# ALGORITHMICS OF MATCHING UNDER PREFERENCES

## SERIES ON THEORETICAL COMPUTER SCIENCE

ISSN: 1793-849X

**Editor-in-Chief:**  Francis Chin *(The University of Hong Kong, Hong Kong)*

**Associate Editors:** Marek Chrobak *(University of California, USA)*
Costas S. Illopoulos *(University of London, UK)*
Kazuo Iwana *(Kyoto University, Japan)*
Ming-Yang Kao *(Northwestern University, USA)*

---

This book series provides an up-to-date and reliable source of information from fundamental knowledge to emerging research areas related to theoretic computer science. The target audiences cover researchers, engineers, scientists, students and professionals.

Series on Vol. 2
Theoretical Computer Science

# ALGORITHMICS OF MATCHING UNDER PREFERENCES

$$M = \begin{cases} (m_1, w_3) \\ (m_2, w_1) \\ (m_3, w_7) \\ (m_4, w_5) \\ (m_5, w_4) \\ (m_6, w_6) \\ (m_7, w_8) \\ (m_8, w_2) \end{cases}$$

**David F. Manlove**
University of Glasgow, UK

**With a foreword by Kurt Mehlhorn**

 **World Scientific**

NEW JERSEY · LONDON · SINGAPORE · BEIJING · SHANGHAI · HONG KONG · TAIPEI · CHENNAI

*Published by*

World Scientific Publishing Co. Pte. Ltd.

5 Toh Tuck Link, Singapore 596224

*USA office:* 27 Warren Street, Suite 401-402, Hackensack, NJ 07601

*UK office:* 57 Shelton Street, Covent Garden, London WC2H 9HE

**British Library Cataloguing-in-Publication Data**
A catalogue record for this book is available from the British Library.

**Series on Theoretical Computer Science — Vol. 2**
**ALGORITHMICS OF MATCHING UNDER PREFERENCES**
Copyright © 2013 by World Scientific Publishing Co. Pte. Ltd.

ISBN 978-981-4425-24-7

Printed in Singapore by B & Jo Enterprise Pte Ltd

To Carolynn, Susan and Paul

# Preface

Matching problems involving preferences occur in widespread applications such as the assignment of children to schools, school-leavers to universities, junior doctors to hospitals, students to campus housing, kidney transplant patients to donors and so on. The common thread is that agents have preferences over the possible outcomes and the task is to find a *matching* (i.e., an assignment of the participants to one another) that is in some sense optimal with respect to these preferences.

These problems are growing in importance in an era in which more and more elements of society are embracing diverse forms of electronic communication, and individuals are increasingly used to making choices via the internet. The ease by which preference information can now be collected has contributed to the growing tendency for matching processes to be centralised. Due to the typical size of applications (for example, in China, over 10 million students apply for admission to higher education annually through a centralised process), trying to construct optimal allocations manually (given a suitable definition of "optimal") is simply not feasible.

Thus algorithms are required to automate the process of constructing optimal matchings. Again, due to the size of typical applications, the efficiency of the algorithms is of paramount importance. The notion of optimality is also a key consideration: many matching processes are conducted by publicly-funded organisations, and there is an increasing tendency for the decisions reached by these organisations to be scrutinised both in the media and by individuals through Freedom of Information requests, for example. Thus the algorithms need to construct matchings that are not just provably optimal, but also are seen to be "fair" by the agents involved.

This book focuses on algorithmic aspects of matching problems involving preferences — our aim is to describe efficient (polynomial-time

algorithms) that produce optimal matchings (under many different notions of optimality) or to highlight complexity results that imply the non-
existence of such algorithms. We also describe some of the many applications in which these algorithms are used.

Our interest, therefore, is in the underlying computational matching
problems that arise in matching markets. The importance of this research
area was recognised by the award, in 2012, of the Sveriges Riksbank Prize
in Economic Sciences in Memory of Alfred Nobel (commonly known as the
Nobel Prize in Economic Sciences) to Alvin Roth and Lloyd Shapley, who
are both leading figures in the research area.

The archetypal matching problem involving preferences is the celebrated
Stable Marriage problem, first introduced by David Gale and Lloyd Shapley
in 1962 [235]. The main contribution of this paper was an algorithm, known
as the Gale–Shapley algorithm, to solve this problem. This algorithm has
been put to practical use in a wide-range of large-scale applications in
countries throughout the world.

Thereafter, Donald Knuth's interest in the problem was reflected in
a collection of lectures published in 1976 [394]. A series of papers written by Dan Gusfield, Rob Irving and co-authors in the 1980s uncovered
deep structural relationships that are present in Stable Marriage problem
instances, and showed how these lead to efficient algorithms for many problems associated with the computation of so-called stable matchings. This
culminated in the publication of their monograph "The Stable Marriage
Problem: Structure and Algorithms" in 1989 [261].

This book has become the standard reference in the literature for structural and algorithmic aspects of the Stable Marriage problem, and indeed its
non-bipartite generalisation, the Stable Roommates problem. It contains
clear and elegant descriptions of a range of stable matching algorithms,
showing how these exploit the fundamental correspondence between the
lattice structure of stable matchings and the so-called rotation poset that
holds in problem instances. It is a much valued resource for those engaged
in stable matching research.

Shortly afterwards, in 1990, a related monograph, entitled "Two-sided
matching: a case study in game-theoretic modeling and analysis", by Al
Roth and Marilda Sotomayor, was published. This book focused mainly on
game-theoretic and strategic questions involving the Stable Marriage problem and its many–one generalisation, the Hospitals / Residents problem,
among others.

Since the publication of these books, the algorithmic interest in the Stable Marriage problem and its variants has shown no sign of diminishing in the literature. Papers since then have also focused on the Stable Roommates problem, the Hospitals / Residents problem and the House Allocation problem.

We classify matching problems with preferences according to three broad classes: (i) bipartite matching problems with two-sided preferences (including the Stable Marriage and Hospitals / Residents problems and their variants), (ii) bipartite matching problems with one-sided preferences (incorporating the House Allocation problem and its variants), and (iii) non-bipartite matching problems with preferences (including the Stable Roommates problem and its variants). Problems in classes (i) and (iii) will mainly be studied with respect to the stability of a given matching, whilst problems in class (ii) involve other optimality criteria such as Pareto optimality, popularity and rank-maximality.

Researchers who are interested in matching problems with preferences are drawn from a range of backgrounds, including the algorithms and complexity, discrete mathematics, game theory and economics, and computational social choice fields. The community is growing and its strength is reflected in the fact that several workshops in the area have taken place in recent years.

The year 2008 saw the first workshop on matching problems with preferences, with an emphasis on algorithmic aspects, co-organised by the author, Magnús Halldórsson, Rob Irving and Kazuo Iwama. This one-day meeting, entitled "MATCH-UP: Matching Under Preferences – Algorithms and Complexity" [266], took place in Reykjavík as a satellite workshop of ICALP 2008 (the 35th International Colloquium on Automata, Languages and Programming). Invited speakers included Kurt Mehlhorn, Al Roth and Marilda Sotomayor. David Gale, who had agreed to open the workshop with an invited lecture, sadly died in the March preceding the workshop; consequently the event was dedicated to his memory and Marilda Sotomayor's invited lecture formed a tribute to his career and their collaboration. Selected papers from the workshop, and several other papers on matching under preferences, appeared in a special issue of the journal *Algorithmica* [417], published in 2010.

A conference entitled "Frontiers of Market Design" was organised by Bettina Klaus in Ascona, Switzerland, in May 2012. Its main focus was on matching markets, and this included matching problems with preferences. Furthermore, a mini-symposium on "Matchings with Preferences" was organised by Christine Cheng at the SIAM Conference on Discrete Mathematics, in Halifax, Canada, in June 2012.

A successor to the 2008 MATCH-UP workshop, entitled "MATCH-UP 2012: the Second International Workshop on Matching Under Preferences" took place, this time over two days, in Budapest in July 2012. This workshop was co-organised by Péter Biró, the author, Tamás Fleiner and Tamás Solymosi, and aimed to involve economists as well as computer scientists. On this occasion, the invited speakers were Nicole Immorlica, Rob Irving, Fuhito Kojima, and Tayfun Sönmez. The workshop was very well-attended, with 68 participants. Again, several papers appearing in the proceedings [78] were invited for submission to a special issue of the journal *Algorithms*.

October 2012 saw the announcement of the award of the Nobel Prize in Economic Sciences to Al Roth and Lloyd Shapley, as previously mentioned, for their work in "the theory of stable allocations and the practice of market design". This reflects both Shapley's contribution to the Stable Marriage algorithm among other theoretical advances, and Roth's application of these results to matching markets involving the assignment of junior doctors to hospitals, pupils to schools and kidney patients to donors (through "kidney exchanges"). The Stable Marriage problem and its variants essentially form the basis of Part 1 of this book, whilst the many applications of matching problems will be highlighted throughout.

These events serve to illustrate the level of interest in the research community in matching under preferences. Indeed at the 2008 MATCH-UP workshop, it became clear from conversations with participants that the time was right for a comprehensive update of the state of the art concerning algorithmic aspects of matching under preferences. This book aims to provide such an update. Whilst in a sense it could be regarded as a "sequel" to Gusfield and Irving's monograph [261], our aim is to expand the range of matching problems that they considered and additionally include alternative optimality criteria besides stability.

The intended readership of this book includes PhD students, postdoctoral researchers and academic staff engaged in research on matching under preferences, senior undergraduate and taught postgraduate students engaged in project work relating to matching under preferences or taking an advanced course on matching theory, and indeed administrators of centralised matching schemes who are interested in the algorithms that underpin these programmes.

The book is largely a comprehensive, classified and guided survey through the literature on matching problems with preferences. Thus the majority of results have been published already in some form. However the book does include some new results due to the author (or his collaborators, reproduced with their permission) that have not been published previously.

Although much effort has been taken to check the correctness of the assertions made in what follows, in a project of this magnitude it is inevitable that there will be errors. For these I offer my apologies now. The careful reader who does discover an error is invited to send the details to david@optimalmatching.com. A list of corrections will be made available via http://www.optimalmatching.com/AMUP.

DFM
Glasgow, October 2012

# Foreword

Matching under preferences is a topic of great practical importance, deep mathematical structure, and elegant algorithmics. The great practical importance stems from the numerous applications such as the assignment of students to universities, families to housing, kidney transplant patients to donors, and so on. In these applications, the participants have preferences over the outcomes and the goal is to find an assignment that optimises the satisfaction of the participants. The importance of the area was recently clearly demonstrated to the world-at-large by the award of the 2012 Nobel Prize in Economics to Alvin Roth and Lloyd Shapley, partly for their work on the stable marriage problem.

Instances may be very large. For example, in China, over 10 million students apply for admission to higher education annually through a centralised process. Thus efficient algorithms are needed. In the past two decades, matching under preferences has became a hot topic in algorithms research with many new structural results, improved algorithms, and new workshops dedicated to the area.

I can attest to the beauty of the subject from my own experience. After hearing a talk by David Manlove on "Strong stability in the Hospitals / Residents Problem" at the British Colloquium for Theoretical Computer Science in 2003, I became interested in matching problems involving preferences. Indeed, some of my results feature in what follows.

This book covers the research area in its full breadth and beauty. It includes most of the recent results in a coherent presentation. Written by one of the foremost experts in the area, it is a timely update to "The Stable Marriage Problem: Structure and Algorithms" (D. Gusfield and R.W. Irving, 1989).

This book will be required reading for anybody working on the subject; it has a good chance of becoming a classic.

Kurt Mehlhorn

# Acknowledgments

There are a lot of people to whom I owe a debt of gratitude in connection with the writing of this book. In some cases, this book simply would not have come about were it not for them.

Firstly, I would like to thank Rob Irving for introducing me to the area of stable matching problems, and subsequently to other matching problems involving preferences. I learned of his expertise in this area as a PhD student under his supervision, but it was not until 1998, when I worked on the "Stable Matching Algorithms" research project (funded by the Engineering and Physical Sciences Research Council) that we started to work together on these problems.

Collaborating with Rob allowed me to quickly discover that studying matching problems with preferences from an algorithmic perspective is a very attractive area to work in for several reasons: (i) many of the problems have direct practical applications; (ii) the problems themselves are engaging and in some cases quite straightforward to specify (and thus accessible to non-specialists), yet challenging to solve; (iii) there are many rich and elegant mathematical structures that can be shown to hold in problem instances, which in turn can be exploited by efficient algorithms; and (iv) although much work has been done on these problems, there are still many interesting open problems to pursue. I gained much from Rob's knowledge, insight and enthusiasm in connection with this research area, which in turn gave me the opportunity to carve out my own research agenda. I have been very fortunate to have had Rob as a colleague and friend, and I hope that our collaboration will continue for many years.

The idea of writing this book was due to Kazuo Iwama. I am very grateful to him for inviting me in early 2008 to make this monograph a contribution to World Scientific's series on Theoretical Computer Science.

Work on the book began in earnest in summer 2008, and therefore it has been a long-running project. I would like to thank Iwama-*san* for his patience over the years whilst I juggled the writing of this book with other research activities, in addition to teaching and administrative duties.

Similarly I would like to thank all those at World Scientific who have been involved in the production of this book for their assistance and patience, including the editors Quan Liang, Yeow-Hwa Quek and Chin Ang Ng. I would also like to thank Rajesh Babu for technical assistance with LaTeX. I have been especially grateful for the publisher's flexibility with deadlines. One of the challenges with this project running over several years is that chapters written some time ago have had to have be constantly updated to take account of the most recent developments.

Within the School of Computing Science at the University of Glasgow, three colleagues in particular have helped me to bring this project to a successful conclusion. Firstly Joe Sventek, the Head of School, kindly granted me a sabbatical from 1 January 2012 to 30 June 2012, which was especially productive. This was due in no small part to Helen Purchase, who took over my duties as Class Head for our Level 4 undergraduate students with great efficiency during this period. Also Rob Irving came out of retirement to cover the Algorithmics course that I usually teach in the second semester.

In terms of the technical writing, there are again many people who have assisted in some way, and who deserve recognition. Rob Irving, once again, has assisted with the task of proof-reading each chapter, which has been very valuable. Also Péter Biró has been a great source of advice in connection with the decentralised algorithms covered in Sec. 2.6 and Sec. 4.3.3. However I stress that any remaining errors are entirely my responsibility.

I would also like to express my gratitude to Kurt Mehlhorn for kindly agreeing to write the Foreword of this book, and for his support of this project in its early stages.

Others who have helped in ways such as providing sources from the literature, answering my queries or commenting on draft sections, or providing new observations (which are acknowledged again at the relevant points) include Haris Aziz, Christine Cheng, Pavlos Eirinakis, Tamás Fleiner, Hal Gabow, Dan Hirschberg, Chien-Chung Huang, Elena Inarra, Augustine Kwanashie, Iain McBride, Eric McDermid, Patrick Prosser, Philip Roscoe, Ildikó Schlotter, Jay Sethuraman and Colin Sng.

The BibTeX references were compiled using "BibTeX Manager", a Java application designed and implemented by Mitesh Furia under my supervision as part of his development project for the MSc in Information Technology in academic session 2008-09. This very useful program helped me greatly to manage a bibliography of the size and complexity that emerged through the writing of this book.

Also on the technical front, other software applications that I have greatly benefited from include MiKTeX, TeXnicCenter and Dropbox. The latter allowed me to seamlessly transfer between my work PC, laptop and home PC in order to work on the book, without the need to copy files using email or pen drive.

During the period of the writing my research was financially supported by the Engineering and Physical Sciences Research Council, through grant EP/E011993/1 ("MATCH-UP: Matching Under Preferences — Algorithms and Complexity", running from 1 June 2007 to 30 June 2010).

I am grateful to Scott Barr, and other friends in Linlithgow, who have spurred me on by asking "how's the book?" at regular intervals.

My father Colin, himself an author of many books (on fantasy literature), has been a source of inspiration, as was my late mother Evelyn who died in 2003.

But lastly, and most importantly, I would like to thank my wife Carolynn for her love, understanding, support and encouragement through all aspects of this project. This book is dedicated to her, and to our children Susan and Paul.

# Contents

## Stable Matching Problems                                     49

# List of Figures

# List of Tables

# List of Algorithms

# Chapter 1

# Preliminary definitions, results and motivation

## 1.1 Introduction

### 1.1.1 *Remit of this book*

#### 1.1.1.1 *Matching under preferences*

This book is about computational problems that involve matching agents to one another, subject to various criteria. Here, the term *agent* is used loosely to mean any participant in a matching process, and could include commodities in addition to human subjects. In many cases the agents form two disjoint sets, and we seek to assign the agents in one set to those in the other. Examples include assigning junior doctors to hospitals, pupils to schools, kidney patients to donors, and so on.

We primarily focus on the case that a subset of the agents have *ordinal preferences* over a subset of the others. That is, there is a notion of first choice, second choice, third choice, etc. For example a school-leaver who is applying for admission to university might rank in order of preference a small subset (say 5) of all available universities. Likewise, the universities might form a ranking of their applicants according to academic merit, and possibly other criteria. We will not always insist that the preference lists be *strictly* ordered: for example a school-leaver might have two universities that are jointly ranked as first choice in her preference list.

Typically there are other constraints in addition to the preference lists: for example, it is reasonable to assume that a school-leaver should not be assigned to more than one university, and likewise a university might have a *capacity*, indicating the maximum number of students that it could admit in a particular academic session.

### 1.1.1.2    *Free-for-all markets*

Applications of matching problems involving ordinal preferences (henceforth the term *ordinal preferences* will be shortened to *preferences*) can be very large in practice. For example in 2011, 140,953 students applied for admission to higher education in Hungary [90]. Economists have identified several problems that arise in free-for-all markets, in which the agents are able to negotiate with one another directly in order to arrange assignments [518, 467].

These problems include *unravelling* in which agents form assignments with one another earlier and earlier in advance of the deadline by which all assignments must be fixed. For example, hospitals wishing to recruit the best applicants might compete with one another by advancing the date when they make their offers. To avoid this issue, agents might be prevented from entering into premature assignments before a certain date.

This can then lead to the problem of *congestion*, in which agents do not have sufficient time to negotiate with one another over potential assignments prior to the deadline. For example a hospital $h$ offering 20 places to applicants might have to make substantially more than 20 offers, to allow for applicants who will turn down $h$'s offer.

In an effort to avoid congestion, a new problem might emerge, namely *exploding offers*. In such a case, agents are given only a short time period to decide whether they are able to form a given assignment, otherwise the potential for making that assignment is removed. For example hospital $h$ might force an applicant $r$ to make a decision swiftly by setting $r$ a deadline, beyond which the offer expires. Unravelling, congestion and exploding offers lead to a situation in which agents might be forced into forming an assignment with one another before they have knowledge of the whole range of potential assignments that may potentially be available to them.

Prior to 1952, the assignment of junior doctors to hospitals in the USA was carried out by a free-for-all market. A detailed description of the problems that this led to (which included unravelling, congestion and exploding offers as described above), with reference to this particular application, has been given in several references in the literature [261, 514, 518, 467].

### 1.1.1.3    *Centralised matching schemes*

*Centralised matching schemes* (referred to as *(centralised) clearinghouses* by economists) can avoid some of the problems that are inherent in free-for-all markets. These work along the following lines: the input data involving the agents and their preferences over one another are collected by a given deadline by a trusted central authority. This third party then computes an optimal matching with respect to the supplied preference lists and capacities, and any other problem-specific constraints. By participating in the process, the agents agree that the outcome is binding. The precise definition of an *optimal matching* has many variations depending on the context, but it could involve, for example, maximising the number of places that are filled at each hospital, or giving the maximum number of school-leavers their first-choice university, or ensuring that no junior doctor and hospital have an incentive to reject their assignees and become matched together, if they were not already assigned to one another.

Centralised matching schemes are often given a name that is an umbrella term for the entire administrative and algorithmic process of data collection, computation of a matching and publication of the outcome. For example the assignment of junior doctors to hospitals in the US is handled by the National Resident[1] Matching Program (NRMP) [498, 602]. In 2012, 38,777 aspiring residents applied via the NRMP for 26,772 available residency positions [602]. The previous subsection indicated that higher education admission in Hungary typically involves an even larger number of applicants.

Due to the potential size of the applications in practice, it is usually infeasible to compute optimal allocations by hand. Centralised matching schemes automate this task by employing algorithms to compute optimal matchings based on the input data supplied by the participants. Our interest is in the design and analysis of efficient (polynomial-time) algorithms for the matching problems that underpin these centralised matching schemes, and in proving results about the non-existence of such algorithms where appropriate. In order to derive these theoretical results it is important to arrive at formal models of the underlying matching problems. This book organises its presentation around a systematic classification of matching problems according to the nature of the agents who are participating, the form of the preference information that they provide, and the criteria that constitute the definition of an optimal matching.

---

[1] *Resident* is the US term for a newly-graduated medical student who is undertaking their first period of supervised postgraduate medical practice before full registration as a junior doctor is granted.

## 1.1.2 The matching problems under consideration

### 1.1.2.1 Classification of matching problems

The matching problems that we consider in this book can be fairly comprehensively classified as follows:

(1) *Bipartite matching problems with two-sided preferences*. Here the participating agents can be partitioned into two disjoint sets, and each member of one set ranks a subset of the members of the other set in order of preference. Example applications include assigning junior doctors to hospitals [596, 602–604], pupils to schools [1, 2, 4] and school-leavers to universities [64, 77, 111, 491].

(2) *Bipartite matching problems with one-sided preferences*. Again the participating agents can be partitioned into two disjoint sets, but this time each member of only one set ranks a subset of the members of the other set in order of preference. Example applications include campus housing allocation [142, 474], DVD rental markets [19] and assigning reviewers to conference papers [242].

(3) *Non-bipartite matching problems with preferences*. Here the participating agents form a single homogeneous set, and each agent ranks a subset of the others in order of preference. Example applications include forming pairs of agents for chess tournaments [401], finding kidney exchanges involving incompatible patient–donor pairs [511, 512, 17, 513, 423] and creating partnerships in P2P networks [232, 233, 403, 429–431].

In the following subsections we describe informally some of the key matching problems that belong to each part of the above classification. Formal definitions of the problem models will be given in later sections of the chapter. We will also expand on the applications mentioned above in greater detail later in this chapter.

### 1.1.2.2 Bipartite matching problems with two-sided preferences

The classical *Stable Marriage problem* (SM) [235, 261] (defined formally in Sec. 1.3.4.1) is the central matching problem in this class. An instance of this problem comprises a set of men and women, and each person ranks each member of the opposite sex in strict order of preference.

A many–one generalisation of SM is the *Hospitals / Residents problem* (HR) [235, 261] (defined formally in Sec. 1.3.2), where each man corresponds to a resident and each woman corresponds to a hospital which can potentially be assigned multiple residents up to some fixed capacity. HR models the assignment of junior doctors to hospitals and many other related applications.

Other generalisations of HR that belong to this class are the *Workers / Firms problem* (WF) and the *Student–Project Allocation problem* (SPA) — see Sec. 5.4 and Sec. 5.5 for definitions of these problems.

In each of the problems in this class, the task is to find a *stable matching*. Informally, a *matching* is a set pairs, each of which represents the assignment of an agent from one set to an agent from the other, such that no agent is assigned more agents than its capacity. A matching is *stable* if no two agents prefer one another to one of their current assignees. Were such a pair of agents to exist, they could undermine the matching by forming a private arrangement outside of it.

Roth and his co-authors [498,503,504,518] have stressed the importance of stability as a solution concept for matching problems in this class. Most of our treatment of bipartite matching problems with two-sided preferences will involve stability as the solution concept, but we will also consider alternative optimality criteria in such settings.

### 1.1.2.3  *Bipartite matching problems with one-sided preferences*

The *House Allocation problem* (HA) (defined in Sec. 1.5.2) [301,595,5] is the variant of SM in which the women do not have preference lists over the men. The men are now referred to as *applicants* and the women are referred to as *houses*. The problem name stems from the application where students are assigned to campus housing, based on their preferences over the available accommodation. This is accomplished using a centralised matching scheme in a number of universities including Carnegie-Mellon University, Duke University, the University of Michigan, Northwestern University and the University of Pennsylvania in the US [142], and the Technion in Israel [474].

A many–one extension of HA, called the *Capacitated House Allocation problem* (CHA) arises when each house can accommodate multiple applicants up to some fixed capacity. CHA can also be regarded as the variant of HR in which hospitals do not have preference lists over residents.

In the context of HA and CHA, only applicants have preferences over houses, so the notion of stability is not relevant. Other optimality criteria have been formulated in the literature, including *Pareto optimality*, *popularity* and *profile-based optimality*. Informally, a matching is *Pareto optimal* if there is no other matching in which some applicant is better off, whilst no applicant is worse off. A matching is *popular* if there is no other matching that is preferred by the majority of the applicants. Finally, the *profile* of a matching $M$ is a vector indicating the number of applicants with their first, second and third choice, etc., in $M$. Optimising the profile of $M$ might, for example, involve maximising the number of applicants with their first choice, and subject to this, maximising the number with their second choice, etc.

### 1.1.2.4  *Non-bipartite matching problems with preferences*

The *Stable Roommates problem* (SR) (defined in Sec. 1.4.2) [235, 306, 261] is the non-bipartite generalisation of SM in which each agent ranks all of the others in strict order of preference. Stability is once again relevant in this context, and the definition of a stable matching is a straightforward extension of the definition in the SM case.

Many–many generalisations of SR have been considered in the literature and are called the *Stable Fixtures problem*, the *Stable Multiple Activities problem* and the *Stable Allocation problem* (see Sec. 4.8.4, Sec. 4.8.5 and Sec. 4.8.6 respectively). Variants of SR have also been considered in which agents can form partnerships into sets of size $> 2$ — this is the *Coalition Formation Game* (see Sec. 4.8.8).

Most of our analysis of non-bipartite matching problems with preferences will involve stability as the solution concept, but there will also be occasions when we will consider optimality criteria other than stability in this context.

### 1.1.2.5  *Further problem variants*

The problem classes above include generalisations, restrictions and other variations of the basic matching problems outlined. For example, in an instance of HR, a large hospital with hundreds of applicants may not be able to distinguish easily between all of them in order to arrive at a strictly-ordered preference list. In particular, the hospital may wish to express *indifference* in its list by grouping together batches of residents into *ties* (see Sec. 1.3.5 for more details). This gives rise to a generalisation of HR.

One example of where a restriction of HR occurs is when we place constraints on the lengths of the preference lists. For example, in the context of assigning school-leavers to universities, typically there are many available universities, and applicants will not want to attend for interview at more than a certain number. This motivates the consideration of restrictions of HR in which the preference lists of the residents are of bounded length.

A variant of HR arises when we alter the stability definition so that we now require that there should be no pair of residents who could exchange one another's assigned hospitals so as to improve their outcome. *Exchange-stability* covers this type of solution concept (see Sec. 5.7 for further details).

One important matching problem with ordinal preferences that does not fit into the above classification is the three-dimensional variant of SM (3DSM) in which there are three sets of agents (Knuth [394] suggested that these could be considered to be men, women and dogs). Strictly speaking, this problem would fit into an additional class entitled *tripartite matching problems with preferences*. See 5.6 for further details concerning this class of stable matching problem.

### 1.1.3 *Existing literature on matching problems*

#### 1.1.3.1 *Algorithms and complexity literature*

Algorithmic aspects of matching problems involving ordinal preferences have been extensively studied in the literature over the last few decades. The seminal paper of Gale and Shapley [235] giving an algorithm for SM was probably the earliest paper in this context. Between 1962, the date of publication of Ref. [235], and 1989, most of the algorithmic effort centred on stable matching problems, and in particular SM, HR, SR, 3DSM and their variants. By contrast, algorithmic results for HA and CHA are more recent, and indeed most of those that we focus on in this book were formulated within the last decade.

Donald Knuth's interest in stable matching problems led to the publication in 1976 of a series of lecture notes in French on the topic, which were later translated into English [394].

1989 saw the publication of Gusfield and Irving's book, entitled "The Stable Marriage problem: Structure and Algorithms" [261]. This gave a deep insight into the underlying structure of the set of stable matchings in instances of SM, HR and SR, and showed how this structure plays a vital role in the derivation of efficient algorithms for problems such as generating all stable matchings, and finding stable matchings with additional useful properties.

The *Encyclopedia of Algorithms* [357], published in 2008, included a range of entries relating to matching problems [120, 314, 315, 333, 361, 416,

450, 555]. Also in the same year, a short survey paper by Iwama and Miyazaki was published [334].

As indicated in the preface, papers accepted to the MATCH-UP workshops held in 2008 and 2012 were published in two proceedings volumes [266, 78], and a special issue of *Algorithmica*, published in 2010 [417], included a selection of papers appearing in the 2008 workshop along with other papers on matching under preferences.

Subsequently to 1989, a number of MSc and PhD theses relating to matching under preferences have been published that involve a substantial algorithmic focus [159, 201, 208, 14, 587, 523, 578, 470, 588, 535, 571, 15, 439, 483, 549].

### 1.1.3.2  Game theory and economics literature

The study of matching problems involving preferences belongs to an area that economists refer to as *matching theory* [541]. Matching theory in turn belongs to a sub-field of microeconomics called *market design* [397]. This is concerned with applying game-theoretic techniques to design economic institutions that avoid the kinds of market failures outlined in Sec. 1.1.1.2.

Economists tend to refer to a matching algorithm as a *mechanism*. They are typically interested in *strategy-proof* or *truthful* mechanisms, which make it a *dominant strategy* for the agents to reveal their true preferences (see Sec. 2.9 where these terms are defined formally).

Roth and Sotomayor's book [514], entitled "Two-sided matching: a case study in game-theoretic modeling and analysis", published shortly after Gusfield and Irving's monograph [261], studied SM and HR in depth, also presenting structural results, but concentrating mainly on a game-theoretic viewpoint. One of the key themes of the book was the investigation of strategy-proof mechanisms for SM and HR.

Several surveys on matching theory have been written by economists following the publication of Roth and Sotomayor's monograph [514]. These study problems such as SM, HR, SR and HA in the context of market design. Survey papers include those by Roth and Sotomayor [515], Gale [234], Roth [505], Niederle *et al.* [467], Sotomayor and Özak [547], Kojima and Troyan [397], Sönmez and Ünver [541], and Abdulkadiroğlu and Sönmez [8].

Increasingly, papers on matching theory and market design are appearing in game theory conferences such as the quadrennial World Congress of the Game Theory Society, and at annual meetings such as the International Summer Festival on Game Theory at Stony Brook and the Spain–Italy–Netherlands series of meetings on Game Theory.

### 1.1.3.3 Algorithmic mechanism design literature

*Computational social choice theory* [150] addresses computational challenges arising in situations where multiple agents must reach a collective decision that affects them all, and which may result in winners and losers.

*Algorithmic mechanism design* [468] is a sub-field of computational social choice theory involving the design of systems that take in users' preferences and produce outcomes based on those preferences. In general such systems may involve monetary transactions, and issues of interest include revenue raised, and social welfare. We are not concerned with monetary payments, and therefore "social welfare" will usually refer to measurable attributes that are derived from the ordinal preferences expressed by the agents.

*Algorithmic game theory* [469] is a very broad field that is concerned with computational questions arising in game-theoretic problems such as finding equilibria, designing mechanisms with or without money, analysing combinatorial auctions, load-balancing machines and routing data through networks.

The algorithmic study of matching under preferences lies at the intersection of algorithmic game theory, computational social choice theory and algorithmic mechanism design [495, 541]. Increasingly, papers that are relevant to the topic of this book are appearing in multi-disciplinary conferences that solicit work at the intersection of these areas, such as EC (ACM Conference on Electronic Commerce), COMSOC (the International Workshop on Computational Social Choice), SAGT (the Symposium on Algorithmic Game Theory) and WINE (the Workshop on Internet and Network Economics).

### 1.1.4 Contribution of this book

#### 1.1.4.1 General overview

Gusfield and Irving's monograph [261] gave a very comprehensive account of the state of the art for SM, HR and SR at the time of its publication. One of the aims of this book is to update the reader on the many structural and algorithmic results for these matching problems that have been published subsequently to 1989. In doing so, we do not assume prior knowledge of any definitions or results from Ref. [261]. All relevant notation, terminology and key results for these problems will be given in this chapter, and hence this book can be read independently of Ref. [261].

This book should not, however, be considered merely as an update to Ref. [261]. Our aim is to broaden the range of matching problems considered to include many additional stable matching problems belonging to Classes (1) and (3) in the classification of Sec. 1.1.2.1. We also cover problems in Class (2), which were not part of the remit of Ref. [261].

### 1.1.4.2   *Chapter outline*

The book is divided into two main parts: Part 1, spanning Chaps. 2–5, deals with stable matching problems in Classes (1) and (3). Part 2, spanning Chaps. 6–8, focuses on other forms of optimality criteria mainly applied to matching problems in Class (2), but also in instances of problems in Classes (1) and (3).

In Part 1, Chap. 2 deals with the central stable matching problem, namely SM, presenting key developments that have appeared in the literature following the publication of Gusfield and Irving's monograph. We provide updates to lists of open problems from Refs. [394, 261], review two important papers by Subramanian and Feder [551, 202] and describe linear and constraint programming approaches to SM, decentralised algorithms for SM and some beautiful results concerning *median* stable matchings. Among the many other results presented, we show how stable matching theory led to a very elegant solution to the Dinitz conjecture [239].

The extensions of SM and HR in which preference lists can include ties and other forms of indifference led to a substantial revival in the study of stable matching problems in the late 1990s and early 2000s. In Chap. 3 we describe algorithmic results for problems involving computing stable matchings in these contexts. One particular problem, namely that of finding a stable matching that matches as many people as possible, given an instance of SM where the preference lists may involve ties and may be incomplete, has led to an interesting "race" to find the tightest, fastest and simplest approximation algorithm.

SR, the non-bipartite version of SM, has traditionally been studied less extensively than SM. However following the publication of [261], some important structural and algorithmic results due to Tan were published [556–559]. These guaranteed the existence of a so-called *stable partition*, a generalisation of a stable matching, even in instances of SR that admit no stable matching. We describe Tan's results, and many other more recent results for SR and its variants, in Chap. 4.

Further results for stable matching problems are presented in Chap. 5. We describe extensions of HR in which hospitals can have lower and/or common quotas, which present additional constraints on the numbers of assignees that they can/must obtain in a stable matching. We also consider the variant of HR in which couples can provide joint preference lists in order to be matched to hospitals that are geographically close to one another. Other problems considered include SPA, WF, 3DSM and exchange-stable matching problems.

Part 2 is concerned with optimality criteria that can be defined for matching problems in Class (2). These include Pareto optimality, popularity and profile-based optimality. These criteria are mainly applied to HA, CHA and their variants, but are also studied in the context of SM, HR and SR. Issues of interest in each case include the existence of an optimal matching, the algorithmic complexity of finding an optimal matching, and the structure of the set of optimal matchings in a given problem instance. Generally speaking, Pareto-optimal and profile-based optimal matchings are bound to exist, but there is no such guarantee in the case of popular matchings. Results for Pareto optimal, popular and profile-based optimal matchings are considered in Chap. 6, Chap. 7 and Chap. 8 respectively.

One of the purposes of this book is to stimulate further research in the area of matching under preferences, and to this end we identify a range of open problems for future investigation. These are presented in the concluding section of each chapter.

Finally, we remark that, given the range of matching problems considered, we do require to define many notational concepts throughout this book. For the reader's convenience we provide a Glossary of Symbols, starting on Page 461.

### 1.1.4.3 *What the book does not contribute*

Due to the magnitude of the range of structural and algorithmic results for matching problems involving preferences that are already within the scope of this book, it has been necessary to omit certain topics. We briefly outline these here.

**Strategic results for matching problems.** Although issues of strategy in matching problems involving preferences are undoubtedly important, we do not have the space in this book to comprehensively update the reader on the wealth of results that have appeared subsequently to the publication of

Roth and Sotomayor's book [514], particularly those from the game theory and economics literature. Instead, we confine our treatment to a short section in Chap. 2 that focuses mainly on results that have appeared in the algorithms and complexity literature (see Sec. 2.9).

**Matching problems with cardinal utilities** The focus of this book is on ordinal preferences, and problems that are primarily based on optimising a matching subject to some constraint involving cardinal utilities are not considered.

In particular we do not consider matching problems with *transferable utility* (sometimes called *monetary transfer* or *payments*) — such problems have been referred to as *TU games* in the literature. These include problems such as the Assignment Game [528], the Matching Game (c.f. [89]), the Stable Roommates problem with Payments [79] (see also Refs. [196, 197, 224]) and the Permutation Game [570].

Matching problems with cardinal utilities such as the Assignment problem [230] and maximum weight matching in general graphs [223, 231] are also not explicitly considered in their own right unless there is some connection with ordinal preferences. More general packing problems such as packing paths and/or cycles into directed or undirected graphs are also beyond the scope of this book. These include problems involving identifying optimal sets of kidney "exchanges" (in such an exchange, a patient with a willing but incompatible donor can "swap" her donor with that of another patient in a similar position) where edge weights are derived from cardinal utilities [511, 512, 17, 513, 423]. (We will however consider kidney exchange problems in the context of ordinal preferences.)

It is often the case that ordinal preferences naturally arise from human subjects who are participants of a centralised matching scheme and able to arrive at a notion of first, second, third choice, etc. However there are some interesting cases where quantitative, objective data do in fact give rise to ordinal preferences, such as the following:

- Junior doctor allocation in Scotland (applicant "scores" based on academic performance and assessment of application forms give rise to ordinal preferences for the hospitals) [309, 604].
- School choice in New York and Boston (children have "priorities" based on factors such as whether they are in the walk zone and whether they have siblings at the school already, and these priorities translate into ordinal preferences for the schools) [2, 1, 4].

• Higher education admission in Hungary (again, the academic performance of the applicants gives rise to ordinal preferences on the part of the universities) [77, 90].

• P2P networking (measures of download / upload bandwidth, latency and storage capacity give rise to ordinal preferences of nodes in a communication network over their peers) [232, 233, 403, 429–431].

We finally remark that there are examples of matching problems that involve both ordinal preferences and cardinal utilities, and these are within the scope of this book. Examples include the *minimum weight stable matching problem* [320] (see Sec. 1.3.4.1).

**Proofs of results.** Our aim is to provide as broad a coverage as possible of the vast literature in the area of algorithmics of matching under preferences. We strive to give an equal balance to results contributed jointly or solely by the author, and those due to other members of the community. In doing so, we will in most cases omit proofs, referring the reader to the relevant references for the full details. However in some cases we do present proofs, generally for one or more of the following reasons: (i) the result is new, and the proof has not been previously published; (ii) the result is known, but an explicit proof has not been given in the literature, and we provide one for completeness; (iii) the proof has already appeared in the literature, but is reproduced (perhaps in a slightly different form) because in the author's opinion, the proof illustrates a general technique and it is instructive to include it.

### 1.1.5 *Outline of this chapter*

In the remainder of this chapter we provide basic definitions of notation and terminology and give key structural and algorithmic results that will be required throughout subsequent chapters. Our presentation is organised according to the problem areas of HR, SM, SR and HA and their variants. In addition we review basic definitions and results concerning matchings in graphs.

The remaining sections are organised as follows. In Sec. 1.2, we begin with general matching concepts in graphs. Then key definitions and results for HR are given in Sec. 1.3, including those for SM in Sec. 1.3.4, and then those for SR in Sec. 1.4. This order allows us to arrive at many definitions and results for SM that follow immediately from the fact that SM is a special case of HR. Key results that were published after 1989 (the date of publication of Ref. [261]) are surveyed in later chapters. Finally, Sec. 1.5 covers important definitions and results for HA.

In each of Sec. 1.3, Sec. 1.4 and Sec. 1.5 we also give further motivation for HR, SR and HA, describing some of their practical applications in more detail.

## 1.2 Matchings in graphs

In this section we briefly review fundamental definitions, together with structural and algorithmic results, concerning matchings in graphs. Let $G = (V, E)$ be an undirected graph, where $n = |V|$ and $m = |E|$. A set $M \subseteq E$ is a *matching* in $G$ if no two edges in $M$ are adjacent in $G$. Let $\mathcal{M}$ denote the set of matchings in $G$. A *maximum cardinality matching* (usually shortened to a *maximum matching* henceforth where there is no ambiguity) is a matching $M \in \mathcal{M}$ that maximises $|M|$. A matching $M$ is *perfect* if every vertex of $G$ is incident to an edge of $M$.

Let $M$ be an arbitrary matching in $G$. For any $v \in V$, if $v$ is incident to some member of $M$ then we say that $v$ is *matched*, otherwise $v$ is *exposed*. If $\{v, w\} \in M$ then $M(v)$ denotes $w$. If $v$ is exposed then $M(v)$ is undefined. An *alternating path* $P$ with respect to $M$ is a simple path in $G$ that contains edges in $M$ and not in $M$ alternately. If the end-vertices of $P$ are exposed then $M$ is an *augmenting path*. A fundamental structural result for maximum matchings is Berge's theorem[2], as follows.

**Theorem 1.1 ([476, 71]).** *Let $G$ be a graph and let $M$ be a matching in $G$. Then $M$ is maximum if and only if $M$ admits no augmenting path.*

Edmonds' *blossom-shrinking* algorithm [183] is (essentially) based on repeatedly finding an augmenting path $P$ relative to a given matching $M$, and *augmenting* along $P$ (that is, replacing the edges in $P \cap M$ by the edges in $P \backslash M$ [3]) until no augmenting path remains. Then $M$ is maximum by Theorem 1.1. The fastest current implementation of Edmonds' algorithm is due to Micali and Vazirani [451, 577], as stated by the following theorem.

---

[2] According to Korte and Vygen [400, p.244], this result had already been observed by Petersen [476] and was rediscovered by Berge.
[3] The matching so obtained is the symmetric difference of $M$ and $P$, denoted by $M \oplus P$.

**Theorem 1.2 ([183, 451, 577]).** *A maximum matching in a graph $G = (V, E)$ can be found in $O(\sqrt{n}m)$ time, where $n = |V|$ and $m = |E|$.*

In the case that $G = (V, E)$ is a bipartite graph, the search for an augmenting path is simpler than for the non-bipartite case, and can be achieved using a variant of breadth-first search (see e.g., Ref. [407] for further details). Hopcroft and Karp [281] established that a maximum matching in $G$ can be found in $O(\sqrt{n}m)$ time, where $n = |V|$ and $m = |E|$ (their approach is simpler than Micali and Vazirani's algorithm for the non-bipartite case, and also pre-dated it).

Hall's Marriage Theorem [265] provides a necessary and sufficient condition for $G$ to admit a perfect matching. To describe this result, we require some additional notation and terminology. A *bipartition* of $G$ is a partition of $V$ into disjoint sets $U$ and $W$ such that, for every edge $e \in E$, $|e \cap U| = |e \cap W| = 1$. We denote such a bipartition of $G$ by $G = (U, W, E)$. Given a vertex $v \in V$, define $N(v) = \{v' \in V : \{v, v'\} \in E\}$ to be the *open neighbourhood* of $v$. Given $V' \subseteq V$, define $N(V') = \bigcup_{v \in V'} N(v)$.

**Theorem 1.3 ([265]).** *Let $G = (V, E)$ be a bipartite graph with bipartition $V = U \cup W$. Then $G$ has a maximum matching in which all vertices in $U$ are matched if and only if $|N(U')| \geq |U'|$ for all $U' \subseteq U$. Moreover, $G$ has a perfect matching if and only if $|U| = |W|$ and $|N(U')| \geq |U'|$ for all $U' \subseteq U$.*

A second fundamental structural result in matching theory is the *Edmonds–Gallai Decomposition* [183, 238] (see also Ref. [407, Theorem 3.2.1]). This is valid for general graphs, but has a particularly nice representation in bipartite graphs, as observed by Dulmage and Mendelsohn [174–176] (see also [407, Theorem 3.2.4])[4]. We will refer to the decomposition in this case as the *Dulmage–Mendelsohn Decomposition*. We first define the decomposition in bipartite graphs and then state its properties.

**Definition 1.4 ([174, 175, 183, 238, 176]).** *Let $G = (V, E)$ be a bipartite graph, and let $M$ be a maximum matching in $G$. The* Dulmage–Mendelsohn Decomposition *is a partition of $V$ into three disjoint[5] sets, namely $\mathcal{E}$, $\mathcal{O}$, and $\mathcal{U}$, as follows. Vertices in $\mathcal{E}$, $\mathcal{O}$, and $\mathcal{U}$ are called* even, odd *and*

---

[4]Lovász and Plummer [407] noted that Dulmage and Mendelsohn formulated their decomposition for bipartite graphs prior to Edmonds and Gallai arriving at their structure theorem for general graphs.

[5]That the sets are disjoint is established by Theorem 1.5.

unreachable *respectively. A vertex $v \in V$ is even (respectively odd) if there exists an alternating path of even (respectively odd) length to $v$ from a vertex that is exposed in $M$. If no such alternating path exists, $v$ is unreachable. We henceforth refer to this vertex labelling as an EOU labelling of $V$.*

**Theorem 1.5 ([183, 238]).** *Let $G = (V, E)$ be a bipartite graph, and let $M$ be a maximum matching in $G$. Let $\mathcal{E}$, $\mathcal{O}$, and $\mathcal{U}$ be the sets defined relative to $M$ in $G$ by Definition 1.4. Then:*

*(i) $\mathcal{E}$, $\mathcal{O}$ and $\mathcal{U}$ are pairwise disjoint. Moreover the EOU labelling of $V$ is independent of the particular choice of $M$.*

*(ii) $M$ satisfies the following properties:*

    *(a) every vertex in $\mathcal{O} \cup \mathcal{U}$ is matched;*

    *(b) every vertex in $\mathcal{O}$ is matched to a vertex in $\mathcal{E}$;*

    *(c) every vertex in $\mathcal{U}$ is matched to another vertex in $\mathcal{U}$;*

    *(d) $|M| = |\mathcal{O}| + |\mathcal{U}|/2$.*

*(iii) $M \cap \{\{u, v\} : u \in \mathcal{O} \wedge v \in \mathcal{O} \cup \mathcal{U}\} = \emptyset$.*

*(iv) $E \cap \{\{u, v\} : u \in \mathcal{E} \wedge v \in \mathcal{E} \cup \mathcal{U}\} = \emptyset$.*

A graph $G = (V, E)$ is *capacitated* if there is a function $c : V \longrightarrow \mathbb{Z}^+$. An *assignment* $M$ in $G$ is a subset of $E$. Given any $v \in V$, define $M(v) = \{w : \{v, w\} \in M\}$. A *matching* $M$ is an assignment such that $|M(v)| \leq c(v)$ for all $v \in V$. A *maximum matching* is a matching $M$ in $G$ that maximises $|M|$. The problem of finding a maximum matching in $G$ is also called the *Upper Degree Constrained Subgraph problem* (UDCS) [226]. Gabow [226] showed that this problem can be solved in $O(\sqrt{C}m)$ time, where $C = \sum_{v \in V} c(v)$.

Now suppose that $G = (V, E)$ is a capacitated bipartite graph. Let $V = U \cup W$ be a bipartition of $G$. In general, the time complexity of Gabow's algorithm is in fact $O(\sqrt{\beta}m)$, where $\beta$ is the size of a maximum matching in $G$ [226]. This leads to the following observation about the algorithm's complexity in a special kind of capacitated bipartite graph.

**Theorem 1.6 ([226]).** *Let $G = (V, E)$ be a capacitated bipartite graph, and let $V = U \cup W$ be a bipartition of $G$. Suppose that $c(u) = 1$ for all $u \in U$. Then a maximum matching in $G$ can be found in $O(\sqrt{n_1}m)$ time[6], where $n_1 = |U|$ and $m = |E|$.*

---

[6]In Refs. [425, 470, 535], the weaker upper bound of $O(\sqrt{C}m)$ was given as the complexity of Gabow's algorithm for this restricted type of capacitated bipartite graph.

A counterpart to a maximum matching is a *minimum maximal matching*, which we now define. Let $G = (V, E)$ be an arbitrary graph. A matching $M$ in $G$ is *maximal* if $M \cup \{e\}$ is not a matching, for every $e \in E \backslash M$. A *minimum maximal matching* is a maximal matching of minimum size. Clearly a maximum matching is maximal. A maximal matching can be found by a simple greedy algorithm (pick an edge at random and delete all of its neighbours, and iterate this process until no edges remain). However the only guarantee regarding the cardinality of the matching $M$ produced by this algorithm is that it is within a factor of 2 of the size of a maximum matching. Indeed, $M$ is trivially within a factor of 2 of the size of a minimum maximal matching also, since any two maximal matchings in $G$ differ in size by at most a factor of 2 [399].

In constrast to the maximisation problem, the problem of finding a minimum maximal matching in $G$ is NP-hard even in some very restricted cases [590,282,165]. To describe these, define MIN MM-D to be the problem of deciding, given a graph $G$ and integer $K$, whether $G$ admits a maximal matching of size at most $K$.

**Theorem 1.7.** MIN MM-D *is NP-complete. The result holds even for bipartite graphs [590]* [7], *for subdivision graphs*[8] *of cubic graphs [282], and for bipartite k-regular graphs, for $k \geq 3$ [165].*

## 1.3 The Hospitals / Residents problem (HR)

### 1.3.1 *Introduction*

In this section we present definitions and fundamental results relating to the Hospitals / Residents problem (HR) and its variants. Key definitions are given in Sec. 1.3.2, and important results appearing in the literature up to 1989 are presented in Sec. 1.3.3. The 1–1 restriction of HR known as the Stable Marriage problem is discussed in Sec. 1.3.4. Variants of HR involving indifference are described in Sec. 1.3.5. Some additional important variants

---

[7] In fact Yannakakis and Gavril proved that MIN EDS-D is NP-complete for this class of graphs. MIN EDS-D is the problem of determining, given a graph $G = (V, E)$ and an integer $K$, whether $G$ contains an *edge dominating set* (i.e., a set of edges $S$ such that every edge in $E \backslash S$ is adjacent to some edge in $S$) of size at most $K$. MIN MM-D and MIN EDS-D are polynomially equivalent; indeed the size of a minimum maximal matching in $G$ is equal to the size of a minimum edge dominating set of $G$ [590].

[8] Given a graph $G$, the *subdivision graph* of $G$, denoted by $S(G)$, is obtained by subdividing each edge $\{u, w\}$ of $G$ in order to obtain two edges $\{u, v\}$ and $\{v, w\}$ of $S(G)$, where $v$ is a new vertex.

of HR are outlined in Sec. 1.3.6. Finally, Sec. 1.3.7 describes practical motivation for HR.

### 1.3.2  Key definitions

The Hospitals / Residents problem (HR) [235, 261, 514, 416] (sometimes referred to as the College (or University or Stable) Admissions problem, or the Stable Assignment problem) was first defined by Gale and Shapley in their seminal paper "College Admissions and the Stability of Marriage" [235], referred to there as the College Admissions problem.

An instance $I$ of HR involves a set $R = \{r_1, \ldots, r_{n_1}\}$ of *residents* and a set $H = \{h_1, \ldots, h_{n_2}\}$ of *hospitals*. Each hospital $h_j \in H$ has a positive integral *capacity*, denoted by $c_j$, indicating the number of *posts* that $h_j$ has. Also there is a set $E \subseteq R \times H$ of *acceptable* resident–hospital pairs. Each resident $r_i \in R$ has an *acceptable* set of hospitals $A(r_i)$, where

$$A(r_i) = \{h_j \in H : (r_i, h_j) \in E\}.$$

Similarly each hospital $h_j \in H$ has an acceptable set of residents $A(h_j)$, where

$$A(h_j) = \{r_i \in R : (r_i, h_j) \in E\}.$$

The *agents* in $I$ are the residents and hospitals in $R \cup H$. Each agent $a_k \in R \cup H$ has a *preference list* in which she/it ranks $A(a_k)$ in strict order. Given any resident $r_i \in R$, and given any hospitals $h_j, h_k \in H$, $r_i$ is said to *prefer* $h_j$ to $h_k$ if $(r_i, h_j) \in E$ and $(r_i, h_k) \in E$, and $h_j$ precedes $h_k$ on $r_i$'s preference list; the *prefers* relation is defined similarly for a hospital. Let $C = \sum_{h_j \in H} c_j$ and let $m = |E|$.

An *assignment* $M$ is a subset of $E$. If $(r_i, h_j) \in M$, $r_i$ is said to be *assigned* to $h_j$, and $h_j$ is *assigned* $r_i$. For each $a_k \in R \cup H$, the set of assignees of $a_k$ in $M$ is denoted by $M(a_k)$. If $r_i \in R$ and $M(r_i) = \emptyset$, $r_i$ is said to be *unassigned*, otherwise $r_i$ is *assigned*. Similarly, a hospital $h_j \in H$ is *undersubscribed*, *full* or *oversubscribed* according as $|M(h_j)|$ is less than, equal to, or greater than $c_j$, respectively.

A *matching* $M$ is an assignment such that $|M(r_i)| \leq 1$ for each $r_i \in R$ and $|M(h_j)| \leq c_j$ for each $h_j \in H$ (i.e., no resident is assigned to an unacceptable hospital, each resident is assigned to at most one hospital, and no hospital is oversubscribed). For notational convenience, given a matching $M$ and a resident $r_i \in R$ such that $M(r_i) \neq \emptyset$, where there is no ambiguity the notation $M(r_i)$ is also used to refer to the single member of the set $M(r_i)$. We will also use this abbreviation occasionally in connection with a hospital $h_j$ of capacity 1. We now present the stability definition for HR.

**Definition 1.8.** *Let $I$ be an instance of* HR *and let $M$ be a matching in $I$. A pair $(r_i, h_j) \in E \backslash M$ blocks $M$, or is a* blocking pair *for $M$, if the following conditions are satisfied relative to $M$:*

*(1) $r_i$ is unassigned or prefers $h_j$ to $M(r_i)$;*

*(2) $h_j$ is undersubscribed or prefers $r_i$ to at least one member of $M(h_j)$ (or both).*

*$M$ is said to be* stable *if it admits no blocking pair.*

Given an instance $I$ of HR, we are interested in the problem of finding a stable matching in $I$.

We will find the following notation and terminology useful: a resident–hospital pair $(r_i, h_j)$ is *stable* if it belongs to some stable matching in $I$; in this case, $r_i$ is a *stable partner* of $h_j$ and vice versa. A resident–hospital pair is called a *fixed pair* if it belongs to every stable matching in $I$. Also the *underlying graph* of an instance $I$ of HR is a capacitated bipartite graph $\langle G, c \rangle$, such that $G = (V, E)$ and $c : R \cup H \longrightarrow \mathbb{Z}^+$, where $V = R \cup H$, $E$ is as defined above[9], $c(r_i) = 1$ for all $r_i \in R$ and $c(h_j) = c_j$ for all $h_j \in H$. Given a resident $r_i \in R$ and a hospital $h_j \in A(r_i)$, we define the *rank* of $h_j$ in $r_i$'s list, denoted by $rank(r_i, h_j)$, to be 1 plus the number of hospitals that $r_i$ prefers to $h_j$. The rank of $r_i$ in $h_j$'s list, denoted by $rank(h_j, r_i)$, is defined similarly.

### 1.3.3 *Key results (up to 1989)*

Gale and Shapley [235] showed that every instance $I$ of HR admits at least one stable matching. Their proof of this result was constructive, i.e., they described a linear-time algorithm for finding a stable matching in $I$.

The algorithm described by Gale and Shapley for HR is known as the *resident-oriented Gale–Shapley algorithm* (or RGS algorithm for short)[10],

---

[9]Following convention in the literature, in a given HR instance, the acceptable pairs, and hence the edges in $G$, are represented by ordered (resident,hospital) pairs, even though $G$ is strictly speaking undirected. The same is true for other bipartite matching problems considered in this book.

[10]Gale and Shapley described their algorithm in terms of *rounds*, during each of which all unassigned residents simultaneously propose (or apply) to the highest hospital on their list that has not already rejected them. McVitie and Wilson [445, 446] observed

since it involves residents applying to hospitals, The RGS algorithm terminates with the unique *resident-optimal* stable matching, in which each assigned resident has the best hospital that she could achieve in any stable matching, whilst each unassigned resident is unassigned in every stable matching [235] (see also [261, Sec. 1.6.3]). Using a suitable choice of data structures (extending those described in Ref. [261, Sec. 1.2.3]), the RGS algorithm can be implemented to run in $O(m)$ time (recall that $m$ is the number of acceptable resident–optimal pairs). We summarise these observations as follows.

**Theorem 1.9 ([235, 261]).** *Given an instance of* HR, *the RGS algorithm constructs, in $O(m)$ time, the unique resident-optimal stable matching, where $m$ is the number of acceptable resident–hospital pairs.*

The resident-optimal stable matching $M_a$ is worst-possible for the hospitals in a precise sense: if $M$ is any other stable matching then every hospital $h_j \in H$ prefers each resident in $M(h_j)$ to each resident in $M_a(h_j) \backslash M(h_j)$ [261, Sec. 1.6.5].

A counterpart of the RGS algorithm, known as the *hospital-oriented Gale–Shapley algorithm*, or HGS algorithm for short, involves hospitals offering posts to residents. The HGS algorithm terminates with the unique *hospital-optimal* stable matching. In this matching, every full hospital $h_j \in H$ is assigned its $c_j$ best stable partners, whilst every undersubscribed hospital is assigned the same set of residents in every stable matching [261, Sec. 1.6.2]. Again, the HGS algorithm can be implemented to run in $O(m)$ time. We obtain the following counterpart of Theorem 1.9.

**Theorem 1.10 ([261]).** *Given an instance of* HR, *the HGS algorithm constructs, in $O(m)$ time, the unique hospital-optimal stable matching, where $m$ is the number of acceptable resident–hospital pairs.*

In the hospital-optimal stable matching $M_z$, each assigned resident has the worst hospital that she could achieve in any stable matching, whilst each unassigned resident is unassigned in every stable matching [261, Theorem 1.6.1]. Note that the RGS / HGS algorithms are often referred to as *deferred acceptance algorithms* by economists [505].

---

that each round can consist of a single resident proposing (or applying) to a hospital. Indeed, most expositions of the Gale–Shapley algorithm in the literature implicitly adopt McVitie and Wilson's modification without explicit reference to Refs. [445, 446].

In general there may be other stable matchings — possibly exponentially many [319] — between the two extremes given by the resident-optimal and hospital-optimal stable matchings. However some key structural properties hold regarding unassigned residents and undersubscribed hospitals with respect to all stable matchings in $I$, as follows.

**Theorem 1.11 ("Rural Hospitals" Theorem [498, 237, 502]).** *For a given instance of* HR, *the following properties hold:*

*(i) the same residents are assigned in all stable matchings;*

*(ii) each hospital is assigned the same number of residents in all stable matchings;*

*(iii) any hospital that is undersubscribed in one stable matching is assigned exactly the same set of residents in all stable matchings.*

Additional background to the Rural Hospitals Theorem for HR is given in Ref. [261, Sec. 1.6.4].

A classical result in stable matching theory states that, for a given instance of the Stable Marriage problem (see Sec. 1.3.4), the set of stable matchings forms a distributive lattice (Knuth [394] attributes this result to John Conway; see also Ref. [261, Sec. 1.3.1]). In fact such a structure is also present for the set of stable matchings in a given instance $I$ of HR [261, Sec. 1.6.5]. Before describing this structure, we require to define some preliminary notation and terminology. Let $S$ denote the set of stable matchings in $I$ and let $M, M' \in S$. We say that $r_i \in R$ *prefers* $M$ to $M'$ if $r_i$ is assigned in both $M$ and $M'$, and $r_i$ prefers $M(r_i)$ to $M'(r_i)$. Also, we say that $r_i$ *is indifferent between* $M$ and $M'$ if either (i) $r_i$ is unassigned in both $M$ and $M'$, or (ii) $r_i$ is assigned in both $M$ and $M'$, and $M(r_i) = M'(r_i)$.

Given these definitions, we can define a natural dominance partial order on the set of stable matchings as follows.

**Definition 1.12.** *Let $I$ be an* HR *instance and let $S$ be the set of stable matchings in $I$. Given $M, M' \in S$, we say that $M$ dominates $M'$, denoted $M \preceq M'$, if each resident either prefers $M$ to $M'$, or is indifferent between them.*

We denote by $M \wedge M'$ (respectively $M \vee M'$) the set of resident-hospital pairs in which either (i) $r_i$ is unassigned if she is unassigned in both $M$ and $M'$, or (ii) $r_i$ is given the better (respectively poorer) of her partners in $M$ and $M'$ if she is assigned in both stable matchings. It turns out that each of $M \wedge M'$ and $M \vee M'$ is a stable matching in $I$, representing the *join* and the *meet* of $M$ and $M'$ respectively [261, Sec. 1.6.5]. These operations give rise to a lattice structure for $\mathcal{S}$, as the following result indicates.

**Theorem 1.13 ([261]).** *Let $I$ be an instance of* HR, *and let $\mathcal{S}$ be the set of stable matchings in $I$. Then $(\mathcal{S}, \preceq)$ forms a distributive lattice, with $M \wedge M'$ representing the meet of $M$ and $M'$, and $M \vee M'$ the join, for two stable matchings $M, M' \in \mathcal{S}$, where $\preceq$ is the dominance partial order on $\mathcal{S}$ as defined in Definition 1.12.*

### 1.3.4    The Stable Marriage problem (SM)

In this section we present fundamental definitions and results for the Stable Marriage problem, in Secs. 1.3.4.1 and 1.3.4.2 respectively. Due to their importance, we provide definitions and results relating to rotations, a key structural concept in the context of SM), in a separate section (namely Sec. 1.3.4.3).

#### 1.3.4.1    Key definitions

The *Stable Marriage problem with Incomplete lists* (SMI) [235,394,261,514, 315] is an important special case of HR in which $c_j = 1$ for all $h_j \in H$ — in this case, the residents and hospitals are more commonly referred to as the *men* and *women*, and we denote these sets by $U = \{m_1, \ldots, m_{n_1}\}$ and $W = \{w_1, \ldots, w_{n_2}\}$ respectively.

Without loss of generality, in an instance of SMI, we assume that $n_1 = n_2 = n$, i.e., the numbers of men and women are equal (for if not, we may simply add additional men or women as appropriate, each with an empty preference list, until the number of men equals the number of women). We define the *size* of a given instance to be the number of men, which is equal to the number of women, given the previous sentence.

The notation and terminology given for HR in Sec. 1.3.2 immediately carry over to the SMI case. For emphasis, however, we define a blocking pair in the SMI case as follows. A pair $(m_i, w_j) \in E \backslash M$ *blocks* a matching $M$, or is a *blocking pair* for $M$, if the following conditions are satisfied relative to $M$:

(1) $m_i$ is unassigned or prefers $w_j$ to $M(m_i)$;
(2) $w_j$ is unassigned or prefers $m_i$ to $M(w_j)$.

A matching $M$ is said to be *stable* if it admits no blocking pair.

The classical *Stable Marriage problem* (SM) [235, 394, 261, 514, 315] is the special case of SMI in which $E = U \times W$ (that is, each man finds every woman acceptable, and vice versa). SM was first studied by Gale and Shapley [235].

Now let $I$ be an instance of SMI and let $M$ be a stable matching in $I$. Define $U_M$ (respectively $W_M$) to be the set of men in $U$ (respectively women in $W$) who are assigned in $M$. We define the *regret* of $M$ to be:

$$r(M) = \max_{a_i \in U_M \cup W_M} rank(a_i, M(a_i)).$$

We say that $M$ is a *minimum regret stable matching* if $r(M)$ is minimum over all stable matchings in $I$.

We now define the *cost of $M$ for the men* to be

$$c^U(M) = \sum_{m_i \in U_M} rank(m_i, M(m_i)).$$

We define the *cost of $M$ for the women*, denoted $c^W(M)$, similarly. Given these two definitions, we define the *cost of $M$* to be $c(M) = c^U(M) + c^W(M)$. We say that $M$ is an *egalitarian stable matching* if $c(M)$ is minimum over all stable matchings in $I$.

We define the *sex-equality measure* of $M$ to be

$$d(M) = c^U(M) - c^W(M).$$

We say that $M$ is a *sex-equal stable matching* if $|d(M)|$ is minimum over all stable matchings in $I$.

Finally suppose that each man $m_i \in U$ has a positive real-numbered weight $wt(m_i, w_j)$ for each woman $w_j$ that he finds acceptable, and suppose that each woman does likewise. We assume that $m_i$ prefers $w_j$ to $w_k$ if and only if $wt(m_i, w_j) < wt(m_i, w_k)$ (and similarly for the women). Now let $a_i \in U_M \cup W_M$. We define the *weight* of $a_i$ with respect to $M$, denoted by $wt_M(a_i)$, to be $wt(a_i, M(a_i))$. We define the *weight* of $M$ to be

$$wt(M) = \sum_{a_i \in U_M \cup W_M} wt_M(a_i).$$

Then a stable matching $M$ in $I$ is of *minimum weight* if $wt(M)$ is minimum, taken over all stable matchings in $I$ (such a matching is referred to as *optimal* in Ref. [261]). Clearly, in the case that $wt(m_i, w_j) = rank(m_i, w_j)$ and $wt(w_j, m_i) = rank(w_j, m_i)$ for each acceptable pair $\{m_i, w_j\}$, a minimum weight stable matching is an egalitarian stable matching.

**1.3.4.2   Key results (up to 1989)**

Clearly the key results mentioned in Sec. 1.3.3 apply to SMI (and hence SM), since SMI is a special case of HR. Here we state some additional results for SMI that do not follow from those in Sec. 1.3.3.

**Theorem 1.14.** *Let $I$ be an instance of* SMI *of size $n$, where $m$ is the number of acceptable pairs. Then:*

(i) *A minimum regret stable matching in $I$ can be found in $O(m)$ time [259].*

(ii) *An egalitarian stable matching in $I$ can be found in $O(m^2)$ time [320, 261].*

(iii) *A minimum weight stable matching in $I$ can be found in $O(m^2 \log n)$ time [320, 261].*

We remark that faster algorithms for (ii) and (iii) are reviewed in Sec. 2.2.10.

The following straightforward result establishes a close relationship between SM and SMI. Its proof is included for completeness since it does not appear to have been explicitly given in the literature previously, although the result itself can probably be attributed to "folklore".

**Proposition 1.15.** *Let $I$ be an instance of* SMI *with $n_1$ men and $n_2$ women. Then there exists an instance $I'$ of* SM *of size $n$, where $n = \max\{n_1, n_2\}$, such that the stable matchings in $I$ are in 1-1 correspondence with the stable matchings in $I'$.*

**Proof.** Suppose that the men and women in $I$ are given by $U = \{m_1, \ldots, m_{n_1}\}$ and $W = \{w_1, \ldots, w_{n_2}\}$ respectively. Without loss of generality assume that $n_1 \leq n_2$ and let $n = n_2$. Let $I'$ comprise the men in $U \cup U'$ and the women in $W$, where $U' = \{m_{n_1+1}, \ldots, m_n\}$. In $I'$, each person in $U \cup W$ initially has his/her preference list in $I$, whilst each man in $U'$ initially has an empty list. Given any man $m_i \in U \cup U'$, if $W_i$ denotes the set of women initially on his list in $I'$, append the women in $W \backslash W_i$ to his list in $I'$ in increasing indicial order. Similarly, given any woman $w_j \in W$, if $U_j$ denotes the set of men initially on her list in $I'$, append the men in $(U \cup U') \backslash U_j$ to her list in $I'$ in increasing indicial order. We claim that the stable matchings in $I$ and $I'$ are in 1-1 correspondence.

For, let $M$ be a stable matching in $I$. Let $U_M = \{m_{j_1}, \ldots, m_{j_p}\}$ and $W_M = \{w_{k_1}, \ldots, w_{k_q}\}$ denote the sets of men and women respectively who are unassigned in $M$ as a matching in $I$. Then $p + n_2 - n_1 = q$. Let $m_{j_r}$

denote $m_{n_1+r-p}$ $(p+1 \leq r \leq q)$, and let $M'' = \{(m_{j_r}, w_{k_r}) : 1 \leq r \leq q\}$. Then $M \cup M''$ is stable in $I'$. Conversely suppose that $M'$ is a stable matching in $I'$. Then $M' \cap (U \times W)$ is stable in $I$. It is also not difficult to see that this correspondence is 1–1. $\qquad \square$

### 1.3.4.3 *Rotations*

Let $I$ be an instance of SMI and let $M$ be a stable matching in $I$. Given any man $m_i \in U_M$ in $I$, let $s_M(m_i)$ denote the most-preferred woman $w_j$ on $m_i$'s list such that $w_j \in W_M$ and $w_j$ prefers $m_i$ to $M(w_j)$. Let $next_M(m_i)$ denote $M(s_M(m_i))$. Then a *rotation (exposed)* in $M$ is a sequence $\rho = (m_{i_0}, w_{j_0}), \ldots, (m_{i_{r-1}}, w_{j_{r-1}})$ of pairs such that, for each $k$ $(0 \leq k \leq r-1)$, $(m_{i_k}, w_{i_k}) \in M$ and $m_{i+1} = next_M(m_i)$, where (here and henceforth in connection with rotations) addition is taken modulo $r$.

Let $\rho$ be a rotation exposed in a matching $M$. The *elimination* of $\rho$ from $M$, denoted $M/\rho$, is defined to be the following matching:

$$M/\rho = (M \backslash \{(m_{i_k}, w_{i_k}) : 0 \leq k \leq r-1\}) \cup \{(m_{i_k}, w_{i_{k+1}}) : 0 \leq k \leq r-1\}.$$

A key result [261, Lemma 2.5.2] is that $M/\rho$ is a stable matching that is dominated by $M$ (under the partial order $\preceq$ defined in Definition 1.12). Moreover if $M$ is any stable matching other than the woman-optimal stable matching $M_z$, then there is at least one rotation exposed in $M$ [261, Lemma 2.5.3].

A pair $(m_i, w_j)$ is a stable pair if and only if $(m_i, w_j) \in M_z$, or $(m_i, w_j)$ belongs to some rotation in $I$ [261, Theorem 2.5.6].

A partial order $\lhd$ is defined on the set of rotations $R(I)$ as follows: $\rho \lhd \sigma$ if and only if, starting from the man-optimal stable matching $M_a$, $\rho$ must be eliminated to give a stable matching in which $\sigma$ is exposed. The rotations under $\lhd$ form the *rotation poset* $(R(I), \lhd)$ for $I$. A subset $S$ of $R(I)$ is *closed* if, whenever $\rho$ is in $S$, so also is every rotation $\sigma$ such that $\sigma \lhd \rho$. The *rotation digraph* $D(I)$ (see Ref. [261, p.105] for the formal definition of $D(I)$) is an efficient representation of $(R(I), \lhd)$ that may be constructed in $O(n^2)$ time, where $n$ is the size of $I$; the transitive closure of the rotation digraph $D(I)$ is precisely the rotation poset $(R(I), \lhd)$.

The following theorem encapsulates the relationship between the rotation poset and the set of all stable matchings in $I$.

**Theorem 1.16 ([319, 261]).** *Let $I$ be an instance of SMI. There is a 1–1 correspondence between the stable matchings in $I$ and the closed subsets of the rotation poset of $I$.*

Exploitation of the rotation poset gives rise to the following algorithmic results for SMI:

**Theorem 1.17** ([259, 261]). *Let $I$ be an instance of* SMI *of size $n$, where $m$ is the number of acceptable pairs in $I$. Then:*

(i) *The stable pairs in $I$ can be found in $O(m)$ time.*

(ii) *The rotations in $I$ can be found in $O(m)$ time.*

(iii) *The stable matchings in $I$ can be listed in $O(n)$ time per matching, after $O(m)$ pre-processing time.*

Many further structural and algorithmic results for SM, especially those that appeared in the literature subsequently to the publication of Gusfield and Irving's book [261] are presented in Chap. 2.

### 1.3.5   The Hospitals / Residents problem with indifference

The National Resident Matching Program in the US (mentioned in Sec. 1.1.1.3), the Canadian Resident Matching Service [603] and the Scottish Foundation Allocation Scheme [604] handle the assignment of graduating medical students to junior doctor appointments in hospitals in their respective countries. At the heart of these matching schemes are efficient algorithms that essentially solve some variant of the Hospitals / Residents problem (HR). In large-scale matching schemes of this kind, participants, particularly large popular hospitals, may not be able to provide a genuine strict preference order over what may be a very large number of applicants, and may prefer to express indifference in their preference lists.

The most general form of indifference can be modelled by the *Hospitals / Residents problem with Partially-ordered lists* (HRP). An instance $I$ of HRP is defined similarly to an instance of HR (see Sec. 1.3.2), except that instead of preference lists, each resident $r_i \in R$ has a *preference poset* $\prec_{r_i} \subseteq A(r_i) \times A(r_i)$, and similarly each hospital $h_j \in H$ has a preference poset $\prec_{h_j} \subseteq A(h_j) \times A(h_j)$. We say that a resident $r_i \in R$ *prefers* $h_j$ to $h_k$ if $h_j \prec_{r_i} h_k$; the *prefers* relation is defined similarly for a hospital. Also we say that a resident $r_i \in R$ is *indifferent between* $h_j$ and $h_k$, denoted by $h_j \sim_{r_i} h_k$, if $\{h_j, h_k\} \subseteq A(r_i)$, $h_j \not\prec_{r_i} h_k$ and $h_k \not\prec_{r_i} h_j$. The $\sim_{h_j}$ relation is defined similarly for a hospital $h_j \in H$. The notation and terminology corresponding to the extension of the *prefers* and *indifferent between* relation to matchings (e.g., $r_i$ prefers $M$ to $M'$, etc.) can be defined in an analogous way to the corresponding definitions in Sec. 1.3.3.

If $\sim_{a_k}$ is transitive for each agent $a_k \in R \cup H$, we obtain an instance of the Hospitals / Residents problem with Ties (HRT). Here each equivalence class under $\sim_{a_k}$ is referred to as a *tie*, and $a_k$ has a linear order over the equivalence classes of $A(a_k)$ under $\sim_{a_k}$. Intuitively, in an instance of HRT, each agent $a_k$ ranks its acceptable partners in tied batches (where a tie can be of length 1), $a_k$ is indifferent between the members of each tie, and prefers each member of a given tie to each member of any successor tie. We more commonly refer to *preference lists* again in this context, rather than preference posets. For any acceptable resident–hospital pair $(r_i, h_j)$, $rank(r_i, h_j)$ and $rank(h_j, r_i)$ are defined as in the HR case.

The concept of indifference can, of course, be applied to instances of SM. The *Stable Marriage problem with Partially ordered and Incomplete lists* (denoted by SMPI) is the restriction of HRP in which each hospital has capacity 1, and the *Stable Marriage problem with Partially ordered lists* (denoted by SMP) is the restriction of SMPI in which all man-woman pairs are acceptable. Analogously, the *Stable Marriage problem with Ties and Incomplete lists* (denoted by SMTI) is the 1–1 restriction of HRT. Finally, the *Stable Marriage problem with Ties* (denoted by SMT) is the special case of SMTI [261, 332, 414, 419, 333] in which every man–woman pair is an acceptable pair.

Three natural stability criteria were defined for SMT [261, 308], and these definitions were generalised first to SMTI [414] and then to HRP [326, 328]. These criteria define three levels of stability in such instances, called *weak stability*, *strong stability* and *super-stability*. For full generality, we now define these terms in the context of HRP.

Given a matching $M$ in an instance $I$ of HRP, a pair $(r_i, h_j) \in E \backslash M$ is said to *block* $M$, or to be a *blocking pair* of $M$ if the following conditions are satisfied depending on the desired level of stability:

- *weak stability*:

  (i) $r_i$ is unassigned or prefers $h_j$ to her assigned hospital in $M$, *and*
  (ii) $h_j$ is undersubscribed or prefers $r_i$ to its worst assigned resident in $M$;

- *strong stability*: *either* (i)

  (a) $r_i$ is unassigned or prefers $h_j$ to her assigned hospital in $M$, *and*
  (b) $h_j$ is undersubscribed or prefers $r_i$ to its worst assigned resident in $M$ or is indifferent between them;

  *or* (ii)

  (a) $r_i$ is unassigned or prefers $h_j$ to her assigned hospital in $M$ or is indifferent between them, *and*

Men's preferences      Women's preferences

$m_1 : (w_2 \; w_3) \; w_1$      $w_1 : m_1 \; (m_2 \; m_3)$

$m_2 : (w_1 \; w_3) \; w_2$      $w_2 : m_2 \; (m_1 \; m_3)$

$m_3 : (w_1 \; w_2) \; w_3$      $w_3 : m_3 \; (m_1 \; m_2)$

Fig. 1.1    An instance of SMT

(b) $h_j$ is undersubscribed or prefers $r_i$ to its worst assigned resident in $M$;

• *super-stability*:

(i) $r_i$ is unassigned or prefers $h_j$ to her assigned hospital in $M$ or is indifferent between them, *and*

(ii) $h_j$ is undersubscribed or prefers $r_i$ to its worst assigned resident in $M$ or is indifferent between them.

Note that, in the context of HRP, a *worst assigned resident* of a hospital $h_j \in H$ is any $r_i \in M(h_j)$ such that there is no $r_k \in M(h_j)$ satisfying $r_i \prec_{h_j} r_k$. $M$ is said to be be *weakly stable, strongly stable* or *super-stable* if it admits no blocking pair with respect to the relevant definition above. Clearly a super-stable matching is strongly stable, and a strongly stable matching is weakly stable. Also, if the preference lists of one set of agents are strictly ordered, clearly the super-stability and strong stability criteria are equivalent. Finally, we remark that the terms *weakly stable pair* and *weakly stable partner* may be defined analogously to *stable pair* and *stable partner* respectively, with a similar comment applying in each of the strong and super-stability cases.

By way of illustration, consider the instance of SMT shown in Fig. 1.1. Here, and henceforth, preference lists are ordered from left to right in decreasing order of preference, and entries in round brackets are tied. The matching

$$M_1 = \{(m_1, w_1), (m_2, w_3), (m_3, w_2)\}$$

is weakly stable, but not strongly stable due to the blocking pair $(m_1, w_2)$ for example. Also

$$M_2 = \{(m_1, w_2), (m_2, w_3), (m_3, w_1)\}$$

is strongly stable, but not super-stable due to the blocking pair $(m_1, w_3)$ for example. Finally the matching

$$M_3 = \{(m_1, w_1), (m_2, w_2), (m_3, w_3)\}$$

is super-stable.

Minimum regret, egalitarian, sex-equal and minimum weight weakly stable, strongly stable and super-stable matchings can be defined in a manner analogous to Sec. 1.3.4.1.

Given an instance $I$ of HRT, a weakly stable matching is bound to exist and can be found in linear time [419]. However the weakly stable matchings in $I$ may be of different sizes and the problem of finding such a matching of maximum size is NP-hard [419]. On the other hand a strongly stable matching need not exist in $I$, though there is a polynomial-time algorithm to find such a matching or report that none exists [328, 364]. A similar remark applies to super-stable matchings [326]. Structural and algorithmic results for HRP under each of the three stability criteria defined above are outlined in much greater detail in Chap. 3.

### 1.3.6 *Other variants of* HR

#### 1.3.6.1 *Couples*

One key extension of HR that has considerable practical importance arises when an instance may involve a set of couples, each of which submits a joint preference list over pairs of hospitals (typically in order that the members of a given couple can be located geographically close to one another, for example). The extension of HR in which couples may be involved is denoted by HRC; the stability definition in HRC is a natural extension of that in HR (see Sec. 5.3 for a formal definition of HRC). It is known that an instance of HRC need not admit a stable matching (see Refs. [498], [261, Sec. 1.6.6] and [514, Sec. 5.4.3]). Moreover, the problem of deciding whether an HRC instance admits a stable matching is NP-complete, even if there are no single residents and each hospital has capacity 1 [493]. See Sec. 5.3 for further details.

#### 1.3.6.2 *Many–many stable matchings*

HR may be regarded as a many–one generalisation of SMI. A further generalisation of SM is to a many–many bipartite stable matching problem, in which both residents and hospitals may be multiply assigned subject to capacity constraints. In this case, residents and hospitals are more commonly referred to as *workers* and *firms* respectively. There are two basic variations of the many–many stable matching problem according to whether (i) workers rank acceptable firms in order of preference and vice versa, or (ii) workers rank acceptable *subsets* of firms in order of preference and vice versa. See Sec. 5.4 for more details concerning both models.

### 1.3.6.3   *Master lists*

In some practical situations, the preference lists on one or both sides of an HR instance may be derived from one or two *master lists*, which may or may not contain ties. To be more precise, a *master list* of residents consists of a single list containing all of the residents, which may or may not involve ties; each hospital's preference list contains its acceptable partners ranked precisely according to the master list. In other words, the preference list of a hospital $h_j$ is precisely the master list of residents, except that each resident $r_i$ that $h_j$ finds unacceptable is deleted (so in general, the deletions that give rise to $h_j$'s preference list could be made from any part of the master list). An analogous meaning is attached to a master list of hospitals. Algorithmic results for variants of HR involving master lists are described throughout Chap. 3.

### 1.3.7   *Motivation*

Practical applications of HR are widespread, most notably arising in the context of centralised automated matching schemes that assign applicants to posts (for example medical students to hospitals, school-leavers to universities, and primary school pupils to secondary schools). Perhaps the best-known example is the NRMP, which handled over 38,000 applicants in 2012 (see Sec. 1.1.1.3). Perhaps the largest existing centralised matching scheme is the one that handles higher education admission in China : there were around 10 million applicants to Chinese higher education institutions in 2007 [593].

Counterparts of the NRMP are in existence in other countries, including Canada [603], Scotland [604] (the *Scottish Foundation Allocation Scheme (SFAS)*) and Japan [596, 354, 355]. These matching schemes essentially employ extensions of the RGS algorithm for HR. In particular, the NRMP has employed a resident-oriented version of the Gale–Shapley algorithm since 1997, having used a hospital-oriented version previously [506].

We describe briefly some practical involvement with SFAS. Rob Irving, of the School of Computing Science, University of Glasgow, has led a collaboration with the author involving the design and implementation of algorithms for SFAS which have been used by NHS Education for Scotland for annual runs of the matching scheme since 2000. In 2012 for example, 710 applicants (as residents are referred to in the SFAS context) were seeking to be assigned to 720 available posts on 52 programmes (*programme* is used rather than *hospital* here, since several programmes might be running at the same hospital). The algorithm allocated 683 applicants to programmes in the first round, giving 470 applicants their first choice. The 27 unassigned applicants were all allocated to programmes in a second run of the algorithm after providing a fresh preference list over the remaining programmes with vacancies.

Centralised matching schemes based largely on HR also occur (or at least have been mooted) in other many contexts, including the following:

- assigning children to daycare places in Denmark [375];
- school placement in Boston [1, 4], Hungary [77, 90], New York [2] and Singapore [567][11];
- higher education admission in China [593, 594], Germany [111], Hungary [77, 90], Spain [491] and Turkey [64, 60, 61]);
- awarding tuition exchange scholarships to dependents of faculty members [177, 601];
- handling the sorority rush in US universities [454];
- university faculty recruitment in France [60, 61];
- placing military cadets in branches [540];
- assigning naval cadets to billets [490, 589, 406];
- hiring federal judicial law clerks [52, 53];
- placement of graduating rabbis [100];
- online dating in the US [277] and online matrimony in India [352];
- auction mechanisms for sponsored search [32];
- supply-chain networks [471];
- metal-only ECO synthesis [349].

A number of additional examples of where centralised matching schemes (with HR as the underlying problem model) are used are given in Ref. [505]. See also Refs. [598–600].

## 1.4 The Stable Roommates problem (SR)

### 1.4.1 *Introduction*

In this section we present key definitions and results for the Stable Roommates problem, in Secs. 1.4.2 and 1.4.3 respectively. As in the SM case, we provide definitions and results relating to rotations separately, in Sec. 1.4.4.

---

[11]Here, the authors argue that school placement should be handled via a matching scheme based on constructing stable matchings.

Preference lists involving forms of indifference are considered in Sec. 1.4.5. Finally, motivation for studying the Stable Roommates problem is given in Sec. 1.4.6.

### 1.4.2  Key definitions

The Stable Roommates problem (SR) [235,394,306,261,514,321] (sometimes referred to as the Stable Matching problem [11, 551, 205, 211]) is a non-bipartite generalisation of SM. We begin by defining an extension of SR, called the *Stable Roommates problem with Incomplete lists* (SRI) [261, Sec. 4.5.2]. Here an instance comprises a single set $A = \{a_1, \ldots, a_n\}$ of *agents*, and $n$ denotes the *size* of the instance. Also there is a set $E$ of unordered pairs of agents, termed the *acceptable* pairs. Each agent $a_i \in A$ has an *acceptable* set of agents $A(a_i)$, where

$$A(a_i) = \{a_j \in E : \{a_i, a_j\} \in E\}.$$

Also, each agent $a_i \in A$ has a *preference list* in which she ranks $A(a_i)$ in strict order. Given any three distinct agents $a_i$, $a_j$ and $a_k$, the definition of $a_i$ *prefers* $a_j$ to $a_k$ can be arrived at in an analogous way to the definition given in Sec. 1.3.2. As before, we let $m = |E|$.

An *assignment* $M$ is a subset of $E$. If $\{a_i, a_j\} \in M$, $a_i$ is said to be *assigned to* $a_j$. With respect to $M$, if $a_i$ is not assigned to any agent then $a_i$ is said to be *unassigned*, otherwise $a_i$ is *assigned*.

A *matching* $M$ is an assignment such that no agent belongs to more than one pair of $M$. If $\{a_i, a_j\} \in M$ then we let $M(a_i)$ denote $a_j$. Let $A_M$ denote the set of agents who are assigned in $M$. A pair $\{a_i, a_j\} \in E \backslash M$ *blocks* a matching $M$, or is a *blocking pair* for $M$, if the following conditions are satisfied relative to $M$:

(1) $a_i$ is unassigned or prefers $a_j$ to $M(a_i)$;
(2) $a_j$ is unassigned or prefers $a_i$ to $M(a_j)$.

A matching $M$ is said to be *stable* if it admits no blocking pair. An instance $I$ of SRI is said to be *solvable* if $I$ admits a stable matching; $I$ is *unsolvable* otherwise.

As in Sec. 1.3.2, for a given instance $I$ of SRI, it is straightforward to define the concepts of a *stable pair*, a *stable partner*, and the *underlying graph* $G = (A, E)$ of $I$. The latter is in general a non-bipartite graph, where $A$ and $E$ are as defined above. Also, for any two distinct agents $a_i, a_j$ such that $\{a_i, a_j\} \in E$, we define $rank(a_i, a_j)$ to be 1 plus the number of agents that $a_i$ prefers to $a_j$.

In the context of an SRI instance, it is straightforward to extend the notions of the *regret* and the *cost* of a given stable matching $M$, in order to define the concepts of a *minimum regret* and an *egalitarian* stable matching. Similarly it is an easy matter to extend the definition of a minimum weight stable matching from the SMI case (as given in Sec. 1.3.4.1) to the SRI context. The classical Stable Roommates problem (SR) is the special case of SRI in which $n$ is even and all pairs are acceptable, i.e., $E$ contains all pairs of distinct agents in $A$.

### 1.4.3  Key results (up to 1989)

Gale and Shapley [235] showed that an instance of SR need not admit a stable matching. The following result, due to Irving [306], indicates that it is possible to determine in polynomial time whether a given instance admits a stable matching, and if so, to find such a matching.

**Theorem 1.18 ([306]).** *Given an instance $I$ of SR with $n$ agents, there in an $O(n^2)$ algorithm that finds a stable matching in $I$ or reports that $I$ is unsolvable.*

It is straightforward to generalise this algorithm to the SRI case [261, Sec. 4.5.2]. In this case, the time complexity is $O(m)$, where $m$ is the number of acceptable pairs of agents.

Clearly SM is a special case of SRI. Moreover, given an instance $I$ of SM, there is an SR instance $J$ (i.e., with complete preference lists) such that there is a 1–1 correspondence between the stable matchings in $I$ and those in $J$ [261, p163].

In a given SRI instance $I$, clearly it is possible that some agents may be unassigned in a stable matching in $I$. However, a counterpart to Theorem 1.11 holds in the SRI context, as follows.

**Theorem 1.19 ([261]).** *Let $I$ be a solvable instance of SRI. The same set of agents are unassigned in every stable matching in $I$.*

A useful structural result for stable matchings in an instance $I$ of SR concerns the *median choice* of a given agent. In order to define this, let $M_1$, $M_2$ and $M_3$ be three stable matchings in $I$, and let $a_i \in A_{M_1}$. By Theorem 1.19, $a_i$ is assigned in each of $M_1$, $M_2$ and $M_3$. If $M_j(a_i) = M_k(a_i) = a_l$

for any $j, k$ ($1 \leq j < k \leq 3$), then $a_i$'s median choice is $a_l$. Otherwise $a_i$ has three distinct stable partners in $M_1$, $M_2$ and $M_3$; denote these by $a_j$, $a_k$ and $a_l$, where, without loss of generality, $a_i$ prefers $a_j$ to $a_k$ to $a_l$. Then $a_i$'s median choice is $a_k$. The following result indicates that giving each agent their median choice among three stable matchings in fact yields another stable matching.

**Theorem 1.20 ([261]).** *Let $I$ be an instance of* SR *and let $M_1$, $M_2$ and $M_3$ be three stable matchings in $I$. Let $M$ be the stable matching obtained by giving each agent in $A_{M_1}$ her median choice among $M_1$, $M_2$ and $M_3$. Then $M$ is a stable matching.*

A counterpart of sorts of Theorem 1.13 holds in the case of SR, namely it is possible to identify a semilattice structure for the set of stable matchings in a given solvable SR instance $I$. In order to state this result, we require to define some notation. Given a stable matching $M$ in $I$, define $P(M)$ to be the set of ordered pairs $(a_i, a_j) \in A_M \times A_M$ such that $\{a_i, a_j\} \in M$ or $a_i$ prefers $a_j$ to $M(a_i)$. Now let $M_0$ be a fixed stable matching in $I$. Define $P_0(M) = P(M) \oplus P(M_0)$, where $\oplus$ denotes the symmetric difference operator.

**Theorem 1.21 ([261]).** *Let $I$ be an instance of* SR *and let $\mathcal{S}$ be the set of stable matchings in $I$. Let $M_0 \in \mathcal{S}$ be a fixed stable matching. Then the set $\{P_0(M) : M \in \mathcal{S}\}$ is closed under intersection and thus forms a meet-semilattice in which $P_0(M_0)$ is the minimal element.*

An alternative semilattice structure, considered to be "more natural" than the above, has been proposed by Cheng and Lin [146] (see Sec. 4.4).

Our final observation of this section is that there is an efficient algorithm for finding a minimum regret stable matching, given an SRI instance.

**Theorem 1.22 ([261]).** *Let $I$ be a solvable* SRI *instance. A minimum regret stable matching in $I$ can be found in $O(m)$ time, where $m$ is the number of acceptable pairs of agents.*

### 1.4.4  Rotations

Let $I$ be an instance of SRI. Irving's algorithm [306, 261] for finding a stable matching in $I$, or reporting that none exists, consists of two phases. The first phase is analogous to an extended form of the classical Gale–Shapley algorithm for SM [235]; it involves a sequence of "proposals" from a given agent $a_i$ to the first agent $a_j$ on her list, where such a proposal results in the deletion of all successors of $a_i$ from $a_j$'s list. (Here, and henceforth, the deletion of $a_k$ from the list of $a_j$ automatically implies the deletion of $a_j$ from the list of $a_k$.) On termination of this phase, the (reduced) preference lists form what is called a *stable table* [261, p.169]; among the properties of such a table are that all first entries are distinct, and that $a_j$ is first in $a_i$'s list if and only if $a_i$ is last in that of $a_j$.

A *rotation* $\rho$ exposed in a stable table $T$ is a sequence $\rho = (a_{i_0}, a_{j_0})$, ..., $(a_{i_{r-1}}, a_{j_{r-1}})$ of pairs such that $a_{j_k}$ is first and $a_{j_{k+1}}$ second in $a_{i_k}$'s list in $T$, for each $k$ $(0 \le k \le r - 1)$, where arithmetic with respect to rotations is taken modulo $r$. *Elimination* of the rotation involves deleting all successors of $a_{i_{k-1}}$ from the list of $a_{j_k}$, for each $k$ $(0 \le k \le r - 1)$. A key result is that, provided no list becomes empty as a consequence, the elimination of an exposed rotation from a stable table gives another (smaller) stable table. Phase 2 of the algorithm consists of the successive elimination of rotations from the current stable table until either some list becomes empty as a result, in which case no stable matching exists, or all lists that were non-empty after phase 1 are reduced to a single entry, in which case these entries constitute a stable matching. In what follows of this section, we assume that $I$ is solvable. At the end of phase 1, we may identify the *fixed pairs* of $I$ — these are the stable pairs that belong to every stable matching in $I$. A pair $\{a_i, a_j\}$ is a fixed pair if and only if $a_i$'s list contains only $a_j$ at the termination of phase 1 [261, Lemma 4.4.1].

Suppose that $\rho = (a_{i_0}, a_{j_0}), \ldots, (a_{i_{r-1}}, a_{j_{r-1}})$ is a rotation that is exposed in some stable table. The *syntactic dual* of $\rho$ is $\bar{\rho} = (a_{j_1}, a_{i_0}), \ldots, (a_{j_0}, a_{i_{r-1}})$. If there is some sequence of rotations that leads to a stable table in which $\bar{\rho}$ is exposed, then $\bar{\rho}$ is also a rotation; in this case $\rho$ and $\bar{\rho}$ are called *non-singular* rotations, and are *duals* of each other, otherwise $\rho$ is *singular*. (Hence the syntactic dual of a singular rotation is not actually a rotation at all.) A partial order $\lhd$ is defined on the set of rotations as follows: $\rho \lhd \sigma$ if and only if $\rho$ must be eliminated to give a stable table in which $\sigma$ is exposed. The rotations under $\lhd$ form the *rotation poset* for $I$. A subset $S$ of this poset is *closed* if, whenever $\rho$ is in $S$, so also is every rotation $\sigma$ such that $\sigma \lhd \rho$. Also $S$ is *complete* if $S$ contains every singular rotation of $I$, together with exactly one of each dual pair of non-singular rotations.

The following theorem encapsulates the relationship between the rotation poset and the set of all stable matchings in $I$.

**Theorem 1.23 ([260, 261]).** *Let $I$ be a solvable* SRI *instance. There is a 1-1 correspondence between the stable matchings in $I$ and the complete closed subsets of the rotation poset of $I$.*

Exploitation of the rotation poset gives rise to the following algorithmic results for SRI:

**Theorem 1.24 ([260, 261]).** *Let $I$ be an instance of* SRI *with $n$ agents, where $m$ is the number of acceptable pairs in $I$. Then:*

(i) *The stable pairs in $I$ can be found in $O(nm \log n)$ time.*

(ii) *The rotations in $I$ can be found and determined as singular or nonsingular in $O(nm \log n)$ time.*

(iii) *The stable matchings in $I$ can be listed in $O(m)$ time per solution, after $O(nm \log n)$ pre-processing time.*

We remark that faster algorithms for the problems identified in (i), (ii) and (iii) of Theorem 1.24 are discussed in Sec. 4.2.4.

Many further structural and algorithmic results for SR, especially those that appeared in the literature subsequently to the publication of Gusfield and Irving's book [261] are presented in Chap. 4.

### 1.4.5 The Stable Roommates problem with indifference

In Sec. 1.3.5 we considered instances of HR and SMI in which preference lists may contain ties, and other forms of indifference. In this section we do likewise for instances of SRI, The abbreviation SRPI (respectively SRP) represents the generalisation of SRI (respectively SR) in which preference lists may be partially ordered. It is a straightforward matter to modify the definitions of the terms *preference poset*, *prefers* and *indifferent between* given in Sec. 1.3.5 so that they apply in the SRPI setting, and we adopt analogous notation to that used in Sec. 1.3.5. As in that section, if $\sim_{a_i}$ is transitive for each agent $a_i$ then we obtain an instance of SRTI (respectively SRT), representing the generalisation of SRI (respectively SR) in which preference lists may contain ties.

As in the HRP context, we may define *weak stability*, *strong stability* and *super-stability* [321]. Given a matching $M$ in an instance $I$ of SRPI, a pair $\{a_i, a_j\} \in E \backslash M$ is said to *block* $M$, or to be a *blocking pair* of $M$ if the following conditions are satisfied depending on the desired level of stability:

- *weak stability*: $a_i$ is unassigned or prefers $a_j$ to $M(a_i)$, and similarly $a_j$ is unassigned or prefers $a_i$ to $M(a_j)$;
- *strong stability*: $a_i$ is unassigned or prefers $a_j$ to $M(a_i)$, and $a_j$ is unassigned or prefers $a_i$ to $M(a_j)$ or is indifferent between them;
- *super-stability*: $a_i$ is unassigned or prefers $a_j$ to $M(a_i)$ or is indifferent between them, and similarly $a_j$ is unassigned or prefers $a_i$ to $M(a_j)$ or is indifferent between them.

The terms *weakly stable pair* and *weakly stable partner* may be defined analogously to *stable pair* and *stable partner* respectively, with a similar comment applying in each of the strong and super-stability cases.

Structural and algorithmic results for SRPI under each of the three stability criteria defined above are contained in Sec. 4.5.

### 1.4.6  Motivation

As the problem name suggests, an application of SR arises in the context of campus housing allocation, where we seek to assign students to share two-person rooms, based on their preferences over one another [474, 50, 475].

Recently, an application of SRI in the medical domain has been studied, involving pairwise kidney exchange markets [511, 512, 17, 513, 423]. Here, a patient with chronic kidney disease who has a willing but incompatible donor may be able to obtain a transplant by swapping their donor with that of another patient in a similar position. Centralised schemes are in existence in many countries, including the US [605], the Netherlands [373, 393] and the UK [597], constructing sets of kidney exchanges among incompatible patient–donor pairs at regular intervals.

We can model the basic market by constructing an agent for each patient, and an undirected edge between any two agents where the incompatible donor for one patient is compatible with the other patient, and vice versa. An edge $\{p_i, p_j\}$ in a matching corresponds to a *pairwise kidney exchange*, in which $p_j$ receives a kidney from $p_i$'s donor in exchange for $p_i$ receiving a kidney from $p_j$'s donor. Preference lists can be constructed on the basis of varying degrees of compatibility between patients and potential donors.

A third application arises in P2P networks [232, 233, 403, 429–431]. For example [403, 431], in a P2P file-sharing network, a given peer may form a preference list over other peers based on the similarity of their interests. In cooperative download applications such BitTorrent, preference functions may be derived from properties such as download / upload bandwidth, latency and storage capacity. The "Tit-for-Tat" strategy of BitTorrent can give rise to preference lists for peers that are based on a master list of peers according to upload capacity.

A final application occurs in the context of forming pairings of players for chess tournaments [401].

## 1.5 The House Allocation problem (HA) and its variants

### 1.5.1 *Introduction*

Many economists and game theorists, and increasingly computer scientists in recent years, have studied the problem of allocating a set $H$ of indivisible goods among a set $A$ of applicants [527, 301, 166, 206]. Each applicant $a_i$ may have ordinal preferences over a subset of $H$ (the *acceptable* goods for $a_i$). Many models have considered the case where there is no monetary transfer. In the literature the situation in which each applicant initially owns one good is known as a *Housing Market*[12] (HM) [527, 508, 497]. When there are no initial property rights, we obtain the *House Allocation problem* (HA) [301, 595, 5]. A mixed model, in which a subset of applicants initially owns a good has also been studied [6].

In this section we begin by defining HA and HM formally in Sec. 1.5.2. We then consider a range of optimality criteria that can be applied to instances of HA and HM, namely *Pareto optimality* (see Sec. 1.5.3), *maximum utility* (see Sec. 1.5.4), *popularity* (see Sec. 1.5.5) and *profile-based optimality* (see Sec. 1.5.6). Extensions of HA arise when houses may be assigned multiple applicants up to some fixed capacity, and/or the preference lists of agents may contain ties. These generalisations are defined in Sec. 1.5.7. Finally, motivation for studying HA and its variants is given in Sec. 1.5.8.

### 1.5.2 *Formal definition of* HA *and* HM

Formally, an instance $I$ of HA comprises a set $A = \{a_1, a_2, \ldots, a_{n_1}\}$ of *applicants* and a set $H = \{h_1, h_2, \ldots, h_{n_2}\}$ of *houses*. The *agents* in $I$ are the applicants and houses in $A \cup H$. There is a set $E \subseteq A \times H$ of *acceptable* applicant–house pairs. Let $m = |E|$. Each applicant $a_i \in A$ has

---

[12]This problem is also referred to as the *House-swapping Game* in the literature [125, 131, 487, 105].

an *acceptable* set of houses $A(a_i)$, where

$$A(a_i) = \{h_j \in H : (a_i, h_j) \in E\}.$$

Similarly each house $h_j \in H$ has an acceptable set of applicants $A(h_j)$, where

$$A(h_j) = \{a_i \in A : (a_i, h_j) \in E\}.$$

Each applicant $a_i \in A$ has a *preference list* in which she ranks $A(a_i)$ in strict order. Given any applicant $a_i \in A$, and given any houses $h_j, h_k \in H$, $a_i$ is said to *prefer* $h_j$ to $h_k$ if $\{h_j, h_k\} \subseteq A(a_i)$, and $h_j$ precedes $h_k$ on $a_i$'s preference list. Houses do not have preference lists over applicants, and it is essentially this feature that distinguishes HA from SMI. For a given acceptable applicant–house pair $(a_i, h_j)$, define $rank(a_i, h_j)$ to be 1 plus the number of houses that $a_i$ prefers to $h_j$. The *underlying graph* of $I$ is the bipartite graph $G = (A \cup H, E)$.

HA is a very general problem model and any application domain having an underlying matching problem that is bipartite, where agents in only one of the sets have preferences over the other, can be viewed as instances of HA. These include the problems of allocating graduates to trainee positions, students to projects, professors to offices, clients to servers, etc. The literature concerning HA has largely described this problem model in terms of assigning applicants to houses, so for consistency we also adopt this terminology.

An *assignment* $M$ is a subset of $E$. If $(a_i, h_j) \in M$, $a_i$ and $h_j$ are said to be *assigned to* one another. For each $p_k \in A \cup H$, the set of assignees of $p_k$ in $M$ is denoted by $M(p_k)$. If $M(p_k) = \emptyset$, $p_k$ is said to be *unassigned*, otherwise $p_k$ is *assigned*. A *matching* $M$ is an assignment such that $|M(p_k)| \leq 1$ for each $p_k \in A \cup H$. For notational convenience, as in the HR case, if $p_k$ is assigned in $M$ then where there is no ambiguity the notation $M(p_k)$ is also used to refer to the single member of the set $M(p_k)$.

An instance $I$ of HM comprises an HA instance $I$ where $n_1 = n_2$, together with a matching $M_0$ in $I$ (the *initial endowment*) such that $|M_0| = n_1$. A matching $M$ in $I$ is *individually rational* if, for each applicant $a_i \in A$, either $a_i$ prefers $M(a_i)$ to $M_0(a_i)$, or $M(a_i) = M_0(a_i)$. Since we are only interested in individually rational matchings, we assume that $M_0(a_i)$ is the last house on $a_i$'s preference list, for each $a_i \in A$. Clearly then, any individually rational matching $M$ in $I$ satisfies $|M| = n_1$.

$$a_1 : h_1 \quad h_2$$
$$a_2 : h_1$$

Fig. 1.2   An instance of HA with Pareto optimal matchings of different sizes

### 1.5.3   *Pareto optimal matchings*

Returning to the HA setting, the preferences of an applicant extend to the set of matchings $\mathcal{M}$ in an HA instance $I$ as follows. Given two matchings $M, M' \in \mathcal{M}$, we say that an applicant $a_i \in A$ *prefers* $M'$ *to* $M$ if either (i) $a_i$ is assigned in $M'$ and unassigned in $M$, or (ii) $a_i$ is assigned in both $M$ and $M'$, and $a_i$ prefers $M'(a_i)$ to $M(a_i)$.

Given this definition, we may define a relation $\lhd$ on $\mathcal{M}$ as follows: if $M, M' \in \mathcal{M}$ then $M' \lhd M$ if no applicant prefers $M$ to $M'$, and some applicant prefers $M'$ to $M$. If $M' \lhd M$ then $M'$ is called a *Pareto improvement* of $M$. It is straightforward to establish that $\lhd$ is a partial order on $\mathcal{M}$. A matching $M \in \mathcal{M}$ is defined to be *Pareto optimal* if $M$ is $\lhd$-minimal. Equivalently, $M$ is Pareto optimal if and only if there is no other matching $M'$ in $I$ such that (i) some applicant prefers $M'$ to $M$, and (ii) no applicant prefers $M$ to $M'$.

Intuitively a matching $M$ is Pareto optimal if no applicant $a_i$ can be better off without requiring another applicant $a_j$ to be worse off. For example, $M$ is not Pareto optimal if two applicants could improve by swapping the houses that they are assigned to in $M$. Fig. 1.2 gives an example HA instance that admits Pareto optimal matchings of different sizes, namely $M_1 = \{(a_1, h_1)\}$ and $M_2 = \{(a_1, h_2), (a_2, h_1)\}$. Further structural and algorithmic results for Pareto optimal matchings are given in Chap. 6.

### 1.5.4   *Maximum utility matchings*

Stronger notions of optimality have been considered in the literature for HA. Suppose that, in a given HA instance $I$, each applicant $a_i \in A$ has a positive integral weight $wt(a_i, h_j)$ for each house $h_j \in A(a_i)$. We assume that $a_i$ prefers $h_j$ to $h_k$ if and only if $wt(a_i, h_j) < wt(a_i, h_k)$. Let $W$ be the largest weight taken over all applicant–house pairs, let $\mathcal{M}^+ \subseteq \mathcal{M}$ denote the set of maximum cardinality matchings in $I$, and let $M \in \mathcal{M}$. We define the *weight* of $M$ to be $wt(M) = \sum_{(a_i, h_j) \in M} wt(a_i, h_j)$. Define the *utility* of an edge $(a_i, h_j) \in E$ to be $ut(a_i, h_j) = W - wt(a_i, h_j)$. The *utility* of $M$ is then $ut(M) = \sum_{(a_i, h_j) \in M} ut(a_i, h_j)$. $M$ is a *maximum utility matching* if $ut(M)$ is maximum, taken over all matchings in $\mathcal{M}$. Clearly a maximum utility matching is Pareto optimal.

$a_1 : h_1 \quad h_2 \quad h_3$
$a_2 : h_1 \quad h_2 \quad h_3$
$a_3 : h_1 \quad h_2 \quad h_3$

Instance $I_1$

$a_1 : h_1 \quad h_4$
$a_2 : h_2 \quad h_5$
$a_3 : h_3 \quad h_4 \quad h_6$
$a_4 : h_1$
$a_5 : h_2$
$a_6 : h_3$

Instance $I_2$

$a_1 : h_1 \quad h_3$
$a_2 : h_2 \quad h_1$
$a_3 : h_2$

Instance $I_3$

Fig. 1.3    Three instances of HA

We also define $M$ to be a *minimum weight maximum cardinality matching* if $M \in \mathcal{M}^+$ and $wt(M)$ is minimum, taken over all matchings in $\mathcal{M}^+$. Each of the problems of finding a maximum utility matching and a minimum weight maximum cardinality matching can be solved in $O(\sqrt{n}m \log(nW))$ time, assuming integral weights [230], where $n = n_1 + n_2$.[13] An important special case arises when $wt(a_i, h_j) = rank(a_i, h_j)$ for all $(a_i, h_j) \in E$, where $rank(a_i, h_j)$ is 1 plus the number of houses that $a_i$ prefers to $h_j$. In this case $W \le n_2$ and the time complexity of each of the problems of finding a maximum utility matching and a minimum weight maximum cardinality matching is $O(\sqrt{n}m \log n)$ [230].

### 1.5.5    *Popular matchings*

Another optimality criterion for an HA instance $I$ is *popularity*. Let $\mathcal{M}$ be the set of matchings in $I$ and let $M, M' \in \mathcal{M}$. Let $P(M, M')$ denote the set of applicants who prefer $M$ to $M'$. Define a "more popular than" relation ▶ on $\mathcal{M}$ as follows: if $M, M' \in \mathcal{M}$, then $M'$ is *more popular than* $M$, denoted $M' \blacktriangleright M$, if $|P(M', M)| > |P(M, M')|$. (Note that ▶ is not in general a partial order on $\mathcal{M}$.) Define a matching $M \in \mathcal{M}$ to be *popular* [21] if $M$ is ▶-maximal (i.e., there is no other matching $M'$ such that $M' \blacktriangleright M$). Thus, put simply, $M$ is popular if there is no other matching that is preferred by a majority of the applicants.

Clearly a matching $M$ is Pareto optimal if there is no other matching $M'$ such that $|P(M, M')| = 0$ and $|P(M', M)| \ge 1$. Hence a popular matching is Pareto optimal. However in contrast to the case for Pareto

---

[13]See Refs. [172, 294] for recent surveys of algorithms for finding minimum weight maximum cardinality matchings in both bipartite and general weighted graphs.

optimal matchings, an HA instance need not admit a popular matching. To
see this, consider the HA instance $I_1$ shown in Fig. 1.3. It is clear that a
matching in $I_1$ cannot be popular unless all applicants are assigned. The
unique matching up to symmetry in which all applicants are assigned is
$M = \{(a_i, h_i) : 1 \leq i \leq 3\}$, however $M' = \{(a_2, h_1), (a_3, h_2)\}$ is preferred
by two applicants, which is a majority. The relation ▶ in this case cycles,
hence the absence of a ▶-maximal solution.

Popular matchings can have different sizes, as illustrated by instance
$I_2$ (due to Rob Irving) in Fig. 1.3. It may be verified that the following
matchings are popular in $I_2$:

$$M_1 = \{(a_1, h_1), (a_2, h_2), (a_3, h_3)\}$$
$$M_2 = \{(a_1, h_1), (a_2, h_2), (a_3, h_4), (a_6, h_3)\}$$
$$M_3 = \{(a_1, h_1), (a_2, h_5), (a_3, h_4), (a_5, h_2), (a_6, h_3)\}$$

However it may also be verified that the unique perfect matching in the
underlying graph of $I_2$ is not popular (as $M_2$ is more popular), and hence
a maximum popular matching can be smaller than a maximum cardinality
matching in the underlying graph.

Clearly a matching $M$ in a given HA instance $I$ is popular if and only
if $|P(M, M')| \geq |P(M', M)|$ for all matchings $M' \in \mathcal{M}$. Given this inter-
pretation, it is then natural to define a *strongly popular* matching, which
is a matching $M$ such that $|P(M, M')| > |P(M, M')|$ for all matchings
$M' \in \mathcal{M}\backslash\{M\}$. It follows that $I$ need not admit a strongly popular matching.

Further structural and algorithmic results for popular matchings are
given in Chap. 7.

### 1.5.6   *Profile-based optimal matchings*

Further notions of optimality are based on the *profile* of a matching. To
define this property, let $I$ be an instance of HA and let $\mathcal{M}$ denote the set of
matchings in $I$. Given a matching $M \in \mathcal{M}$, define the *regret* of $M$, denoted
$r(M)$, to be the maximum rank of an applicant's partner in $M$. Formally
$r(M)$ is defined as follows:

$$r(M) = \max\{rank(a_i, h_j) : (a_i, h_j) \in M\}.$$

$$a_1 : h_1$$
$$a_2 : h_1 \quad h_2$$
$$a_3 : h_1 \quad h_3$$
$$a_4 : h_1 \quad h_4$$
$$a_5 : h_2 \quad h_5$$
$$a_6 : h_3 \quad h_6$$
$$a_7 : h_4 \quad h_7$$

$$M_1^7 = \{(a_1, h_1), (a_5, h_2), (a_6, h_3), (a_7, h_4)\}$$
$$M_2^7 = \{(a_i, h_i) : 1 \le i \le 7\}$$

Fig. 1.4   Instance $I^7$ of HA with two particular matchings due to Irving [313]

The *profile*[14] of $M$, denoted by $p(M)$, is a vector $\langle p_1, \ldots, p_{r^*} \rangle$, where $r^* = r(M)$ and for each $k$ $(1 \le k \le r^*)$,

$$p_k = |\{(a_i, h_j) \in M : rank(a_i, h_j) = k\}|.$$

Intuitively, $p_k$ is the number of applicants who have their $k$th-choice house in $M$.

A matching $M$ is *rank-maximal* [318] if $p(M)$ is lexicographically maximum, taken over all matchings in $\mathcal{M}$. Intuitively, in such a matching, the maximum number of applicants are assigned to their first-choice house, and subject to this condition, the maximum number of applicants are assigned to their second-choice house, and so on. A rank-maximal matching need not be of maximum cardinality. To see this, consider the HA instance $I_3$ shown in Fig. 1.3. Define the following matchings in $I_3$:

$$M_1 = \{(a_1, h_1), (a_2, h_2)\}$$
$$M_2 = \{(a_1, h_3), (a_2, h_1), (a_3, h_2)\}$$

Clearly $M_1$ is a rank-maximal matching in $I_3$ of size 2, whereas $M_2$ is a maximum matching in $I_3$ of size 3.

Irving [313] gave a family of instances of HA, denoted by $I^n$ $(n \ge 3)$, each with $n$ applicants and $n$ houses, in which a rank-maximal matching has size $1 + \lfloor n/2 \rfloor$ and a maximum matching has size $n$.

This family of instances can be described as follows. In $I^n$, $A(a_1) = \{h_1\}$. Also for each $i$ $(2 \le i \le \lceil n/2 \rceil)$, $A(a_i) = \{h_1, h_i\}$ and $a_i$ prefers $h_1$ to $h_i$. Finally for each $i$ $(\lceil n/2 \rceil + 1 \le i \le n)$, $A(a_i) = \{h_{i-\lceil n/2 \rceil+1}, h_i\}$ and $a_i$ prefers $h_{i-\lceil n/2 \rceil+1}$ to $h_i$. Instance $I^7$ is illustrated in Fig. 1.4.

---

[14]The profile of a matching $M$ has also been referred to the *signature* of $M$ in the literature [318, 371, 27, 242, 453, 295].

It may be verified that

$$M_1^n = \{(a_1, h_1)\} \cup \{(a_i, h_{i-\lceil n/2 \rceil + 1}) : \lceil n/2 \rceil + 1 \le i \le n\}$$

is a rank-maximal matching in $I^n$, where $p(M_1^n) = \langle \lfloor n/2 \rfloor + 1 \rangle$ and $|M_1^n| = \lfloor n/2 \rfloor + 1$. On the other hand $M_2^n = \{(a_i, h_i) : 1 \le i \le n\}$ is the unique maximum matching in $I^n$ and $|M_2^n| = n$. Matchings $M_1^7$ and $M_2^7$ are illustrated relative to $I^7$ in Fig. 1.4.

In many applications we seek to assign as many applicants as possible. With this in mind, consider $\mathcal{M}^+$, the set of maximum matchings in a given HA instance $I$, and let $r$ be the maximum rank of a house in an applicant's list. A *greedy maximum matching*[15] is a matching $M \in \mathcal{M}^+$ such that $p(M)$ is lexicographically maximum, taken over all matchings in $\mathcal{M}^+$. Both rank-maximal and greedy maximum matchings maximise the number of applicants with their $s$th-choice house as a higher priority than maximising the number of those with their $t$th-choice house, for any $1 \le s < t \le r$. As a consequence, both of these types of matchings could end up assigning applicants to houses relatively low down on their preference lists.

Consequently, define a *generous maximum matching*[16] to be a matching $M \in \mathcal{M}^+$ such that $p^R(M)$ is lexicographically minimum, taken over all matchings in $\mathcal{M}^+$, where $p^R(M)$ is the reverse of $p(M)$. That is, $M$ is a maximum cardinality matching that assigns the minimum number of applicants to their $r$th-choice house, and subject to this, the minimum number to their $(r-1)$th-choice house, and so on.

We collectively refer to rank-maximal, greedy maximum and generous maximum matchings as *profile-based optimal matchings*. Returning to instance $I_3$ shown in Fig. 1.3, the matching $M_2$ defined above is the unique maximum matching and is therefore both a greedy maximum matching and a generous maximum matching.

To give an illustrative comparison of a greedy maximum matching, a generous maximum matching and a minimum weight maximum cardinality matching (relative to the weight function $wt$ where $wt(a_i, h_j) = rank(a_i, h_j)$ for each acceptable pair $(a_i, h_j)$), consider the HA instance $I_4$, together with the three matchings in $I_4$, shown in Fig. 1.5. Table 1.1 lists the matchings and indicates which is a greedy maximum / generous maximum / minimum weight maximum cardinality matching, showing also the profile and weight of each matching.

---

[15] A greedy maximum matching has also been referred to as a *maximum rank maximal matching* in the literature [447, 452, 295].

[16] A generous maximum matching has also been referred to as a *fair* matching in the literature [19, 447, 367, 442].

$$a_1 : h_1 \quad h_2 \quad h_3 \quad h_4 \quad h_5$$
$$a_2 : h_1 \quad h_2 \quad h_3 \quad h_4 \quad h_5$$
$$a_3 : h_1 \quad h_2 \quad h_3 \quad h_4 \quad h_5$$
$$a_4 : h_1 \quad h_3 \quad h_5 \quad h_4 \quad h_2$$
$$a_5 : h_2 \quad h_5 \quad h_4 \quad h_3 \quad h_1$$

$$M_3 = \{(a_1, h_1), (a_2, h_4), (a_3, h_5), (a_4, h_3), (a_5, h_2)\}$$
$$M_4 = \{(a_1, h_1), (a_2, h_2), (a_3, h_3), (a_4, h_5), (a_5, h_4)\}$$
$$M_5 = \{(a_1, h_1), (a_2, h_2), (a_3, h_4), (a_4, h_3), (a_5, h_5)\}$$

Fig. 1.5 Instance $I_4$ of HA with three particular matchings

Table 1.1 Three matchings in the HA instance $I_4$ of Fig. 1.5, together with their profiles and weights

| Matching | Profile | Weight |
|---|---|---|
| $M_3$ : greedy maximum | $\langle 2, 1, 0, 1, 1 \rangle$ | 13 |
| $M_4$ : generous maximum | $\langle 1, 1, 3 \rangle$ | 12 |
| $M_5$ : minimum weight maximum | $\langle 1, 3, 0, 1 \rangle$ | 11 |

Further structural and algorithmic results for profile-based optimal matchings are described in Chap. 8.

### 1.5.7 *Extensions of* HA

A natural extension of HA arises when applicants are permitted to have ties in their preference lists (as in the case of SMTI, but without preferences of women over men). This gives rise to the *House Allocation problem with Ties* (HAT). In this case each of the definitions of a Pareto optimal, popular and profile-based optimal matching, as defined for the HA case in Sec. 1.5.3, Sec. 1.5.5 and Sec. 1.5.6 respectively, carry over to the HAT case without alteration.

The *Capacitated House Allocation problem* (CHA) is the generalisation of HA in which houses can be assigned more than one applicant, up to some fixed limit. That is, it is similar to HR except that hospitals do not have preference lists.

Formally, an instance $I$ of CHA is an instance of HA together with a *capacity* $c_j \in \mathbb{Z}^+$ for each $h_j \in H$. All of the definitions relating to an HA instance as given in Sec. 1.5.2 carry over to the CHA context without change. Furthermore, the definition of a matching is as given for HR in Sec. 1.3.2. Also *full* and *undersubscribed* houses relative to a matching are defined in an analogous way to the corresponding definitions in the HR case.

The *Capacitated House Allocation problem with Ties* (CHAT) is then the hybrid of HAT and CHA, namely the extension of CHA in which the applicants' preference lists can include ties.

Again, each of the definitions of a Pareto optimal, popular and profile-based optimal matching, as defined for the HA case in Sec. 1.5.3, Sec. 1.5.5 and Sec. 1.5.6 respectively, carry over to the CHAT case without change.

In instances of CHAT, we remark that, for an acceptable applicant–house pair, $rank(a_i, h_j)$ is defined in the same way as for the HA case (that is, it is 1 plus the number of houses that $a_i$ prefers to $h_j$).

## 1.5.8 *Motivation*

A number of applicatons can be found in different countries and contexts that involve centralised matching schemes based on HA or some variant of this problem.

We begin with campus housing allocation. In a number of universities, centralised matching schemes allocate students to campus accommodation, taking into account student preferences over available housing. Examples include Carnegie-Mellon University, Duke University, the University of Michigan, Northwestern University, and the University of Pennsylvania in the US [142] and the Technion in Israel [474].

The HA problem model also arises in the context of allocating families to government-subsidised housing in China [592]. However this application was not described in terms of a centralised matching scheme. Rather a description of a "residence exchange fair" was given, in which families can meet and arrange to exchange their houses with one another. In a residence exchange fair held in Beijing in 1991, involving 80,000 person-attendances, one of the swaps arranged involved 9 families!

Assigning students to projects and elective courses in an academic department also gives rise to applications of HA and its extensions. For example at the University of Glasgow, School of Computing Science, students are assigned to final-year projects using a profile-based optimal matching algorithm (implemented by Rob Irving and the author) that operates on the basis of the students' preferences over available projects. In the University of Glasgow, School of Medicine, a similar profile-based optimal matching algorithm is used to assign students to elective courses, taking into account student preferences over courses and course capacities. For example in academic year 2006-07, 246 final-year medical students were seeking to be matched to 17 elective courses. Each student ranked 6 courses in order of preference; of these 246, every student was assigned in a generous maximum matching to their fourth-choice course or better, with 228 students obtaining their third choice or better.

Two further applications involve assigning customers to DVDs in the context of a DVD rental market operated by regular mail taking into account the preferences of customers over available DVDs [19], and assigning reviewers to conference papers via a conference management software system, based on the preferences of reviewers over the submitted papers [242]. Each of these applications is described in more detail in Sec. 8.5.1 and Sec. 8.5.2.

Finally, we mention the Teacher Induction Scheme run by the General Teaching Council for Scotland. As part of this scheme, an eligible student who is graduating with a teaching qualification from a Scottish University is guaranteed a one-year probationary teaching post in a Scottish school. Graduating students are asked to rank in strict order 5 out of Scotland's 32 local authorities, to indicate their preferences over potential school locations. Alternatively they can (in return for a financial inducement) indicate that they are willing to work anywhere in Scotland (meaning that they are likely to be assigned to more rural areas that typically have a shortfall in probationers).

# PART 1
# Stable Matching Problems

Chapter 2

# The Stable Marriage problem: An update

## 2.1 Introduction

The Stable Marriage problem has had something of an illustrious history in the fifty or so years since the publication of Gale and Shapley's seminal paper in 1962 [235]. Some of the most significant structural and algorithmic developments were published by Gusfield and Irving in a series of papers in the late 1980s [319, 259, 262, 320], culminating in their book, published in 1989 [261]. For a newcomer to the area, after reading this book, it would perhaps be tempting to believe that all of the most interesting problems had been solved, and that apart from the open problems posed in the appendix, little new remained to be proved.

It is perhaps surprising, then, just how much progress has been made on problems relating to SM and its variants subsequent to the publication of Ref. [261]. The purpose of this chapter is to update the reader on some of the most important developments that have been made since then, with an emphasis on structural and algorithmic results. In describing these results, we aim to overlap with material already presented in Ref. [261] as little as possible, though in some cases a certain amount of scene-setting may be required in order for the context to be clear. Some of the research "highlights" that we cover include the ground-breaking papers of Subramanian and Feder [551, 202], the linear programming characterisations of SM and its variants, decentralised algorithms for constructing stable matchings, and the beautiful (and unexpected) structural results concerning generalised median stable matchings.

This chapter is organised as follows. In Sec. 2.2, we begin the technical discussion by updating the reader on the status of the 12 open problems posed by Gusfield and Irving in their book [261] that relate to SM and its

51

variants. Some of these problems have been fully solved, some others are partially solved, and a few remain open. We next describe in Sec. 2.3 the papers of Subramanian and Feder. Very broadly, these papers relate stable matchings in instances of SMI and SRI to so-called stable configurations of a network composed of a certain type of gate. This characterisation is taken further by Feder in order to relate stable matchings in SRI to satisfying truth assignments in a certain 2-SAT instance. The consequences of these transformations are wide-ranging, both in terms of yielding new structural results and faster algorithms for problems concerned with computing stable matchings.

In Sec. 2.4 we focus on the various papers that have used linear programming to characterise stable matchings in instances of SM and its generalisations. These techniques again contribute new structural results, whilst also yielding alternative algorithms for computing types of optimal stable matchings, such as egalitarian, minimum regret and minimum weight stable matchings. Constraint programming approaches to SM and its variants are described in Sec. 2.5. These demonstrate that the action of the Gale–Shapley algorithm can be simulated using so-called arc consistency propagation. One of the benefits of encoding stable matching problems as Constraint Satisfaction Problems is that it becomes easy to then model extensions that are not obviously solvable in polynomial time, by adding "side constraints" to the basic model.

The Gale–Shapley algorithm for SMI can be viewed as a centralised matching algorithm. Decentralised algorithms for producing stable matchings are surveyed in Sec. 2.6. These typically start from a given matching (which may be empty) and iteratively satisfy blocking pairs in order to arrive at a stable matching. Interestingly, for an arbitrary SM instance, there may be some stable matchings that can *never* be reached by starting from the empty set, no matter what sequence of blocking pairs is followed.

Some of the most beautiful (and unexpected) structural results concerning stable matchings in SMI that have been discovered since 1989 involve so-called *generalised median stable matchings*. To give an idea of the concept, suppose that each man in an SM instance arranges his stable partners in preference order, allowing repetitions. Then it turns out that, for each $k$, assigning each man the $k$th element in this ordered list gives rise to not only a matching, but one that is stable, called the $k$th *generalised median stable matching*. Results concerning the computation of generalised median stable matchings in SMI are surveyed in Sec. 2.7.

In many practical matching applications where the underlying theoretical model is based on a bipartite matching problem with preferences on both sides, stability is the key criterion to be satisfied. However, when preference lists are incomplete, a stable matching might be smaller (up to 50% smaller in the worst case) than a maximum cardinality matching. In some applications, a limited number of blocking pairs may be tolerated if that enables a larger matching to be found. In Sec. 2.8 we describe results connected with finding maximum cardinality matchings with the minimum number of blocking pairs, given an instance of SMI.

Issues of strategy in stable matching problems concern the question of whether an agent can misrepresent his/her true preferences in order to obtain a better outcome with respect to a given mechanism. Such questions have been the focus of much research by economists traditionally, and an extensive coverage of results up to 1990 appears in Ref. [514]. In the subsequent years, increasingly this line of research has been taken up by computer scientists. We review the post–1989 research on strategic issues, with a particular focus on algorithmic results, in Sec. 2.9.

Some further extensions of SM are discussed in Sec. 2.10. These include variants where certain acceptable pairs may be forced or forbidden (Sec. 2.10.1), where we seek a *balanced* stable matching (Sec. 2.10.2) — this minimises the maximum of the sum of the ranks of the men's partners and the sum of the ranks of the women's partners, and where we are given a set of matchings, and we wish to determine whether there is an SM instance in which all of the given matchings are stable (Sec. 2.10.3). We also discuss how the theory of stable matchings led to a very elegant proof of the Dinitz conjecture (Sec. 2.10.4) — this concerns list colouring the edges of a complete bipartite graph. The proof relates stable matchings to so-called *kernels* in directed graphs. This connection was further developed in several papers on the *marriage digraph*, which we also survey (Sec. 2.10.5). We also cover the problems of counting and sampling stable matchings (Sec. 2.10.6), online algorithms for SM (Sec. 2.10.7), and a general framework for finding stable matchings with additional "useful" properties in instances of SMI and HR (Sec. 2.10.8). So-called *locally stable matchings* which arise from *social network graphs* are studied in Sec. 2.10.9. Further miscellaneous results are also gathered together in a single subsection (Sec. 2.10.10).

We close the chapter in Sec. 2.11 by gathering together a selection of those open problems mentioned in the preceding sections that are, in the author's opinion, among the most notable and most deserving of further investigation.

## 2.2    The 12 open problems of Gusfield and Irving

### 2.2.1    Introduction

At the end of their book, Gusfield and Irving [261] gave a list of 12 open problems relating to the stable marriage and stable roommates problems. This was intended to follow the format of a similar list of 12 open problems given by Knuth [394]. The first four of Knuth's open problems concern the mean number of partner changes / proposals by men or women during an execution of the Gale–Shapley algorithm, and are different in nature to the structural and algorithmic material contained in Ref. [261]. Gusfield and Irving did not consider them further, and neither do we. Another six of Knuth's open problems have been solved, as noted by Gusfield and Irving. That leaves two of Knuth's open problems, which form problems 1 and 2 in Gusfield and Irving's list.

In this section we give updates to the open problems from Ref. [261] that relate to SM — these correspond to Problems 1–7 and half of each of Problems 10 and 11. Problems 8, 9, 12 and half of each of Problems 10 and 11 from Ref. [261] specifically relate to SR, and are dealt with in Sec. 4.2.

### 2.2.2    1. Maximum number of stable matchings

Given an SM instance $I$ of size $n$, we let $\mathcal{S}_I$ denote the set of stable matchings in $I$ (we omit the subscript if the instance is clear from the context). This problem in Ref. [261] relates to constructing an SM instance of size $n$, for each $n \geq 1$, that admits the maximum number of stable matchings taken over all SM instances of size $n$. We let $x_n$ denote this number; formally:

$$x_n = \max\{|\mathcal{S}_{I_n}| : I_n \text{ is an SM instance of size } n\}.$$

This problem is still open. However some progress has been made, which we now summarise.

Knuth [394, p.56] gave an example SM instance of size 4 with 10 stable matchings. For completeness, this instance is illustrated in Fig. 2.1. Eilers [187] showed (by exhaustive computer search) that indeed $x_4 = 10$, and in fact that Knuth's example is the unique SM instance of size 4 (up to isomorphism) with 10 stable matchings.

More generally, Knuth [394, p.4] also showed that, for each $n \geq 1$, there is an instance of SM of size $n$ that admits at least $2^{n/2}$ stable matchings. This result was later strengthened by Irving and Leather [319] who showed that, for $n = 2^k$ for some $k \geq 0$, there is an instance $J_n$ of SM of size $n$ that

| Men's preferences | Women's preferences |
|---|---|
| $m_1 : w_1 \quad w_2 \quad w_3 \quad w_4$ | $w_1 : m_4 \quad m_3 \quad m_2 \quad m_1$ |
| $m_2 : w_2 \quad w_1 \quad w_4 \quad w_3$ | $w_2 : m_3 \quad m_4 \quad m_1 \quad m_2$ |
| $m_3 : w_3 \quad w_4 \quad w_1 \quad w_2$ | $w_3 : m_2 \quad m_1 \quad m_4 \quad m_3$ |
| $m_4 : w_4 \quad w_3 \quad w_2 \quad w_1$ | $w_4 : m_1 \quad m_2 \quad m_3 \quad m_4$ |

Fig. 2.1 An instance of SM with 10 stable matchings due to Knuth [394, p.56]

admits least $2^{n-1}$ stable matchings. (A neat proof of this is given in Ref. [261, Sec. 1.3.2], where it is shown that, given SM instances of sizes $p$ and $q$, admitting $r$ and $s$ stable matchings respectively, there is an SM instance of size $pq$ with at least $\max\{rs^p, r^q s\}$ stable matchings.) In fact Irving and Leather [319] proved that $y_n = |\mathcal{S}_{J_n}|$ satisfies the following recurrence relation for $n \geq 4$:

$$y_n = 3y_{n/2}^2 - 2y_{n/4}^4$$

subject to $y_1 = 1$ and $y_2 = 2$. Knuth (personal communication, reported in Ref. [261]) showed that the solution of the recurrence relation satisfies

$$y_n > 2.28^n / (1 + \sqrt{3}). \tag{2.1}$$

Gusfield and Irving [261] conjectured that the family of instances so constructed satisfies $x_n = y_n$ for all such $n$.

Benjamin et al. [70] showed that, for each $n \geq 1$, there is an instance of SM of size $2n$ that admits $(n+1)\binom{2n}{n} - 2^{2n-1}$ stable matchings. In general this is a weaker bound than Inequality (2.1), but gives a lower bound for $x_n$ for even values of $n$ (recall that (2.1) is valid only when $n$ is a power of 2).

Hwang [300] showed that, for two positive integers $n_1$ and $n_2$, the inequality $x_{n_1+n_2} \geq x_{n_1} x_{n_2}$ holds. Since $x_1 = 1$, it follows that $x_n$ is a non-decreasing function of $n$. Thurber [569] showed that in fact $x_n$ is an increasing function of $n$. He also showed that, for each $n \geq 1$, $x_n > 2.28^n/(1+\sqrt{3})^{(\log n + 1)}$, generalising Inequality 2.1 which was shown to hold only for $n$ a power of 2.

Stathopoulos [549] showed that, if $n = 2^k$ for some $k \geq 2$, $x_n \leq (\frac{5}{12})^{\frac{n}{4}} n!$. He also gave an upper bound for the maximum number of stable matchings in an SR instance.

### 2.2.3   2. The "divorce digraph"

Let $I$ be an SM instance and let $\mathcal{M}_I$ be the set of matchings in $I$. The divorce digraph of $I$ is a digraph $D_I = (V, A)$, where $D_I$ contains a vertex for each matching in $\mathcal{M}_I$ (so $|V| = n!$, where $n$ is the size of $I$), and the edges in $D_I$ are defined as follows. Given two matchings $M, M'$ in $\mathcal{M}_I$, we say that $M'$ can be obtained from $M$ by a *divorce operation* (referred to as a *b-interchange* in Ref. [554]) if

$$M' = (M \backslash \{(m, M(m)), (M(w), w)\}) \cup \{(m, w), (M(w), M(m))\}$$

for some blocking pair $(m, w)$ of $M$. (Thus in $M'$, the man and woman involved in the blocking pair are matched together, and the "divorcees" are also matched together.) Given two vertices $v_M$, $v_{M'}$ in $V$, corresponding to matchings $M$ and $M'$ in $\mathcal{M}_I$ respectively, $(v_M, v_{M'}) \in A$ if and only if $M'$ can be obtained from $M$ by a divorce operation. It follows that $v_M \in V$ is a sink vertex of $D_I$ if and only if the corresponding matching $M \in \mathcal{M}_I$ is stable. The problem posed by Gusfield and Irving relates to exploring the structure of $D_I$, and in particular determining whether, given an arbitrary vertex $v_{M_0} \in V$, we can always find a path from $v_{M_0}$ to a sink vertex. Knuth [394, pp.2–3] showed that $D_I$ could contain cycles, and therefore it is not the case that, given an arbitrary vertex $v_{M_0} \in V$, *every* path from $v_{M_0}$ leads to a sink vertex.

Tamura [554] solved this problem by constructing an SM instance $I_4$ and identifying a set $\mathcal{M}_0 \subseteq \mathcal{M}_{I_4}$ such that (i) $I_4$ has size 4, (ii) five of the $4! = 24$ matchings in $\mathcal{M}_{I_4}$ are stable, (iii) $|\mathcal{M}_0| = 16$, and (iv) given any matching $M_0 \in \mathcal{M}_0$, there is no path in $D_{I_4}$ from $v_{M_0}$ to a sink vertex. Tamura showed how to generalise this construction to form an arbitrarily large SM instance $I_n$ of size $n$, for each $n \geq 4$, with a similar property.

Returning to an arbitrary SM instance $I$, in the case that we start from a given matching $M_0$ and follow a sequence of divorce operations which leads to a cycle in $D_I$, Tamura showed that all is not lost. More formally, he gave an algorithm (a so-called *b-interchange algorithm*) that finds, starting from $v_{M_0}$, a path on vertices $v_{M_0}, v_{M_1}, \ldots, v_{M_s}$ for some $s \geq 1$, where either $v_{M_s}$ is a sink vertex, or $v_{M_s} = v_{M_r}$ for some $r$ ($0 \leq r < s$), in which case the path starting from $v_{M_0}$ leads to a cycle $C$. In the latter case, Tamura showed how to construct a matching $M$ from $C$ such that $bp(M) \subset bp(M_0)$ (note that this step does not involve divorce operations). Matching $M$ is then the next starting point for a further iteration of the b-interchange algorithm. Thus, iterating this approach will ultimately lead to a stable matching being constructed, starting from an arbitrary matching. Note that Tamura was not able to conclude whether this process is guaranteed to terminate in a polynomial number of steps. The difficulty is not the number of times that the b-interchange algorithm is invoked (which must be bounded above by $n^2$, since $|bp(M)| \leq n^2$ for any matching $M$), but rather the potential length of a cycle output by the algorithm.

Tan and Su [560] independently solved this problem by providing an SM instance $I_4'$ (not the same as Tamura's instance $I_4$) of size 4 and identifying a set $\mathcal{M}_0' \subseteq \mathcal{M}_{I_4'}$ such that (i) $I_4'$ has size 4, (ii) two of the 4! = 24 matchings in $\mathcal{M}_{I_4'}$ are stable, (iii) $|\mathcal{M}_0'| = 16$, and (iv) given any matching $M_0 \in \mathcal{M}_0'$, there is no path in $D_{I_4}$ from $v_{M_0}$ to a sink vertex. The authors also proved that any SM instance with the property that some matching cannot be transformed to a stable matching via a sequence of divorce operations must be of size at least 4. Further, they showed that $I_4'$ could be generalised to produce a family of instances of arbitrarily large size with the desired property.

Tan and Su additionally gave an algorithm for transforming an arbitrary matching $M_0$ into a stable matching in an SM instance $I$ using a sequence of divorce operations. However, as in the case of Tamura's algorithm, additional types of operations may be necessary. The approach of Tan and Su is to regard $I$ as an instance of SRI and invoke the theory of stable partitions in $I$ (see Sec. 4.3). Without loss of generality, suppose that $M_0 = \{(m_i, w_i) : 1 \leq i \leq n\}$. The idea is to use an incremental approach, constructing for each $k$ ($1 \leq k \leq n$) a matching of size $k$ that is stable in $I_k$, where $I_k$ is the sub-instance of $I$ obtained by deleting each man $m_j$ and each woman $w_j$ such that $k < j \leq n$. The starting point is the matching $\{(m_1, w_1)\}$, which is trivially stable in $I_1$. Inductively, for $k \geq 2$, given a matching $M_{k-1}$ (of size $k - 1$), that is stable in $I_{k-1}$, the authors show that, within a sequence of $O(n^2)$ divorce operations, either $M_{k-1}$ can be transformed into a matching $M_k$ (of size $k$) that is stable in $I_k$, or else the process cycles in $D_I$, in which case a stable partition $\Pi_k$ in $I_k$ is produced. As $I_k$ is an instance of SM, every party in $\Pi_k$ has even length, and therefore by Theorem 4.4, $\Pi_k$ can be transformed into a stable matching $M_k$ in $I_k$ (however this step does not of course involve divorce operations). At each iteration of Tan and Su's algorithm, the next divorce operation takes $O(n^2)$ time to locate[1], and therefore the algorithm has $O(n^5)$ complexity overall.

We make two remarks about the approaches of Tamura and of Tan and Su. Firstly, Step 2 of Tamura's b-interchange algorithm closely resembles Tan and Su's proposal-rejection alternating sequence which is used to construct $M_n$ iteratively from $M_0$. Secondly, it is possible that the algorithms in both of these papers might not exclusively use divorce operations even

---

[1] With the aid of appropriate data structures, it is possible that $O(n^2)$ could be improved to $O(n)$.

when there is a path from $v_{M_0}$ to a sink vertex in $D_I$, due to a cycle being traversed in $D_I$ instead. Hence the complexity of the following fundamental decision problem is still open: given an SM instance $I$ and a matching $M_0$, is there a path in $D_I$ from $v_{M_0}$ to a sink vertex?

Note that, if we drop the insistence, as in this subsection, that the "divorcees" marry one another as part of the divorce operation, then the landscape changes dramatically — see Sec. 2.6 for more details.

### 2.2.4    3. Parallel algorithms for stable marriage

This problem asks whether SM belongs to NC.[2] The problem is still open. The only significant development following the publication of Ref. [261] is the paper of Feder *et al.* [205], where the following was proved, indeed for the more general case where we are given an instance $I$ of SRI:

- the problem of finding a stable matching in $I$ or reporting that none exists can be solved in $O(\sqrt{m} \log^3 m)$ time on an $m^4$-processor CRCW PRAM (here, $m$ is the number of acceptable pairs);
- the agents who are matched in all stable matchings in $I$ can be found in $O(\sqrt{m} \log^3 m)$ time on an $m^3$-processor CRCW PRAM;
- the set of stable matchings in $I$ can be characterised in terms of a 2-SAT instance (see Sec. 4.2.4 for more details regarding this characterisation) in $O(\sqrt{m} \log^3 m)$ time on an $m^4$-processor CRCW PRAM.

The approach of Feder *et al.* is based on the primal–dual interior path-following method for linear programming; the difficulty of parallelising the McVitie–Wilson algorithm for SM [445,446] due to its inherently sequential nature in parts had already been observed [486].

Additional results regarding the parallel complexity of SRI were given by Subramanian [551]. These are reviewed in Sec. 2.3.1, as it is more appropriate to present these results in the wider context of the discussion of the framework presented by that paper.

The original question of Gusfield and Irving, and the results of Feder *et al.* [205] and Subramanian [551] apply to a shared memory architecture. Alternatively, Lu and Zheng [408] considered parallel algorithms for an SM instance $I$ in three computational models based on a message-passing architecture, namely a hypercube, a mesh of trees (MOT) and an array

---

[2]NC is the class of decision problems that admit a parallel algorithm running in poly-logarithmic time on a polynomial number of processors. It is a major open problem as to whether P=NC.

with multiple broadcasting buses. Their algorithms each consist of two alternating phases, namely an *Initiation Phase* and an *Iteration Phase*, which consists of multiple iterations. The authors show that an execution of the Initiation Phase, and an iteration of an Iteration Phase, have $O(\log n)$ complexity on an array with multiple broadcasting buses, and $O(\log^2 n)$ complexity on either a hypercube or an MOT, where each architecture comprises $n^2$ processors (here $n$ is the size of $I$). Their simulations indicate that each algorithm converges within $n$ rounds with high probability.

Although not directly relevant to the original problem posed by Gusfield and Irving, it is appropriate to briefly review distributed algorithms for SM here. Here the underlying model of distributed computation is based on a bipartite graph, where each node corresponds to a processor, and each one ranks nodes in the opposite set of the bipartition in order of preference.

Amira *et al.* [44] focused on a special case of SMI where the preferences are derived from a global ranking function of the edges in the underlying bipartite graph (see Sec. 4.7 for more details of this model). They considered two different models of communication. In the so-called *billboard* model, they showed that any algorithm for the problem requires at least $n - 1$ steps, where $n$ is the size of the SMI instance, and they provided an algorithm that achieves this bound. In the so-called *distributed weighted model*, the authors gave an $O(\sqrt{n})$ algorithm for the problem.

For a general SMI instance $I$ of size $n$, Kipnis and Patt-Shamir [383] proved that any distributed algorithm for $I$ requires $\Omega(\sqrt{n/B}\log n)$ communication rounds in the worst case, where $B$ is the number of bits per message. They also gave an $O(D + m)$ distributed algorithm for the finding a stable matching, where $D$ is the diameter and $m$ is the number of acceptable pairs.

Other studies of distributed algorithms for SM and its variants include Refs. [382, 152, 532, 112, 113, 151, 220].

### 2.2.5   4. Batch stability testing

This problem concerns whether a set of matchings in an SM instance $I$ can be checked for stability within a time bound that substantially improves on the naïve approach which involves checking each matching for stability in $O(n^2)$ time, where $n$ is the size of $I$. In particular, Gusfield and Irving [261] asked whether each matching could be checked for stability in time sub-quadratic in $n$ for each matching, following some "reasonable" pre-processing time ($O(n^4)$ was suggested as a measure of "reasonable").

This problem has been largely solved by Dabney and Dean [158], even in the SMI case. To describe their results, let $I$ be an instance of SMI and let $\mathcal{M}_0$ be a set of $k$ matchings in $I$ that we require to check for batch stability. Clearly it is possible to verify the stability of the matchings in $\mathcal{M}_0$ in $O(km) = O(kn^2)$ overall time, where $m$ is the number of acceptable pairs in $I$. However the authors give a new characterisation of the stability of a matching in $I$ in terms of the connectivity of the so-called *expanded rotation graph*, which is built up from the rotations in $I$ together with precedence relations between them. Using this characterisation, together with existing results concerning efficient data structures for fully dynamic connectivity in graphs, they show that the stability of the matchings in $\mathcal{M}_0$ can be checked in $O((m + kn)\log^2 n)$ overall time. This essentially equates to the verification of each matching in $O(n\log^2 n)$ amortised time, following $O(m\log^2 n)$ pre-processing time. This does not, however, imply that *each* matching in $\mathcal{M}_0$ can be checked in $O(n\log^2 n)$ time in the worst case, following $O(m\log^2 n)$ pre-processing time.

It remains open as to whether the "amortised" qualification on the time complexity for checking a single matching in $\mathcal{M}_0$ for stability can be dropped, and moreover whether each matching in $\mathcal{M}_0$ can be checked for stability in $o(m)$ time in the worst case.

### 2.2.6    *5. Structure of stable marriage with ties*

Gusfield and Irving [261] asked whether there is a characterisation or compact representation of the set of stable matchings for an instance of SMTI. To answer this question properly, we need to be clear as to which stability definition is being used. As mentioned in Sec. 1.3.5, Irving [308] defined three levels of stability for SMT instances, namely weak stability, strong stability and super-stability, and these definitions were later generalised to the SMTI case [414].

It is known that, given an instance of SMT, the set of weakly stable matchings need not form a lattice (see Sec. 3.2.2 for more details). In the case of SM, the link between elegant structural characterisations of stable matchings and efficient algorithms for a range of problems concerned with computing stable matchings was a recurring theme in Ref. [261]. However as noted in Sec. 3.2, many of the corresponding problems turn out to be NP-hard in SMT or SMTI under weak stability. Hence it seems unlikely that we can expect any kind of natural structure to be present for weakly stable matchings.

On the other hand, for super-stability, it is known that the set of super-stable matchings for an instance $I$ of SMTI forms a distributive lattice [548, 415] (see Sec. 3.4.3 for more details). The concept of a *meta-rotation* (consisting of a set of rotations that must be eliminated in turn) has been defined for SMTI under super-stability [523, Chapter 6] and leads to a characterisation of the super-stable matchings in $I$ in terms of the closed subsets of a related digraph. However a simpler characterisation may be obtained by transforming from $I$ to an instance of the Stable Marriage problem with Forbidden Pairs (see Sec. 2.10.1). The transformation itself is described in the proof of Theorem 4.39 for the more general case that $I$ is an instance of SRTI.

We now turn to the third stability criterion for SMTI, namely strong stability, which lies "in between" weak stability and super-stability. It is known [415] that the set of equivalence classes of the strongly stable matchings for a given SMTI instance (under a natural equivalence relation) forms a distributive lattice — see Sec. 3.3.3 for more details. To date, no definition of a rotation has been given that is applicable in the strong stability case. Thus a range of interesting algorithmic problems remain open concerning the computation of various kinds of strongly stable matchings — see Sec. 3.6 for further details.

As part of their Open Problem 5, Gusfield and Irving [261] asked whether there is an LP representation of an instance of SMTI under weak, strong or super-stability. This problem is still open, however formulations of SMTI and HRT under weak stability as a Constraint Satisfaction Problem were given in Refs. [251, 252] and [470] respectively (see Sec. 2.5).

### 2.2.7  6. Sex-equal matching

Gusfield and Irving [261] observed that, with respect to an egalitarian stable matching, it is possible for the members of one sex to fare much better than those of the other sex. As noted by Romero-Medina [491], this is illustrated very well by Knuth's example SM instance of size 4 [394, p.56], illustrated in Fig. 2.1. Recall that this instance has 10 stable matchings; moreover each has cost 20, and thus each stable matching is egalitarian. However the sex-equality measure measures of the 10 stable matchings constitute the set $\{-12, -8, -4, 0, 4, 8, 12\}$. This motivates the *Sex-Equal Stable Matching problem* (SESM), the problem of finding a sex-equal stable matching, given an SMI instance. Gusfield and Irving asked whether there is a polynomial-time algorithm for SESM.

Kato [360] was the first to show that SESM is NP-hard, by reducing from PARTIALLY ORDERED KNAPSACK [350], [241, pp.247–248]. McDermid and Irving [443] gave a shorter reduction from CLIQUE [241, p.194] to SESM, which was inspired by Johnson and Niemi's reduction from CLIQUE to PARTIALLY ORDERED KNAPSACK. Moreover McDermid and Irving proved that SESM is NP-hard even if each preference list in the constructed SMI instance is of length at most 3. In fact they proved that the problem of determining whether, given an SMI instance where each preference list is of length at most 3, there exists a stable matching $M$ where $d(M) = 0$, is NP-complete.

On the other hand, when preference lists are of length at most 2 on one side (and there is no upper bound on the preference list lengths on the other side), they showed that SESM is solvable in $O(n^3)$ time, where $n$ is the instance size, using dynamic programming.

Exact algorithms for NP-hard cases of SESM have been given. If $I$ is an SMI instance in which the preference lists on one side are of length at most $k$ (for some $k \geq 3$), and there is no upper bound on the preference list lengths on the other side, McDermid and Irving [443] showed that, given any $\varepsilon > 0$, SESM can be solved in $O^*(2^{\alpha n} + 2^\beta)$ time[3], where $\alpha = (5 - 2\sqrt{4})(k - 2 + \varepsilon)$, $\beta = (k-1)/2\varepsilon$ and $n$ is the size of $I$. For a sufficiently small choice of $\varepsilon$, this equates to $O^*(1.0726^n)$ for $k = 3$, $O^*(1.1504^n)$ for $k = 4$ and $O^*(1.2339^n)$ for $k = 5$.

Romero-Medina [491] gave an exact algorithm for SESM where there are no restrictions on the preference list lengths. The author did not analyse the complexity of his algorithm, though claimed in his concluding section that the algorithm runs in polynomial time. However this would obviously contradict Kato's result [360] (which is not referenced in Ref. [491]) unless P=NP, and therefore it is more likely that the algorithm's complexity is exponential in the worst case. Moreover the running time is likely to be poorer than that of McDermid and Irving for the case that preference lists on one side are of bounded length.

At a very general level, both algorithms are based on computing an appropriate closed subset of rotations whose elimination leads to a sex-equal stable matching. McDermid and Irving's algorithm, however, employs a very novel technique along the following lines: if the number of rotations in the rotation digraph $D(I)$ of $I$ is "small" enough, it is sufficient to enumerate all subsets of the vertices in $D(I)$ in order to solve the problem.

---

[3]A function $f$ satisfies $f(n) = O^*(g(n))$ if $f(n) = O(p(n)g(n))$, where $p$ is a polynomial function of $n$.

Otherwise, it is shown that $G(I)$ (which is obtained from $D(I)$ by replacing every arc by an undirected edge) must have bounded average degree, in which case $G(I)$ admits a sufficiently large induced subgraph $G'(I)$ that is series–parallel [184]. The algorithm enumerates each subset $S_1$ of the rotations in $G(I)\backslash G'(I)$ (again this subgraph is sufficiently "small"), and, for each such subset, an optimal closed subset $S_2$ of rotations from $G'(I)$ is identified in polynomial time (by exploiting the fact that $G'(I)$ is series-parallel), such that $S_1 \cup S_2$ is an optimal closed subset of rotations in $I$.

Iwama *et al.* [344] also studied SESM, mainly from an approximability point of view. Let $I$ be an instance of SMI, and let $M_a$ and $M_z$ denote the man-optimal and woman-optimal stable matchings in $I$ respectively. Let $\Delta = \min\{|d(M_a)|, |d(M_z)|\}$. The authors gave a polynomial-time algorithm that finds a *near-optimal* solution to SESM. That is, given some fixed $\varepsilon > 0$, in polynomial time the algorithm returns a matching $M$ such that $-\varepsilon\Delta \leq d(M) \leq \varepsilon\Delta$, or reports that no such matching exists (the complexity of the algorithm is $O(n^{3+\frac{1}{\varepsilon}})$, where $n$ is the size of $I$).

Recall from Sec. 1.3.4.1 that the *cost* of a stable matching $M$ in $I$, denoted by $c(M)$, is a measure that is minimised by an egalitarian stable matching. Iwama *et al.* [344] gave an example SM instance in which two near-optimal stable matchings have very different cost values. In fact, the example demonstrates that this is the case for two sex-equal stable matchings (in a sense, this is an "opposite" example to the one described earlier in this subsection, in which two egalitarian stable matchings had very different values of $d(M)$). This motivates the *Minimum Egalitarian Sex-Equal Stable Marriage problem* (MESESM): among all near-optimal stable matchings in a given SM instance, find a matching $M$ such that $c(M)$ is minimum (or report that no such matching exists). The authors showed that MESESM is NP-hard and gave an approximation algorithm with performance guarantee $(2 - (\varepsilon - \delta)/(2 + 3\varepsilon))$ for any fixed $\delta$ such that $0 < \delta < \varepsilon$ (the running time of the algorithm is $O\left(n^{3+2\left(\frac{1+\varepsilon}{\delta}\right)}\right)$, where $n$ is the size of $I$).

Genetic and ant colony-based algorithms for SESM have also been considered [459, 582].

A concept that is superficially similar to a sex-equal stable matching is that of a *balanced* stable matching — see Sec. 2.10.2 for more details.

### 2.2.8 7. Lying and egalitarian matchings

This question relates to the investigation of forms of strategic behaviour that could benefit an individual or several members of a coalition when we are concerned with a mechanism (i.e., an algorithm) for SM that produces a stable matching other than the man-optimal or the woman-optimal solutions. In particular, in relation to this question, Gusfield and Irving [261] refer to a mechanism based on computing an egalitarian stable matching. We are not aware that this specific question has been addressed explictly in the literature, however there have been several studies of strategic issues relating to SM mechanisms following the publication of Ref. [261] — see Sec. 2.9 for more details.

### 2.2.9    *10. Succinct certificates*

The part of this problem that relates to SM asks whether there is a succinct (i.e., $o(n^2)$ size) certificate of the stability of a matching in a given SM instance of size $n$, given that there is a very simple (i.e., $O(1)$ size) witness that a matching is *not* stable, namely a blocking pair. This question was answered in the negative by Dougherty and Selkow [170], who proved that the *certificate complexity* [114] of determining whether a given matching is stable in an SM instance of size $n$ is $\Omega(n^2)$. They proved that similar lower bounds hold for determining whether a man–woman pair is (i) stable, (ii) a fixed pair, or (iii) unstable (i.e., not a stable pair). Note that certificate complexity is a measure of the *size* of a witness of a given property, and does not correspond to the time taken to compute that witness. Thus the results of Dougherty and Selkow (and result (i) in particular) are subtly different from those of Ng and Hirschberg (see Sec. 2.10.10).

### 2.2.10    *11. Algorithmic improvements*

Gusfield and Irving conjectured that an egalitarian stable matching can be found for a given SMI instance $I$ in $O(m)$ time, where $m$ is the number of acceptable pairs in $I$. No such algorithm has been found to date. However Feder [202,203] gave an $O(m^{1.5})$ algorithm for the problem, which improved on the $O(m^2)$ algorithm due to Irving *et al.* [320,261], and an $O(n^3\sqrt{\log n})$ algorithm due to Ng [462]. More generally, Feder showed that the problem of finding a minimum weight stable matching in $I$ can be solved in $O(\min(n, \sqrt{K})m\log(K/m + 2))$ time, where $n$ is the size of $I$ and $K$ is the weight of a minimum weight stable matching. This improved on the $O(m^2 \log n)$ algorithm described by Gusfield and Irving [261]. Note that in Ref. [203], the running time of Feder's algorithm for minimum weight stable matching is given as $O(m\sqrt{K})$ if $K = O((m/\log^2 m)^2)$, and $O(nm \log K)$ for arbitrary $K$. See also Ref. [314] for a discussion of these results.

Feder [202] showed that the problem of finding an egalitarian stable matching in $I$ is at least as difficult as the UDCS problem in a bipartite multigraph with $n$ vertices and at most $m$ edges, where the vertex bounds add up to at most $m$. The fastest algorithm for this problem has $O(m^{1.5})$ complexity [226] (see Sec. 1.2), and therefore any improvement for the egalitarian stable matching problem below a bound of $O(m^{1.5})$ would imply a similar speed-up for UDCS.


## 2.3 The Subramanian and Feder papers

Around the time that Gusfield and Irving were finalising their manuscript [261], two significant papers relating stable matching theory to network stability in the context of circuit design were about to be published as extended abstracts in the proceedings of two international conferences [433, 200]. The complete versions of these papers, authored by Subramanian [551] and Feder [202], appeared in journals some years later. As the full details of these papers were not available to Gusfield and Irving at the time of writing, their book [261] contained only brief descriptions of the results of Subramanian and Feder. We update the reader by summarising more fully the contributions of these papers here, and in particular, outlining the impact of their results on the theory of stable matching.

Subramanian's paper [551], which builds on earlier work of Mayr and Subramanian [433, 434], is described in Sec. 2.3.1. It is based on modelling stable matchings in an instance of SRI in terms of so-called *stable configurations* in a network composed of so-called *scatter-free* gates. Subramanian's construction has a variety of structural and algorithmic consequences in the SRI context, and perhaps most notably, leads to results concerning the parallel complexity of a range of problems concerned with finding stable matchings in SRI instances. In the SMI case, Subramanian applies a fixed-point approach to his characterisation to establish that the set of stable matchings is non-empty and forms a lattice.

Feder's paper [202] also extends earlier work of Mayr and Subramanian [433, 434, 551], and is dealt with in Sec. 2.3.2. This paper is the journal version of the author's earlier conference paper [200] and much of the material contained therein is drawn from the author's PhD thesis [201]. Feder demonstrates, using so-called *stable configurations* in an *adjacency-preserving network*, and in turn using fixed points of an associated function defined on a hypercube, that there is a correspondence between stable matchings in an instance of SRI and satisfying truth assignments in an instance of 2-SAT. This characterisation, combined with the algorithmic results for 2-SAT contained in Ref. [203], leads to a range of algorithmic consequences for problems concerned with computing stable matchings in the SRI context. That Feder's characterisation is a profound structural result is without question, however it is arguable that if one is only interested in the algorithmic consequences for SRI, an easier way to achieve these complexity results is to use the simpler 2-SAT characterisation of SRI contained in Ref. [261], together with the 2-SAT algorithms from Ref. [203]. We explore this point in more detail in Sec. 2.3.2.

Finally, we review other fixed-point approaches to characterising SM in Sec. 2.3.3.

### 2.3.1   *Subramanian:* SRI *and network stability*

Subramanian [551] demonstrated that there is a deep relationship between stable matchings in SRI and so-called *network!stability*. In this context, a *network* is a boolean circuit with feedback: that is, it is a digraph where source nodes and sink nodes correspond to input and output bits respectively, and all other nodes represent *gates*, where a *gate* corresponding to a vertex $v$ with indegree $r_1$ and outdegree $r_2$ is essentially a boolean function $f : \{0,1\}^{r_1} \longrightarrow \{0,1\}^{r_2}$. The *Network Stability problem* (NS) is to determine whether a given network $N$ and a given input bit string (applied to the source nodes of $N$) has a *stable configuration*, which is an assignment of boolean values to the arcs of the network that respects the input string and the gate constraints. In general NS is NP-complete [434]. However X-NS is more accessible — this is the special case of NS where each gate is a so-called *X-gate* (consisting of two inputs and two outputs) [434, 551]; these satisfy the so-called *scatter-free* property, which can be exploited by efficient algorithms. Subramanian's framework leads to a range of concise proofs of existing results, including simpler algorithms, and also new structural and algorithmic consequences. We review these in this subsection.

The first main result of the paper is a linear-time reduction from an instance $I$ of SRI to an instance $N$ of X-NS, with the property that the stable matchings in $I$ are in 1–1 correspondence with the stable configurations of $N$. Subramanian gave a linear-time algorithm for X-NS, and hence it follows that there is a linear-time algorithm for finding a stable matching in $I$ or reporting that none exists (thus providing an alternative method to that in Ref. [306]). From properties of the stable configurations of $N$, Subramanian was also able to deduce that the same set of agents are assigned in all stable matchings in $I$ (thus giving an alternative proof of Theorem 4.5.2 in Ref. [261]). Furthermore, the linear-time algorithm for finding a stable configuration in $N$ may be adapted in order to construct a minimum regret stable matching in $I$ (thus providing an alternative approach to that in Ref. [261, Sec. 4.4.3]).

Subramanian was also able to use his characterisation in order to deduce, for an unsolvable SRI instance $I$, the existence of disjoint cycles in the corresponding network $N$, each containing an odd number of NOT gates. This structural result is reminiscent of the existence of an odd party in a stable partition for $I$ (see Sec. 4.3). Furthermore, Subramanian used the #P-completeness of the problem of counting the number of stable configurations for a given instance of X-NS [551] in order to show that the problem of counting the number of stable matchings for a given instance of SRI is #P-complete. This gives an alternative proof of the existing result of Irving and Leather [319] for this problem (although Irving and Leather had established hardness for the counting problem in the SM case).

In the case that $I$ is an instance of SMI, Subramanian showed that $N$ may be formulated as a *comparator network*, i.e., it can be constructed from (the simpler two-input, two-output) *comparators* rather than X-gates. In this case $N$ is an instance of C-NS (*Comparator Network Stability*). This observation leads to a simple (non-constructive) proof that $I$ admits a stable matching. We remark that whilst Gale and Shapley's proof of this result [235] was of course constructive, Sotomayor [542] had already provided an alternative non-constructive proof. Moreover, the lattice structure for the set of stable matchings $S$ in $I$ (see Theorem 1.13) follows from the fact that the set of stable configurations of $N$ is a distributive lattice [551]. Alternatively, fixed-point theory can be invoked to deduce that $S$ is non-empty and forms a lattice. That is, in the SMI context, the relationship between the stable matchings in $I$ and the stable configurations of $N$ can be expressed in terms of a monotone function. As a consequence of Tarski's fixed-point theorem [562], it follows that $S$ is non-empty and forms a lattice.

We now turn to parallel complexity. Subramanian [551] also showed that the X-NS and the *Comparator Circuit Value* (C-CV) problems are equivalent under many–one logspace reductions [351]. C-CV is the problem of determining whether a given output node has value 1, given a comparator network $N$ and an input bit string (here $N$ must be a *circuit*, i.e., the underlying digraph must be acylic). C-CV can be solved in $O(\sqrt{m} \log^c m)$ on an $O(\sqrt{m})$-processer PRAM, for some $c > 0$, where $m$ is the size of the circuit [434].

With the use of appropriate reductions, this result about the parallel complexity of C-CV sheds light on the parallel complexity of various stable matching problems. In particular, the results are stated in terms of *CC-completeness*. The class CC consists of those decision problems that are reducible to C-CV [434] (again under many–one logspace reductions). CC-complete problems are therefore equivalent to C-CV under many–one logspace reductions.

It is known that L$\subseteq$ NL $\subseteq$ NC $\subseteq$ P.[4] The inclusions L $\subseteq$ NL $\subseteq$ CC $\subseteq$ P hold [434], and it is conjectured [434] that the inclusions are strict, and that CC and NC are incomparable.

The following problems have been shown to be CC-complete [434, 551]:

- Does a given man–woman pair belong to the man-optimal stable matching for a given SMI instance?
- Does a given SRI instance have a stable matching?
- Is a given pair of agents $\{a_i, a_j\}$ in a given SRI instance $I$ a *fixed pair* (i.e., does $\{a_i, a_j\}$ belong to every stable matching in $I$)?
- Is a given pair of agents in a given SRI instance a stable pair?
- Does a stable matching in a given SRI instance have regret at most $K$, for a given integer $K$?

Subramanian [551] noted that SRI is in NC for the case that all preference lists are of length at most 2.

Another (logspace) reduction in Subramanian's paper is from an SMI instance $I$ to an instance $J$ of the Assignment problem (i.e., maximum weight bipartite matching) — this reduction allows the man-optimal stable matching in $I$ to be computed from an optimal solution in $J$. Note that this is the *only* stable matching in $I$ that is preserved under this reduction; it is an open problem to formulate a reduction from an SMI instance $I$ to the Assignment problem that preserves the structure of all solutions in $I$.

Lê *et al.* [402] proposed an alternative definition of the class CC, based on a weaker form of reducibility to C-CV. The authors showed that SMI is complete for CC under this weaker reduction, claiming that their proof is simpler than Subramanian's corresponding proof for (many–one) logspace reductions. See also Ref. [155].

The final result in Subramanian's paper is the NP-completeness of determining whether a stable matching exists, given an instance of 3GSM (see Sec. 5.6.1.1). This problem was already known to be NP-complete [465].

---

[4]L and NL are classes of decision problems that can be solved using logarithmic space on a deterministic and non-deterministic Turing machine respectively.

Subramanian's alternative proof is based on a reduction from Y-NS, the restriction of the Network Stability problem in which the network comprises so-called *Y-gates* (involving three inputs and three outputs).

### 2.3.2 *Feder:* SRI *and* 2-SAT

Feder [202] extended Subramanian's network stability characterisation of stable matching problems by considering networks with so-called *adjacency preserving* gates (to be defined below). A special case of these are scatter-free gates, which formed the basis of Subramanian's framework (Mayr and Subramanian [434] showed that the problem of finding a stable configuration or reporting that none exists, for a given network with scatter-free gates, is solvable in linear time, whereas the problem is NP-hard for arbitrary networks). Feder then characterised networks with adjacency-preserving gates in terms of 2-SAT instances, which in turn led to a new framework for representing SRI instances. The end result was a series of theorems yielding new structural and algorithmic results for a whole range of problems associated with computing types of stable matchings. In many cases, these implied improved algorithms for problems that were already known to be polynomial-time solvable.

As the title of Feder's paper alludes to, the approach that he took is based on viewing stable configurations in a network as fixed points of a so-called *edge-preserving* function on a hypercube, which in turn leads to a 2-SAT representation. We now give an overview of the framework; many of the key definitions closely follow the treatment in Ref. [202].

Recall that a configuration is an assignment of boolean values to the arcs of a network $N$, which can be represented by an $m$-bit string $x = x_1 \ldots x_m$, where $x_i$ is the value assigned to arc $i$ and $m$ is the number of arcs in $N$. We can associate with $N$ a *transition function* $f_N : \{0,1\}^m \longrightarrow \{0,1\}^m$ on the set of configurations. Function $f_N$ maps a configuration $x = x_1 \ldots x_m$ to the configuration $y = y_1 \ldots y_m$ that is obtained by evaluating in parallel all of the gates in $N$ with the $x_i$ values as inputs and the $y_i$ values as outputs. Network $N$ can be regarded as a single gate such that the output on the $i$th arc feeds into the input on the $i$th arc. The configuration $x$ is then stable if and only if it is a fixed point under $f_N$, i.e., $f_N(x) = x$.

We now define the concepts of an *adjacency-preserving* network and an *edge-preserving* mapping. Two bit strings are defined to be *adjacent* if they differ in at most one bit. A gate is defined to be *adjacency-preserving* if, for any input bit string $x$, the corresponding output bit string $y$ is adjacent to $x$. A network $N$ is *adjacency-preserving* if all gates in $N$ are adjacency-preserving. This is equivalent to requiring that $f_N$, viewed as a single gate, be adjacency-preserving. The graph with $2^m$ vertices corresponding to the bit strings of length $m$ and with edges between adjacent bit strings is a *reflexive hypercube*, the $m$-cube (*reflexive* here refers to the fact that each vertex has a self-loop). If $f_N$ is adjacency-preserving, it maps adjacent vertices to adjacent vertices in the hypercube and is therefore an *edge-preserving* mapping.

Feder gave an $O(m^3)$ algorithm which, given an edge-preserving mapping $f$ on the $m$-cube, finds a fixed point of $f$, or reports that none exists. If the algorithm is applied to $f_N$ then it therefore yields a stable configuration in the adjacency-preserving network $N$ or reports that none exists.

Feder then went on to explore structural properties of edge-preserving mappings on the hypercube. The so-called *median* structure of the hypercube is used to show that (a) the set of stable configurations has a simple characterisation as a 2-SAT instance on the boolean variables associated with the edges of the network, and (b) the behaviour of the network is closely related to a certain permutation on these boolean variables. In particular, the 2-SAT clauses characterise the set of all fixed points of $f$. In a network of *gatewidth* $c$ (this is the maximum, taken over each gate $g$, of the minimum of the number of inputs and outputs of $g$), an instance of 2-SAT with $O(cm)$ clauses that characterises all the stable configurations can be found in $O(c^2 m)$ time.

Putting all of these reductions together, we are thus able to characterise stable matchings in an SRI instance by satisfying truth assignments for a 2-SAT instance. Specifically, given an SRI instance $I$, Feder constructed a 2-SAT instance $J$ with $O(m)$ variables and clauses in $O(m)$ time, where $m$ is the number of acceptable pairs in $I$, such that the stable matchings in $I$ are in 1–1 correspondence with the satisfying truth assignments in $J$. Feder [202] also gave a reduction in the opposite direction, which transforms a minimum weight 2-SAT instance[5] $I$ into an SRI instance $J$ such that a satisfying truth assignment of minimum weight in $I$ corresponds to an egalitarian stable matching in $J$ and vice versa.

In a different paper [203], Feder used a network flow-based approach to derive efficient algorithms for a range of problems relating to 2-SAT, namely minimising the weight of a solution, finding the transitive closure,

---

[5]Here, each variable has a non-negative weight, the weight of a truth assignment is the sum of the weights of the variables that are true under $f$, and the objective is to find a satisfying truth assignment with minimum weight.

recognising partial solutions and enumerating all solutions. Using the above reduction, these then yield a range of structural and algorithmic results for problems relating to SRI, including an efficient algorithm for finding a minimum weight stable matching, given an SMI instance (see Sec. 2.2.10), a 2-approximation algorithm for the problem of finding a minimum weight stable matching, given an SRI instance (see Sec. 4.2.5), and efficient algorithms for finding all stable pairs and listing all stable matchings, given an SRI instance (see Theorem 4.1).

Feder [202] also considered the parallel complexity of SRI and proved that, given an SRI instance $I$, the problem of finding a stable matching or reporting that none exists is reducible to (and therefore no harder than) the problem of deciding whether $I$ admits a stable matching, showing that the former problem is CC-complete (recall that the latter problem was already known to be CC-complete [434, 551]).

Feder's approach [202] involves creating a 2-SAT instance $J$ corresponding to an SRI instance $I$ without appealing directly to the structural results from Ref. [261, Chapter 4] concerning the rotation poset in $I$. Even so, the notion of a rotation is implicit in his construction [202, p.264]. Feder argued that his framework is mathematically appealing and sheds light on the parallel complexity of SRI. However if one is only interested in sequential algorithms, and is familiar with the structure of rotations in an SRI instance, it is arguable that the transformation from SRI to 2-SAT given by Theorem 4.3.4 in Ref. [261] is much simpler. The algorithms for 2-SAT provided by Feder in Ref. [203] and used in Ref. [202] can equally be applied to the 2-SAT instances constructed by Gusfield and Irving [261, pp.194–195] in order to obtain the same structural and algorithmic consequences as observed by Feder in Refs. [202, Sec. 8] and [203, Sec. 9]. A fuller description of Gusfield and Irving's construction, together with the application of Feder's 2-SAT algorithms [203], appears in Sec. 4.2.4 (see also Ref. [216, Sec. 5]).

### 2.3.3 Other fixed-point approaches

Two additional characterisations of SM based on a fixed-point approach have been formulated by Adachi [31] and Fleiner [208–211].

Adachi [31] characterised the set of stable matchings $S$ in an SM instance in terms of the fixed points of an increasing function (distinct from those of Subramanian and Feder). That $S$ is non-empty and forms a lattice are then consequences of Tarski's fixed-point theorem [562], in view of the monotonicity of the mapping.

Fleiner's approach involves characterising stable matchings via the fixed points of a function that is defined in terms of so-called *co-monotone* set functions. His model is valid not just for SM, but for the many–many stable marriage problem (see Sec. 5.4). Again, he invokes Tarski's fixed-point theorem [562] to deduce the non-emptyness and lattice structure of the set of stable matchings in a given problem instance. In fact, his framework draws together a whole range of fundamental results from the field of combinatorics, which might previously have been considered to be completely unrelated to the theory of stable matchings. See Refs. [208–210] for further details. One result that follows from his framework is a generalisation of stable matching theory to the matroid context [209]. We conclude by remarking that Eguchi *et al.* [185] extended the matroidal model to the framework of discrete convex analysis (see also Ref. [224]).

## 2.4 Linear programming approaches

Two Linear Programming (LP) formulations of SM are given in Ref. [261, Sec. 3.7]. Each of these takes an SM instance $I$ and constructs a set of linear inequalities $J$ such that the set of stable matchings in $I$ is in 1–1 correspondence with the extreme points of the polytope of solutions to the LP defined by $J$. The first LP model [261, Sec. 3.7.1] is based on expressing a stable matching in terms of a closed subset of rotations that are to be eliminated. The second [261, Sec. 3.7.2] is a more direct set of inequalities that refers only to the preference lists in $I$ and does not require prior knowledge of the rotation poset of $I$. The first approach is due to Gusfield and Irving, whilst the second is due to Vande Vate [576]. When proving the correctness of his model, Vande Vate implicitly transformed his system of inequalities into those given in Sec. 3.7.1 of Ref. [261].

Rothblum [520] extended Vande Vate's results to the SMI case, and indeed to the HR case too, and claimed that his proofs (showing that the extreme points of the polytopes of solutions corresponding to his LP models are integral) are simpler and more transparent than those of Vande Vate. Roth *et al.* [510] used duality theory in the linear programming context to obtain new results and to derive new proofs of known results for SM. Again, they claimed a simpler proof (compared to those of Vande Vate and Rothblum) of correctness for their LP formulation of SM. They also defined a *fractional stable matching* to be a solution to their linear inequalities (although not necessarily an extreme point of the polytope of such solutions) and showed that the set of fractional stable matchings for a given SM instance forms a distributive lattice.

Abeledo and Rothblum [12] showed that the Gale–Shapley algorithm for SM can be considered as an application of the dual–simplex method. Also, Abeledo et al. [10] showed that a fractional stable matching has a unique representation as a convex combination of (integral) stable matchings, and used this construction in order to give an alternative proof of the existence of a lattice structure for fractional stable matchings, as earlier observed by Roth et al. [510]. The results of Abeledo et al. [10] were also obtained independently by Teo and Sethuraman [565], whose correctness proofs were claimed to be simpler.

LP formulations of SM have been extended in order to find egalitarian, minimum regret and minimum weight stable matchings [261, Sec. 3.7.1][576, 565, 141].

Baïou and Balinski [58] formulated an LP model for HR, whose solutions correspond to the so-called *stable admissions polytope*. They also extended their technique to the case that a minimum weight stable matching is required (given a weight function on the edges of the underlying bipartite graph). Sethuraman et al. [526] studied the linear inequalities as given by Baïou and Balinski [58] for HR from a geometric point of view, and showed that, in the HR context, a fractional stable matching has a decomposition into a convex combination of (integral) stable matchings. This leads to an alternative proof that the extreme points of the polytope of solutions corresponding to the LP model are integral.

Fleiner [212] further generalised the LP characterisation of HR due to Baïou and Balinski to the many–many bipartite stable matching context.

## 2.5  Constraint programming approaches

### 2.5.1  *Introduction*

Over the last 15 years, stable matching problems have been the focus of much interest from the Constraint Programming (CP) community (see Refs. [48, 494] for a general introduction to CP). In particular, several authors have modelled SM and SMI in terms of a *Constraint Satisfaction Problem* (CSP) [39, 250, 409, 257, 420, 573, 572, 571] (see Ref. [48, pp.9–10] for a definition of a CSP). Modifications and extensions of these models have also been considered for a range of variants of SMI, including SMTI [251, 252], HR [424, 188], HRT under weak stability [470], many–many bipartite stable matching [189] and student–project allocation [179, 568]. Models for distributed versions of SM and SR have also been proposed [112, 113, 532, 530], and additionally, a so-called *soft* CSP formulation of the problem of finding a minimum weight stable matching in SM has been constructed [96].

In this section we illustrate the techniques involved by presenting a simple $(n + 1)$-valued binary[6] CSP encoding [420] for an instance $I$ of SMI. This model bears some resemblance to the encoding of SM given in Ref. [409] and develops the "conflict matrices" model of Ref. [250]. For our model we demonstrate that Arc Consistency (AC) propagation [72] (see Ref. [48, Sec. 5.2] for a general introduction to AC) can be carried out in $O(n^3)$ time, where $n$ is the size of $I$. Furthermore, we demonstrate that AC propagation achieves the same results as an extended version of the Gale–Shapley algorithm in a precise sense.

This section is organised as follows. We begin by giving some preliminary definitions and results in Sec. 2.5.2, prior to presenting the CSP model from Ref. [420] in Sec. 2.5.3. Then, in Sec. 2.5.4, we explore the structural relationship between the *GS-lists* (defined in the next subsection) in the SMI instance and the effect of establishing AC in this model.

### 2.5.2 Preliminaries

Gusfield and Irving [261, Sec. 1.2.4] described an extended version of the Gale–Shapley algorithm for SM that avoids some unnecessary steps by deleting from the preference lists certain man–woman pairs that cannot belong to a stable matching. Henceforth we refer to this as the EGS algorithm; it is straightforward to extend this to the SMI case [261, Sec. 1.4.2]. We refer to the man-oriented (respectively woman-oriented) version of the EGS algorithm as the MEGS (respectively WEGS) algorithm.

Upon termination of the MEGS (respectively WEGS) algorithm for a given SMI instance, the reduced preference lists that arise following the deletions are referred to as the *MGS-lists* (respectively *WGS-lists*). The intersection of the MGS-lists with the WGS-lists yields the *GS-lists* [261, p.16]. Some important structural properties of the GS-lists are given by the following theorem.

---

[6]The term *binary* refers to the fact that each constraint has arity 2, while the term *boolean* refers to the case that each variable's domain has size 2.

**Theorem 2.1 ([261, Theorem 1.2.5]).** *For a given instance of* SMI:

(i) *all stable matchings are contained in the GS-lists;*

(ii) *no matching of size $k$ that is contained in the GS-lists can be blocked by a pair that is not in the GS-lists, where $k$ is the number of men who have a non-empty GS-list;*

(iii) *in the man-optimal (respectively woman-optimal) stable matching, each person who has an empty GS-list is unassigned, each man with a non-empty GS-list is partnered by the first (respectively last) woman on his GS-list, and each woman with a non-empty GS-list is partnered by the last (respectively first) man on hers.*

In the next subsection we will construct a CSP encoding $J$ of an SMI instance $I$, proving that the GS-lists in $I$ correspond to the domains remaining after establishing AC in $J$. Furthermore, we will show that we are guaranteed a failure-free enumeration of all stable matchings in $I$ using AC propagation combined with a value-ordering heuristic in $J$.

### 2.5.3 *Overview of the* CSP *model*

Let $I$ be an SMI instance in which $U = \{m_1, m_2, \ldots, m_n\}$ is the set of men and $W = \{w_1, w_2, \ldots, w_n\}$ is the set of women. For each man $m_i \in U$ and woman $w_j \in W$, the lengths of $m_i$'s and $w_j$'s preference lists are denoted by $l(m_i)$ and $l(w_j)$ respectively. We let $m$ denote the number of acceptable pairs in $I$. Also, for any person $a_i \in U \cup W$, we let $PL(a_i)$ denote the set of persons on $a_i$'s original preference list in $I$, and we let $GS(a_i)$ denote the set of persons on $a_i$'s GS-list in $I$. For avoidance of ambiguity, throughout this section, for any acceptable pair $(m_i, w_j)$, each of $rank(m_i, w_j)$ and $rank(w_j, m_i)$ is defined with respect to the *original* preference lists in $I$ (i.e., prior to any potential deletions by the MEGS / WEGS algorithms).

We define a CSP encoding $J$ for an instance $I$ of SMI [420] by introducing $2n$ variables to represent the men and women in the original instance $I$. For each man $m_i \in U$, we introduce a variable $x_i$ in $J$ whose domain, denoted by $dom(x_i)$, is initially defined as $dom(x_i) = \{1, 2, \ldots, l(m_i)\} \cup \{n+1\}$. Similarly, for each woman $w_j \in W$, we introduce a variable $y_j$ in $J$ whose domain, denoted by $dom(y_j)$, is initially defined as $dom(y_j) = \{1, 2, \ldots, l(w_j)\} \cup \{n+1\}$.

An intuitive meaning of the variables is now given. Informally, if $x_i = p$ ($1 \leq p \leq l(m_i)$), then $m_i$ marries the woman $w_j$ such that $rank(m_i, w_j) = p$, and similarly for the case that $y_j = q$ ($1 \leq q \leq l(w_j)$).

1.  $x_i \geq p \Rightarrow y_j \leq q$   $(1 \leq i \leq n, 1 \leq p \leq l(m_i))$
2.  $y_j \geq q \Rightarrow x_i \leq p$   $(1 \leq j \leq n, 1 \leq q \leq l(w_j))$
3.  $y_j \neq q \Rightarrow x_i \neq p$   $(1 \leq j \leq n, 1 \leq q \leq l(w_j))$
4.  $x_i \neq p \Rightarrow y_j \neq q$   $(1 \leq i \leq n, 1 \leq p \leq l(m_i))$

Fig. 2.2   The constraints for the $(n + 1)$-valued encoding of an instance of SMI [420]

More formally, if $\min dom(x_i) \geq p$ $(1 \leq p \leq l(m_i))$, then the pair $(m_i, w_s)$ has been deleted as part of the MEGS algorithm applied to $I$, for all $w_s$ such that $rank(m_i, w_s) < p$. Hence if $w_j$ is the woman such that $rank(m_i, w_j) = p$, then either $m_i$ proposes to $w_j$ during the execution of the MEGS algorithm or the pair $(m_i, w_j)$ will be deleted before the proposal occurs. Similarly if $\min dom(y_j) \geq q$ $(1 \leq q \leq l(w_j))$, then the pair $(m_r, w_j)$ has been deleted as part of the WEGS algorithm applied to $I$, for all $m_r$ such that $rank(m_r, w_j) < q$. Hence if $m_i$ is the man such that $rank(w_j, m_i) = q$, then either $w_j$ proposes to $m_i$ during the execution of the WEGS algorithm or the pair $(m_i, w_j)$ will be deleted before the proposal occurs. If $x_i = n + 1$ (respectively $y_j = n + 1$) then $m_i$ (respectively $w_j$) is unassigned upon termination of each of the MEGS or WEGS algorithms applied to $I$.

The constraints used for the $(n + 1)$-valued encoding are shown in Fig. 2.2. In the context of Constraints 1 and 4, $j$ is the integer such that $rank(m_i, w_j) = p$; also $q = rank(w_j, m_i)$. In the context of Constraints 2 and 3, $i$ is the integer such that $rank(w_j, m_i) = q$; also $p = rank(m_i, w_j)$.

An interpretation of Constraints 1 and 3 is now given (a similar interpretation can be attached to Constraints 2 and 4 with the roles of the men and women reversed). First consider Constraint 1, a stability constraint. This ensures that if a man $m_i$ obtains a partner no better than his $p^{th}$-choice woman $w_j$, then $w_j$ obtains a partner no worse than her $q^{th}$-choice man $m_i$. Now consider Constraint 3, a consistency constraint. This ensures that if man $m_i$ is removed from $w_j$'s list, then $w_j$ is removed from $m_i$'s list.

### 2.5.4   Arc consistency in the CSP model

We now show that, given the above CSP encoding $J$ of an SMI instance $I$, the domains of the variables in $J$ following AC propagation correspond to the GS-lists of $I$. That is, we prove that, after AC is established, for any $i, j$ $(1 \leq i, j \leq n)$, $w_j \in GS(m_i)$ if and only if $p \in dom(x_i)$, and similarly

$m_i \in GS(w_j)$ if and only if $q \in dom(y_j)$, where $rank(m_i, w_j) = p$ and $rank(w_j, m_i) = q$.

The proof is presented using two lemmas. The first lemma shows that the arc consistent domains are equivalent to subsets of the GS-lists. This is done by proving that the deletions made by the MEGS and WEGS algorithms applied to $I$ are correspondingly made during AC propagation. The second lemma shows that the GS-lists correspond to a subset of the domains remaining after AC propagation. This is done by proving that the GS-lists for $I$ give rise to arc consistent domains for the variables in $J$.

**Lemma 2.2 ([420]).** *For a given $i$ ($1 \leq i \leq n$), let $p$ be an integer ($1 \leq p \leq l(m_i)$) such that $p \in dom(x_i)$ after AC propagation. Then the woman $w_j$ such that $rank(m_i, w_j) = p$ belongs to the GS-list of $m_i$. A similar correspondence holds for the women.*

**Lemma 2.3 ([420]).** *For each $i$ ($1 \leq i \leq n$), define a domain of values $dom(x_i)$ for the variable $x_i$ as follows: if $GS(m_i) = \emptyset$, then $dom(x_i) = \{n+1\}$; otherwise $dom(x_i) = \{rank(m_i, w_j) : w_j \in GS(m_i)\}$. The domain of each $y_j$ ($1 \leq j \leq n$) is defined analogously. Then the domains so defined are arc consistent in $J$.*

The two lemmas above, together with the fact that AC algorithms find the unique maximal set of arc consistent domains[7], lead to the following theorem.

**Theorem 2.4 ([420]).** *Let $I$ be an instance of SMI, and let $J$ be a CSP instance obtained by the (n+1)-valued encoding. Then the domains remaining after AC propagation in $J$ correspond to the GS-lists of $I$ in the following sense: for any $i, j$ ($1 \leq i, j \leq n$), $w_j \in GS(m_i)$ if and only if $p \in dom(x_i)$, and similarly $m_i \in GS(w_j)$ if and only if $q \in dom(y_j)$, where $rank(m_i, w_j) = p$ and $rank(w_j, m_i) = q$.*

The constraints shown in Fig. 2.2 may be revised in $O(1)$ time during propagation, assuming that upper and lower bounds for the variables' domains are maintained. Hence the time complexity for establishing AC is $O(ed)$, where $e$ is the number of constraints and $d$ is the domain size [574]. For this encoding we have $e = O(m)$ and $d = O(n)$, therefore AC may be

---

[7] This follows because (i) AC algorithms return a maximal set of arc consistent domains, and (ii) the union of any two sets of arc consistent domains gives rise to a set of arc consistent domains.

established in $O(nm)$ time; also the space complexity is $O(m)$. These complexities represent an improvement on the "conflict matrices" encoding in Ref. [250], whose time and space complexities are $O(n^4)$ and $O(m^2)$ respectively. Moreover we claim that the model presented in this section is a very natural and intuitive encoding for SMI.

Theorem 2.4 and Part (iii) of Theorem 2.1 show that we can find a solution to the CSP giving the man-optimal stable matching $M_a$ without search: for each man $m_i \in U$, we let $p = \min dom(x_i)$. If $p = n+1$ then $m_i$ is unassigned in $M_a$, otherwise the partner of $m_i$ is the woman $w_j \in W$ such that $rank(m_i, w_j) = p$. Considering the $y_j$ variables in a similar fashion gives the woman-optimal stable matching $M_z$.

In fact we may go further and show that the CSP encoding yields all stable matchings in $I$ without having to backtrack due to failure. That is, we may enumerate all solutions of $I$ in a failure-free manner using AC propagation in $J$ combined with a value-ordering heuristic. The following theorem describes the enumeration procedure.

**Theorem 2.5 ([420]).** *Let $I$ be an instance of SMI and let $J$ be a CSP instance obtained using the $(n + 1)$-valued encoding. Then the following search process enumerates all solutions in $I$ without repetition and without ever failing due to an inconsistency:*

- *AC is established as a preprocessing step, and after each branching decision, including the decision to remove a value from a domain;*
- *if all domains are arc consistent and some variable $x_i$ has two or more values in its domain, then the search proceeds by setting $x_i$ to the minimum value $p$ in its domain. On backtracking, the value $p$ is removed from the domain of $x_i$;*
- *when a solution is found, it is reported and backtracking is forced.*

The above results show that, provided the model is chosen carefully, AC propagation within a CSP formulation of SMI captures the structure produced by the EGS algorithm. Furthermore, in many practical situations there may be additional constraints that cannot be accommodated by a straightforward modification of the EGS algorithm. Such constraints could however be built on top of the model that we present here. Possible extensions could arise from variants of SMI that are NP-hard (see e.g., Secs. 2.2.7, 3.2.4, 5.3.3 and 5.6.1) — see Ref. [571, Chapter 6] for further details.

## 2.6 Paths to stability

### 2.6.1 Introduction

In Sec. 2.2.3 we considered the problem of transforming an arbitrary matching $M_0$ in an SM instance $I$ to a stable matching in $I$ using a sequence of divorce operations, each of which involves satisfying a blocking pair and assigning the "divorcees" to one another. As described, Tamura [554], and independently Tan and Su [560], each gave an example SM instance and initial matching $M_0$ such that it is not possible to transform $M_0$ to a stable matching using only divorce operations. At the end of that section we remarked that the situation changes dramatically if the divorcees are allowed to remain single.

In this case a number of papers have shown that, for several stable matching problems, we can always find a sequence of matchings $M_0, M_1, \ldots, M_t$ such that $M_t$ is stable, and for each $i$ ($1 \leq i \leq t$), $M_i$ can be obtained from $M_{i-1}$ by *satisfying a blocking pair* $(m_p, w_q)$ of $M_{i-1}$, i.e., $M_i$ is obtained from $M_{i-1}$ by adding $(m_p, w_q)$ and letting $M_{i-1}(m_p)$ and $M_{i-1}(w_q)$ (as applicable in either case) be unassigned. Decentralised algorithms for constructing such sequences will be reviewed in this section. We begin in Sec. 2.6.2 by reviewing the Roth–Vande Vate Mechanism for SMI [516]. Then in Sec. 2.6.3, we describe results that relate to the question of which stable matchings can be reached by the a special case of this algorithm, called the Random Order Mechanism. Finally in Sec. 2.6.4 we survey other decentralised algorithms for stable matching problems.

### 2.6.2 The Roth–Vande Vate Mechanism

Let $I$ be an instance of SMI and let $M_0$ be an arbitrary matching in $I$. Roth and Vande Vate [516] considered a random sequence of matchings $M_0, M_1, \ldots$, where for each $i \geq 1$, $M_i$ is obtained from $M_{i-1}$ by satisfying a blocking pair. The blocking pair that is satisfied at each step is chosen at random, subject to the constraint that there is a positive probability that any particular blocking pair (from among those that exist at a given step) is chosen. Roth and Vande Vate [516] showed that this random sequence converges to a stable matching with probability 1.

Nevertheless, it is still possible that the process can cycle if the "wrong" choice of blocking pair is made at each step. Roth and Vande Vate [516] illustrated this using an example of Knuth [394]. Abeledo and Rothblum

[13] gave a stronger example to show that cycling is still possible even in the case that the blocking pair that is satisfied at each step involves a pair of agents $(m_p, w_q)$ where $w_q$ is the most-preferred woman in $m_p$'s list who forms a blocking pair with $m_p$.

Roth and Vande Vate gave a constructive proof of their convergence result which contained an algorithm for building a sequence $M_1, \ldots, M_t$ such that $M_t$ is stable, and for each $i$ $(1 \leq i \leq t)$, $M_i$ can be obtained from $M_{i-1}$ by satisfying a blocking pair. This algorithm has become known in the literature as the *Roth–Vande Vate Mechanism*.

We refer to the Roth–Vande Vate Mechanism as Algorithm RVV, and a pseudocode description of this algorithm can be found in Algorithm 2.1 (the presentation of the pseudocode is influenced by the descriptions of Cechlárová [119], Biró *et al.* [81], and Biró and Norman [95]).

The algorithm uses the following notation. Suppose that $U = \{m_1, \ldots, m_n\}$ is the set of men and $W = \{w_1, \ldots, w_n\}$ is the set of women in $I$. The algorithm will modify a matching $M$, which is initially equal to $M_0$. The algorithm maintains a set $S$ of agents, which is initially empty. At any point during the algorithm's execution $M|_S$ denotes $M \cap (S \times S)$ and $I|_S$ denotes the sub-instance of $I$ obtained by deleting every member of $(U \cup W)\backslash S$, and deleting each such agent from the preference list of each member of $S$. Throughout this section we will use the notation $a_i$ and $b_j$ to denote arbitrary agents (i.e., $a_i$ can either be a man or a woman, and similarly for $b_j$). We will also use $(a_i, b_j)$ to denote a man–woman pair; thus if $a_i \in W$ and $b_j \in U$ then $(a_i, b_j)$ actually corresponds to the man–woman pair $(b_j, a_i)$.

We will now describe an execution of Algorithm RVV. The loop in this algorithm iterates as long as $M$ is not stable in $I$. It will maintain two loop invariants, as follows: (i) no member of $S$ is assigned in $M$ to a member outside of $S$, and (ii) $M|_S$ is stable in $I|_S$. During a loop iteration, if there is a blocking pair $(a_i, b_j)$ of $M$ in $I$ such that $a_i \notin S$ and $b_j \in S$ then Algorithm add is called with parameter $a_i$. Otherwise, any blocking pair $(m_i, w_j)$ of $M$ in $I$ must satisfy the property that $m_i \notin S$ and $w_j \notin S$, by loop invariant (ii). In this case, Algorithm satisfy is called with paramters $m_i$ and $w_j$.

Algorithm add and Algorithm satisfy are described in Algorithms 2.2 and 2.3 respectively. Note that these algorithms, along with Algorithm RVV, assume that $I$, $M$, $S$ and $t$ are declared as global variables.

At the start of an execution of Algorithm add with parameter $a_i$, $a_i$'s partner in $M$ (if applicable) is set to be unassigned. Also $a_i$ is added to

---

**Algorithm 2.1** Algorithm RVV [516]

**Require:** SMI instance $I$ and a matching $M_0$ in $I$
**Ensure:** $M$ is stable in $I$ and can be obtained from $M_0$ by iteratively satisfying blocking pairs
1: $M := M_0$;
2: $S := \emptyset$;
3: $t := 0$;
4: **while** $M$ is not stable in $I$ **do**
5:    **if** there exists $(a_i, b_j) \in bp(I, M)$ such that $a_i \notin S$ and $b_j \in S$ **then**
6:       add($a_i$);
7:    **else**
8:       choose $(m_i, w_j) \in bp(I, M)$;    {then $m_i \notin S$ and $w_j \notin S$}
9:       satisfy($m_i, w_j$);
10:   **end if**
11: **end while**

---

$S$. The task is to ensure that, following the arrival of $a_i$, we can restabilise the matching so that $M|_S$ is again stable in $I|_S$. This is carried out by the while loop, whose execution is referred to as a *proposal–rejection sequence*.

If $a_i$, the *proposer*, is a *blocking agent* (i.e., is involved in a blocking pair) of $M|_S$ in $I|_S$, we let $(a_i, b_j)$ be the best blocking pair of $M|_S$ in $I|_S$ according to $a_i$'s preference list. Then $b_j \in S$. If $b_j$ is assigned in $M$, to $a_s$ say, the pair $(a_s, b_j)$ is removed from $M$, and $a_s$ is recorded as the next proposer (note that $a_s \in S$, since $(a_s, b_j) \in M|_S$). The pair $(a_i, b_j)$ is then added to $M$. The while loop terminates when the proposer is not a blocking agent of $M|_S$ in $I|_S$.

Algorithm satisfy is used to satisfy a blocking pair $(m_i, w_j)$ of $M$, where $m_i \notin S$ and $w_j \notin S$. Both of these agents are added to $S$. Also each of the partners of $m_i$ and $w_j$ (if applicable) is set to be unassigned in $M$. Finally $(m_i, w_j)$ is added to $M$.

We will prove via the following lemma and theorem that the algorithm produces a finite sequence of matchings $M_0, M_1, \ldots, M_t$, where $M_t$ is stable in $I$. Some of the exposition in the proof of Lemma 2.6 follows the approach taken in Ref. [81].

**Lemma 2.6.** *Each loop iteration $\ell$ of Algorithm* RVV *terminates. Moreover, at the end of $\ell$, (i) $M$ contains no pair $(a_i, b_j)$ such that $a_i \notin S$ and $b_j \in S$, and (ii) $M|_S$ is stable in $I|_S$.*

**Proof.** We prove this by induction on the number of iterations $r$ of the main loop of Algorithm RVV. Clearly at the beginning of the first loop

---

**Algorithm 2.2** Algorithm add (method for Algorithm RVV) [516]

**Require:** agent $a_i \notin S$
**Ensure:** $a_i \in S$ and $M|_S$ is stable in $I|_S$
1: **if** $a_i$ is assigned in $M$ **then**
2:     $M := M\backslash\{(a_i, M(a_i))\}$;
3: **end if**
4: $S := S \cup \{a_i\}$;
5: **while** isBlockingAgent($I|_S, M|_S, a_i$) **do**
6:     $\{a_i$ is the "proposer"$\}$
7:     $(a_i, b_j) := $ bestBlockingPair($I|_S, M|_S, a_i$);
8:     $a_z := a_i$;
9:     **if** $b_j$ is assigned in $M$ **then**
10:         $a_s := M(b_j)$;
11:         $M := M\backslash\{(a_s, b_j)\}$;
12:         $a_i := a_s$;
13:     **end if**
14:     $M := M \cup \{(a_z, b_j)\}$;
15:     $t$++;
16:     $M_t := M$;
17: **end while**

---

**Algorithm 2.3** Algorithm satisfy (method for Algorithm RVV) [516]

**Require:** agents $m_i \notin S$ and $w_j \notin S$
**Ensure:** $m_i \in S$, $w_j \in S$ and $M|_S$ is stable in $I|_S$
1: $S := S \cup \{m_i, w_j\}$;
2: **if** $m_i$ is assigned in $M$ **then**
3:     $M := M\backslash\{(m_i, M(m_i)\}$;
4: **end if**
5: **if** $w_j$ is assigned in $M$ **then**
6:     $M := M\backslash\{(M(w_j), w_j\}$;
7: **end if**
8: $M := M \cup \{(m_i, w_j)\}$;
9: $t$++;
10: $M_t := M$;

---

iteration $\ell$, $S = \emptyset$. During $\ell$, Algorithm satisfy is called with agents $m_i$ and $w_j$ such that $(m_i, w_j) \in bp(M)$. It is immediate that this method terminates, and once it does, $S = \{m_i, w_j\}$ and $(m_i, w_j) \in M$, so clearly (i) and (ii) from the lemma statement are satisfied.

Now suppose that $r \geq 1$, and let $M^*$ and $S^*$ denote the contents of $M$ and $S$ at the very end of the $r$th loop iteration of Algorithm RVV. Then the induction hypothesis is that (i) $M^*$ contains no pair $(a_i, b_j)$ such that

$a_i \notin S$ and $b_j \in S$, and (ii) $M^*|_{S^*}$ is stable in $I|_{S^*}$. Let $\ell$ denote the $(r+1)$th loop iteration of Algorithm RVV. During $\ell$, either (a) Algorithm add is called, or (b) Algorithm satisfy is called.

In case (a), Algorithm add is called with parameter $a_i$, for some $a_i \notin S^*$. Let $\ell'$ denote this execution of Algorithm add, and suppose that the proposers during $\ell'$ are $a_{k_0}, a_{k_1}, a_{k_2}, \ldots$, where $a_{k_0} = a_i$. Note that the proposers need not be distinct, however we claim that the proposal–rejection sequence is finite, i.e., there is some $z \geq 0$ such that $a_{k_z}$ is the last proposer.

To establish the claim, we note that there is at least one iteration of the while loop during $\ell'$, because there is some $b_j \in S^*$ such that $(a_i, b_j) \in bp(M)$, which led to the call of Algorithm add during $\ell$. Let the sequence $b_{k_p}$ ($p \geq 1$) be defined such that $(a_{k_{p-1}}, b_{k_p})$ is the best blocking pair for $a_{k_{p-1}}$, identified in line 7 of the while loop iteration in which $a_{k_{p-1}}$ is the proposer. For each $p \geq 1$, by the induction hypothesis and by the choice of best blocking pair for $a_{k_{p-1}}$ (in line 7 of the while loop iteration in which $a_{k_{p-1}}$ is the proposer), it is clear that $a_{k_p} \in S$ and $b_{k_p} \in S$. Also $a_{k_0}$ is added to $S$ during $\ell'$.

It is immediate that, for each $p \geq 1$, $b_{k_p}$ prefers $a_{k_{p-1}}$ to $a_{k_p}$. Also, for each $p \geq 1$, $a_{k_p}$ prefers $b_{k_p}$ to $b_{k_{p+1}}$ (for otherwise $(a_{k_p}, b_{k_{p+1}})$ blocks $M^*|_{S^*}$ in $I|_{S^*}$, a contradiction. Thus for each $p \geq 0$, if $a_{k_p}$ (respectively $b_{k_p}$) appears in the sequence more than once, their partner in $M$ will be successively worse (respectively better). Hence each pair $(a_{k_p}, b_{k_{p+1}})$ can occur in the sequence at most once. Thus, since the number of acceptable pairs is finite, it follows that there is some $q \geq 1$ such that either (I) $b_{k_q}$ is unassigned in $M$, or (II) $a_{k_q}$ is not a blocking agent of $M|_S$ in $I|_S$. In each case, the proposal–rejection sequence terminates (in case (I), $a_{k_{q-1}}$ is the last proposer, whilst in case (II), $a_{k_q}$ is the last proposer).

As already noted, at the termination of $\ell'$, each member of the proposal–rejection sequence is in $S$, and thus condition (i) from the statement of the lemma is satisfied. At this point, suppose that $(a_{k_p}, b_c)$ blocks $M|_S$ in $I|_S$, for some $p$ ($0 \leq p \leq q-1$) and $b_c \in S$. Consider the while loop iteration $z$ which corresponds to the last occurrence of $a_{k_p}$ as a proposer during $\ell'$. Let $M'$ be the matching at the beginning of loop iteration $z$. As this while loop iteration corresponds to the last occurrence of $a_{k_p}$ as a proposer during $E$, it follows that $(a_{k_p}, M(a_{k_p}))$ is the best blocking pair of $M'|_S$ in $I|_S$ identified during $z$. Now either $b_c$ prefers $M$ to $M'$ or is indifferent between them. Hence we obtain a contradiction to the choice of best blocking pair for $a_{k_p}$ during loop iteration $z$.

Also $(a_c, b_d)$ cannot block $M|_S$ in $I|_S$, for any $b_d \in S$ and for any $a_c \in S$ such that $a_c \neq a_{k_p}$ for each $0 \leq p \leq q - 1$, for (I) $a_c$ is indifferent between $M$ and $M^*$, (II) either $b_d$ prefers $M$ to $M^*$ or is indifferent between them, and (III) $M^*|_{S^*}$ is stable in $I|_{S^*}$.

Hence condition (ii) in the statement of the lemma is true.

In case (b), Algorithm satisfy is called during $\ell$. This method clearly terminates. By the induction hypothesis, if $a_i$ is assigned in $M^*$ then $M^*(a_i) \notin S$. A similar remark holds for $M^*(b_j)$, if $b_j$ is assigned in $M^*$. Hence condition (i) in the statement of the lemma holds. Also if $(a_i, b_k)$ blocks $M|_S$ in $I|_S$, for some $b_k \in S\backslash\{b_j\}$, then we contradict the fact that Algorithm satisfy was called during $\ell$, rather than Algorithm add. Similarly $(a_k, b_j)$ cannot block $M|_S$ in $I|_S$, for some $a_k \in S\backslash\{a_i\}$. Hence $M|_S$ is stable in $I|_S$ at the termination of $\ell$.

Thus the $(r + 1)$th loop iteration $\ell$ of Algorithm RVV terminates, and at the end of $\ell$, conditions (i) and (ii) in the statement of the lemma both hold. Hence the overall result follows by induction.  $\square$

**Theorem 2.7 ([516]).** *Let $I$ be an instance of* SMI *of size $n$, let $m$ be the number of acceptable pairs, and let $M_0$ be an arbitrary matching in $I$. Algorithm* RVV *applied to $I$ and $M_0$ produces a finite sequence of matchings $M_0, M_1, \ldots, M_t$, where $M_t$ is stable in $I$, and for each $k$ ($1 \leq k \leq t$), $M_k$ is obtained from $M_{k-1}$ by satisfying a blocking pair of $M_{k-1}$. Moreover $M_t$ and an abbreviated representation of the intermediate matchings can be obtained in $O(nm)$ overall time, whilst the complexity is $O(n^2 m)$ for an explicit representation of all matchings. Finally, $t \leq 2nm$.*

**Proof.** During each iteration of Algorithm RVV, $S$ increases in size by either one or two elements. At the end of each such iteration, the invariant $M|_S$ is stable in $I|_S$ holds by Lemma 2.6. Hence we are bound to ultimately reach the outcome that $M$ is stable in $I$, in which case Algorithm RVV terminates. Moreover it is clear from the operation of the algorithm that, for each $k$ ($1 \leq k \leq t$), $M_k$ is obtained from $M_{k-1}$ by satisfying a blocking pair of $M_{k-1}$.

The complexity of the algorithm is obtained by observing that $S$ increases in size by a minimum of one element at each loop iteration of Algorithm RVV. Since $|S| \leq 2n$, it follows that the same upper bound applies to the number of loop iterations of an execution of Algorithm RVV.

As noted in the proof of Lemma 2.6, each proposal–rejection sequence during an execution of Algorithm add can involve at most $m$ agents, given

---

**Algorithm 2.4** Algorithm ROM [410]

---

**Require:** SMI instance $I$
**Ensure:** $M$ is stable in $I$ and can be obtained from $\emptyset$ by iteratively satisfying
    blocking pairs
1: $M := \emptyset$;
2: $S := \emptyset$;
3: $t := 0$;
4: **while** $S \neq U \cup W$ **do**
5:    choose $a_i \in (U \cup W) \backslash S$;    {agent $a_i$ "arrives"}
6:    add($a_i$);
7:    $t$++;
8:    $M_t := M$;
9: **end while**

---

that each acceptable pair can occur in a sequence at most once. With a suitable choice of data structures, each proposal–rejection sequence can be implemented to run in $O(m)$ time if we discount the time required for the assignment to $M_t$ at line 16 during an iteration of the while loop. Also each call to Algorithm satisfy takes $O(1)$ time if we ignore for the moment the time required for the assignment to $M_t$ in line 10.

Thus an abbreviated representation of the sequence of constructed matchings can be given in $O(nm)$ overall time if we simply output the blocking pairs that have been satisfied when transforming $M_0$ to a stable matching. The upper bound for the final value of $t$ also follows. If an explicit description of each matching in the sequence is required then the complexity of Algorithm add becomes $O(nm)$, that of Algorithm satisfy becomes $O(n)$, and hence the overall time complexity bound for Algorithm RVV increases to $O(n^2 m)$. □

### 2.6.3 The Random Order Mechanism

Ma [410] argued that, in a decentralised market, it is more natural to assume that agents arrive one at a time, rather than in pairs (as in Algorithm RVV). He described a modification of Algorithm RVV that reflects this observation for the special case that $M_0 = \emptyset$. Ma referred to his modification as the *Random Order Mechanism*. Essentially, to obtain this algorithm from Algorithm RVV, ensure that $M = \emptyset$ initially, the loop should iterate as long as $S \neq U \cup W$, change line 5 to read "choose some $a_i \in U \cup W) \backslash S$" and delete lines 7-10. For completeness, pseudocode for the Random Order Mechanism is shown as Algorithm ROM, contained in Algorithm 2.4.

Men's preferences    Women's preferences

$m_1 : w_1 \ w_3 \ w_2$        $w_1 : m_3 \ m_2 \ m_1$

$m_2 : w_2 \ w_1 \ w_3$        $w_2 : m_1 \ m_3 \ m_2$

$m_3 : w_3 \ w_2 \ w_1$        $w_3 : m_2 \ m_1 \ m_3$

Fig. 2.3    An instance of SM of size 3 due to Gale and Shapley [235]

It is straightforward to modify Lemma 2.6 and Theorem 2.7 so that they also hold for Algorithm ROM (ensuring that $M_0$ is taken to be the empty set in the statement of each of these results).

In the remainder of this section we consider the question as to which stable matchings, for a given SMI instance, can be reached by all possible executions of Algorithm ROM. Here, the possible executions relate to the "arrival orders" of the agents, i.e., the order in which they are chosen at line 5 of Algorithm ROM.

As observed by Roth and Vande Vate [516], during Algorithm ROM, if all the women arrive first followed by all the men, Algorithm ROM is equivalent to the MEGS algorithm. Obviously if the men arrive first followed by the women, we obtain the WEGS algorithm. Ma [410] showed that, for an example SM instance of size 4 due to Knuth (see Fig. 2.1), 4 out of the 10 stable matchings can never be reached, regardless of the arrival order of the agents. In fact, in order to show that not all stable matchings can be reached in general, a simpler proof can be obtained by considering the SM instance of size 3 due to Gale and Shapley [235], as shown in Fig. 2.3. The fact that not all stable matchings can be reached for this instance is established by the following result.

**Theorem 2.8.** *Let $I$ be the* SM *instance shown in Fig. 2.3. There is a stable matching $M_e$ in $I$ such that no execution of Algorithm* ROM *leads to $M_e$.*

**Proof.**    It is not difficult to verify that $I$ has three stable matchings, namely $M_a$, $M_e$ and $M_z$, in which each man has his first, second and third choice respectively. We will show that Algorithm ROM fails to arrive at $M_e$ starting from the empty matching, regardless of the arrival order of the agents.

Let $Q$ be some initial ordering of the agents, and let $E$ be the execution of Algorithm ROM with respect to $Q$. Let $M$ be the stable matching obtained upon termination of $E$. Suppose for a contradiction that $M = M_e$. Suppose that $m_1$ is last in $Q$; by the symmetry of the preference lists, the argument is similar if any other agent is last in $Q$. Let $M^*$ denote the stable matching just prior to $m_1$'s arrival during $E$. Then as $(m_1, w_3) \in M$, it follows that either $(m_2, w_1) \in M^*$ or $(m_3, w_1) \in M^*$, for otherwise $(m_1, w_1) \in M$. Also either $w_3$ is unassigned in $M^*$ or $(m_3, w_3) \in M^*$. In the former case, $(m_3, w_3)$ blocks $M^*$, and hence $(m_3, w_3) \in M^*$. Thus $(m_2, w_1) \in M^*$. Then $w_2$ is unassigned in $M^*$, and hence $(m_2, w_2)$ blocks $M^*$ just prior to $m_1$ arriving, a contradiction. $\square$

We remark that, in the context of the above theorem, the fact that $M_e$ cannot be reached is no surprise, given the following observation of Blum *et al.* [98] and Cechlárová [118]. Independently these authors showed that, for the SM case, in the stable matching output by any execution of Algorithm ROM, some agent obtains his/her best stable partner. With respect to $M_e$, since each agent obtains his/her second choice in that matching, whilst each agent's best stable partner is his/her first choice, it follows that $M_e$ cannot be reached by Algorithm ROM.

In fact, Blum *et al.* [98] proved a more general result. To describe this, following Biró *et al.* [81], we define an *active phase* of Algorithm ROM to be an execution of Algorithm add in which the agent $a_i$ passed to the method is involved in a blocking pair of $M|_S$ in $I|_S$. For a given SMI instance, Blum *et al.* [98] showed that, in the stable matching output by Algorithm ROM, if $a_i$ (respectively $b_j$) is a proposer (received a proposal) and became assigned during the last active phase, then $a_i$ ($b_j$) receives his/her best (worst) stable partner. Specifically, we can conclude that the last agent to arrive during an active phase obtains his/her best stable partner, as observed by Blum and Rothblum [99]. These authors also showed that, for a given agent $a_i$ and for a fixed ordering of the other agents, $a_i$ can only improve his/her outcome in the stable matching output by Algorithm ROM by deferring his/her arrival time.

As Cechlárová [118] remarked, even if $M$ is a stable matching in which some agent has their best stable partner, it is possible that no execution of Algorithm ROM produces $M$. Ma's observation that 4 out of the 10 stable matchings in Knuth's SM instance (see Fig. 2.1) are unreachable does not illustrate this, since no agent obtains their best stable partner in any of these 4 stable matchings.

However, this phenomenon can be illustrated by considering the SM instance shown in Fig. 2.4, and the following stable matching (both due to Iain McBride [435]):

$$M = \{(m_1, w_4), (m_2, w_3), (m_3, w_2), (m_4, w_1)\}$$

Men's preferences            Women's preferences
$m_1 : w_1 \quad w_4 \quad w_2 \quad w_3$        $w_1 : m_3 \quad m_4 \quad m_2 \quad m_1$
$m_2 : w_2 \quad w_3 \quad w_1 \quad w_4$        $w_2 : m_1 \quad m_3 \quad m_4 \quad m_2$
$m_3 : w_3 \quad w_4 \quad w_2 \quad w_1$        $w_3 : m_1 \quad m_2 \quad m_4 \quad m_3$
$m_4 : w_1 \quad w_3 \quad w_4 \quad w_2$        $w_4 : m_2 \quad m_4 \quad m_1 \quad m_3$

Fig. 2.4   An instance of SM of size 4 due to McBride [435]

Clearly $m_4$ has his best stable partner in $M$, however computer simulation indicates that no execution of Algorithm ROM produces $M$. Experimentation also reveals [435] that there is no SM instance of size 3 that admits a stable matching $M$ such that (i) some agent has their best stable partner in $M$, and (ii) no execution of Algorithm ROM produces $M$.

It is an open question to obtain a complete characterisation of the set of stable matchings that can be reached by Algorithm ROM, for a given SMI instance. Indeed, related open problems are to determine the complexity of the following two decision problems (the second of which is due to Péter Biró) :

(1) given an SM instance $I$ and a stable matching $M$, is there an execution of Algorithm ROM that terminates with $M$?
(2) given an SM instance $I$, a matching $M_0$ and a stable matching $M$, is there an execution of Algorithm RVV that transforms $M_0$ to $M$?

### 2.6.4   *Other decentralised algorithms*

In this subsection we discuss counterparts of Algorithm RVV that have been formulated for other stable matching problems. We begin by mentioning that, for the SMI case, Abeledo and Rothblum [13] formulated a family of decentralised algorithms for transforming an arbitrary matching to a stable matching based on iteratively satisfying blocking pairs. This family includes the Gale–Shapley algorithm, Algorithm RVV and Algorithm ROM as special cases.

Ackermann *et al.* [30] categorised decentralised algorithms for SMI into *better response dynamics* and *best response dynamics*. The former description applies to mechanisms that are based on satisfying blocking pairs, whilst the latter refers to a more specific mechanism where, should a blocking pair be satisfied, it is the best blocking pair for the *active* agent (i.e., the agent who makes the proposal). It is tempting to believe that Algorithm RVV is an example of a best response dynamics, however this is not the case because the best blocking pair for a proposer is only selected from among those agents who belong to $S$ at that point in time. Ackermann *et al.* [30] also considered *random better response dynamics* and *random best response dynamics*. In the former case, a blocking pair is chosen uniformly at random, whilst in the latter case, a blocking pair that corresponds to the best blocking pair for a given proposer is selected uniformly at random. The authors gave exponential lower bounds for the convergence time (i.e., the time to reach a stable matching, starting from a given matching) of both the random better response dynamics and the random best response dynamics in SMI. Note that this result for the random better response dynamics does not contradict Theorem 2.7, since Algorithm RVV is not strictly speaking a *random* better response dynamics (it involves a random choice of blocking pair at certain points, but not during an execution of Algorithm add, for example).

Generalisations of Roth and Vande Vate's results [516] to the SRI case have been given in various papers — see Sec. 4.3.6 for more details. Furthermore, Biró *et al.* [81] extended many of the results obtained for the SMI case by Blum *et al.* [98], and by Blum and Rothblum [99], as described in the previous subsection, to the SRI setting.

Algorithm ROM may be extended to the HR case by using the fact that, for a given instance $I$ of HR, there is an instance $J$ of SMI such that there is a 1–1 correspondence between the stable matchings in $I$ and $J$. This "cloning" operation is described in Ref. [261, p.38] (see also Ref. [514, pp.131–132]). This approach was taken by Boyle and Echenique [110], who showed that in the HR setting, the last agent to arrive obtains the best outcome they could obtain in any stable matching (thus generalising an earlier result of Blum *et al.* [98], and of Blum and Rothblum [99]). Klaus and Klijn [390] showed that Algorithm RVV can be extended to the variant of HR in which couples are allowed to submit joint preference lists (see Sec. 5.3 for more details), as long as the couples have so-called *weakly responsive* preferences. Similarly Kojima and Ünver [398] showed that the same is true for a many–many bipartite stable matching problem, as long as the preferences of the agents on one side satisfy the so-called *responsive* property, whilst agents on the other side have so-called *substitutable* preferences (see Sec. 5.4.4 for more details). See also Refs. [116,389].

Biró *et al.* [81] referred to the Roth–Vande Vate Mechanism [516] (and by implication Algorithm ROM also) and the Tan–Hsueh algorithm [559] (see Sec. 4.3.3) as *incremental* in nature, because they are applicable in dynamic versions of bipartite and non-bipartite matching markets

respectively, in which agents arrive and leave over time, and we require to
restore the stability of a given matching after each such change.

## 2.7 Median stable matchings

Recall from Chap. 1 that many important structural results are already
known for stable matchings. For example, Theorem 1.13 states that, in
a given SM instance, if each man picks the better of his partners in two
given stable matchings, the result is also a stable matching. Furthermore,
Theorem 1.20 indicates that, in a given SR instance, if we assign each agent
their "median choice" over their partners in three given stable matchings,
we obtain another stable matching. These two results are intriguing because
in each case it is by no means obvious that the set of pairs so constructed
should be a matching, let alone stable. However, it turns out that these
results can be generalised in a very unexpected way, giving rise to an even
more elegant structure. We describe recent developments concerning so-
called "median" stable matchings in this section, but first we require to
define some notation.

Let $I$ be an SM instance where $U$ is the set of men and $W$ is the set
of women. Denote by $\mathcal{S}$ the set of stable matchings in $I$, and let $\mathcal{T} \subseteq \mathcal{S}$.
Suppose that $s = |\mathcal{S}|$ and $t = |\mathcal{T}|$. For each agent $a_i \in U \cup W$, denote
by $P_{\mathcal{T}}(a_i)$ the multiset $\langle M(a_i) : M \in \mathcal{T} \rangle$ of $a_i$'s partners (with possible
repetitions) in the stable matchings in $\mathcal{T}$, where $P_{\mathcal{T}}(a_i)$ has been sorted
according to $a_i$'s preference list. Let $p_{j,\mathcal{T}}(a_i)$ denote the $j$th element in this
sorted multiset ($1 \leq j \leq t$).

To give an example of the notation defined so far, consider the SM
instance $I_8$ of size 8 (due to Irving *et al.* [320]) as shown in Fig. 2.5. This
instance has 23 stable matchings, $M_1, \ldots, M_{23}$, as shown in Table 2.1 (the
partner of man $m_i$ in matching $M_j$ is shown in the $(i+1)$th row and $(j+1)$th
column)[8]. The sorted multiset of partners of each man $m_i$ among all these
stable matchings is shown in the $(i + 2)$th row of Table 2.2. Note that
Tables 2.1 and 2.2 are due to Teo and Sethuraman [565][9].

Teo and Sethuraman proved the following surprising and beautiful
result.

**Theorem 2.9 ([565]).** *Let $I$ be an* SM *instance and let $\mathcal{T}$ be a set of stable
matchings in $I$. Assuming the notation defined above, let $\alpha_{j,\mathcal{T}}$ (respectively*

---

[8]The rows involving the $D(M)$ and $\Delta(M)$ values should be ignored for the time being.
[9]However we correct two small errors in their presentation of these tables (one in $M_{19}$ and one in $\alpha_{14}$.

| Men's preferences | | | | | | | | | Women's preferences | | | | | | | | |
|---|---|---|---|---|---|---|---|---|---|---|---|---|---|---|---|---|---|---|
| $m_1$ : | $w_3$ | $w_1$ | $w_5$ | $w_7$ | $w_4$ | $w_2$ | $w_8$ | $w_6$ | $w_1$ : | $m_4$ | $m_3$ | $m_8$ | $m_1$ | $m_2$ | $m_5$ | $m_7$ | $m_6$ |
| $m_2$ : | $w_6$ | $w_1$ | $w_3$ | $w_4$ | $w_8$ | $w_7$ | $w_5$ | $w_2$ | $w_2$ : | $m_3$ | $m_7$ | $m_5$ | $m_8$ | $m_6$ | $m_4$ | $m_1$ | $m_2$ |
| $m_3$ : | $w_7$ | $w_4$ | $w_3$ | $w_6$ | $w_5$ | $w_1$ | $w_2$ | $w_8$ | $w_3$ : | $m_7$ | $m_5$ | $m_8$ | $m_3$ | $m_6$ | $m_2$ | $m_1$ | $m_4$ |
| $m_4$ : | $w_5$ | $w_3$ | $w_8$ | $w_2$ | $w_6$ | $w_1$ | $w_4$ | $w_7$ | $w_4$ : | $m_6$ | $m_4$ | $m_2$ | $m_7$ | $m_3$ | $m_1$ | $m_5$ | $m_8$ |
| $m_5$ : | $w_4$ | $w_1$ | $w_2$ | $w_8$ | $w_7$ | $w_3$ | $w_6$ | $w_5$ | $w_5$ : | $m_8$ | $m_7$ | $m_1$ | $m_5$ | $m_6$ | $m_4$ | $m_3$ | $m_2$ |
| $m_6$ : | $w_6$ | $w_2$ | $w_5$ | $w_7$ | $w_8$ | $w_4$ | $w_3$ | $w_1$ | $w_6$ : | $m_5$ | $m_4$ | $m_7$ | $m_6$ | $m_2$ | $m_8$ | $m_3$ | $m_1$ |
| $m_7$ : | $w_7$ | $w_8$ | $w_1$ | $w_6$ | $w_2$ | $w_3$ | $w_4$ | $w_5$ | $w_7$ : | $m_1$ | $m_4$ | $m_5$ | $m_6$ | $m_2$ | $m_8$ | $m_3$ | $m_7$ |
| $m_8$ : | $w_2$ | $w_6$ | $w_7$ | $w_1$ | $w_8$ | $w_3$ | $w_4$ | $w_5$ | $w_8$ : | $m_2$ | $m_5$ | $m_4$ | $m_3$ | $m_7$ | $m_8$ | $m_1$ | $m_6$ |

Fig. 2.5   An instance of SM of size 8 due to Irving, Leather and Gusfield [320]

$\beta_{j,\mathcal{T}}$) denote the set of pairs obtained by assigning each man $m_i$ (woman $w_i$) to $p_{j,\mathcal{T}}(m_i)$ ($p_{j,\mathcal{T}}(w_i)$), the $j$th element in the sorted multiset $P_{\mathcal{T}}(m_i)$ ($P_{\mathcal{T}}(w_i)$)). Then each of $\alpha_{j,\mathcal{T}}$ and $\beta_{j,\mathcal{T}}$ is a stable matching, and moreover $\alpha_{j,\mathcal{T}} = \beta_{t-j+1,\mathcal{T}}$.

In the above theorem, if $t = 2$ and $j = 1$, we obtain Theorem 1.13 in the context of SM. Similarly if $t = 3$ and $j = 2$, we arrive at Theorem 1.20 in the SM setting. Finally, if $\mathcal{T} = \mathcal{S}$ and $j = 1$, we obtain the man-optimal stable matching in $I$. More generally, in the context of Theorem 2.9, $\alpha_{j,\mathcal{T}}$ (and indeed $\beta_{t-j+1,\mathcal{T}}$) is referred to as the $j$th generalised median stable matching of $\mathcal{T}$. Note that the number of generalised median stable matchings of $\mathcal{T}$ could be substantially smaller than $t$. For example, with respect to instance $I_8$, there are 9 generalised median stable matchings of $\mathcal{S}$, as follows:

$$\{M_1, M_3, M_8, M_{15}, M_{18}, M_{19}, M_{20}, M_{21}, M_{22}\}.$$

A very special case of Theorem 2.9 occurs when $j$ is chosen so that each agent $a_i$ obtains his/her partner from the "middle" of $P_{\mathcal{T}}(m_i)$ in a precise sense — such an $\alpha_{j,\mathcal{T}}$ is called a *median stable matching of* $\mathcal{T}$. Specifically, if $t$ is odd then $j = (t + 1)/2$, i.e., $\alpha_{(t+1)/2,\mathcal{T}}$ is the unique median stable matching of $\mathcal{T}$. If $t$ is even then in $\alpha_{t/2,\mathcal{T}}$ (respectively $\alpha_{t/2+1,\mathcal{T}}$), each man obtains his lower (upper) median stable partner, whilst each woman obtains her upper (lower) median stable partner in $\mathcal{T}$. Cheng [144] defined the median stable matchings of $\mathcal{T}$ in this case to be $[\alpha_{t/2,\mathcal{T}}, \alpha_{t/2+1,\mathcal{T}}]$, where

$$[M_{p,\mathcal{T}}, M_{q,\mathcal{T}}] = \{M \in \mathcal{T} : M_{p,\mathcal{T}} \preceq M \preceq M_{q,\mathcal{T}}\},$$

and $\preceq$ is the dominance partial order on $\mathcal{S}$ from Definition 1.12.

Table 2.1  The 23 stable matchings for the SM instance of size 8 shown in Fig. 2.5

| | $M_1$ | $M_2$ | $M_3$ | $M_4$ | $M_5$ | $M_6$ | $M_7$ | $M_8$ | $M_9$ | $M_{10}$ | $M_{11}$ | $M_{12}$ | $M_{13}$ | $M_{14}$ | $M_{15}$ | $M_{16}$ | $M_{17}$ | $M_{18}$ | $M_{19}$ | $M_{20}$ | $M_{21}$ | $M_{22}$ | $M_{23}$ |
|---|---|---|---|---|---|---|---|---|---|---|---|---|---|---|---|---|---|---|---|---|---|---|---|
| $m_1$ | $w_3$ | $w_1$ | $w_3$ | $w_3$ | $w_1$ | $w_1$ | $w_3$ | $w_1$ | $w_5$ | $w_1$ | $w_3$ | $w_5$ | $w_5$ | $w_1$ | $w_5$ | $w_5$ | $w_5$ | $w_5$ | $w_5$ | $w_7$ | $w_5$ | $w_1$ | $w_1$ |
| $m_2$ | $w_1$ | $w_3$ | $w_1$ | $w_1$ | $w_3$ | $w_3$ | $w_1$ | $w_3$ | $w_3$ | $w_4$ | $w_1$ | $w_4$ | $w_3$ | $w_4$ | $w_4$ | $w_4$ | $w_8$ | $w_8$ | $w_4$ | $w_8$ | $w_8$ | $w_4$ | $w_3$ |
| $m_3$ | $w_7$ | $w_7$ | $w_4$ | $w_7$ | $w_4$ | $w_7$ | $w_4$ | $w_4$ | $w_4$ | $w_3$ | $w_4$ | $w_3$ | $w_4$ | $w_3$ | $w_3$ | $w_1$ | $w_3$ | $w_1$ | $w_1$ | $w_1$ | $w_2$ | $w_3$ | $w_4$ |
| $m_4$ | $w_5$ | $w_5$ | $w_5$ | $w_8$ | $w_5$ | $w_8$ | $w_8$ | $w_8$ | $w_8$ | $w_5$ | $w_6$ | $w_8$ | $w_6$ | $w_6$ | $w_6$ | $w_8$ | $w_6$ | $w_6$ | $w_6$ | $w_6$ | $w_1$ | $w_8$ | $w_6$ |
| $m_5$ | $w_4$ | $w_4$ | $w_2$ | $w_4$ | $w_2$ | $w_4$ | $w_2$ | $w_2$ | $w_2$ | $w_2$ | $w_8$ | $w_2$ | $w_8$ | $w_8$ | $w_8$ | $w_2$ | $w_7$ | $w_7$ | $w_8$ | $w_3$ | $w_6$ | $w_2$ | $w_8$ |
| $m_6$ | $w_6$ | $w_6$ | $w_6$ | $w_5$ | $w_6$ | $w_5$ | $w_5$ | $w_5$ | $w_7$ | $w_6$ | $w_5$ | $w_7$ | $w_7$ | $w_5$ | $w_7$ | $w_7$ | $w_4$ | $w_4$ | $w_7$ | $w_4$ | $w_4$ | $w_5$ | $w_5$ |
| $m_7$ | $w_8$ | $w_8$ | $w_8$ | $w_6$ | $w_8$ | $w_6$ | $w_6$ | $w_6$ | $w_6$ | $w_8$ | $w_2$ | $w_6$ | $w_2$ | $w_2$ | $w_2$ | $w_6$ | $w_2$ | $w_2$ | $w_2$ | $w_2$ | $w_3$ | $w_6$ | $w_2$ |
| $m_8$ | $w_2$ | $w_2$ | $w_7$ | $w_2$ | $w_7$ | $w_2$ | $w_7$ | $w_7$ | $w_1$ | $w_7$ | $w_7$ | $w_1$ | $w_1$ | $w_7$ | $w_1$ | $w_3$ | $w_1$ | $w_3$ | $w_3$ | $w_5$ | $w_5$ | $w_7$ | $w_7$ |
| $D(M)$ | 98 | 85 | 83 | 85 | 70 | 72 | 70 | 57 | 60 | 71 | 73 | 61 | 63 | 61 | 64 | 74 | 79 | 92 | 77 | 111 | 132 | 58 | 60 |
| $\Delta(M)$ | 10 | 9 | 9 | 9 | 8 | 8 | 8 | 7 | 6 | 7 | 7 | 5 | 5 | 5 | 6 | 6 | 7 | 8 | 7 | 9 | 10 | 6 | 6 |

Table 2.2  The stable partners of each man in preference order, allowing repetitions

| | 1 | 2 | 3 | 4 | 5 | 6 | 7 | 8 | 9 | 10 | 11 | 12 | 13 | 14 | 15 | 16 | 17 | 18 | 19 | 20 | 21 | 22 | 23 |
|---|---|---|---|---|---|---|---|---|---|---|---|---|---|---|---|---|---|---|---|---|---|---|---|
| | $M_1$ | $M_1$ | $M_1$ | $M_1$ | $M_3$ | $M_8$ | $M_8$ | $M_8$ | $M_8$ | $M_8$ | $M_8$ | $M_8$ | $M_{22}$ | $M_{15}$ | $M_{15}$ | $M_{15}$ | $M_{15}$ | $M_{15}$ | $M_{19}$ | $M_{18}$ | $M_{18}$ | $M_{20}$ | $M_{21}$ |
| $m_1$ | $w_3$ | $w_3$ | $w_3$ | $w_3$ | $w_3$ | $w_1$ | $w_1$ | $w_1$ | $w_1$ | $w_1$ | $w_1$ | $w_1$ | $w_1$ | $w_5$ | $w_5$ | $w_5$ | $w_5$ | $w_5$ | $w_5$ | $w_5$ | $w_5$ | $w_7$ | $w_7$ |
| $m_2$ | $w_1$ | $w_1$ | $w_1$ | $w_1$ | $w_1$ | $w_3$ | $w_3$ | $w_3$ | $w_3$ | $w_3$ | $w_3$ | $w_3$ | $w_4$ | $w_4$ | $w_4$ | $w_4$ | $w_4$ | $w_4$ | $w_4$ | $w_8$ | $w_8$ | $w_8$ | $w_8$ |
| $m_3$ | $w_7$ | $w_7$ | $w_7$ | $w_7$ | $w_7$ | $w_4$ | $w_4$ | $w_4$ | $w_4$ | $w_4$ | $w_4$ | $w_4$ | $w_3$ | $w_3$ | $w_3$ | $w_3$ | $w_3$ | $w_3$ | $w_1$ | $w_3$ | $w_3$ | $w_1$ | $w_2$ |
| $m_4$ | $w_5$ | $w_5$ | $w_5$ | $w_5$ | $w_5$ | $w_8$ | $w_8$ | $w_8$ | $w_8$ | $w_8$ | $w_8$ | $w_8$ | $w_8$ | $w_6$ | $w_6$ | $w_6$ | $w_6$ | $w_6$ | $w_6$ | $w_6$ | $w_6$ | $w_6$ | $w_1$ |
| $m_5$ | $w_4$ | $w_4$ | $w_4$ | $w_4$ | $w_4$ | $w_2$ | $w_2$ | $w_2$ | $w_2$ | $w_2$ | $w_2$ | $w_2$ | $w_2$ | $w_8$ | $w_8$ | $w_8$ | $w_8$ | $w_8$ | $w_8$ | $w_7$ | $w_7$ | $w_3$ | $w_6$ |
| $m_6$ | $w_6$ | $w_6$ | $w_6$ | $w_6$ | $w_6$ | $w_5$ | $w_5$ | $w_5$ | $w_5$ | $w_5$ | $w_5$ | $w_5$ | $w_5$ | $w_7$ | $w_7$ | $w_7$ | $w_7$ | $w_7$ | $w_7$ | $w_4$ | $w_4$ | $w_4$ | $w_4$ |
| $m_7$ | $w_8$ | $w_8$ | $w_8$ | $w_8$ | $w_8$ | $w_6$ | $w_6$ | $w_6$ | $w_6$ | $w_6$ | $w_6$ | $w_6$ | $w_6$ | $w_2$ | $w_2$ | $w_2$ | $w_2$ | $w_2$ | $w_2$ | $w_2$ | $w_2$ | $w_2$ | $w_3$ |
| $m_8$ | $w_2$ | $w_2$ | $w_2$ | $w_2$ | $w_2$ | $w_7$ | $w_7$ | $w_7$ | $w_7$ | $w_7$ | $w_7$ | $w_7$ | $w_7$ | $w_1$ | $w_1$ | $w_1$ | $w_1$ | $w_1$ | $w_1$ | $w_3$ | $w_3$ | $w_5$ | $w_5$ |

When $\mathcal{T} = \mathcal{S}$, we abbreviate $\alpha_{j,\mathcal{T}}$ by $\alpha_j$ for any $j$ $(1 \leq j \leq s)$. We also simply refer to the $j$th generalised median stable matching of $\mathcal{T}$ as the $j$th *generalised median stable matching*, and the median stable matching(s) of $\mathcal{T}$ as the *median stable matching(s)* (where the use of the plural depends on whether $s$ is odd or even). Moreover, we say that a given matching $M$ is a *generalised median stable matching* in $I$ if $M$ is the $j$th generalised median stable matching in $I$ for some $j$ $(1 \leq j \leq s)$. For example in $I_8$, for $\mathcal{T} = \mathcal{S}$, the median stable matching is $\alpha_{12}$, which is $M_8$.

As the median stable matching assigns each agent to his/her (lower or upper) median stable partner among a set of $t$ stable matchings, it can be considered to be a type of "fair" stable matching, giving a counterpart to the notions of egalitarian, minimum regret, sex-equal and balanced stable matchings (see Secs. 1.3.4.1 and 2.10.2). However, these latter four concepts are very different from the notion of a median stable matching. Note, for example, that in $I_8$, there are two egalitarian stable matchings, namely $M_{10}$ and $M_{12}$ (each with cost 54), neither of which is even a generalised median stable matching, let alone the median stable matching. On the other hand $M_8$, the median stable matching, has cost 58.

Cheng [143, 144] gave the following remarkable characterisation of generalised median stable matchings in terms of the rotation poset of a given SM instance.

**Theorem 2.10 ([143, 144]).** *Let $I$ be an SM instance, let $R(I)$ be the set of rotations in $I$, let $\mathcal{T}$ be a set of stable matchings in $I$, and let $t = |\mathcal{T}|$. For each $M \in \mathcal{T}$, let $S_M$ denote the unique closed subset of rotations in $R(I)$ that corresponds to $M$ (the existence of $S_M$ is established by Theorem 1.16). For each $\rho \in R(I)$, let $\bar{n}_{\rho,\mathcal{T}} = |\{M \in \mathcal{T} : \rho \notin S_M\}|$. That is, $\bar{n}_{\rho,\mathcal{T}}$ is the number of stable matchings in $\mathcal{T}$ whose corresponding closed subset of rotations does not contain $\rho$. Then, for each $j$ $(1 \leq j \leq t)$, $\bar{R}_{j,\mathcal{T}}(I) = \{\rho \in R : \bar{n}_{\rho,\mathcal{T}} < j\}$ is a closed subset of $R(I)$, and $\alpha_{j,\mathcal{T}}$ is precisely the stable matching obtained by starting from the man-optimal stable matching in $I$ and eliminating every rotation in $\bar{R}_{j,\mathcal{T}}(I)$.*

Parts of Theorem 2.10 were proved independently by Nemoto [461], for the case that $\mathcal{T} = \mathcal{S}$; Nemoto referred to $\bar{R}_{j,\mathcal{T}}(I)$ as the *$j$th level ideal* of $R(I)$.

We illustrate Theorem 2.10 with respect to $I_8$ when $\mathcal{T} = \mathcal{S}$ (recall that $s = |\mathcal{S}| = 23$). The rotations for $I_8$, together with their corresponding $\bar{n}_{\rho,\mathcal{T}}$ values, are shown in Table 2.3. The rotations are taken from [320] where the rotation poset is also illustrated; for brevity we do not repeat it here.

Table 2.3   The rotations for the SM instance $I_8$ shown in Fig. 2.5, together with their corresponding $\bar{n}_{\rho,\mathcal{T}}$ values

| | |
|---|---|
| $\rho_1 = (m_1, w_3), (m_2, w_1)$ | $\bar{n}_{\rho_1,\mathcal{T}} = 5$ |
| $\rho_2 = (m_3, w_7), (m_5, w_4), (m_8, w_2)$ | $\bar{n}_{\rho_2,\mathcal{T}} = 4$ |
| $\rho_3 = (m_4, w_5), (m_7, w_8), (m_6, w_6)$ | $\bar{n}_{\rho_3,\mathcal{T}} = 5$ |
| $\rho_4 = (m_1, w_1), (m_6, w_5), (m_8, w_7)$ | $\bar{n}_{\rho_4,\mathcal{T}} = 13$ |
| $\rho_5 = (m_2, w_3), (m_3, w_4)$ | $\bar{n}_{\rho_5,\mathcal{T}} = 12$ |
| $\rho_6 = (m_4, w_8), (m_7, w_6), (m_5, w_2)$ | $\bar{n}_{\rho_6,\mathcal{T}} = 13$ |
| $\rho_7 = (m_3, w_3), (m_8, w_1)$ | $\bar{n}_{\rho_7,\mathcal{T}} = 18$ |
| $\rho_8 = (m_2, w_4), (m_5, w_8), (m_6, w_7)$ | $\bar{n}_{\rho_8,\mathcal{T}} = 19$ |
| $\rho_9 = (m_1, w_5), (m_7, w_7), (m_8, w_3)$ | $\bar{n}_{\rho_9,\mathcal{T}} = 21$ |
| $\rho_{10} = (m_3, w_1), (m_7, w_2), (m_5, w_3), (m_4, w_6)$ | $\bar{n}_{\rho_{10},\mathcal{T}} = 22$ |

The correspondence between the closed subsets of the rotation poset and the stable matchings in $I_8$ is illustrated via the Hasse diagram $\mathcal{H}_{I_8}$ for the lattice of stable matchings in $I_8$ shown in Fig. 2.6. Here an arc $(M_i, M_j)$ between two stable matchings $M_i$ and $M_j$ indicates that $M_i \prec M_j$ (where $\prec$ is the partial order on stable matchings from Definition 1.12) and there is no $M_k \in \mathcal{S}$ such that $M_i \prec M_k \prec M_j$ (that is, $M_i$ is an *immediate predecessor* of $M_j$). Each arc $(M_i, M_j)$ is annotated by the rotation $\rho$ that is exposed in $M_i$ whose elimination from $M_i$ yields $M_j$. Thus the closed subset of rotations corresponding to a given $M_i \in \mathcal{S}$ is the set of rotations adjacent to the arcs on the path from $M_1$ to $M_i$ in $\mathcal{H}_{I_8}$. Recall from Table 2.2 that $\alpha_{12} = M_8$. Theorem 2.10 states that the closed subset of rotations corresponding to $\alpha_{12}$ is $\{\rho : \bar{n}_{\rho,\mathcal{T}} < 12\}$. Table 2.3 in turn indicates that this set is $\{\rho_1, \rho_2, \rho_3\}$, which does indeed correspond to $M_8$ as confirmed by Fig. 2.6.

Cheng [143, 144] also studied the algorithmic complexity of computing the $j$th generalised median stable matching, answering an open question from Refs. [565] and [526]. Note that, when trying to formalise related problem definitions, it is not immediately obvious how to specify the bounds of $j$, given an SM instance $I$. It is not possible to merely insist that $1 \leq j \leq s$, where $s$ is the number of stable matchings in $I$, since the problem of computing $s$ is itself #P-complete [319]. Instead, it may be noted that $|R(I)| \leq n(n-1)/2$ [261, Corollary 3.2.1], and hence the number of closed subsets of $R(I)$ (and by implication the number of stable matchings in $I$) is at most $2^{n(n-1)/2}$. We follow Cheng's convention and assume that, for $s < j < 2^{n(n-1)/2}$, $\alpha_j$ is defined to be the woman-optimal stable matching in $I$. With this in mind, Cheng [144] defined the following problems:

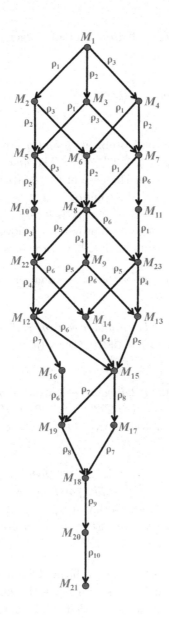

Fig. 2.6   The lattice of stable matchings for the SM instance $I_8$ shown in Fig. 2.5

GEN-MEDIAN-1

*Input*: an SM instance $I$ of size $n$ and an integer $j$ $(1 \leq j \leq 2^{n(n-1)/2})$.
*Solution*: $\alpha_j$.

GEN-MEDIAN-$(p,q)$

*Input*: an SM instance $I$ of size $n$.
*Solution*: $\alpha_{\lceil ps/q \rceil}$.

In the case of the second problem, $p$ and $q$ are two relatively prime integers such that $1 \leq p < q \leq 2^{n(n-1)/2}$. The *duals* of each problem can also be defined as follows: in the case of GEN-MEDIAN-1-DUAL, the input is as for GEN-MEDIAN-1 and the solution is $\alpha_{s-j+1}$. Similarly in the case of GEN-MEDIAN-$(p,q)$-DUAL, the input is as for GEN-MEDIAN-$(p,q)$ and the solution is $\alpha_{s-\lceil ps/q \rceil+1}$. Cheng [144] proved the following results concerning the computational complexity of the above problems:

**Theorem 2.11 ([144]).** GEN-MEDIAN-1 *and* GEN-MEDIAN-1-DUAL *are* #*P-hard.*

**Theorem 2.12 ([144]).** *Let $p$ and $q$ be two relatively prime positive integers such that $p < q$. Then* GEN-MEDIAN-$(p,q)$ *and* GEN-MEDIAN-$(p,q)$-DUAL *are* #*P-hard.*

**Theorem 2.13 ([144]).** *Let $I$ be an SM instance of size $n$ and let $j$ be an integer such that $1 \leq j \leq 2^{n(n-1)/2}$. Then if $j = O(\log n)$, each of* GEN-MEDIAN-1 *and* GEN-MEDIAN-1-DUAL *can be solved in polynomial time.*

**Theorem 2.14 ([144]).** *Let $I$ be an SM instance of size $n$. If the rotation poset in $I$ is a series–parallel poset, an interval order or a two-dimensional poset, each of* GEN-MEDIAN-$(p,q)$ *and* GEN-MEDIAN-$(p,q)$-DUAL *can be solved in polynomial time, where $p$ and $q$ are two relatively prime integers such that $1 \leq p < q \leq 2^{n(n-1)/2}$.*

Theorem 2.12 indicates that the problem of finding $\alpha_j$ in a given SM instance $I$ is #P-hard even if $j = \Theta(s)$, where $s$ is the number of stable matchings in $I$. Also Theorem 2.13 implies that, by contrast, if $j = O(\log \log s)$, then $\alpha_j$ can be found in polynomial time. Kijima and Nemoto [380] strengthened these results by establishing the algorithmic complexity of computing $\alpha_j$ for cases "in between" $j = O(\log \log s)$ and $j = \Theta(s)$. Specifically, they proved the following.

**Theorem 2.15 ([380]).** *Let $I$ be an* SM *instance with $s$ stable matchings.* GEN-MEDIAN-1 *and* GEN-MEDIAN-1-DUAL *can be solved in polynomial time if $j = O((\log s)^c)$ for some arbitrary constant $c > 0$.*

**Theorem 2.16 ([380]).** *Let $I$ be an* SM *instance with $s$ stable matchings.* GEN-MEDIAN-1 *and* GEN-MEDIAN-1-DUAL *are #P-hard, even if $j = \Theta(s^{1/c})$ for an arbitrary constant $c \geq 1$.*

**Theorem 2.17 ([380]).** *Let $I$ be an* SM *instance and let $M$ be a matching in $I$. The problem of deciding whether $M$ is a generalised median stable matching in $I$ is #P-hard.*

Kijima and Nemoto [380] also give randomised approximation schemes for the problem of computing $\alpha_j$ $(1 \leq j \leq s)$, given an SM instance with $s$ stable matchings.

Theorems 2.11, 2.12 and 2.16 indicate the hardness of computing generalised median stable matchings in general. This prompted Cheng [143,144] to consider the complexity of approximating generalised median stable matchings. She arrived at the following result, which also indicates what is meant by an "approximate" generalised median stable matching. In what follows, if $\mathcal{S}$ is the set of stable matchings in a given SM instance and $M_1, M_2 \in \mathcal{S}$, then $[M_1, M_2]$ denotes the set $\{M \in \mathcal{S} : M_1 \preceq M \preceq M_2\}$.

**Theorem 2.18 ([144]).** *Let $I$ be an* SM *instance of size $n$, and let $s$ be the number of stable matchings in $I$. Then, for any $\varepsilon > 0$ such that $\varepsilon = O(\log s/s)$, we may construct in $O(n^2)$ time a stable matching $M$ in $\left[\alpha_{\lceil \frac{\varepsilon s}{2} \rceil}, \alpha_{\lceil s - \frac{\varepsilon s}{2} + \frac{1}{2} \rceil}\right]$. Such a matching $M$ is referred to as an $\varepsilon$-approximation to a median stable matching in $I$.*

Cheng [144] also characterised median stable matchings in terms of medians in the distributive lattice of stable matchings in a given SM instance $I$. Formally, suppose that $\mathcal{S}$ is the set of stable matchings in $I$, and $\mathcal{H}_I$ is the Hasse diagram of $\mathcal{S}$ under $\preceq$. Given two stable matchings $M$ and $M'$ in $\mathcal{S}$, define $d(M, M')$ to be the length of a shortest path between $M$ and $M'$ in the undirected version of $\mathcal{H}$. Let $D(M) = \Sigma_{M' \in \mathcal{S}} d(M, M')$. Then a *median* of $(\mathcal{S}, \preceq)$ is a stable matching $M \in \mathcal{S}$ such that $D(M)$ is minimum, taken over all stable matchings in $I$. Cheng [144] proved the following:

**Theorem 2.19 ([144]).** *Let $I$ be an instance of* SM *where $\mathcal{S}$ is the set of stable matchings, and let $s = |\mathcal{S}|$. If $s$ is odd then $\alpha_{(s+1)/2}$ is precisely the* *median stable matching in* $(S, \preceq)$. *If s is even then the stable matchings in the interval* $[\alpha_{s/2}, \alpha_{s/2+1}]$ *are precisely the median stable matchings in* $(S, \preceq)$.

Cheng [144] remarked that a median of $(S, \preceq)$ is fair in a global sense, whilst each agent being matched to either their lower or upper median stable partner among a set of stable matchings is fair in a local sense. Theorem 2.19 thus states that stable matchings are "globally median" if and only if they are "locally median" in the senses indicated. This is a quite remarkable observation, which Cheng and Lin [146] referred to as the *local/global median phenomenon* of stable matchings. By way of illustration, Table 2.1 shows the $D(M)$ values for the stable matchings in the SM instance $I_8$ given in Fig. 2.5, which confirms that $D(M_8)$ is minimum (recall that $\alpha_{12} = M_8$ is the median stable matching in $I_8$).

Theorem 2.19, combined with Theorems 2.11, 2.12 and 2.16, show that computing this type of "globally fair" stable matching in $(S, \preceq)$ is hard. With this in mind, Cheng *et al.* [149] defined another type of "globally fair" stable matching as follows. A *center stable matching* is a stable matching $M \in S$ that minimises $\Delta(M)$, the maximum distance (in the undirected version of $\mathcal{H}_I$) from $M$ to all other stable matchings in $S$. Cheng *et al.* [149] proved that a center stable matching (which need not be unique) can be found in $O(n^5)$ time, where $n$ is the size of $I$. Moreover, they showed that this type of globally fair stable matching is an approximation to a locally median stable matching in the following sense: there exists a set $\mathcal{T} \subseteq S$, where the stable matchings in $\mathcal{T}$ form a maximum-length chain in $(S, \preceq)$, such that, if $t = |\mathcal{T}|$, $\alpha_{(t+1)/2,\mathcal{T}}$ is a center stable matching of $I$ if $t$ is odd, and $\alpha_{t/2,\mathcal{T}}$ and $\alpha_{t/2+1,\mathcal{T}}$ are center stable matchings of $I$ if $t$ is even[10]. Again, to illustrate the results described here, Table 2.1 shows the $\Delta(M)$ values for the stable matchings in the SM instance $I_8$ given in Fig. 2.5. They imply that $M_{12}$, $M_{13}$ and $M_{14}$ are center stable matchings in $I_8$. Let $\mathcal{T}$ be the following maximum-length chain of size 11 in $(S, \preceq)$:

$$\{M_1, M_4, M_7, M_{11}, M_{23}, M_{14}, M_{15}, M_{17}, M_{18}, M_{20}, M_{21}\}.$$

Then it may be verified that $\alpha_{6,\mathcal{T}} = M_{14}$, and $M_{14}$ is indeed a center stable matching of $I$ as previously observed.

Note that additional structural and algorithmic results relating to generalised median stable matchings in the SM context appear in Refs.

---

[10]Cheng *et al.* did not include the second subscript (involving $\mathcal{T}$) on $\alpha_{(t+1)/2,\mathcal{T}}$, $\alpha_{t/2,\mathcal{T}}$ and $\alpha_{t/2+1,\mathcal{T}}$ when stating these results in Ref. [149], however those subscripts are in fact necessary [145].

[143,144,380]. Generalised median stable matchings have been considered in the context of other stable matching problems. Fleiner [210] generalised Theorem 2.9 to the many–many bipartite case. Also, independently, Sethuraman *et al.* [526], and Klaus and Klijn [388,391], generalised the same theorem to the HR case. Cheng [144, Sec. 6] discussed the extension of her results to other stable matching problems, including SMI and HR. Median stable matchings have also been considered in the context of SM with side payments [522].

Results concerning median stable matchings in the SR setting are described in Sec. 4.4. We conclude with two open problems, due to Kijima and Nemoto [380]. They asked whether the following two decision problems belong to the class NP: (1) given an SM instance $I$ and a stable matching $M$ in $I$, determine whether $M$ is a generalised median stable matching in $I$; (2) given an SM instance $I$ with $s$ stable matchings, a stable matching $M$ in $I$, and an integer $j$ $(1 \leq j \leq s)$, determine whether $M$ is the $j$th generalised median stable matching in $I$.

## 2.8  Size versus stability

In the 2006–07 run of SFAS, the Scottish medical matching scheme (see Sec. 1.3.7) [604], there were 781 students and 53 hospitals, the latter having total capacity 789. The matching algorithm (designed and implemented at the the School of Computing Science, University of Glasgow) found a stable matching of size 744, thus leaving 37 students unassigned. Clearly stability is the key property to be satisfied, and it is this that restricts the size of the resultant matching. Nevertheless the administrators asked whether, were the stability criterion to have been relaxed, a larger matching could have been found. We found that a matching of size 781 did exist, but the matching we computed admitted 400 blocking pairs.

In practical situations, a blocking pair of a given matching $M$ need not always lead to $M$ being undermined, since the agents involved might be unaware of their potential to improve relative to $M$. For example, in situations where preference lists are not public knowledge, there may be limited channels of communication that would lead to the awareness of blocking pairs in practice. Nevertheless, it is reasonable to assert that the greater the number of blocking pairs of a given matching $M$, the greater the likelihood that $M$ would be undermined by a pair of agents in practice. In particular, a maximum cardinality matching for the 2006–07 SFAS data that admits only 10 blocking pairs might be considered to be "more stable" than one with 400 blocking pairs. This motivates the problem of finding a maximum matching that admits the smallest number of blocking pairs (and is therefore, in the sense described above, "as stable as possible"). Eriksson and Häggström [195] also argued that counting the number of blocking pairs of a matching can be an effective way to measure its degree of instability; this approach had already been taken in earlier references [378, 466, 199]. An alternative method is to count the number of blocking agents (i.e., the number of agents who are involved in a blocking pair) [519, 195].

Further practical applications of "almost stable" maximum matchings arise in similar bipartite settings, where the size of the matching may be considered to be a higher priority than its stability in a particular matching market. Examples include school placement [7] and the allocation of students to projects in a university department [23]. Furthermore, the US Navy has a bipartite matching problem involving the assignment of sailors to billets [490, 589] in which every sailor should be matched to a billet, and meanwhile there are some critical billets that cannot be left vacant.

In non-bipartite contexts, applications arise in the context of constructing pairwise kidney exchanges (see Sec. 1.4.6), where preference lists are constructed on the basis of compatibility profiles between donors and patients. In most matching schemes, the main goal is to maximise the number of transplants (i.e., the first priority is to find a maximum matching) [512]. However the stability of the matching could also be an issue [511]. Another example in a non-bipartite setting involves pairing up chess players [401].

Given an instance $I$ of SMI, let $\mathcal{M}$ denote the set of matchings in $I$ and let $\mathcal{M}^+$ denote the set of maximum matchings in $I$. Given a matching $M \in \mathcal{M}$, let $bp(I, M)$ (respectively $ba(I, M)$) denote the set of blocking pairs (respectively blocking agents) with respect to $M$ in $I$ (we omit the first argument $I$ when the instance is clear from the context). Clearly $M \in \mathcal{M}$ is stable in $I$ if and only if $bp(I, M) = \emptyset$, which in turn is true if and only if $ba(I, M) = \emptyset$. Let

$$bp^+(I) = \min\{|bp(I, M)| : M \in \mathcal{M}^+\}$$

and let

$$ba^+(I) = \min\{|ba(I, M)| : M \in \mathcal{M}^+\}.$$

Define MAX SIZE MIN BP SMI (respectively MAX SIZE MIN BA SMI) to be the problem of finding, given an SMI instance $I$, a matching $M \in \mathcal{M}^+$ such that $|bp(I, M)| = bp^+(I)$ (respectively $|ba(I, M)| = ba^+(I)$).

Men's preferences   Women's preferences
$m_1 : w_2 \ w_1$        $w_1 : m_1$
$m_2 : w_2$              $w_2 : m_1 \ m_2$

Fig. 2.7   An instance of SMI of size 2

To illustrate this problem, consider the SMI instance $I$ shown in Fig. 2.7. Clearly $M = \{(m_1, w_2)\}$ is the unique stable matching and $M^+ = \{(m_1, w_1), (m_2, w_2)\}$ is the unique maximum cardinality matching. Hence $bp^+(I) = 1$ and $ba^+(I) = 2$.

It turns out that each of MAX SIZE MIN BP SMI and MAX SIZE MIN BA SMI are NP-hard and very difficult to approximate, as indicated by the following result.

**Theorem 2.20 ([93]).** *Each of* MAX SIZE MIN BP SMI *and* MAX SIZE MIN BA SMI *is not approximable within* $n^{1-\varepsilon}$, *where n is the size of a given instance, for any* $\varepsilon > 0$, *unless P=NP.*

We can also define MAX SIZE EXACT BP SMI (respectively MAX SIZE EXACT BA SMI) to be the problem of finding, given an SMI instance $I$ and an integer $K$, a matching $M \in \mathcal{M}^+$ such that $|bp(I, M)| = K$ (respectively $|ba(I, M)| = K$). It turns out that both problems are NP-hard if $K$ is part of the problem input, but solvable in polynomial time if $K$ is a fixed integer [93].

We now consider preference lists of fixed length. Given two integers $p$ and $q$, let MAX SIZE MIN BP $(p, q)$-SMI (respectively MAX SIZE MIN BA $(p, q)$ -SMI) denote the restriction of MAX SIZE MIN BP SMI (respectively MAX SIZE MIN BA SMI) in which each man's preference list is of length at most $p$, and each woman's list is of length at most $q$. We use $p = \infty$ or $q = \infty$ to denote the possibility that the men's lists or women's lists are of unbounded length, respectively.

Biró *et al.* [93] showed that each of MAX SIZE MIN BP $(3,3)$-SMI and MAX SIZE MIN BA $(3,3)$ -SMI is not approximable within $\delta$, for some constant $\delta > 1$, unless P=NP. Hamada *et al.* [274] extended the reduction used to prove these results in order to strengthen the inapproximability lower bound for the first problem, as stated by the following theorem.

**Theorem 2.21 ([274]).** MAX SIZE MIN BP $(3,3)$-SMI *is not approximable within* $n^{1-\varepsilon}$, *where n is the size of a given instance, for any* $\varepsilon > 0$, *unless P=NP.*

For preference lists of length at most 2 on one side, it turns out that the above problems are solvable in polynomial time. In fact, this is even true for instances of SMTI with respect to weak stability, as stated by the following result.

**Theorem 2.22 ([93]).** *Each of* MAX SIZE MIN BP $(2,\infty)$-SMTI *and* MAX SIZE MIN BA $(2,\infty)$-SMTI *is solvable in* $O(n^3)$ *time, where n is the size of a given instance.*

Floréen *et al.* [220] considered a slightly different definition of an "almost stable" matching in a distributed model of computation. They defined a matching $M$ in an SMI instance $I$ to be $\varepsilon$-stable, for some $\varepsilon > 0$, if $|bp(I, M)| \leq \varepsilon |M|$. The authors showed that an $\varepsilon$-stable matching can be found in at most $4 + 2k^2/\varepsilon$ synchronous communication rounds, assuming that each preference list is of length at most $k$, for some constant $k$.

We close this section by noting that Biermann [75] argued that counting the number of blocking pairs as a measure of the instability of a matching may not necessarily be appropriate. He reasoned that, for example, a matching $M$ such that $|bp(M)| = 20$ and where $bp(M)$ itself forms a matching, could be said to be more "unstable" than a matching $M'$ where $|bp(M')| = 50$ and a single agent belongs to every pair of $bp(M')$. This is simply due to the fact that all the blocking pairs in $bp(M)$ can be satisfied simultaneously, whereas only one from $bp(M')$ can be. Biermann defined the notion of a *permissible* set of blocking pairs and proved that a matching $M$ is stable if and only if its set of permissible blocking pairs is empty. He proposed that the cardinality of the set of permissible blocking pairs relative to $M$ be used as the measure of $M$'s degree of instability.

## 2.9   Strategic issues

In this section we consider various strategic results relating to SM and SMI — these mainly centre around the question as to whether an individual agent, or some coalition of agents, can falsify their preference list/s (typically by permuting some entries and/or truncating their list) so as to obtain a better partner (with respect to the true preferences) than they would obtain in either the man-optimal or woman-optimal stable matchings. Chapter 4 of Ref. [514] and Sec. 1.7 of Ref. [261] describe the main results under this heading up to around 1990. We update the reader on subsequent results in this section, concentrating mainly on those that have an algorithmic flavour. However in order to describe more recent work, it is necessary to briefly recap on some of the key older strategic results.

In a matching market modelled by SM or SMI, ideally there would exist a mechanism for constructing a matching in which it is a *dominant strategy* [514, p.84] for each agent to report his/her true preferences, regardless of whether the other agents are doing likewise. (That is, the best outcome for each agent would be obtained by telling the truth, no matter whether the other agents are doing so.) Such a mechanism is called a *strategy-proof mechanism* (also known as a *truthful mechanism*). Roth [496] showed that, with respect to the man-oriented Gale–Shapley algorithm, it is a dominant strategy for the men to tell the truth. On the other hand he showed that, more generally, there is no mechanism for SM for which it is a dominant strategy for *all* agents to be truthful, and hence there is no strategy-proof mechanism for SM.

To describe strategic results for SM in more detail, let $I$ be an SM instance representing the true preferences of the agents, and let $M_a$ (respectively $M_z$) denote the man-optimal (respectively woman-optimal) stable matching in $I$. Let $C$ be a coalition of agents who falsify their preferences, and denote by $I'$ the preference lists that result (each agent not in $C$ has the same preference list in $I$ and $I'$).

Dubins and Freedman [173] proved that there is no coalition $C$ of men who could falsify their preferences so as to yield a matching $M'$ that is stable in $I'$ such that *every* man in $C$ has a better (with respect to $I$) partner in $M'$ than in $M_a$. Roth [496] independently proved this for the special case that $C$ comprises a single man. Demange, Gale and Sotomayor [164] extended Dubins and Freedman's result to the case where $C$ can include both men and women as follows. They proved that there is no stable matching $M'$ in $I'$ such that every member of $C$ prefers (in $I$) their partner in $M'$ to their partner in *every* stable matching in $I$. If incomplete lists are permitted, Gale and Sotomayor [236] proved that, as long as two stable matchings in $I$ exist, then we can choose $C$ to contain a single woman who could falsify her preferences so as to yield a better (with respect to $I$) partner than in $M_a$. They also showed that if $C$ is the set of all women, then each woman can truncate her preference list so as to force the man-oriented Gale–Shapley algorithm to yield $M_z$ in $I'$ (rather than yielding $M_a$ in $I$). Further truncation strategies are described in Ref. [509, 186].

Gusfield and Irving [261, p.65] observed that, up to the time of writing, cheating strategies for women with respect to the man-oriented Gale–Shapley algorithm had been restricted to preference list truncation.

However they gave an example to show that, in an SM instance, a single woman could permute her preference list so as to obtain a better (in $I$) partner with respect to the man-oriented Gale–Shapley algorithm. That is, with respect to the above notation, $C$ contains a single woman who permutes her list so as to force the man-oriented Gale–Shapley algorithm to yield $M_z \neq M_a$ in $I'$ (rather than yielding $M_a$ in $I$)[11].

We now turn to results published after Ref. [261]. Roth and Vande Vate [517] considered the incentives to cheat for agents in an instance of SM when a decentralised mechanism such as the Roth–Vande Vate Mechanism [516] (see Sec. 2.6) is used.

Teo et al. [567] considered the SM setting and, in particular, the case where there is a single woman $w$ who knows the preferences of all other agents, which are declared truthfully. They showed how to construct, in polynomial time, an optimal cheating strategy for $w$ relative to the man-oriented Gale–Shapley algorithm. However, interestingly, they showed in simulations that it is relatively unlikely that a woman could benefit by cheating. In particular, they generated 1000 random instances of size 8 and found that, for 74% of these, the deceitful woman did not improve from the partner that she would obtain in $M_a$ (the man-optimal stable matching with respect to the true preferences), and on average, only 5.1% of women did improve by cheating. The authors also presented a discussion of school placement in Singapore, arguing that a mechanism based on stable matching would be more appropriate than the algorithm that was in place at the time of writing.

Immorlica and Mahdian [302] showed that, for an SMI instance in which the men's preferences are of length at most $k$, for some constant $k$, and each is drawn from an arbitrary probability distribution of the women, and the women's preferences are arbitrary and complete, the number of participants with more than one stable partner is vanishingly small. As a consequence of this, for a given agent, his/her best strategy is to be truthful assuming that all other agents have been truthful.

Huang [285] considered instances of SM and exploited a loophole in the Dubins–Freedman theorem [173]. He defined a cabal $C_0$ to be a $k$-tuple of men $\langle m_0, m_1, \ldots, m_{k-1} \rangle$, for some $k$, such that, for each $i$ ($0 \leq i \leq k-1$),

---

[11]The example given by Gusfield and Irving on page 65 of Ref. [261] is a contradiction to Theorem 1.7.2 in their book (for the case that the chosen coalition contains a single woman and no men). On page 65 of Ref. [261], Theorem 1.7.2 is attributed to Demange et al. [164], but in fact this is an incorrect statement of the main result in the latter paper.

$m_i$ prefers $M_a(m_{i-1})$ to $M_a(m_i)$, where addition is taken modulo $k$ (and $M_a$ is the man-optimal stable matching). Huang showed that if a cabal exists relative to $M_a$, then there exists a coalition $C \supseteq C_0$ of men who can falsify their preferences such that (i) each man in the cabal is better off (with respect to $I$) in $M'$ relative to $M_a$ and (ii) each man outside the cabal has the same partner in $M'$ and $M_a$, where $M'$ is the man-optimal stable matching in the SM instance $I'$ so obtained. Huang [284] gave a polynomial-time algorithm for constructing such a coalition if it exists. He also showed that this strategy is the only one in which no deceitful agent is worse off. Huang then suggested a randomised coalition strategy in which every man in $C$ has a chance to obtain a better partner, whilst no man is worse off, though proved that such a strategy is unrealisable. If, however, some of the men in $C$ are allowed to be worse off, then Huang constructed a randomised strategy in which every man in $C$ can expect to obtain a better partner.

Kobayashi and Matsui [395] studied problems concerned with enabling a coalition $C$ of women to construct a cheating strategy that would force the man-oriented Gale–Shapley algorithm to return a matching in which the women in $C$ obtain a desired set of partners. The first problem that the authors considered has the following input: (i) a set of $n$ men and $n$ women, (ii) for each man, a strict preference list over all $n$ women, (iii) a partial matching $M$ of the men and women, and (iv) for each woman who is unassigned in $M$, a strict preference list over all $n$ men. Let $C$ be the set of women who are assigned in $M$. The problem is to find, for each woman in $C$, a strict preference list over all $n$ men such that, for the overall SM instance (including the preference lists described in the input) obtained, the man-oriented Gale–Shapley algorithm constructs a stable matching $M'$ containing $M$, or else reports that no such set of preference lists exists. Kobayashi and Matsui showed that this problem is solvable in $O(n^2)$ time.

Additionally, the problem is solvable in $O(n^2)$ time if, in (iii), $M$ is a perfect matching of the $n$ men and $n$ women, and in (iv), there is a set $C'$ of women (possibly empty), each of whom has a preference list over the $n$ men. Letting $C = W \backslash C'$ (where $W$ is the set of women), the remainder of the problem is as before. By contrast, in the description of the first problem in the previous paragraph, if (iv) is no longer part of the input, and $C$ comprises *all* women, the authors showed that the corresponding problem is NP-hard. For further work along these lines, see Ref. [552].

Matsui [432] considered the following strategic game involving an SM instance. We are given a set of $n$ men with true preferences over a set of $n$

women. In a given round of the game, each woman chooses a preference list over the $n$ men. The outcome for a woman is the partner that she receives in the man-optimal stable matching with respect to the men's (true) preferences and the women's preferences as determined by their joint strategy. Matsui gave an $O(n^2)$ algorithm to determine whether, given a matching $M$, there is a joint strategy for the women that forms an equilibrium such that $M$ is the overall outcome of the game. He also gave an $O(n^4)$ algorithm to determine whether a given joint strategy for the women forms an equilibrium or not.

Pini *et al.* [478] defined a stable matching mechanism for SM to be *manipulable* if there exists an agent who could falsify their preferences so as to obtain a better partner than they would obtain, were the mechanism to be executed with respect to the true preferences. They argued that, despite the earlier observation by Roth [496] that every stable matching mechanism for SM is manipulable, it might in fact be NP-hard to compute a manipulation. Such a complexity result could then indicate that there is at least some difficulty in arriving at a cheating strategy, even if we know that one is possible.

Teo *et al.*'s result [567] described earlier indicates that the man-oriented Gale–Shapley algorithm can be manipulated in polynomial time. More generally, Pini *et al.* [478] defined instances of SM that are *universally manipulable* by some woman $w$. Such instances can also be manipulated in polynomial time. In particular, given an SM instance $I$ and a woman $w$ such that $I$ is universally manipulable by $w$, and given any stable matching mechanism $A$ for $I$, there is a polynomial-time strategy that allows $w$ to permute her preference list so as to arrive at an SM instance $I'$, such that $A$ applied to $I'$ gives $w$ the partner that she obtains in the woman-optimal stable matching in $I$. On the other hand, the authors showed that, for general SM instances, there is a stable matching mechanism for which it is NP-hard to find a manipulation.

Inoshita *et al.* [305] observed that, with respect to the man-optimal stable matching $M_a$ in a given SM instance $I$, some men may have partners relatively far down their preference lists. They considered the problem of finding a single man $m_i$ who is allowed to falsify his preference list, so as to produce an SM instance $I'$, such that (i) relative to $I$, no man has a worse partner in $M'_a$ than in $M_a$, and (ii) $c^U(M_a) - c^U(M'_a)$ is maximum, taken over all possible preference list falsifications by a single man, where $M'_a$ is the man-optimal stable matching in $I'$, and each of $c^U(M_a)$ and $c^U(M'_a)$ are measured relative to $I$. That is, we wish to find some permutation of the preference list of a single man $m_i$ which leads to an instance $I'$ in which the man-optimal stable matching $M'_a$ in $I'$ gives the maximum overall improvement for the men (in terms of the sum of the ranks of their partners, with respect to the true preferences) compared to the man-optimal stable matching $M_a$ in $I$, subject to the constraint that no man is worse off in $M'_a$ than in $M_a$ (again, relative to the true preferencs). Note that, by the result of Dubins and Freedman, $M_a(m_i) = M'_a(m_i)$ for any such man $m_i$.

The authors gave an $O(n^3)$ algorithm for this problem which outputs a single man $m_i$ and his modified list that leads to the largest overall improvement for the men (which may be 0), where $n$ is the size of $I$. However if we only want to know whether there exists a man $m_i$ who can lead to a positive overall improvement for the men, then this problem can be solved in $O(n^2)$ time.

We also remark that Sönmez [538, 539] considered methods of manipulation by hospitals in an instance of HR. He showed in Ref. [538] that any stable matching mechanism for HR is open to manipulation by hospitals under-reporting their capacities. In a later paper [539] he dealt with another form of strategic behaviour by hospitals and proved that no stable matching mechanism for HR is resistant to manipulation by a hospital pre-arranging some set of desired assignees prior to the execution of the mechanism.

The study of strategic issues was extended to SR by Huang [286]. For a solvable SR instance $I$, Huang proved that there exists no SR instance $I'$, obtained from $I$ by a single deceitful agent $a$ falsifying her preferences, that admits a stable matching $M'$ in which $a$ obtains a better partner (with respect to $I$) than her best stable partner in $I$. More generally, this result extends to coalitions of agents in the following way. Let $M$ be a stable matching in $I$ in which each member of a non-empty coalition of agents $C$ obtains her best stable partner. Then there exists no SR instance $I'$, obtainable from $I$ by the agents in $C$ falsifying their preferences, that admits a stable matching $M'$ in which *every* agent in $C$ obtains a better partner (with respect to $I$) in $M'$ than in $M$.

## 2.10 Further results

### 2.10.1 The Stable Marriage problem with Forbidden pairs

A natural generalisation of SM was considered by Dias *et al.* [168], namely the *Stable Marriage problem with Forced and Forbidden pairs* (SMFF). An instance $I$ of SMFF comprises a standard SM instance, together with a set $P$ of *forbidden* pairs and a set $Q$ of *forced* pairs. In $I$, a matching $M$ is defined to be *stable* if (i) $M \cap P = \emptyset$ and $Q \subseteq M$, and (ii) $M$ is stable in the underlying SM instance obtained by ignoring the forbidden and forced pairs. Such a matching may not exist, of course.

We denote the special case of SMFF in which $P = \emptyset$ by the *Stable Marriage problem with Forced pairs* (SMFD). Similarly the restriction of SMFF in which $Q = \emptyset$ is called the *Stable Marriage problem with Forbidden pairs* (SMF). Knuth [394] described an $O(n^2)$ algorithm for SMFD, where $n$ is the size of the underlying SM instance. Gusfield and Irving [261, Sec. 3.4.2] gave a polynomial-time checkable necessary and sufficient condition for a set $Q$ of forced pairs to be part of a stable matching in an SM instance. This leads to an $O(|Q|^2)$ algorithm for SMFD, following $O(n^4)$ pre-processing time.

For SMFF, Dias *et al.* [168] showed that the problem of finding a stable matching or reporting that none exists, given an instance of SMFF, is solvable in $O((|P| + |Q|)^2)$ time, following $O(n^4)$ pre-processing time. The authors also gave a simple reduction from SMFF to SMF as follows: given an instance $I$ of SMFF in which $W$ is the set of women, $P$ is the set of forbidden pairs and $Q$ is the set of forced pairs, we construct an instance $I'$ of SMF in which $P'$ is the set of forbidden pairs, as follows:

$$P' = \{(m_i, w_k) : (m_i, w_j) \in Q \land w_k \in W \backslash \{w_j\}\}.$$

Clearly a matching $M$ is a solution for $I$ if and only if $M$ is a solution for $I'$. However, $|P'| = |P| + (n-1)|Q|$, where $n = |W|$, and hence the blow-up in the number of forbidden pairs produced by this reduction justifies applying the $O((|P| + |Q|)^2)$ time (following $O(n^4)$ pre-processing time) algorithm, as described above, to $I$, rather than to $I'$.

The algorithm described in the previous paragraph can be useful if many sets of forced and forbidden pairs are given with respect to the same SM instance, as the pre-processing step need only be carried out once. However suppose we are only interested in a single set of forbidden pairs in an instance $I$ of SMF. Dias *et al.* [168] gave an $O(n^2)$ algorithm for finding a stable matching or reporting that none exists in $I$. They also showed that all the stable pairs in $I$ can be found within the same time bound. Finally, the authors proved that all stable matchings in $I$ can be listed in $O(n^2 + n|\mathcal{S}|)$ time and $O(n^2)$ space, where $\mathcal{S}$ is the set of stable matchings in $I$. That is, the first stable matching can be output in $O(n^2)$ time, and each subsequent stable matching can be output in $O(n)$ time. For further results concerning SMF, see Sec. 4.8.2.

We give some motivation for SMFF. In the context of a centralised matching scheme, a forced pair might represent an arrangement that has been made for two participants prior to the matching run — perhaps due to certain special circumstances — that must be respected in a final solution. By contrast, a forbidden pair may be supplied by an administrator in order to prevent a given pairing from appearing in a constructed matching, for whatever reason. Note that, although SMF superficially resembles SMI (that is, we may be tempted to simply delete the forbidden pairs from the relevant agents' preference lists, since they cannot form part of a stable matching), there is a clear distinction. Namely, in the SMF context, a forbidden pair can still be a blocking pair of a matching. Again, this is reasonable in practical applications, for even if an administrator tries to prevent a pairing $(m_i, w_j)$ from being part of a constructed matching, it could still be the case that $m_i$ and $w_j$ would prefer to make a private arrangement outside of the matching than to remain with their partners.

### 2.10.2  Balanced stable matchings

A stable matching $M$ in an SMI instance $I$ is said to be *balanced* [201] if $b(M) = \max\{c^U(M), c^W(M)\}$ is minimum, taken over all stable matchings in $I$, where $U$ is the set of men and $W$ is the set of women in $I$. (Here we are assuming notation defined in Sec. 1.3.4.1.) Thus the notion of a balanced stable matching gives a further concept of a stable matching that is "fair" to both sexes; previously in this respect we have considered egalitarian and minimum regret stable matchings (see Sec. 1.3.4.1, [261] and Sec. 2.2.10), sex-equal stable matchings (see Secs. 1.3.4.1 and 2.2.7) and median stable matchings (see Sec. 2.7).

It is non-trivial to construct an SM instance in which no balanced stable matching is a sex-equal stable matching, and vice versa. In fact we construct an instance $I$ of size $5r + 3$, due to Eric McDermid [438], that satisfies this property, for any $r \geq 1$. In $I$, the set of men is

$$U = \{m_0, \ldots, m_{5r-1}, p_1, p_2, p_3\},$$

and the set of women is

$$W = \{w_0, \ldots, w_{5r-1}, q_1, q_2, q_3\}.$$

The preference list of each person is shown in Fig. 2.8. In a given person's list, the symbol "..." denotes all remaining members of the opposite sex listed in arbitrary strict order; also addition is taken modulo $5r$. Clearly

Men's preferences

$m_i : w_i \quad q_1 \quad q_2 \quad q_3 \quad w_{i+1} \quad \cdots \qquad (0 \le i \le 5r - 1)$

$p_i : q_i \quad \cdots \qquad\qquad\qquad\qquad\quad (1 \le i \le 3)$

Women's preferences

$w_j : p_1 \quad p_2 \quad m_{j-1} \quad m_j \quad \cdots \qquad (0 \le j \le 5r - 1)$

$q_j : p_j \quad \cdots \qquad\qquad\qquad\qquad\quad (1 \le j \le 3)$

Fig. 2.8 An instance of SM due to McDermid [438]

$(p_i, q_i) \in M$ for $1 \le i \le 3$, where $M$ is any stable matching in $I$. As a result it is not difficult to see that there are only two stable matchings in $I$, namely the man-optimal stable matching $M_a$ and the woman-optimal stable matching $M_z$, where

$$M_a = \{(m_i, w_i) : 0 \le i \le 5r - 1\} \cup \{(p_i, q_i) : 1 \le i \le 3\}$$
$$M_z = \{(m_i, w_{i+1}) : 0 \le i \le 5r - 1\} \cup \{(p_i, q_i) : 1 \le i \le 3\}$$

It may be verified that $|d(M_a)| = 15r$ and $b(M_a) = 20r + 3$, whilst $|d(M_z)| = 10r$ and $b(M_z) = 25r + 3$. Thus $M_z$ is the unique sex-equal stable matching, whilst $M_a$ is the unique balanced stable matching.

Feder [201] showed that the problem of finding a balanced stable matching is NP-hard, though approximable within a factor of 2. McDermid [438] showed that, in the SMI context, this performance guarantee can be improved when preference lists are of bounded length, as indicated by the following result.

**Theorem 2.23 ([438]).** *Let $I$ be an instance of* SMI *where $k$ is the length of the longest preference list. Then the problem of computing a balanced stable matching is approximable within $2 - \frac{1}{k}$.*

**Proof.** Let $U$ and $W$ be the sets of men and women in $I$ respectively, and let $n$ be the size of $I$. Let $M$ be an egalitarian stable matching and let $M'$ be a balanced stable matching in $I$. Without loss of generality suppose that $c^U(M) \ge c^W(M)$.

We firstly note that $b(M') \le c^U(M)$, for otherwise $b(M) < b(M')$, a contradiction. We next note that $b(M') \ge (c^U(M) + c^W(M))/2$, for otherwise

$$c(M') = c^U(M') + c^W(M') \le 2b(M') < c^U(M) + c^W(M) = c(M),$$

a contradiction.

Hence $c^U(M) \leq 2b(M') - c^W(M)$. Recall from our earlier assumption that $c^U(M) \geq c^W(M)$. Hence

$$b(M) = c^U(M) \leq 2b(M') - c^W(M) = b(M')(2 - c^W(M)/b(M')).$$

Trivially, we note that $c^W(M) \geq n$. If all preference lists are of length at most $k$, then $b(M') \leq c^U(M) \leq nk$. Hence $c^W(M)/b(M') \geq \frac{1}{k}$ and the result follows. $\qquad \square$

### 2.10.3  Rationalizing matchings

Echenique [181] considered the problem of determining whether a set of matchings $\mathcal{M}$ is *rationalizable*. Informally, this concerns whether we can find an SMI instance $I$ in which each matching in $\mathcal{M}$ is stable. Formally, $\mathcal{M}$ is rationalizable if and only if there is an SMI instance $I$ such that $\mathcal{M} \subseteq \mathcal{S}$, where $\mathcal{S}$ is the set of stable matchings in $I$.

Suppose that $U = \{m_1, \dots, m_{n_1}\}$ and $W = \{w_1, \dots, w_{n_2}\}$ are the sets of men and women collectively assigned in the matchings in $\mathcal{M}$, respectively. We may assume that each member of $U \cup W$ is assigned in every matching in $\mathcal{M}$, for if the matchings in $\mathcal{M}$ are to be stable in $I$, the Rural Hospitals Theorem (Theorem 1.11) implies that any agent who is unassigned in one matching in $\mathcal{M}$ is unassigned in all matchings in $\mathcal{M}$. Thus if this property is not satisfied, we can simply report that $\mathcal{M}$ is not rationalizable. Otherwise, we can thus assume that $n_1 = n_2 = n$. Moreover, by Proposition 1.15, we lose no generality in insisting that $I$ is an SM instance of size $n$.

Echenique showed that if $n \geq 3$ and $\mathcal{M}$ is the set of all possible bipartite matchings between $U$ and $W$, then $\mathcal{M}$ is not rationalizable. However he gave a simple example of a set of matchings $\mathcal{M}$ that *is* rationalizable, namely the case where $M(m_i) \neq M'(m_i)$ for all $i$ $(1 \leq i \leq n)$ and for all distinct matchings $M, M' \in \mathcal{M}$. Echenique illustrated the construction used in the proof of this result; in doing so he pointed out that the SM instance $I$ constructed has a set of stable matchings $\mathcal{S}$ where $\mathcal{M} \subset \mathcal{S}$. He remarked that if one requires that $\mathcal{M} = \mathcal{S}$, different techniques to those employed in his paper seem to be required, and he did not pursue this variant of the problem further. A related problem, as also described by Echenique, is to find an SM instance $I$ of size $n$ such that $\mathcal{M} \subseteq \mathcal{S}$, where $|\mathcal{S}|$ is minimum, or else report that no such $I$ exists.

Echenique gave a polynomial-time checkable necessary condition for a set of matchings to be rationalizable. He also gave a necessary and sufficient condition, however it is unlikely that this can be verified in polynomial time, given the following result of Kalyanaraman and Umans [353]. Define RATIONALIZABILITY to be the decision problem which takes as input a set $\mathcal{M}$ of matchings involving $n$ men and $n$ women, and asks whether there exists an SM instance of size $n$ such that $\mathcal{M} \subseteq \mathcal{S}$, where $\mathcal{S}$ is the set of stable matchings in $I$. Kalyanaraman and Umans proved that RATIONALIZABILITY is NP-complete.

Given the NP-completeness of RATIONALIZABILITY, Kalyanaraman and Umans defined two related optimisation problems along the lines of trying to find an SM instance that rationalizes the given set of matchings "as much as possible". The first of these problems, MAX STABLE MATCHINGS, takes as an instance a set of matchings $\mathcal{M}$ involving $n$ men and $n$ women, and a solution is an SM instance of size $n$ such that $|\mathcal{M} \cap \mathcal{S}|$ is maximised, where $\mathcal{S}$ is the set of stable matchings in $I$. The NP-hardness of this problem follows by the NP-completeness of RATIONALIZABILITY. The authors showed that MAX STABLE MATCHINGS is not approximable within $\delta$, for some $\delta > 1$, unless P=NP.

The other optimisation problem related to rationalizability that the authors defined concerns maximising the number of pairs in $\mathcal{M}$ that are not blocking in $I$, taken over all matchings in $\mathcal{M}$. However this problem needs to be defined carefully. Let $U = \{m_1, \ldots, m_n\}$ and $W = \{w_1, \ldots, w_n\}$ be the sets of men and women respectively who are assigned in the matchings in $\mathcal{M}$. Define a man–woman pair $(m_i, w_j) \in U \times W$ to be *active* if $(m_i, w_j) \in M$ for some $M \in \mathcal{M}$, and *inactive* otherwise. An SM instance $I$ of size $n$ is said to be *valid* if, whenever $(m_i, w_j)$ is an active pair and $(m_i, w_k)$ is an inactive pair, $m_i$ prefers $w_j$ to $w_k$, and similarly, whenever $(m_i, w_k)$ is an active pair and $(m_j, w_k)$ is an inactive pair, $w_k$ prefers $m_i$ to $m_j$. Intuitively, if $\mathcal{M}$ can be rationalized by some SM instance, then it can be rationalized by some valid SM instance.

Given these definitions, Kalyanaraman and Umans defined MAX STABILITY to be the problem of finding, given a set of matchings $\mathcal{M}$ involving the set $U = \{m_1, \ldots, m_n\}$ of men and the set $W = \{w_1, \ldots, w_n\}$ of women, a valid SM instance of size $n$ that maximises

$$\left| \left\{ (m_i, w_j, M) : \begin{array}{l} M \in \mathcal{M} \wedge (m_i, w_j) \in U \times W \text{ is an} \\ \text{active pair and does not block } M \text{ in } I \end{array} \right\} \right|.$$

Again, the NP-hardness of this problem follows by the NP-completeness of RATIONALIZABILITY. The authors show that MAX STABILITY is approximable within $4/3$, though not approximable within $\delta$, for some $\delta > 1$, unless P=NP.

One intriguing problem that the authors leave open is whether RATIO-NALIZABILITY remains NP-complete even if $|\mathcal{M}|$ is a constant. It is also of interest to extend the study of rationalizability to the SR context.

### 2.10.4 The Dinitz conjecture and stable marriage theory

The famous Dinitz conjecture (see Ref. [194]) asserted that $L(K_{n,n})$, the line graph of the complete bipartite graph with $n$ vertices on each side $(n \geq 1)$, is $n$-choosable. A graph $G = (V, E)$ is $k$-choosable for some integer $k \geq 1$ if, given any function $f$ that assigns a set of integers (or "colours") of size $k$ to each vertex of $G$, there exists a colouring $c : V \longrightarrow \{1, 2, \ldots, k\}$ such that $c(v) \in f(v)$ for each $v \in V$, and $c(v) \neq c(w)$ for each $\{v, w\} \in E$. The Dinitz conjecture is equivalent to the assertion that $\chi'_l(K_{n,n}) = n$, i.e., the *list chromatic index* of $K_{n,n}$ is equal to $n$, where the *list chromatic index* of a graph $G$ is the minimum $k$ such that $L(G)$ is $k$-choosable.

The Dinitz conjecture was proved by Galvin [239] using a beautiful connection with the theory of stable marriage[12]. Rather than describing Galvin's proof in its entirety, the purpose of this section is to indicate where the connection with stable matchings arises. We firstly require to establish some further terminology.

A directed graph $D = (V, A)$ can be viewed as an *orientation* of the edges in the underlying undirected graph $G = (V, E)$, where $E = \{\{u, w\} : (u, w) \in A\}$. Given an arc $(u, v) \in A$, we say that $v$ is a *successor* of $u$. A set $S \subseteq V$ is said to be *absorbant* if each vertex in $V \backslash S$ has a successor in $S$. Also $S$ is *independent* if no two members of $S$ are adjacent in $D$. A *kernel* of $D$ is a set $S \subseteq V$ that is both absorbant and independent.

A *clique* in $D$ is a subset $S \subseteq V$ such that, for every pair of vertices $u, v \in S$, either $(u, v) \in A$ or $(v, u) \in A$. If a clique $S$ has a kernel, then it comprises a single node $v$ that is a *sink* of $S$ (i.e., $v$ is a successor of every vertex in $S \backslash \{v\}$). If every clique in $D$ has a kernel, then $D$ is said be *normal*, and $D$ is a *normal orientation* of $G$. If every normal orientation of $G$ has a kernel, then $G$ is said to be *solvable*.

In order to prove that $G = L(K_{n,n})$ is $n$-choosable, Galvin constructed a normal orientation $D$ of $G$, and required to prove that every induced subdigraph of $D$ has a kernel[13]. However it is enough to show

---

[12]Galvin's proof was selected as an example of a "book proof" by the authors of Ref. [34].

[13]The fact that it follows from this that $G$ in $n$-choosable is established by Corollary 2.2 in Ref. [239].

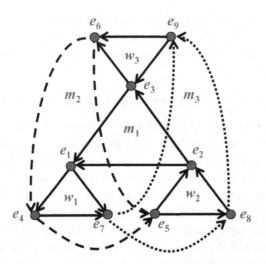

Fig. 2.9   The line graph of $K_{3,3}$ with a normal orientation.

that $G$ is solvable, since any induced subgraph of a solvable graph is also solvable.

Maffray [411] proved the the line graph $G$ (of a multigraph $H$) is solvable if and only if $G$ is perfect. Note that if $H$ is a bipartite multigraph then $H$ is perfect, and so is $G = L(H)$. Hence $G$ is solvable in this case. This result can be applied to the special case that $H$ is $K_{n,n}$. However Maffray's proof that $G = L(K_{n,n})$ is solvable can also be established using the theory of stable marriage.

Let $D$ be an arbitrary normal orientation of $G = L(H)$, where $H$ is $K_{n,n}$. We show that $D$ gives rise to an SM instance $I$ of size $n$. For, let $U = \{m_1, \ldots, m_n\}$ and $W = \{w_1, \ldots, w_n\}$ denote the sets of vertices on each side of $H$. Each man $m_i \in U$ corresponds to a clique $C_{m_i}$ in $G$: this comprises the set of edges that are incident to $m_i$ in $H$. Similarly each woman $w_j \in W$ corresponds to a clique $C_{w_j}$ in $G$. Conversely each clique in $G$ is either a $C_{m_i}$ for some $m_i \in U$ or a $C_{w_j}$ for some $w_j \in W$.

Given $m_i \in U$, we form $m_i$'s preference list in $I$ as follows. $C_{m_i}$ is a clique in $G$ whose kernel is a sink vertex $e_r = \{m_i, w_j\}$; $w_j$ is $m_i$'s first choice. Similarly $C_{m_i} \backslash \{e_r\}$ is a clique in $G$ having a sink $e_s = \{m_i, w_k\}$ as its kernel; $w_k$ is $m_i$'s second choice. Continuing in this way, we deduce the preference list of $m_i$; we do likewise for each other man and woman.

Gale and Shapley [235] showed that $I$ admits a stable matching $M$. We claim that $M$ is a kernel of $G$. For, clearly the elements of $M$ are independent as vertices in $G$. Now suppose that $e_r = \{m_i, w_j\} \notin M$. Then by the stability of $M$, either (i) $m_i$ prefers $M(m_i) = w_k$ to $w_j$, or (ii) $w_j$ prefers $M(w_j) = m_l$ to $m_i$. In case (i), $e_s = \{m_i, w_k\}$ is a successor of $e_r$ in $M$, whilst in case (ii), $e_t = \{m_l, w_j\}$ is a successor of $e_r$ in $M$. It follows that $G$ is solvable, and hence Galvin's proof is complete. The connection between stable matchings in SM and kernels in line graphs of bipartite graphs was first made by Maffray [411], and it was his observation that provided some of the inspiration for Galvin's subsequent proof of the Dinitz conjecture.

Indeed, given any SM instance, there is a corresponding line graph of a bipartite graph with a normal orientation. For, suppose $I$ is an SM instance of size $n$. Let $H = (U, W, E)$ be the underlying bipartite graph (that is, $H$ is isomorphic to $K_{n,n}$), and let $G = L(H)$.

Form a digraph $D$ from $G$ by orienting the edges in $G$ as follows: if $\{e_i, e_j\} \in V(G)$, then either (i) $e_i = \{m_p, w_q\}$ and $e_j = \{m_p, w_r\}$ for some $m_p \in U$ and $w_q, w_r \in W$, or (ii) $e_i = \{m_p, w_r\}$ and $e_j = \{m_q, w_r\}$ for some $m_p, m_q \in U$ and $w_r \in W$. In case (i), without loss of generality suppose that $m_p$ prefers $w_q$ to $w_r$; orient $\{e_i, e_j\}$ as $(e_j, e_i)$ in $D$. Do likewise for case (ii). Then $D$ is a normal orientation of $G$, since every clique $S$ in $D$ satisfies either (i) $S \subseteq C_{m_i}$ for some $m_i \in U$, or (ii) $S \subseteq C_{w_j}$ for some $w_j \in W$. In case (i), $S$ has a sink vertex $e_p = \{m_i, w_j\}$, where $w_j$ is the most-preferred woman according to $m_i$'s preferences among the women incident to an edge in $S$. The argument for case (ii) is similar.

To illustrate the various connections described in this subsection, a normal orientation of the line graph corresponding to the SM instance shown in Fig. 2.3 is indicated in Fig. 2.9. In that instance, $e_k = \{m_{\lceil k/3 \rceil}, w_{((k-1) \bmod 3)+1}\}$. Also, for each $m_i \in U$, $C_{m_i}$ comprises vertices $\{e_{3(i-1)+k} : 1 \le k \le 3\}$, whilst for each $w_j \in W$, $C_{w_j}$ comprises vertices $\{e_{3k+j} : 0 \le k \le 2\}$. Each man and woman is indicated within the group of three vertices that forms his/her corresponding clique. In order to assist with visualising $C_{m_2}$ and $C_{m_3}$, the edges joining vertices in these cliques are shown with dashed and dotted lines respectively. The stable matching $M_e = \{e_3, e_4, e_8\}$ is a kernel in $G$.

### 2.10.5 The marriage graph

Ratier [489] defined the *marriage graph* (strictly speaking a digraph) $D_I = (V, E)$ for an SMI instance $I$, where $U$ is the set of men and $W$ is the set of women, as follows:

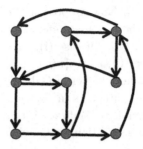

Fig. 2.10    The marriage graph for the SM instance of Fig. 2.3.

(1) There is a vertex $s_{m,w} \in V$ for each acceptable pair $(m, w)$ in $I$;

(2) There is an edge $(s_{m,w}, s_{m',w'}) \in E$ if and only if either

- $m' = m$ and $m$ prefers $w'$ to $w$, or
- $w' = w$ and $w$ prefers $m'$ to $m$.

Recall from Sec. 2.10.4 that $I$ has a corresponding digraph $D_I'$, which is a normal orientation of the line graph of the underlying bipartite graph of $I$. Clearly $D_I$ is isomorphic to $D_I'$. Henceforth, we will assume that $D_I$ contains only arcs corresponding to the transitive reductions of the preference orders belonging to the men and women in $I$.

The marriage graph for the SM instance of Fig. 2.3 is shown in Fig. 2.10. In the latter figure, the vertex in the $i$th row and $j$th column corresponds to the acceptable pair $(m_i, w_j)$.

Following on from the observations of Maffray [411] as described in the previous subsection, Ratier [489] noted that a matching $M$ is stable in $I$ if and only if its corresponding vertices in $D_I$ form a kernel. Ratier defined two marriage graphs to be *equivalent* if they admit the same set of stable matchings (i.e., they have the same kernels). He showed that $D_I$ is equivalent to a marriage graph in which certain vertices, corresponding to pairs that cannot belong to a stable matching in $I$, are deleted. It turns out that these vertices correspond to precisely those acceptable pairs in $I$ that do not belong to the GS-lists in $I$ (i.e., they are deleted during an execution of either the MEGS or WEGS algorithms). Ratier's observation is thus immediate by Theorem 1.2.5 of [261].

Ratier noted that, following these deletions, $D_I$ is a *principal marriage graph*, i.e., it can be decomposed into *principal circuits* (see Ref. [489] for further details). Ratier went on to characterise the polytope of solutions to the LP for SMI [576,520] (see Sec. 2.4) in terms of the principal marriage graph.

Two further papers by Balinski and Ratier [62,63] also explore the marriage graph and use it to reprove a range of existing results concerning the theory of stable matchings in SMI. Baïou and Balinski extended the marriage graph to the case of HR in Refs. [58] (see Sec. 2.4 for more details) and [60], and to the case of the many–many stable marriage problem in Ref. [57] (see Sec. 5.4 for more details).

### 2.10.6 Sampling and counting

Bhatnagar et al. [73] and Chebolu et al. [133,134] considered the problems of sampling and counting stable matchings in various restricted models of SM and SR. It turns out that, in general, these are all hard problems. We begin with problems involving sampling stable matchings. Here, results have been obtained with respect to several restricted models of SM. We begin by defining these models; the definitions have natural extensions to the SR context.

In the *k-attribute* model ($k \geq 1$), each agent has $k$ scores, each according to $k$ different attributes (e.g., attractiveness, intelligence, wealth, etc.) and can therefore be associated with a point in $\mathbb{R}^k$. Each man $m_i$ has a linear function of these attributes (representing his opinion of the relative importance of these characteristics) which essentially projects the women's points onto a line — this gives rise to $m_i$'s preference list. The preference list for each woman $w_j$ is arrived at in a similar way.

In the *k-range* model ($k \geq 1$), the preference lists satisfy the property that, for each agent $a_i$, there is some $j$ such that $a_i$ appears between positions $j, j + 1, \ldots, j + k - 1$ on the preference list of each member of the opposite sex. This could correspond to the case that the preference lists of the men and women conform, to within a measure of closeness given by $k$, to master lists of the women and men respectively. Such master lists can occur on both sides in practical applications: consider for example HR instances, where common academic examinations give rise to a master list of residents, whilst a national league table gives rise to a master list of hospitals. For $k = 1$, all preference lists are identical and there is a unique stable matching.

In the *k-list* model ($k \geq 1$), the men and women are each partitioned into $k$ sets, and all members of the same set have an identical preference list. This model reflects the possibility that the agents may be partitioned into groupings which characterise their rankings of the members of the opposite sex.

In the *k-Euclidean model* ($k \geq 1$), each agent is represented by two points in $\mathbb{R}^k$, a *preference point* and a *position point*. Denote these points by $\hat{m}_i$ and $\bar{m}_i$ for each man $m_i$, and by $\hat{w}_j$ and $\bar{w}_j$ for each woman $w_j$, respectively. Then $m_i$ prefers $w_j$ to $w_k$ if and only if $d(\hat{m}_i, \bar{w}_j) < d(\hat{m}_i, \bar{w}_k)$, where $d(x,y)$ represents the Euclidean distance between two points $x, y \in \mathbb{R}^k$. A similar condition for preference holds in the case of the women.

Bogomolnaia and Laslier [103] gave an example of an SM instance of size $n$ that cannot be represented using the $k$-attribute model for $k \leq n-2$. Bhatnagar *et al.* [73] strengthened this by showing that there exists a lattice of stable matchings involving $n$ men and $n$ women that is not the lattice of stable matchings for any SM instance that conforms to the $k$-attribute model, for any $k < n/2$. They also showed that there are SM instances conforming to the $k$-attribute model ($k \geq 2$) or the $k$-range model ($k \geq 2$) that admit an exponential number of stable matchings.

Bhatnagar *et al.* [73] studied the problem of sampling a stable matching uniformly at random. To this end, they defined a Markov Chain based on a simple random walk on the lattice of stable matchings for a given SM instance conforming to one of the above models. The mixing time of this Markov Chain corresponds to the time that the random walk takes to converge to an equilibrium. The authors showed that there are SM instances $I$ belonging to the $k$-attribute model ($k \geq 2$) for which the mixing time is exponential (in the size of $I$). An analogous result holds for the $k$-range model ($k \geq 5$) and the $k$-list model ($k \geq 4$). However the mixing time is polynomial for the $k$-range model when $k = 2$.

Gelain *et al.* [246] presented a local search approach to sampling the stable marriage lattice.

We now turn to the problem of counting the number of stable matchings, given an instance of SM, which we refer to as #SM. It is known that #SM is #P-complete [319]. Thus, it is of interest as to whether a *Fully Polynomial Randomised Approximation Scheme* (FPRAS) [358, 359] exists for this problem.

Dyer *et al.* [180] defined a complexity class #RHΠ₁ of counting problems, and identified a subclass of problems in this class that are complete with respect to *approximation-preserving (AP) reductions* (henceforth problems in this subclass are referred to as being #RHΠ₁-complete). The #RHΠ₁-complete problems are equivalent to one another in the sense that if one of these problems admits an FPRAS, then they all do. At present it is not known whether an FPRAS exists for any of the #RHΠ₁-complete problems, however it is felt to be unlikely [180]. One of the #RHΠ₁-complete problems is that of counting the number of closed subsets of a given poset $\mathcal{P} = (P, \blacktriangleleft)$ [180]. Given that Irving and Leather [319] proved that there is a corresponding SM instance $I$, which can be constructed from $\mathcal{P}$ in polynomial time, such that the subsets of $P$ that are closed under $\blacktriangleleft$ are in 1–1 correspondence with the stable matchings in $I$, it follows that #SM is also #RHΠ₁-complete.

Chebolu *et al.* [133] showed that #SM is #RHΠ₁-complete even for instances that belong to the $k$-attribute model for $k \geq 3$. They also showed that the same is true for the $k$-Euclidean model ($k \geq 2$). On the other hand, for the 1-attribute model, the authors gave a polynomial-time algorithm for #SM.

In more recent work, Chebolu *et al.* [134] extended their study to SR. We denote by #SR the problem of counting the number of stable matchings, given an SR instance. The authors showed that #SR is complete for #P under AP-reductions, even for instances conforming to the $k$-attribute model ($k \geq 4$) or the $k$-Euclidean model ($k \geq 3$). This means that there is no FPRAS for either restriction of #SR unless NP=RP. Additionally, the authors showed that each of #SR under the 3-attribute model, and #SR under the 2-Euclidean model, is #RHΠ₁-complete. Finally, for #SR under the 1-attribute model, Chebolu *et al.* showed that #SR is solvable in polynomial time (in fact the number of stable matchings is either 1 or 2 in this case).

### 2.10.7  *Online algorithms*

Khuller *et al.* [378] considered online algorithms for SM. In their model, it is assumed that the set of $n$ women and their preference lists over the $n$ men are known in advance. The men arrive one by one, and when a man $m_i$ arrives, his preference list is revealed. At this point, $m_i$ is to be assigned to some woman $w_j$, in such a way that no other assigned woman changes her partner. The authors measure the "competitiveness" of an online algorithm $A$ that conforms to this model by measuring the number of blocking pairs of a matching $M$ output by $A$ in the final SM instance $I$ whose definition becomes complete when the last man arrives.

Note that the authors do not consider the notion of a *competitive ratio* for $A$ (which is the ratio of the performance of $A$ to the performance of an optimal offline algorithm) since the "performance" measure of the optimal offline algorithm (namely the Gale–Shapley algorithm) is 0 for this problem. Similarly, it would not lead to an interesting problem if the authors were to have allowed a previously matched woman to change her partner during the execution of $A$ — in this case it is clear that $A$ will always produce a stable matching (since the arriving man $m_i$ will trigger a proposal–rejection sequence that transforms a stable matching in the SMI instance immediately prior to $m_i$'s arrival into an updated matching that is stable in the SMI instance immediately after $m_i$'s arrival).

Khuller *et al.* proved that if $A$ is the "obvious" online algorithm (i.e., assign the arriving man to the most-preferred single woman on his preference list) then $A$ produces a matching $M$ with $O(n \log n)$ blocking pairs in the average case. They also showed that no randomised algorithm can produce a matching with fewer than $\Omega(n^2)$ blocking pairs in the worst case.

See also Ref. [404] for further work regarding online algorithms for SM.

### 2.10.8 *Unified approach to finding "good" stable matchings*

Cheng *et al.* [148] developed a unified approach to solving variants of SM and HR that involve finding stable matchings that satisfy some additional criteria. Specifically, these additional criteria are separated into two distinct forms: (i) where there is a set of constraints, and a solution is a stable matching that satisfies each of these; and (ii) where a cost function is defined over the set of stable matchings, and a solution is a stable matching that is optimal with respect to this function. We refer to a solution as a *feasible stable matching* or an *optimal stable matching* depending on whether the additional criteria fall into category (i) or (ii), respectively. We remark that a feasible stable matching need not exist.

An example set of constraints for category (i) could be the condition that no two men can swap their partners so as to both improve (this is the notion of *man-exchange-stability* — see Sec. 5.7). For category (ii), cost functions can be defined in order that an optimal solution is an egalitarian or minimum regret stable matching, for example.

The authors' treatment is in fact generalised to the HR setting rather than the more restricted SMI context, because practical applications of bipartite matching problems with two-sided preferences most commonly correspond to the many–one setting, and moreover in the authors' experience, HR gives rise to a richer variety of feasibility constraints and cost functions that can be defined.

The approach that Cheng *et al.* adopted that enables these feasible or optimal stable matchings to be found efficiently (assuming such a matching exists) is based on navigating through the lattice of stable matchings in an HR instance with the aid of the *meta-rotation poset*. Bansal *et al.* [67] defined the concept of a *meta-rotation* in the context of the many–many stable marriage problem (see Sec. 5.4). A meta-rotation is a natural generalisation of the notion of a rotation from the 1–1 SMI setting (see Sec. 1.3.4.3). Cheng *et al.* [148] then specialised Bansal *et al.*'s definition to a given HR instance $I$. The traversal of the lattice of stable matchings in $I$ is then facilitated by a fundamental result of Bansal *et al.* [67], namely that the closed subsets of the meta-rotation poset are in 1–1 correspondence with the stable matchings in $I$.

Let $I$ be an HR instance where $R$ is the set of residents and $H$ is the set of hospitals, and suppose that we seek a feasible stable matching in $I$ relative to a set $\mathcal{X}$ of constraints (i.e., corresponding to category (i) above). Cheng *et al.* gave an algorithm that will find such a matching or report that none exists, so long as each constraint $X \in \mathcal{X}$ satisfies the so-called *identification property*. A constraint $X$ satisfies this property if, whenever $M$ is a stable matching in $I$ that does not satisfy $X$, where $M$ is not the hospital-optimal stable matching, there exists a resident $r^*$ such that, for each stable matching $M'$ where $M \preceq M'$ ($\preceq$ is the dominance partial order defined in Definition 1.12) and $M(r^*) = M'(r^*)$, $M'$ does not satisfy $X$ either. Such a resident $r^*$ is called a *candidate resident*. The running time of the algorithm is $O(n_1 n_2 f(\mathcal{X}))$, where $n_1 = |R|$ and $n_2 = |H|$, and $f(\mathcal{X})$ is the worst-case time taken to check whether each constraint $X \in \mathcal{X}$ is satisfied, or else identify a candidate resident with respect to $X$ and the current matching.

Now suppose that we seek a stable matching in $I$ that is optimal with respect to a given cost function (i.e., corresponding to category (ii) above). Cheng *et al.* assumed that there is a cost function $s : (R \cup H) \times \mathcal{S} \longrightarrow \mathbb{R}_0^+$, where $\mathcal{S}$ is the set of stable matchings in $I$, and the larger the value of $s$ with respect to an agent $a_i \in R \cup H$ and a matching $M \in \mathcal{S}$, the happier $a_i$ is with $M$. The authors considered functions $s$ that satisfy the *independence property*: that is, for all $a_i \in R \cup H$, $s(a_i, M)$ is a function of $M(a_i)$ only. They define a *generalised minimum regret stable matching* to be a matching $M \in \mathcal{S}$ such that

$$r'(M) = \min_{a_i \in R \cup H} s(a_i, M)$$

is maximum, and a *generalised egalitarian stable matching* to be a matching $M \in \mathcal{S}$ such that

$$c'(M) = \sum_{a_i \in R \cup H} s(a_i, M)$$

is maximum. Cheng *et al.* gave algorithms for computing generalised minimum regret and generalised egalitarian stable matchings with respect to cost functions satisfying the independence property. These algorithms have time complexity $O(n_1 n_2 f(r'))$ and $O(n_1 n_2 f(c') + n_1^4)$, where $f(r')$ and $f(c')$ are the worst-case time taken to evaluate $r'$ and $c'$ respectively.

### 2.10.9    *Locally stable matchings*

Arcaute and Vassilvitskii [49] introduced the notion of a *locally stable matching* in instances of HR that are augmented with a *social network graph*. Formally, let $I$ be an instance of HR and let $G = (R, E)$ be an undirected graph, where $R$ is the set of residents in $I$. Intuitively $\{r_i, r_j\} \in E$ if and only if $r_i$ and $r_j$ know one another. When instance $I$ is equipped with $G$, we have an instance of HR+SN.

Given a matching $M$ in an instance $\langle I, G \rangle$ of HR+SN (where $I$ is the underlying HR instance and $G = (R, E)$ is the social network graph), a *blocking pair* of $M$ is a resident–hospital pair $(r_i, h_j)$ such that (i) $(r_i, h_j)$ is a blocking pair of $M$ in $I$, and (ii) there exists some $r_k \in M(h_j)$ such that $\{r_i, r_k\} \in E$. Intuitively, in order to be a blocking pair in an HR+SN instance, $(r_i, h_j)$ is a blocking pair in the classical sense such that $r_i$ knows someone in $M(h_j)$. The motivation is that, in reality, often an employer's awareness of the merits of an applicant depends on the recommendation of an existing employee. $M$ is *locally stable* if $M$ admits no blocking pair with respect to this revised definition.

Clearly if $M$ is stable in $I$ then $M$ is locally stable in $\langle I, G \rangle$. Hence the set of locally stable matchings in $\langle I, G \rangle$ is a superset of the set of stable matchings in $I$. In fact it is possible for the former set to be much larger than the latter [49].

Arcaute and Vassilvitskii proved that locally stable matchings do not form a lattice structure in general. They also considered a dynamic version of the problem, giving a decentralised algorithm, which they referred to as the "local Gale-Shapley algorithm", that, given an initial matching, converges almost surely to a locally stable matching.

Hoefer [279] studied locally stable matchings in instances of SMI and SRI augmented with a social network graph (it is straightforward to extend the definition of a locally stable matching to the SRI case). He extended Arcaute and Vassilvitskii's study of decentralised algorithms by considering *best response* and *better-response dynamics* (see Sec. 2.6.4).

Cheng and McDermid [147] considered the HR+SN case and showed that locally stable matchings can have different sizes in a given problem instance $\langle I, G \rangle$, though a maximum locally stable matching is at most twice the size of a stable matching. This gives rise to the problem MAX HR+SN: given an instance of HR+SN, the objective is to find a maximum locally stable matching.

Cheng and McDermid considered some special cases of $G$ (such as the cases when $G$ is empty, or has a constant number of edges, or is the complete graph) for which MAX HR+SN is solvable in polynomial time. On the other hand, in general, they showed that MAX HR+SN is NP-hard and not approximable within $\frac{21}{19} - \delta$, for any $\delta > 0$, unless P=NP. The result holds even if each hospital has capacity 1. For a particular class of social network graphs, the authors showed that MAX HR+SN is approximable within $\frac{3}{2}$, using corresponding results for MAX HRT (see Sec. 3.2.6). See also Ref. [280].

### 2.10.10   *Miscellaneous results*

In this section we present a number of additional results for SM that have not already been covered in previous sections, but are still worthy of noting.

**Lower bounds for stable matching.** The ground-breaking paper of Ng and Hirschberg [464] established lower bounds of $\Omega(n^2)$ for each of the problems of determining whether a given man–woman pair is stable and finding a stable matching, given an SM instance of size $n$. Although the year of publication of Ng and Hirschberg's paper was 1990, these results were described in Ref. [261, Sec. 1.5]. Dias *et al.* [168] showed that listing all stable matchings requires $\Omega(n)$ amortised time per solution. Dougherty and Selkow [170] also gave lower bounds for the certificate complexity of various problems related to finding stable matchings in SM (see Sec. 2.2.9).

**Rank profiles of stable matchings.** Let $n \geq 2$ be an integer. Boros *et al.* [107] defined a *rank profile* to be a pair of $n$-tuples $\langle p, q \rangle$, where $p = \langle x_1, \ldots, x_n \rangle$, $q = \langle y_1, \ldots, y_n \rangle$ and $1 \leq x_i, y_j \leq n$ $(1 \leq i, j \leq n)$. Further, they defined $\langle p, q \rangle$ to be *stable* if there is an instance $I$ of SM of size $n$ and a stable matching $M$ in $I$ such that $x_i = rank(m_i, M(m_i)$ $(1 \leq i \leq n)$ and $y_j = rank(w_j, M(w_j))$ $(1 \leq j \leq n)$. For example, the rank profile $\langle\langle 2, 2\rangle, \langle 2, 2\rangle\rangle$ is not stable. Among other results, they gave a characterisation of stable rank-profiles, leading to an $O(n^5)$ algorithm to determine if a given rank-profile $\langle p, q\rangle$ is stable. If $\langle p, q\rangle$ is stable then the algorithm constructs an SM instance $I$ of size $n$ such that the matching that is uniquely determined by $I$ and $\langle p, q\rangle$ is stable.

**Genetic algorithms for stable matching.** Aldershof and Carducci [38] described two genetic algorithms for finding stable matchings: one is for the classical SMI case, and the other is for the SMI variant where couples can submit joint preference lists (see Sec. 5.3). In each case, a set of inequalities is given — these constraints are based on the LP inequalities for SM and SMI due to Vande Vate and Rothblum [576,520]. Chromosomes are generated in order to satisfy a subset of these inequalities (these correspond to feasible matchings). The remaining inequalities correspond to blocking pairs, and the fitness function is the number of these constraints that are satisfied. This is useful in the case of SMI with couples, because a stable matching need not exist, and therefore a solution will be a matching with the minimum number of blocking pairs. As mentioned in Sec. 2.2.7, genetic and ant colony-based algorithms for finding a sex-equal stable matching, given an SMI instance, have also been formulated [459,582]. Refs. [581,381,159,405] also discuss genetic algorithms for SM.

**Private stable matching.** Golle [256] argued that in many practical scenarios, the preference list of an agent might contain sensitive information, which should not be shared with other participants in a given matching scheme. Algorithms for finding stable matchings typically take as input the entire set of preferences of all agents. In mitigation of the need for an agent to reveal their preferences to all other participants, these lists are often passed to a trusted third party that is responsible for administering the matching scheme, but that third party is then vulnerable to corruption as a consequence of pressure from more powerful participants. Golle proposed an alternative stable matching algorithm for SM that aims to maintain the privacy of each agent's preference list throughout the matching process. See also Refs. [532,530,222,69,531].

**$\alpha$-stability and $k$-stability.** Arkin *et al.* [50] defined the notions of an $\alpha$-*stable matching* $(\alpha \geq 1)$ and a $k$-*stable matching*, each of which generalises the concept of a (classical) stable matching. In fact, Arkin *et al.* defined these terms in the SRI setting. Nevertheless, given the clear applicability of these concepts to SMI, we describe them in this chapter.

Let $I$ be an instance of SRI. Relative to $\alpha$-stability, a blocking pair of a given matching $M$ in $I$ is an acceptable pair of agents $\{a_i, a_j\}$ such that either $a_i$ is unmatched in $M$ or $rank(a_i, M(a(i)))/rank(a_i, a_j) > \alpha$, and similarly for $a_j$. Thus, in order to be in a blocking pair, each agent must either be unmatched or improve their rank by a factor more than $\alpha$. Clearly an $\alpha$-stable matching is $\alpha'$-stable, for any $\alpha' > \alpha$.

Clearly classical stability corresponds to $\alpha$-stability where $\alpha = 1$. The authors show that, for a fixed $\alpha > 1$, finding an $\alpha$-stable matching is at least as hard as finding a classical stable matching. They also consider $\alpha$-stability in the context of 3D variants of SR (see Sec. 5.6). We note that $\alpha$-stability was also considered by Anshelevich $et\ al.$ [46] and by Emek $et\ al.$ [191] (albeit in a slightly different form in the latter case).

Arkin $et\ al.$ [50] also remarked that it is possible to define an additive counterpart of $\alpha$-stability, which we will refer to as $k$-stability. A matching $M$ is $k$-stable if there is no acceptable pair of agents $\{a_i, a_j\}$ in $I$ such that either $a_i$ is unmatched in $M$ or $rank(a_i, M(a(i))) - rank(a_i, a_j) \geq k$, and similarly for $a_j$. That is, in order to be in a blocking pair, each agent must either be unmatched or improve their partner by at least $k$ places in their preference list. Clearly a $k$-stable matching is $k'$-stable, for any $k' > k$. Again, classical stability corresponds to $k$-stability where $k = 1$. Pini $et\ al.$ [477,479,480] also considered the analogue of $k$-stability (unfortunately they referred to this concept as $\alpha$-stability) in the context of SMI with weighted preferences (see Sec. 1.3.4.1).

## 2.11   Conclusions and open problems

Although SM and its variants have been the focus of much attention in the literature since the publication of Gusfield and Irving's book in 1989, a number of intriguing problems remain open. From among the list of 12 research problems posed by Gusfield and Irving in the appendix of their book that relate to SM and its variants, perhaps the most noteworthy of those that remain open are Problems 1 and 3: these concern characterising SM instances that admit the maximum number of stable matchings, and determining whether SM belongs to NC (see Secs. 2.2.2 and 2.2.4 respectively).

Although Problem 2 has been solved (this concerns the structure of the divorce digraph, and determining whether an arbitrary matching can always be transformed into a stable matching via a sequence of divorce operations), one intriguing problem remains open, as noted in Sec. 2.2.3. This relates to establishing the algorithmic complexity of the decision problem which asks, given an SM instance $I$ and a matching $M_0$, whether there is a path from $v_{M_0}$ to a sink vertex in the divorce digraph $D_I$ (that is, whether an arbitrary stable matching can be obtained from $M_0$ by a sequence of exclusively divorce operations).

As noted in Sec. 2.3.1, Subramanian [551] gave a logspace reduction from an SMI instance $I$ to an instance $J$ of the Assignment problem that enables the man-optimal stable matching in $I$ to be constructed from an optimal solution in $J$. It remains open as to whether there is a polynomial-time reduction from $I$ to an instance $J$ of the Assignment problem that gives a correspondence between *all* the stable matchings in $I$ and the optimal solutions in $J$.

Determining which stable matchings can be reached, by starting from a given matching (which may be empty) and iteratively satisfying blocking pairs, was discussed in Sec. 2.6. As noted there, the complexity of the following decision problems remains open:

(1) given an SM instance $I$ and a stable matching $M$, is there an execution of Algorithm ROM that terminates with $M$?

(2) given an SM instance $I$, a matching $M_0$ and a stable matching $M$, is there an execution of Algorithm RVV that transforms $M_0$ to $M$?

The study of each of the problems of finding a balanced stable matching and rationalizing matchings (see Secs. 2.10.2 and 2.10.3 respectively) is at a relatively early stage, and there is scope for progress to be made, despite the NP-hardness of both problems in general. For example, special cases of each problem might be more accessible, such as restrictions where preference lists are of bounded length and/or master lists are in place involving either the men or the women.

It is also of interest to determine the complexity of problems relating to finding $\alpha$-stable and $k$-stable matchings in instances of SMI (see Sec. 2.10.10).

# Chapter 3

# The Stable Marriage and Hospitals / Residents problems with indifference

## 3.1 Introduction

In Sec. 1.3.5, we motivated the study of variants of HR that involve forms of indifference, and we defined HRT, the Hospitals / Residents problem with Ties. We also defined three stability criteria, namely weak stability, strong stability and super-stability, that are appropriate in this setting. Following the first paper to define these criteria in SMT [308], the study of SMTI and HRT under these forms of stability has been a very active area of research from 1999 to date, triggered by Refs. [414, 332].

Among the three stability criteria mentioned in the preceding paragraph, it is weak stability that has received by far the most attention in the literature (Refs. [419, 267, 271, 272, 340, 322, 323, 344, 385] represent just some of the papers published on this topic). It is likely that one of the main reasons for this is the guaranteed existence of a weakly stable matching, given an instance of HRT, as we will show in Sec. 3.2. By contrast, as revealed in Secs. 3.3 and 3.4, the same is not true in general in the cases of strong stability and super-stability, respectively.

One of the most exciting areas of research in this context has been the search for approximation algorithms for finding large weakly stable matchings in the context of SMTI (it turns out that, for such a problem instance, weakly stable matchings can have different sizes, and the problem of finding the largest is NP-hard) [333]. After the initial straightforward upper bound of 2 was established [419], a series of papers derived successively smaller upper bounds for the general SMTI case [336, 341, 340, 385], culminating in the current best bound of $\frac{3}{2}$ [437, 386, 472]. It is likely that we have not yet heard the last word on this.

This chapter is organised as follows. Sections 3.2, 3.3, 3.4 detail structural and algorithmic results for HRT under each of the weak, strong and super-stability criteria respectively. In the majority of practical applications, indifference takes the form of ties in the preference lists, hence this chapter mainly focuses on HRT. However in Sec. 3.5 we describe some related stable matching problems involving indifference, including HRP in its full generality in Sec. 3.5.3. Finally Sec. 3.6 contains some conclusions and open problems.

## 3.2 Weak stability

In this section we focus on the weakest of the three stability criteria. In Sec. 3.2.1 we show that every instance of HRT admits a weakly stable matching, and we give a simple linear-time algorithm for finding one. Despite the guaranteed existence of a weakly stable matching, it turns out that many structural properties enjoyed by stable matchings in instances of SM or HR are absent in the case of weakly stable matchings in SMTI instances. Sections 3.2.2 and 3.2.3 indicate some properties that no longer hold for weakly stable matchings. In particular, a fundamental observation is that, in contrast to the case for SMI, weakly stable matchings need not be of the same size, for a given instance of SMTI.

This leads to the question of whether there exist efficient algorithms for the problem of finding a maximum cardinality weakly stable matching (henceforth a maximum weakly stable matching) in instances of SMTI and HRT, denoted by MAX SMTI and MAX HRT respectively. In Sec. 3.2.4, we show that MAX SMTI is NP-hard. We briefly survey parameterized complexity results for this problem in Sec. 3.2.5. Section 3.2.6 deals with the approximability of MAX SMTI and MAX HRT. We give an overview of previous results in the literature, focusing on one particular approximation algorithm due to Király [385]. We also describe two heuristics for MAX HRT whose implementations have been compared empirically with that of Király's algorithm. We then give a lower bound for the approximability of MAX HRT, and we further show how instances of HRT can be "cloned" to form instances of SMTI, in many cases enabling approximation algorithms for the latter to be applied to instances of the former without affecting the performance guarantee.

Finally, in Sec. 3.2.7, we discuss some other problems involving weak stability in instances of SMTI and HRT.

### 3.2.1 Existence of a weakly stable matching

The following result, first proved in Ref. [418], gives a necessary and sufficient condition for a matching to be weakly stable in an instance of HRT. Before stating the result, we make the following definitions. Given an instance $I$ of HRT and a matching $M$ in $I$, let $\mathcal{R}_M(I)$ denote the set of HR instances formed by breaking the ties in $I$ in some way, subject to the requirement that if $t$ is a tie on a resident $r_i$'s list containing $M(r_i)$, then $t$ must be broken so that $r_i$ prefers $M(r_i)$ to each member of $t \backslash M(r_i)$. Similarly let $\mathcal{H}_M(I)$ denote the set of HR instances formed by breaking the ties in $I$ in some way, subject to the requirement that if $t$ is a tie on a hospital $h_j$'s list containing some member of $M(h_j)$, then $t$ must be broken so that $h_j$ prefers each member of $t \cap M(h_j)$ to each member of $t \backslash M(h_j)$.

**Lemma 3.1 ([418]).** *Let $I$ be an instance of* HRT, *and let $M$ be a matching in $I$. Then $M$ is weakly stable in $I$ if and only if $M$ is stable in some instance $I'$ of* HR *obtained by breaking the ties in $I$.*

*Proof.* Let $R = \{r_1, \ldots, r_n\}$ be the residents in $I$, and let $H = \{h_1, \ldots, h_m\}$ be the hospitals in $I$. Suppose $M$ is a weakly stable matching in $I$. Let $I'$ be any member of $\mathcal{R}_M(I) \cap \mathcal{H}_M(I)$. (Then $I'$ is an HR instance obtained by breaking the ties in $I$.) Suppose that $(r_i, h_j)$ blocks $M$ in $I'$. Then in $I'$, either $r_i$ is unassigned or prefers $h_j$ to $M(r_i)$, and either $h_j$ is undersubscribed or prefers $r_i$ to at least one member of $M(h_j)$. But the same is also true in $I$, in view of the way that ties were broken to form $I'$. Hence $(r_i, h_j)$ blocks $M$ in $I$, a contradiction.

Conversely suppose that $M$ is stable in some HR instance $I'$ obtained by breaking the ties in $I$. It is then straightforward to verify that if $(r_i, h_j)$ blocks $M$ in $I$, then the same pair blocks $M$ in $I'$, a contradiction. $\square$

Lemma 3.1 and Theorem 1.9 therefore indicate that a weakly stable matching in an HRT instance $I$ can be found by breaking the ties arbitrarily in $I$ to obtain an HR instance $I'$, and then applying the RGS algorithm. We thus obtain:

**Theorem 3.2 ([418, 419]).** *Every instance $I$ of* HRT *admits a weakly stable matching, and such a matching can be found in $O(m)$ time, where $m$ is the number of acceptable resident–hospital pairs in $I$.*

Men's preferences        Women's preferences
$m_1 : w_1 \ w_2$        $w_1 : (m_1 \ m_2)$
$m_2 : w_1 \ w_2$        $w_2 : m_1 \ m_2$

Fig. 3.1   An instance of SMT with no man-optimal weakly stable matching

## 3.2.2   *Absence of a lattice structure*

As mentioned in Sec. 1.3.3, the set of stable matchings for a given instance of SM forms a distributive lattice. However, in the case of weak stability, this structure is absent (under the "usual" definitions of meet and join as described in Sec. 1.3.3) even for SMT. This was first observed by Roth [498], who gave an example SMT instance of size 3 that admits no man-optimal weakly stable matching. Figure 3.1 shows an SMT instance $I$ of size 2 with the same property (in a preference list, agents within parentheses are tied). Here, $M_1 = \{(m_1, w_1), (m_2, w_2)\}$ and $M_2 = \{(m_1, w_2), (m_2, w_1)\}$ are the two weakly stable matchings in $I$. Since man $m_1$ has his first-choice partner in $M_1$ and his second-choice partner in $M_2$, whereas man $m_2$ has his second-choice partner in $M_1$ and his first-choice partner in $M_2$, no man-optimal weakly stable matching in $I$ exists.

The absence of a lattice structure for weakly stable matchings in instances of SMT is a strong indicator that other structural results (such as Theorem 1.11) and efficient algorithms that apply in the case of SMI do not carry over to SMT and SMTI. As we shall see in the forthcoming subsections, given an instance of SMTI, although a weakly stable matching always exists and can be computed in linear time, whenever any additional constraints are placed on the weakly stable matching to be found (such as requiring a maximum cardinality, minimum regret or egalitarian weakly stable matching), NP-hardness, and in some cases strong inapproximability results, prevail.

## 3.2.3   *Sizes of weakly stable matchings*

We begin by noting that weakly stable matchings can have different sizes, given an instance of SMTI. Figure 3.2 shows an SMTI instance in which there are two weakly stable matchings, namely $M_1 = \{(m_1, w_2), (m_2, w_1)\}$ of size 2 and $M_2 = \{(m_1, w_1)\}$ of size 1. Clearly this instance may be replicated to yield an arbitrarily large instance of SMTI having two weakly stable matchings $M$ and $M'$, where $|M| = 2|M'|$. (In fact this bound of 2 is the worst possible — see Sec. 3.2.6.1 for further details.)

Men's preferences      Women's preferences
$m_1 : w_1 \quad w_2$        $w_1 : (m_1 \quad m_2)$
$m_2 : w_1$               $w_2 : m_1$

Fig. 3.2   An instance of SMTI with weakly stable matchings of sizes 1 and 2

We now give a structural result that does hold for weakly stable matchings in SMTI, namely the interpolation of weakly stable matchings. That is, we show that, given an SMTI instance and weakly stable matchings of sizes $p$ and $r$, we may find in polynomial time a weakly stable matching of size $q$, for each $p < q < r$. Our starting point is the following lemma, first proved in Ref. [418]. Henceforth, we denote by $s(I)$ the size of the stable matchings in an SMI instance $I$ (recall from Theorem 1.11 that all stable matchings in $I$ have the same size).

**Lemma 3.3 ([418]).** *Let $I$ and $I'$ be two instances of* SMI *with the same set of men and women, such that exactly one agent's preference list in $I$ differs in $I'$. Then $|s(I) - s(I')| \leq 1$.*

**Proof.** Let $a_k$ be the agent whose preference list in $I$ differs in $I'$. Let $M$ and $M'$ be stable matchings in $I$ and $I'$ respectively. Let $G = M \oplus M'$. The connected components of $G$ are paths and cycles whose edges alternate between $M$ and $M'$. Suppose that $G$ has a component that is an odd-length path which does not contain $a_k$ — say it is $(m_1, w_1), (w_1, m_2), \ldots, (m_r, w_r)$, where without loss of generality $(m_i, w_i) \in M$ $(1 \leq i \leq r)$ and $(m_{i+1}, w_i) \in M'$ $(1 \leq i \leq r - 1)$.

Clearly, because of the way in which $G$ was constructed, both $m_1$ and $w_r$ are unassigned in $M'$. If $w_1$ prefers $m_1$ to $m_2$ then $(m_1, w_1)$ blocks $M'$, so $w_1$ prefers $m_2$ to $m_1$. If $m_2$ prefers $w_1$ to $w_2$ then $(m_2, w_1)$ blocks $M$. Similarly, for each $i$ $(1 \leq i \leq r - 1)$, $w_i$ prefers $m_{i+1}$ to $m_i$, and $m_{i+1}$ prefers $w_{i+1}$ to $w_i$. It follows that $(m_r, w_r)$ blocks $M'$, a contradiction.

Hence $G$ contains at most one alternating path of odd length, and an easy counting argument establishes the lemma.                                □

Lemmas 3.3 and 3.1 may be used to demonstrate our interpolation result, first proved in Ref. [418].

**Theorem 3.4 ([418]).** *Weak stability is an interpolating invariant, i.e., if a given instance $I$ of* SMTI *has weakly stable matchings of sizes $p$ and $r$, and $p < q < r$, then $I$ also has a weakly stable matching of size $q$, and such a matching can be constructed in polynomial time.*

**Proof.** Let $M$ and $M'$ be weakly stable matchings of sizes $p$ and $r$ respectively in $I$, and let $I_M$ and $I_{M'}$ be instances of SMI obtained by breaking the ties in $I$ so that $M$ and $M'$ are stable in $I_M$ and $I_{M'}$ respectively (note that $I_M$ and $I_{M'}$ exist by Lemma 3.1). Suppose that the preference lists of $t$ agents in $I_M$ differ in $I_{M'}$. Let $a_1, \ldots, a_t$ be these agents, and let $P_i$ be the preference list of $a_i$ in $I_{M'}$ ($1 \leq i \leq t$).

There exists a sequence $I_M = I_0, I_1, I_2, \ldots, I_t = I_{M'}$ of instances of SMI such that, for each $i$ ($1 \leq i \leq t$), $I_i$ is obtained from $I_{i-1}$ by giving agent $a_i$ the preference list $P_i$, and by giving every other agent the same preference list as in $I_{i-1}$. Let $s_i = s(I_i)$ ($0 \leq i \leq t$). Then by Lemma 3.3, successive entries in the sequence $s_0, s_1, \ldots, s_t$ differ by at most 1, and hence there is some $i$ ($1 \leq i \leq t - 1$) such that $s_i = r$.

Note that $t \leq 2n$ (where $n$ is the size of $I$), and thus a weakly stable matching in $I$ of size $r$ can be found in $O(m \log n)$ time using a binary search, where $m$ is the number of acceptable man–woman pairs in $I$. $\quad\square$

We finally remark that Theorem 3.4 carries over to HRT by Theorem 3.11 (see Sec. 3.2.6.5).

### 3.2.4   *NP-hardness of* MAX SMTI

As described in Sec. 3.2.3, finding a weakly stable matching $M$ in an instance $I$ of SMTI is equivalent to finding some instance $I'$ of SMI, obtained from $I$ by breaking the ties in some way, in which $M$ is stable. However, as the example in Fig. 3.2 shows, different ways of breaking the ties in $I$ can give rise to instances of SMI that admit stable matchings of different sizes.

In almost all practical situations, a larger stable matching is preferable to a smaller one. Typically, in a centralised matching scheme, an unassigned resident may well be disappointed, and possibly disillusioned, at being unassigned, and will have to enter some secondary process that allocates them to unfilled places, such as the so-called "scramble" that follows the NRMP match [602]. Hence it is desirable to find some strategy of breaking the ties that gives rise to stable matchings that are as large as possible. Unfortunately, as we show in this section, such a strategy is unlikely to have polynomial-time complexity.

Let COM SMTI denote the problem of deciding whether a given instance of SMTI admits a *complete* weakly stable matching (i.e., a weakly stable matching in which all men and women are assigned.) We now show that COM SMTI is NP-complete, as first proved in Ref. [332]. To do this, we use a reduction from a problem relating to maximal matchings in graphs. Recall from Sec. 1.2 the definition of MIN MM-D. A related decision problem is EXACT MM, which asks whether, given a graph $G$ and an integer $K$, $G$ admits a maximal matching of size exactly $K$. It turns out that MIN MM-D and EXACT MM are polynomially equivalent, which yields NP-completeness for EXACT MM in bipartite graphs, as we now demonstrate.

**Lemma 3.5 ([332, 419]).** EXACT MM *is NP-complete, even for bipartite graphs.*

**Proof.** Clearly EXACT MM belongs to NP. To show NP-hardness, we reduce from MIN MM-D restricted to bipartite graphs, which is NP-complete by Theorem 1.7. Let $G$ (a bipartite graph) and $K$ (a positive integer) be an instance of MIN MM-D. Without loss of generality we may assume that $K \leq \beta^+(G)$, where $\beta^+(G)$ denotes the size of a maximum matching of $G$. Suppose that $G$ admits a maximal matching $M$, where $|M| = k \leq K$. If $k = K$, we are done. Otherwise suppose that $k < K$. We note that maximal matchings satisfy the interpolation property [276] (i.e., $G$ has a maximal matching of size $j$, for $k \leq j \leq \beta^+(G)$) and hence $G$ has a maximal matching of size $K$. The converse is clear. $\qquad \square$

**Theorem 3.6 ([332, 419]).** COM SMTI *is NP-complete.*

**Proof.** Clearly COM SMTI belongs to NP. To show NP-hardness, we transform from EXACT MM restricted to bipartite graphs, which is NP-complete by Lemma 3.5. Hence let $G = (U, W, E)$ (a bipartite graph) and $K$ (a positive integer) be an instance of EXACT MM. Let $U = \{m_1, m_2, \ldots, m_s\}$ and $W = \{w_1, w_2, \ldots, w_t\}$. Without loss of generality assume that $K \leq \min\{s, t\}$ (for otherwise the EXACT MM instance trivially has a "no" answer).

We construct an instance $I$ of COM SMTI as follows: let $U \cup X$ be the set of men, and let $W \cup Y$ be the set of women, where $X = \{x_1, x_2, \ldots, x_{t-K}\}$ and $Y = \{y_1, y_2, \ldots, y_{s-K}\}$. For any $m_i \in U$, let $W_i \subseteq W$ denote the vertices adjacent to $m_i$ in $G$. Similarly for any $w_j \in W$, let $U_j \subseteq U$ denote the vertices adjacent to $w_j$ in $G$. Create preference lists for each agent as follows:

$$
\begin{aligned}
m_i &: (W_i) \ [Y] & (1 \leq i \leq s) \\
x_i &: [W] & (1 \leq i \leq t - K) \\
w_j &: (U_j) \ [X] & (1 \leq j \leq t) \\
y_j &: [U] & (1 \leq j \leq s - K)
\end{aligned}
$$

In a given preference list, the symbol $(S)$ denotes a tie containing all members of $S$, and the symbol $[S]$ denotes all members of the set $S$ listed in strict order, in increasing subscript order, from the point at which the symbol appears. We claim that $G$ has a maximal matching of size $K$ if and only if $I$ has a complete weakly stable matching.

For, suppose that $G$ has a maximal matching $M$ where $|M| = K$. We construct a matching $M'$ in $I$ as follows. Initially let $M' = M$. In $I$, there remain $s - K$ men in $U$ who are unassigned in $M'$, and $t - K$ women in $W$ who are unassigned in $M'$. Denote these men and women respectively by

$$m_{p_1}, m_{p_2}, \ldots, m_{p_{s-K}}, \text{ where } p_1 < p_2 < \cdots < p_{s-K},$$

and

$$w_{q_1}, w_{q_2}, \ldots, w_{q_{t-K}}, \text{ where } q_1 < q_2 < \cdots < q_{t-K}.$$

Add $(m_{p_i}, y_i)$ to $M'$ $(1 \leq i \leq s-K)$ and add $(x_j, w_{q_j})$ to $M'$ $(1 \leq j \leq t-K)$. Clearly $M'$ is a complete matching in $I$. We claim that $M'$ is weakly stable in $I$. For, as $M$ is maximal in $G$, clearly no member of $U \times W$ can block $M'$. Suppose $(m_i, y_j) \in U \times Y$ blocks $M'$. Then $(m_i, y_k) \in M'$ for some $y_k \in Y$ such that $j < k$. But then $(m_r, y_j) \in M'$ for some $m_r \in U$ such that $r < i$, so $(m_i, y_j)$ does not block $M'$ after all. Similarly no member of $X \times W$ blocks $M'$. Hence $M'$ is weakly stable in $I$.

Conversely suppose that $M'$ is a complete weakly stable matching in $I$. Let $M = M' \cap E$. Since each of the $t - K$ men in $X$ is assigned in $M'$ to a woman in $W$, and each of the $s - K$ women in $Y$ is assigned in $M'$ to a man in $X$, it follows that

$$|M| = |M'| - (t - K) - (s - K) = (s + t - K) - (s + t - 2K) = K.$$

Finally, suppose that $M$ is not maximal in $G$. Then $M \cup \{\{m_i, w_j\}\}$ is a matching in $G$, for some $\{m_i, w_j\} \in E$, where $m_i \in U$ and $w_j \in W$. Hence $(m_i, y_l) \in M'$ for some $y_l \in Y$ and $(x_k, w_j) \in M'$ for some $x_k \in X$. It follows that $(m_i, w_j)$ blocks $M'$ in $I$, a contradiction.                                  $\square$

**Corollary 3.7.** MAX SMTI *is NP-hard.*

Using more intricate reductions, the NP-completeness of COM SMTI (and hence the NP-hardness MAX SMTI) has been demonstrated for some highly restricted cases. For example COM SMTI is NP-complete even if each man's list is strictly ordered, and even if each woman's list is either strictly ordered or is a tie of length 2 [419].

The case where preference lists are of bounded length is of practical interest; typically in applications, the members of at least one set of agents have "short" preference lists (for example in the case of SFAS, until recently, each resident was asked to rank up to 6 hospitals in order of preference.) COM SMTI remains NP-complete even if each preference list is of length at most 3, and each man's list is strictly ordered [325, 444]. By contrast, MAX SMTI is solvable in polynomial time when the preference lists of one sex are of length at most 2 [325]. It is currently open as to whether MAX HRT is polynomial-time solvable the preference list of each resident is of length at most 2, and the preference lists of the hospitals are unbounded.

Also of practical significance is the case where the preference lists on one or both sides of an SMTI instance are derived from one or two master lists. It turns out that COM SMTI is NP-complete, even if each man's preference list is derived from a strictly-ordered master list of women, and each woman's preference list is derived from a master list of men that contains only one tie [329]. NP-completeness also holds in the case that the master list of women is strictly ordered, and the master list of men contains ties of length 2 only (though in general more than one tie) [329].

A further NP-complete case is where the SMTI instance has *symmetric preferences* (that is, for any acceptable man–woman pair $(m_i, w_j)$, $rank(m_i, w_j) = rank(w_j, m_i)$) [470, 27].


### 3.2.5   *Parameterized complexity of* MAX SMTI

Marx and Schlotter [427] studied the parameterized complexity of MAX SMTI under various parameterizations of a given SMTI instance $I$, as follows:

* $\kappa_1$: the number of ties in $I$;
* $\kappa_2$: the maximum length of a tie in $I$;
* $\kappa_3$: the total length of the ties in $I$.

The authors showed that MAX SMTI with parameterization $\kappa_3$ belongs to FPT. By contrast its decision version with parameterization $\kappa_1$ is W[1]-hard, even if ties belong to the women's lists only. The authors also proved that if W[1]$\neq$FPT, there is no FPT local search algorithm for MAX SMTI with parameterization $\ell$, the size of the neighbourhood to be searched, even if $\kappa_2 = 2$ and ties occur in the women's lists only.

### 3.2.6 Approximability of MAX SMTI and MAX HRT

#### 3.2.6.1 Overview of approximability results for MAX SMTI

The NP-hardness of MAX SMTI implies that the approximability of this problem is of interest. It is straightforward to show that the problem admits an approximation algorithm with a performance guarantee of 2 [419]: namely, break the ties in a given instance $I$ of SMTI arbitrarily and run the Gale–Shapley algorithm in the resulting instance of SMI to obtain a stable matching $M$. The performance guarantee of 2 follows from the fact that if $M'$ is an arbitrary weakly stable matching, each of $M$ and $M'$ is maximal in the underlying bipartite graph $G$ of $I$, and any two maximal matchings in $G$ differ in size by at most a factor of 2 [399][1]. A number of improved approximation algorithms for versions of MAX SMTI have recently been proposed.

For the general case, Iwama *et al.* [336] gave an algorithm with a performance guarantee of $2 - (c \log n)/n$, where $n$ is the size of the given instance and $c$ is a positive constant. This was later improved to $2 - c'/\sqrt{n}$ [341], where $c'$ is a positive constant such that $c' \leq 1/4\sqrt{6}$. Iwama *et al.* [340] gave the first approximation algorithm for the general case with a constant performance guarantee better than 2, namely $\frac{15}{8}$. This performance guarantee was improved to $\frac{5}{3}$ by Király [385], and then further improved to $\frac{3}{2}$ by McDermid [437]. Recently Király [386] and Paluch [472] independently derived approximation algorithms for MAX SMTI with performance guarantee $\frac{3}{2}$, which were faster (running in $O(m)$ time as opposed to the $O(n^{3/2}m)$ running time of McDermid's algorithm, where $m$ is the number of acceptable pairs in a given instance), and claimed by the authors to be simpler, than McDermid's.

As far as special cases are concerned, Halldórsson *et al.* [272] gave a $(2/(1 + t^{-2}))$-approximation algorithm for the case where all ties are on one side, and are of length at most $t$ — so, for example, this gives a bound of $\frac{8}{5}$ when all ties are of length 2. If ties are on both sides and restricted to be of length 2, a bound of $\frac{13}{7}$ is shown in Ref. [272]. Halldórsson *et al.* [271] also described a randomised algorithm with an expected guarantee of $\frac{10}{7}$ for the same special case. For the case where ties are on one side only,

---

[1]For a short proof of this, observe that each of $S_1 = V(M)$ and $S_2 = V(M')$ is a vertex cover in $G$, where $V(M)$ is the set of vertices that are matched by $M$. Let $S$ be a minimum vertex cover in $G$. Then $S$ contains at least one vertex from each edge of $M$, whilst $S_1$ contains two vertices corresponding to each edge of $M$, so $|S| \geq |S_1|/2$. Hence $|S_1| \leq 2|S_2|$, for otherwise $|S| \geq |S_1|/2 > |S_2|$, a contradiction. Since $|S_1| = 2|M|$ and $|S_2| = 2|M'|$, the result follows.

there is at most one tie per list, and each tie occurs at the *tail* of some list (the *tail* of an agent $a_i$'s list is the set of one or more agents, tied in its list, to whom it prefers all other agents in its list), Irving and Manlove [322] described an approximation algorithm with a performance guarantee of $\frac{5}{3}$. This bound was later improved to $\frac{3}{2}$ by Király [385], and even more recently to $\frac{25}{17}$ by Iwama *et al.* [347], in both cases for the more general SMTI restriction that ties are on one side only, but the number and location of ties in a given list on that side is unrestricted.

There is also an approximability result for MAX SMTI that gives rise to an additive error bound. If $I$ is an instance of MAX SMTI and $s^+(I)$ denotes the maximum size of a weakly stable matching in $I$, any weakly stable matching $M$ in $I$ satisfies $|M| \geq s^+(I) - t(I)$, where $t(I)$ is the number of preference lists in $I$ that contain at least one tie [267].

Many of these positive results also carry over to the MAX HRT case — see Sec. 3.2.6.5 for further details.

From the inapproximability point of view, Halldórsson *et al.* [267] showed that MAX SMTI is not approximable within $\delta$, for some $\delta > 1$, unless P=NP (see also Ref. [268]). This result holds even if each man's list is of length at most 7 and each woman's list is of length 4 [267], and in addition to these restrictions, even if each preference list is derived from two master lists of the men and women [329].

Irving *et al.* [325] proved that MAX SMTI is hard to approximate within some constant factor for shorter length preference lists on the men's side.

**Theorem 3.8 ([325]).** MAX SMTI *is not approximable within $\delta$, for some $\delta > 1$, unless P=NP. The result holds even if each man's list is of length at most 3, and each woman's list is of length at most 4.*

However in the context of each of the aforementioned inapproximability results, the constant $\delta$ is very close to 1. Halldórsson *et al.* [272] strengthened these results for the case of unbounded length preference lists and gave a lower bound of $\frac{21}{19} - \varepsilon$ on any approximation algorithm, for any $\varepsilon > 0$ (assuming P$\neq$NP). This result holds even if each man's list is strictly ordered, and each women's list is strictly ordered or is a tie of length 2. Yanagisawa [588] improved the lower bound (albeit for the weaker case that ties can occur on both sides), as follows.

**Theorem 3.9 ([588]).** MAX SMTI *is not approximable within $\frac{33}{29}$ unless P=NP. The result holds even if each tie is of length 2.*

---

**Algorithm 3.1** Algorithm Király [386]

---

1: $M := \emptyset$;
2: **for each** man $m_i \in U$ **do**
3:     $secondChance(m_i) :=$ false;
4:     $exhausted(m_i) :=$ false;
5: **end for**
6: **while** some man $m_i \in U$ is unassigned in $M$ **and** $!exhausted(m_i)$ **do**
7:     $w_j :=$ most-preferred woman on $m_i$'s list;     {any one, if more than one}
8:     **if** $w_j$ is unassigned in $M$ **then**
9:         $M := M \cup \{(m_i, w_j)\}$;
10:    **else**
11:        $m_k := M(w_j)$;
12:        **if** $w_j$ prefers $m_i$ to $m_k$ **or** $precarious(w_j)$ **then**
13:            $reject(m_k, w_j)$;
14:            $M := M \cup \{(m_i, w_j)\}$;
15:        **else**
16:            $reject(m_i, w_j)$;
17:        **end if**
18:    **end if**
19: **end while**
20: **return** $M$;

---

Yanagisawa [588] also showed that if MAX SMTI is approximable within $\frac{4}{3} - \varepsilon$, then MIN VERTEX COVER (the problem of finding a vertex cover of minimum size, given a graph $G$) is approximable within $2 - \varepsilon$, for any $\varepsilon > 0$.[2]

#### 3.2.6.2 *Király's approximation algorithm*

Recently, Király [385, 386] described ingenious approximation algorithms for MAX SMTI and MAX HRT. In this subsection we describe his algorithm for general MAX SMTI (i.e., ties can occur on both sides) with performance guarantee $\frac{3}{2}$ [386].

Suppose we are given an instance of SMTI, where $U = \{m_1, m_2, \ldots, m_n\}$ is the set of men and $W = \{w_1, w_2, \ldots, w_n\}$ is the set of women. Algorithm Király, shown in Algorithm 3.1, is a variant of the MEGS algorithm for SMI [261, Sec. 1.2.4]. The algorithm operates as follows. It will ultimately return a matching $M$. Initially $M$ is the empty set; consequently each

---

[2]The truth of the Unique Games Conjecture (UGC) [376] would imply that, if P$\neq$NP, MIN VERTEX COVER is not approximable within $2 - \varepsilon$, for any $\varepsilon > 0$ [377]. Some authors therefore state that it is *UGC-hard* to approximate MIN VERTEX COVER within $2 - \varepsilon$, for any $\varepsilon > 0$.

---

**Algorithm 3.2** Algorithm reject (method for Algorithm Király) [386]

**Require:** man $m_i \in U$ and woman $w_j \in W$
1: $M := M \backslash \{(m_i, w_j)\}$;
2: **if** !precarious($w_j$) **then**
3:    delete($m_i, w_j$);
4:    **if** $m_i$'s list is empty **then**
5:       **if** $secondChance(m_i)$ **then**
6:          $exhausted(m_i) :=$ true;
7:       **else**
8:          $secondChance(m_i) :=$ true;
9:          recover($m_i$);
10:       **end if**
11:    **end if**
12: **end if**

---

man and woman is initially unassigned in $M$. Also, each man $m_i$ has two booleans, $secondChance(m_i)$ and $exhausted(m_i)$, each of which is false initially. Intuitively, if $secondChance(m_i)$ is true (in which case we say that $m_i$ *has a second chance*), this means that $m_i$ has proposed to (and has been rejected by) every woman on his preference list, in which case he is given a second chance to propose to every woman on his list. If $exhausted(m_i)$ is true (in which case we say that $m_i$ *is exhausted*), this means that $m_i$ has already had such a second chance, but has again been rejected by every woman on his list.

In what follows, the definition of *prefers* needs to be adapted for the purposes of this algorithm. For a man $m_i$ and for two women $w_j$ and $w_k$, we say that $m_i$ *prefers* $w_j$ to $w_k$ if either (i) $rank(m_i, w_j) < rank(m_i, w_k)$ (i.e., $m_i$ prefers $w_j$ to $w_k$ in the usual sense) or (ii) $rank(m_i, w_j) = rank(m_i, w_k)$ and $w_j$ is unassigned in $M$ whilst $w_k$ is assigned in $M$. Thus, in case (ii), if $m_i$ is indifferent (in the usual sense) between two women, one unassigned and the other assigned in $M$, then he gives priority to the unassigned woman. For a woman $w_j$ and for two men $m_i$ and $m_k$, we say that $w_j$ *prefers* $m_i$ to $m_k$ if either (i) $rank(w_j, m_i) < rank(w_j, m_k)$ (i.e., $w_j$ prefers $m_i$ to $m_k$ in the usual sense) or (ii) $rank(w_j, m_i) = rank(w_j, m_k)$ and $secondChance(m_i)$ is true whilst $secondChance(m_k)$ is false. Thus, in case (ii), if $w_j$ is indifferent (in the usual sense) between two men, one who has a second chance and one who does not, then she gives priority to the former man.

The while loop of the algorithm iterates as long as we can find an unassigned man $m_i$ who is not exhausted. Assuming such an $m_i$ exists, we let $w_j$ be the most-preferred woman on his preference list (according to his new definition of *prefers*). Intuitively $m_i$ proposes to $w_j$. If $w_j$ is unassigned in $M$ then she accepts the proposal and $(m_i, w_j)$ is added to $M$. Otherwise she is assigned in $M$ — let $m_k$ be her partner in $M$. The algorithm tests whether $w_j$ prefers $m_i$ to $m_k$ (according to her new definition of *prefers*) or whether precarious($w_j$) holds (i.e., whether $w_j$ is *precarious*), which holds if $M(w_j) = m_k$ prefers some woman to $w_j$ (again, according to his new definition of *prefers*)[3]. If either of these is true, $w_j$ rejects $m_k$, which is carried out by Algorithm reject, shown in Algorithm 3.2, and then $(m_i, w_j)$ is added to $M$. Otherwise, $w_j$ rejects $m_i$, which is also carried out by Algorithm reject.

Algorithm reject works as follows. Assume man $m_i$ and woman $w_j$ are passed as parameters. The pair $(m_i, w_j)$ is removed from $M$ (assuming it was in $M$). If $w_j$ is precarious then nothing further happens (there are no deletions from the preference lists in this case). Otherwise $w_j$ is deleted from $m_i$'s list and vice versa. If $m_i$'s list becomes empty and $m_i$ had already had a second chance, then $m_i$ becomes exhausted. Otherwise $m_i$ is given a second chance to propose to all the women on his preference list — in particular, the method recover($m_i$) is called, which reinstates every deleted woman $w_k$ to $m_i$'s list (and equivalently reinstates $m_i$ to the list of each such woman $w_k$).

Once the while loop of Algorithm Király terminates, the final matching $M$ is returned. The following result, proved in Ref. [386], indicates that $M$ is a weakly stable matching of size at least two-thirds of that of a maximum weakly stable matching.

**Theorem 3.10 ([386]).** *Algorithm* Király *is a $\frac{3}{2}$-approximation algorithm for* MAX SMTI.

### 3.2.6.3 *Comparison of approximation algorithms for* MAX SMTI

Podhradský [483] compared empirically a range of approximation algorithms for MAX SMTI for the following cases: (i) ties are on both sides, (ii) ties belong to the women's lists only, and (iii) men's lists are strictly ordered, and each woman's list is either strictly ordered or has one tie at the tail. The third case corresponds to the practical scenario (which used to be relevant in the context of the Scottish Foundation Allocation Scheme [604]) in which women rank a subset of men in strict order of preference, and then express indifference among the remaining men that they find acceptable.

---

[3]This possibility can arise if, among the most-preferred tie comprising undeleted women on $m_k$'s list, there are two single women, and $m_k$ became provisionally assigned to one of them.

The approximation algorithms that Podhradský featured in his experiments were as follows:

- Case (i):

  - IMY (performance guarantee $\frac{15}{8}$) [340];
  - Király–GSA2 (performance guarantee $\frac{5}{3}$) [385];
  - McDermid (performance guarantee $\frac{3}{2}$) [437];
  - Paluch (performance guarantee $\frac{3}{2}$) [472];
  - RandBrk [271] [4];
  - ShiftBrk (performance guarantee $2/(1 + t^{-2})$ where $t$ is the longest tie length) [272];

- Case (ii):

  - all of the algorithms from Case (i);
  - GSA–LP (performance guarantee $\frac{25}{17}$) [343];
  - Király–GSA1 (performance guarantee $\frac{3}{2}$) [385];

- Case (iii):

  - all of the algorithms from Cases (i) and (ii);
  - SSMTIApprox (performance guarantee $\frac{5}{3}$) [322].

The author also implemented an algorithm for finding a maximum stable matching based on an integer programming formulation of MAX SMTI. The algorithms were compared against one another on a range of randomly-generated SMTI instances of varying sizes corresponding to the above restrictions, where the lengths of the ties were bounded in some cases. The lengths of the preference lists do not appear to have been bounded, though this detail is unclear from the accompanying discussion.

The results for case (i) indicate that McDermid's algorithm produced the largest stable matchings and was the fastest on average. For case (ii), no single algorithm was the clear-cut winner, although in the experiments performed, ShiftBrk, GSA–LP, Király–GSA1 and McDermid (the latter two algorithms are essentially the same in this case) performed well. In case (iii), it is interesting to note that, although SSMTIApprox performed reasonably well, it was generally beaten by other approximation algorithms for the instances generated, despite being specifically designed for this restriction of SMTI.

---

[4]This approximation algorithm was in fact mainly analysed in Ref. [271] for the restriction of Case (ii) in which each tie is of length 2, however Podhradský analysed the algorithm in the general case in which ties are on both sides and their length is unbounded.

Király's newest approximation algorithm for general MAX SMTI [386] came too late to be included in this study.

### 3.2.6.4  *Heuristics for* MAX HRT

Two heuristics for the special case of MAX HRT in which ties belong to hospitals' lists only were presented by Irving and Manlove in Ref. [323]. The first of these (Algorithm R) is an extension of the RGS algorithm for HR and employs network flow to attempt to optimise the size of the constructed weakly stable matching. The second (Algorithm H) is a variant of the HGS algorithm for HR and utilises maximum matching in bipartite graphs to attempt to maximise the cardinality of its weakly stable matching. Both of these heuristics were compared empirically with an implementation of Király's $\frac{3}{2}$-approximation algorithm for this restricted version of HRT (Algorithm HRGSA1 from Ref. [385]) and with two simple random tie-breaking heuristics using both real-world and randomly-generated data [322].

For example, data arising from the 2006 SFAS run involved 759 residents and 53 hospitals with a total capacity of 801. When each algorithm was run for 5 minutes, the maximum sizes of weakly stable matchings found by each of Algorithm R, Algorithm H, Király's algorithm and the two random tie-breaking heuristics were 755, 753, 753, 746 and 744 respectively. Empirical results for real data from the 2007 and 2008 SFAS runs, and from randomly-generated instances, confirmed a pattern of behaviour, namely that the largest weakly stable matching found by Algorithm R was consistently larger than, or at least as large as, those found by the other heuristics.

We remark that Algorithms R and H were not considered by Podhradský in his empirical investigation as summarised in the previous subsection.

Local search heuristics for MAX SMTI were presented by Gelain *et al.* [246, 247]. The authors generated random SMTI instances of size 100 with varying density of ties in the preference lists, and varying levels of incompleteness of the lists. They showed that a heuristic based on satisfying so-called *undominated* blocking pairs almost always finds a complete stable matching. See also Refs. [251, 252, 244, 248].

### 3.2.6.5  *"Cloning" hospitals*

Clearly the NP-hardness and inapproximability results proved in Secs. 3.2.4 and 3.2.6 carry over to MAX HRT, the problem of finding a maximum weakly stable matching, given an instance of HRT. We now show that, under certain conditions, the approximation algorithms for MAX SMTI described in Sec. 3.2.6.1 can also be applied to instances of MAX HRT, achieving the same performance guarantee.

It is known that, by identifying residents with men, and "cloning" each hospital into a number of women equal to its capacity, an instance $I$ of HR may be transformed in polynomial time to an instance $I^*$ of SMI such that there is a bijective function between the set of stable matchings in $I$ and those in $I^*$ [261, p.38] (see also Ref. [514, pp.131–132]). As we now show, a similar reduction holds from HRT to MAX SMTI in the case of weakly stable matchings, preserving matching cardinality, however the correspondence is no longer a bijective function in general.

**Theorem 3.11.** *Given an instance $I$ of HRT, we may construct in $O(n_1 + c_{\max}m)$ time an instance $I'$ of SMTI such that a weakly stable matching $M$ in $I$ can be transformed in $O(c_{\max}m)$ time to a weakly stable matching $M'$ in $I'$ where $|M| = |M'|$, and conversely, where $n_1$ is the number of residents, $c_{\max}$ is the maximum hospital capacity and $m$ is the number of acceptable resident–hospital pairs in $I$.*

**Proof.** Let $I$ be an instance of HRT in which $R = \{r_1, r_2, \ldots, r_{n_1}\}$ is the set of residents and $H = \{h_1, h_2, \ldots, h_{n_2}\}$ is the set of hospitals. Let $c_j$ be the capacity of hospital $h_j \in H$. We form an instance $I'$ of SMTI as follows. Each resident in $I$ corresponds to a man in $I'$. Each hospital $h_j \in H$ gives rise to $c_j$ women (hospital "clones") in $I'$, denoted by $h_j^1, h_j^2, \ldots, h_j^{c_j}$, each of whom has the same preference list as $h_j$ in $I'$. Each man $r_i \in R$ starts off with the same preference list in $I'$ as he has in $I$. We then replace each entry $h_j$ on his list by the $c_j$ women $h_j^1, h_j^2, \ldots, h_j^{c_j}$. These women are listed in strict order (with increasing superscripts) in the case that $h_j$ is not involved in a tie in $r_i$'s preference list in $I$, otherwise the women are simply added to that tie in $I'$.

Now let $M$ be a weakly stable matching in $I$. We form a matching $M'$ in $I'$ as follows. For each $h_j \in H$, let $r_{j,1}, r_{j,2}, \ldots, r_{j,x_j}$ be the set of residents assigned to $h_j$ in $M$, where $x_j \leq c_j$, and $k < l$ implies that $h_j$ prefers $r_{j,k}$ to $r_{j,l}$ or is indifferent between them. Add $(r_{j,k}, h_j^k)$ to $M'$ ($1 \leq k \leq x_j$). Clearly $M'$ is a matching in $I'$ such that $|M'| = |M|$, and it is straightforward to verify that $M'$ is weakly stable in $I'$.

Conversely let $M'$ be a weakly stable matching in $I'$. We form a matching $M$ in $I$ as follows. For each $(r_i, h_j^k) \in M'$, add $(r_i, h_j)$ to $M$. Clearly $M$ is a weakly stable matching in $I$ such that $|M| = |M'|$.

The stated time complexities follow from the fact that $I'$ has $O(n_1 + C)$ agents and $O(c_{max}m)$ acceptable man–woman pairs, where $C$ is the total capacity of the hospitals in $I$.                                                                    □

An immediate consequence of Theorem 3.11 is that, given an approximation algorithm $A$ for MAX SMTI with performance guarantee $\delta$, for some constant $\delta > 1$, we may obtain (except in certain cases, as we will describe shortly) an approximation algorithm for MAX HRT with the same performance guarantee as follows. Starting from an instance $I$ of MAX HRT, simply apply $A$ to the instance $I'$ of MAX SMTI as constructed by the proof of Theorem 3.11, and map the obtained weakly stable matching $M'$ in $I'$ to a weakly stable matching $M$ in $I$ such that $|M| = |M'|$. The special cases that constitute an exception to this arise when $A$ depends on certain properties of the preference lists that are not preserved under "cloning": an example of such a property is the length of the ties in the residents' lists, which are in general inflated under such a transformation. This implies that the approximation algorithms for MAX SMTI mentioned in the previous subsection, except for those with performance guarantees $2/(1 + t^{-2})$, $\frac{10}{7}$ and $\frac{13}{7}$ [271, 272], also yield approximation algorithms for corresponding versions of MAX HRT with the same performance guarantees.

### 3.2.7   Other problems involving weak stability

In this section we describe some additional problems involving the computation of weakly stable matchings — many of these problems turn out to be NP-hard.

**Minimum weakly stable matchings.**   Define MIN SMTI to be the problem of finding a minimum cardinality weakly stable matching (henceforth a minimum weakly stable matching), given an instance of SMTI. MIN SMTI is NP-hard, even if each tie occurs at the tail of some woman's list, there is at most one tie per list, and each tie is of length 2 [419]. Thus instances of SMTI give rise to minimisation and maximisation problems that are NP-hard for the same simultaneous restrictions — there are relatively few examples in the literature where this phenomenon occurs. Lower bounds for the approximability of MIN SMTI are given in Refs. [267, 587].

As noted in Sec. 3.2.6.1, a maximal matching is at most twice the size of a minimum maximal matching, and hence MIN SMTI is approximable within 2. Also, if $I$ is an instance of MIN SMTI and $s^-(I)$ denotes the size of a minimum weakly stable matching in $I$, any weakly stable matching $M$ in $I$ satisfies $|M| \leq s^-(I) + t(I)$, where $t(I)$ is the number of preference lists in $I$ that contain at least one tie [267]. This gives a form of approximation algorithm for MIN SMTI with an additive error bound.

**Weakly stable pairs.** It is unlikely that there is an efficient algorithm for finding all weakly stable pairs in $I$, for, it turns out that the problem of deciding whether a man–woman pair is weakly stable in $I$ is NP-complete. This holds even if $I$ is an instance of SMT in which the ties occur at the tails of lists and on one side only, there is at most one tie per list, and each tie is of length 2 [419].

Three additional NP-complete cases are where (i) $I$ is an instance of SMT with symmetric preferences [470], (ii) $I$ is an instance of SMT in which the preference lists on one side are identical, and ties occur on one side only, and (iii) $I$ is an instance of SMTI in which the preference lists on both sides are derived from two master lists of men and women, one of which is strictly ordered [329].

However if $I$ is an instance of SMT in which the preference lists on both sides are derived from two master lists of men and women (both of which may contain ties), the weakly stable pairs in $I$ can be found in $O(n + s)$ time, where $n$ is the size of $I$, and $s$ is the number of weakly stable pairs in $I$ [329].

**Minimum regret weakly stable matchings.** The problem of finding a minimum regret weakly stable matching, given an instance $I$ of SMT, is NP-hard and not approximable within $n^{1-\varepsilon}$, for any $\varepsilon > 0$, unless P=NP [419]. Here $n$ is the size of $I$. The result holds even if the ties occur on one side only, there is at most one tie per list, and each tie is of length 2 [419]. The lower bound for the inapproximability of this problem was strengthened to $\Omega(n)$ in Ref. [267] though without the restrictions involving the ties.

NP-hardness also holds even if $I$ is an instance of SMT in which the men's preference lists are derived from a single master list of women, which contains a tie at the tail, even if the women's lists are strictly ordered [329]. However, if there is no tie at the tail of this master list, the problem is solvable in $O(n^2)$ time (even if women's lists contain ties) [329]. Given an instance of SMT in which the preference lists on both sides are derived from two master lists of men and women, a minimum regret weakly stable matching can be found in $O(n)$ time [329]. NP-hardness also holds for SMT with symmetric preferences [470, 27].[5]

Marx and Schlotter [427] also showed that the problem belongs to FPT with parameterization $\kappa_3$, though there is no FPT approximation algorithm for the problem with parameterization $\kappa_1$ that has a performance guarantee of $n^{1-\varepsilon}$, for any $\varepsilon > 0$, unless W[1]=FPT (see Sec. 3.2.5 for the definitions of $\kappa_1$ and $\kappa_3$ in this context).

**Egalitarian weakly stable matchings.** The problem of finding an egalitarian weakly stable matching, given an instance $I$ of SMT, is NP-hard and not approximable within $n^{1-\varepsilon}$, for any $\varepsilon > 0$, unless P=NP [419]. Here $n$ is the size of $I$. The result holds even if the ties occur on one side only, there is at most one tie per list, and each tie is of length 2 [419]. The lower bound for the inapproximability of this problem was strengthened to $\Omega(n)$ in Ref. [267] though without the restrictions involving the ties.

NP-hardness also holds even if $I$ is an instance of SMT in which the men's preference lists are derived from a single master list of women, even if ties occur on one side only [329]. This particular restriction is, however, approximable within a factor of 3 [329]. Given an instance of SMT in which the preference lists on both sides are derived from two master lists of men and women, an egalitarian weakly stable matching can be found in $O(n)$ time [329]. NP-hardness also holds for SMT with symmetric preferences [470, 27] (see Footnote 5 on Page 146).

Marx and Schlotter [427] also show that the problem belongs to FPT with parameterization $\kappa_3$, though there is no FPT approximation algorithm for the problem with parameterization $\kappa_1$ that has a performance guarantee of $\delta n$, for some $\delta > 0$, unless W[1]=FPT, even if ties belong to the women's lists only (see Sec. 3.2.5 for the definitions of $\kappa_1$ and $\kappa_3$ in this context).

**Sex-equal weakly stable matchings.** The problem of finding a sex-equal weakly stable matching, given an instance $I$ of SMT, is NP-hard and not approximable within $\Omega(n)$ unless P=NP [267].

**Generation of weakly stable matchings.** It is not known whether there is an efficient algorithm for listing all weakly stable matchings, given an instance $I$ of SMT of size $n$. By *efficient*, we mean that the algorithm should have complexity $O(p(n) + kq(n))$, where $p$ and $q$ are polynomial

---

[5]This NP-hardness result holds for a slightly different definition of $rank$, namely $rank(m_i, w_j) = k$ if and only if $w_j$ belongs to the $k$th tie in $m_i$'s preference list, and similarly for $rank(w_j, m_i)$, for any acceptable pair $(m_i, w_j)$.

functions and $k$ is the number of weakly stable matchings in $I$. A partial result along these lines is, however, provided by Scott [523]. He showed that, given a weakly stable matching $M$ in $I$, we can, in polynomial time, find a weakly stable matching $M' \neq M$ if one exists, or else report that $M$ is unique. If $I$ is an instance of SMT in which the preference lists on both sides are derived from two master lists of men and women, all the weakly stable matchings in $I$ can be generated in $O(n + s + k \log n)$ time, where $s$ is the number of weakly stable pairs in $I$ [329].

**Pareto stable matchings.** In instances of SMTI or SRTI, Sotomayor [546] defined a *Pareto stable matching* to be a matching that is both weakly stable and Pareto optimal (note that it is straightforward to extend the Pareto optimality definition from Sec. 1.5.3 to the SMTI and SRTI contexts). In particular, in the SMTI case, a Pareto stable matching is Pareto optimal with respect to both men *and* women. Sotomayor remarked that a strongly stable matching is Pareto stable. She also gave an example SRTI instance with a weakly stable matching that is not Pareto stable, and a Pareto stable matching that is not strongly stable.

Erdil and Ergin [192] considered Pareto stable matchings in HRT. In order to define such matchings, we must first define Pareto optimality in the context of an HRT instance $I$. The definition of *prefers* relative to two matchings in $I$ for a resident $r_i \in R$ is the same as that given in Sec. 1.5.3 for an applicant in the context of an HA instance.

In order to define *prefers* for a hospital $h_j \in H$, we assume that hospitals' preferences over subsets are *responsive* [500] to their preferences over individuals (see also Sec. 5.4.4). That is, (i) for any $R' \subseteq A(h_j)$ such that $|R'| < c_j$ and for any $r_i \in A(h_j) \backslash R'$, $h_j$ prefers $R' \cup \{r_i\}$ to $R'$, and (ii) for any two subsets $R', R'' \subseteq A(h_j)$ such that $R'' = (R' \backslash \{r_i\}) \cup \{r_k\}$ for two distinct residents $r_i \in R'$ and $r_k \in R''$, $h_j$ prefers $R''$ to $R'$ if and only if $h_j$ prefers $r_i$ to $r_k$. (That is, if one set of assignees is obtained from another by adding an acceptable resident, then the hospital prefers the larger set of assignees, and if two sets of assignees differ by replacing one resident $r_i$ by another, $r_k$, then the hospital prefers the set with the most-preferred resident from among $r_i$ and $r_k$). In addition, we take the transitive closure of this definition to arrive at the final notion of *prefers* for $h_j$ over subsets of residents. Thus, given two matchings $M$ and $M'$ in $I$, $h_j$ *prefers* $M$ to $M'$ if and only if $h_j$ prefers $M(h_j)$ to $M'(h_j)$.

A matching $M$ in $I$ is then *Pareto optimal* if there is no other matching $M'$ in $I$ such that some agent in $I$ prefers $M'$ to $M$ and no agent in $I$

prefers $M$ to $M'$. $M$ is *Pareto stable* if it is both weakly stable and Pareto optimal in $I$.

Erdil and Ergin [192] defined the concepts of a *Pareto improvement cycle* and a *Pareto improvement chain* with respect to a given matching $M$ in an HRT instance $I$ as follows.

**Definition 3.12 ([192]).** *Let $I$ be an instance of* HRT *and let $M$ be a matching in $I$. A Pareto improvement cycle with respect to $M$ is a sequence of residents $r_0, \ldots, r_{k-1}$, for some $k \geq 2$, each assigned in $M$, such that*

*(1) for each $i$ ($0 \leq i \leq k-1$), $r_i$ prefers $M(r_{i+1})$ to $M(r_i)$ or is indifferent between them, and $M(r_{i+1})$ prefers $r_i$ to $r_{i+1}$ or is indifferent between them;*

*(2) there is some $j$ ($0 \leq i \leq k-1$) such that either $r_j$ prefers $M(r_{j+1})$ to $M(r_j)$ or $M(r_{j+1})$ prefers $r_j$ to $r_{j+1}$;*

*where addition is taken modulo $k$. Similarly a Pareto improvement chain with respect to $M$ is a sequence of residents $r_0, \ldots, r_{k-1}$, for some $k \geq 2$, and a hospital $h$ such that*

*(1) $r_0$ is unassigned in $M$, whilst for each $i$ ($1 \leq i \leq k-1$), $r_i$ is assigned in $M$;*

*(2) $r_0$ finds $M(r_1)$ acceptable;*

*(3) $h$ is undersubscribed in $M$;*

*(4) for each $i$ ($1 \leq i \leq k-2$), $r_i$ prefers $M(r_{i+1})$ to $M(r_i)$ or is indifferent between them;*

*(5) for each $i$ ($0 \leq i \leq k-2$) $M(r_{i+1})$ prefers $r_i$ to $r_{i+1}$ or is indifferent between them;*

*(6) $r_{k-1}$ prefers $h$ to $M(r_{k-1})$ or is indifferent between them.*

Erdil and Ergin showed that a weakly stable matching is Pareto stable if and only if it admits no Pareto improvement chain or Pareto improvement cycle. They gave an $O(n_1^3 C)$ algorithm for finding a Pareto stable matching, where $n_1$ is the number of residents and $C$ is the total capacity of the hospitals in $I$. This algorithm is based on repeatedly finding and applying Pareto improvement cycles and chains, starting from an arbitrary weakly stable matching; note that the weak stability of the matching is preserved after each such augmentation.

Chen [138] (see also Ref. [140]) described a strongly polynomial-time algorithm for finding a Pareto stable matching in an instance of the many–many stable marriage problem with ties and incomplete lists. An alternative algorithm for the same problem was given by Kamiyama [356].

Returning to the HRT setting, Erdil and Ergin [192] also considered the problem of finding a weakly stable matching that is Pareto optimal for the residents only — we call such a matching a *resident-Pareto stable matching*. That is, a weakly stable matching $M$ is resident-Pareto stable if and only if there is no matching $M'$ in which some resident prefers $M'$ to $M$, and no resident prefers $M$ to $M'$. They identified analogues of the notions of Pareto improvement cycles and chains for this context, which we refer to as *resident-Pareto improvement cycles* and *resident-Pareto improvement chains*[6] respectively. Note that, in contrast to the case for Pareto stable matchings, a hospital could be worse off after a resident-Pareto improvement cycle or chain is applied. The authors showed that a weakly stable matching is resident-Pareto stable if and only if it admits no resident-Pareto improvement chain or resident-Pareto improvement cycle. They gave an $O(n_1^3 n_2)$ algorithm for finding a resident-Pareto stable matching in $I$, where $n_2$ is the number of hospitals in $I$. As in the case of Pareto stable matchings, this algorithm is based on repeatedly finding and applying resident-Pareto improvement cycles and chains, starting from an arbitrary weakly stable matching; again, the weak stability of the matching is preserved after each such augmentation.

Erdil and Ergin [193] studied resident-Pareto stable matchings in instances of HRT where resident preference lists are strictly ordered. In such a setting, a weakly stable matching is resident-Pareto stable if and only if it admits no resident-Pareto improvement cycle (that is, resident-Pareto improvement chains need not be considered).[7] Erdil and Ergin showed that, for such an instance of HRT, a resident-Pareto stable matching can be found in $O(n_1 n_2 m)$ time, where $m$ is the number of acceptable pairs. Abdulkadiroğlu *et al.* [3] reported that, had Erdil and Ergin's algorithm for finding and applying resident-Pareto improvement cycles been executed on the preference data arising from the 2003–04 New York City High School Match, 6,854 students (equating to 10.5% of the 63,795 matched students) would have been matched with schools higher on their preference lists without any other student receiving a poorer school.

---

[6]Erdil and Ergin [192] used the terms *workers* and *firms*, rather than *residents* and *hospitals*, and they referred to *resident-Pareto improvement cycles* and *resident-Pareto improvement chains* as *stable worker improvement cycles* and *stable worker improvement chains* respectively.

[7]Erdil and Ergin [193] referred to *resident-Pareto improvement cycles* as *stable improvement cycles*.

Men's preferences          Women's preferences

$m_1 : w_1 \ w_2$                  $w_1 : (m_1 \ m_2)$

$m_2 : w_1 \ w_2$                  $w_2 : m_1 \ m_2$

Fig. 3.3    An instance of SMT with no strongly stable matching

## 3.3    Strong stability

This section focuses on strong stability in the context of SMT, SMTI and HRT instances. There is a sense in which strong stability can be viewed as the most appropriate criterion for a practical matching scheme when there is indifference in the preference lists, and that in cases where a strongly stable matching exists, it should be chosen instead of a matching that is merely weakly stable. Consider a weakly stable matching $M$ for an instance of HRT, and suppose that a resident $r_i$ prefers a hospital $h_j$ to her assigned hospital in $M$, whilst $h_j$ is full and is indifferent between $r_i$ and its worst assignee $r_k$ in $M$. Such a pair $(r_i, h_j)$ would not constitute a blocking pair for weak stability. However, $r_i$ might have such an overriding preference for $h_j$ over $M(r_i)$ that she is prepared to engage in persuasion, even bribery, in the hope that $h_j$ will reject $r_k$ and accept $r_i$ instead. Hospital $h_j$, being indifferent between $r_i$ and $r_k$, may yield to such persuasion, and, of course, a similar situation could arise with the roles reversed. However, the matching cannot be potentially undermined in this way if it is strongly stable.

We begin by providing in Sec. 3.3.1 an example SMT instance that admits no strongly stable matching. We also give a necessary and sufficient condition for a matching to be strongly stable in an HRT instance in terms of instances of HR obtained by tie-breaking. In Sec. 3.3.2 we show that the Rural Hospitals Theorem holds for strongly stable matchings in the context of HRT. We also demonstrate that a strongly stable matching, if one exists, is at least two-thirds of the size of a maximum weakly stable matching. Section 3.3.3 demonstrates that the set of strongly stable matchings in an SMT instance forms a distributive lattice. Then, in Section 3.3.4, we describe an $O(Cm)$ algorithm due to Kavitha *et al.* [364] for finding a strongly stable matching or reporting that none exists, given an HRT instance where $C$ is the total capacity of the hospitals and $m$ is the number of acceptable pairs.

### 3.3.1    *Existence of a strongly stable matching*

An instance of SMT need not admit a strongly stable matching. To see this, consider the SMT instance $I$ shown in Fig. 3.3. The matching

$\{(m_1, w_1), (m_2, w_2)\}$ is blocked by $(m_2, w_1)$, whilst the matching $\{(m_1, w_2), (m_2, w_1)\}$ is blocked by $(m_1, w_1)$.

We now give an equivalent criterion for strong stability in terms of tie-break instances, which utilises the notation defined just before Lemma 3.1.

**Theorem 3.13.** *Let $I$ be an instance of* HRT, *and let $M$ be a matching in $I$. Then $M$ is strongly stable in $I$ if and only if $M$ is stable in every member of $\mathcal{R}_M(I) \cup \mathcal{H}_M(I)$.*

**Proof.** Let $R = \{r_1, \ldots, r_{n_1}\}$ be the residents in $I$, and let $H = \{h_1, \ldots, h_{n_2}\}$ be the hospitals in $I$. Suppose $M$ is a strongly stable matching in $I$. Let $I' \in \mathcal{H}_M(I)$. Suppose that $(r_i, h_j)$ blocks $M$ in $I'$. Then $h_j$ is undersubscribed or prefers $r_i$ to at least one member of $M(h_j)$ in $I'$. But this is also true in $I$, in view of the way that ties were broken to form $I'$. Also either $r_i$ is unassigned or prefers $h_j$ to $M(r_i)$ or is indifferent between them. Hence $(r_i, h_j)$ blocks $M$ in $I$, a contradiction. Using a similar proof it follows that $M$ is stable in every member of $\mathcal{R}_M(I)$.

Conversely suppose that $M$ is stable in every member of $\mathcal{R}_M(I) \cup \mathcal{H}_M(I)$. Suppose that $(r_i, h_j)$ blocks $M$ in $I$. Suppose firstly that $r_i$ is unassigned or prefers $h_j$ to $M(r_i)$ in $I$. Then in $I$, $h_j$ is undersubscribed or prefers $r_i$ to at least one member of $r_k \in M(h_j)$ or is indifferent between them. Then $(r_i, h_j)$ blocks $M$ in any member of $\mathcal{R}_M(I)$ in which $h_j$ prefers $r_i$ to $r_k$, a contradiction. A similar contradiction involving $\mathcal{H}_M(I)$ holds if $h_j$ is undersubscribed or prefers $r_i$ to some member of $M(h_j)$ in $I$. $\square$

### 3.3.2 Rural Hospitals Theorem for strongly stable matchings in HRT

In Sec. 1.3.3 we presented the Rural Hospitals Theorem for stable matchings in an instance of HR. We now state a counterpart of this result for strongly stable matchings in the context of HRT.

**Theorem 3.14 ([327]).** *Let $I$ be a given instance of* HRT. *Then:*

(i) *the same residents are assigned in all strongly stable matchings;*

(ii) *each hospital is assigned the same number of residents in all strongly stable matchings;*

(iii) *any hospital that is undersubscribed in one strongly stable matching is assigned exactly the same set of residents in all strongly stable matchings.*

$$m_1 : w_1 \qquad\qquad w_1 : m_2 \ \ m_1$$
$$m_2 : (w_1 \ w_2) \qquad\quad w_2 : (m_2 \ m_3)$$
$$m_3 : w_2 \ \ w_3 \qquad\qquad w_3 : m_3$$
$$m_4 : (w_4 \ w_5) \qquad\quad w_4 : (m_4 \ m_5)$$
$$m_5 : (w_4 \ w_6) \qquad\quad w_5 : (m_4 \ m_6)$$
$$m_6 : (w_5 \ w_6) \qquad\quad w_6 : (m_5 \ m_6)$$

Men's preferences        Women's preferences

Fig. 3.4   An instance of SMTI

By contrast to Theorem 3.14, it has already been observed in Sec. 3.2.3 that weakly stable matchings can be of different sizes, given an instance of SMTI. Indeed it turns out that it is possible to find weakly stable matching both smaller than, and larger than, the size of a strongly stable matching in a given instance of SMTI, as the following example shows.

Consider the SMTI instance $I$ shown in Fig. 3.4, and the following matchings in $I$:

$$M_1 = \{(m_2, w_1), (m_3, w_2), (m_4, w_4), (m_6, w_6)\}$$
$$M_2 = \{(m_2, w_1), (m_3, w_2), (m_4, w_4), (m_5, w_6), (m_6, w_5)\}$$
$$M_3 = \{(m_2, w_1), (m_3, w_2), (m_4, w_5), (m_5, w_4), (m_6, w_6)\}$$
$$M_4 = \{(m_1, w_1), (m_2, w_2), (m_3, w_3), (m_4, w_4), (m_5, w_6), (m_6, w_5)\}$$

It may be verified that each of $M_2$ and $M_3$ is a strongly stable matching of size 5, whilst $M_1$ is a weakly stable matching of size 4 and $M_4$ is a weakly stable matching of size 6. In an HRT instance $I$, although a weakly stable matching can be larger than the size of a strongly stable matching in $I$, it turns out that the former can never be more than $\frac{3}{2}$ times as large as the latter, as the following result indicates.

**Theorem 3.15 ([523]).** *Let $I$ be an instance of* HRT, *let $M$ be a strongly stable matching in $I$ and let $M'$ be a maximum weakly stable matching in $I$. Then $|M'| \leq \frac{3}{2}|M| - \frac{1}{2}u_M$, where $u_M = \sum_{h_j \in H'} f_j$, where $H'$ is the set of hospitals that are undersubscribed in $M$, and $f_j$ is the number of posts that a hospital $h_j \in H'$ fills in $M$.*

### 3.3.3   Strongly stable matchings form a lattice

Let $S$ denote the set of strongly stable matchings for a given SMT instance $I$. In this section we show that, with the aid of a suitable equivalence relation defined on $\mathcal{S}$, we may deduce that $\mathcal{S}$ forms a distributive lattice. For brevity we describe the lattice for the SMT case, however the structural results of this section may be extended to HRT in a manner analogous to the extension of the lattice results for stable matchings in SM to the HR case [261, Sec. 1.6.5].

We begin with the following result, which is central to establishing the lattice structure.

**Theorem 3.16 ([415]).** *Let $I$ be an instance of* SMT, *and let $M, M'$ be two strongly stable matchings in $I$. Suppose that $(m_i, w_j) \in M \backslash M'$. Then either*

*(i) one of $m_i, w_j$ prefers $M$ to $M'$, and the other prefers $M'$ to $M$, or*
*(ii) both $m_i$ and $w_j$ are indifferent between $M$ and $M'$.*

We now define the equivalence relation $\sim$ on $\mathcal{S}$, on whose equivalence classes the lattice dominance relation will be defined.

**Definition 3.17.** *Let $\mathcal{S}$ denote the set of strongly stable matchings for a given* SMT *instance, and let $M, M' \in \mathcal{S}$. Define an equivalence relation $\sim$ on $\mathcal{S}$ as follows: $M \sim M'$ if and only if each man is indifferent[8] between $M$ and $M'$. Denote by $\mathcal{C}$ the set of equivalence classes of $\mathcal{S}$ under $\sim$, and denote by $[M]$ the equivalence class containing $M$, for $M \in \mathcal{S}$.*

We now define the dominance relation for the lattice.

**Definition 3.18.** *Let $\mathcal{S}$ denote the set of strongly stable matchings for a given* SMT *instance, and let $M, M' \in \mathcal{S}$. Then $M$ dominates $M'$, denoted by $M \preceq M'$, if each man either prefers $M$ to $M'$, or is indifferent between them.[9] Now let $\mathcal{C}$ and $[M]$ be as defined in Definition 3.17. We may extend $\preceq$ to a partial order $\preceq^*$ defined on equivalence classes as follows: for any two equivalence classes $[M], [M'] \in \mathcal{C}$, $[M] \preceq^* [M']$ if and only if $M \preceq M'$.*

It turns out that $(\mathcal{C}, \preceq^*)$ forms a lattice, as the following results indicate.

In order to define the "meet" and "join" operations for two strongly stable matchings in $I$, we require the following notation. Given two strongly stable matchings $M, M' \in \mathcal{S}$, let $\tilde{U}(M, M')$ be the set of men in $I$ who are

---

[8]The same equivalence classes arise if we define $M \sim M'$ if and only if each woman is indifferent between $M$ and $M'$.
[9]The same lattice structure prevails if we define $M \preceq M'$ if and only if each woman either prefers $M$ to $M'$, or is indifferent between them.

indifferent between $M$ and $M'$ (note that possibly $\tilde{U}(M, M') = \emptyset$.) The following lemma is key to defining the "meet" operation.

**Lemma 3.19 ([415]).** *Let $I$ be an instance of* SMT, *and let $M, M'$ be two strongly stable matchings in $I$. Let $M^*$ be a set of man–woman pairs defined as follows: for each man $m_i \in \tilde{U}(M, M')$, $m_i$ has in $M^*$ the same partner as in $M$, and for each man $m_i \notin \tilde{U}(M, M')$, $m_i$ has in $M^*$ the better of his partners in $M$ and $M'$. Then $M^*$ is a strongly stable matching.*

We denote by $M \wedge M'$ the strongly stable matching $M^*$ defined by Lemma 3.19. We remark that, in general, it need not be the case that $M \wedge M' = M' \wedge M$, however it certainly is true that $[M \wedge M'] = [M' \wedge M]$.

We now present a result along the same lines as Lemma 3.19, which will be key to our definition of a join operation between two equivalence classes.

**Lemma 3.20 ([415]).** *Let $I$ be an instance of* SMT, *and let $M, M'$ be two strongly stable matchings in $I$. Let $M^*$ be a set of man–woman pairs defined as follows: for each man $m_i \in \tilde{U}(M, M')$, $m_i$ has in $M^*$ the same partner as in $M$, and for each man $m_i \notin \tilde{U}(M, M')$, $m_i$ has in $M^*$ the poorer of his partners in $M$ and $M'$. Then $M^*$ is a strongly stable matching.*

We denote by $M \vee M'$ the strongly stable matching $M^*$ defined by Lemma 3.20. As in the case of the meet operation, we remark that, in general, it need not be the case that $M \vee M' = M' \vee M$, however it does follow that $[M \vee M'] = [M' \vee M]$.

The following theorem presents our main result of this section.

**Theorem 3.21 ([415]).** *Let $I$ be an instance of* SMT, *and let $S$ be the set of strongly stable matchings in $I$. Let $C$ be the set of equivalence classes of $S$ under $\sim$ (as defined by Definition 3.17), and let $\preceq^*$ be the dominance partial order on $C$ (as defined in Definition 3.18). Then $(C, \preceq^*)$ forms a distributive lattice, with $[M \wedge M']$ representing the meet of $[M]$ and $[M']$, and $[M \vee M']$ the join, for two equivalence classes $[M], [M'] \in C$.*

### 3.3.4 Finding a strongly stable matching

The problem of finding a strongly stable matching, or reporting that none exists, was shown to be solvable in $O(n^4)$ time, given an instance $I$ of SMT of size $n$ [308]. This algorithm was extended to the SMTI case with the same time complexity [414] and to the HRT case with complexity $O(m^2)$, where $m$

is the number of acceptable pairs [327,328]. Later, an $O(nm)$ improvement was obtained by Kavitha *et al.* [364] for SMTI. They also extended their algorithm to the HRT context — in this case the time complexity is $O(Cm)$, where $C$ is the sum of the hospital capacities. In this section we present the algorithm of Kavitha *et al.* [364] for the HRT case.

For a given instance $I$ of HRT, Algorithm HRT-Strong-Res, shown in Algorithm 3.3, finds a strongly stable matching or reports that none exists. This algorithm is resident-oriented (as noted in Sec. 1.3.7, practical matching schemes based on the HR problem model tend to employ resident-oriented versions of the Gale–Shapley algorithm). We now describe informally an execution of Algorithm HRT-Strong-Res.

The algorithm maintains two assignment relations, $A$ and $M$, where $M$ is a matching in $I$, though $A$ in general is not, as hospitals can be oversubscribed in $A$. We shall refer to the $z$th iteration of the outer while loop as *phase $z$* $(z \geq 1)$. During a particular phase, residents *apply* to hospitals, and the preference lists are reduced by potential deletions of pairs that cannot belong to a strongly stable matching. By a resident $r_i$'s *current list* at a particular point during the algorithm's execution, we mean $r_i$'s list after any deletions have been carried out up to this point.

The while loop on line 3 iterates as long as there is a resident $r_i \in R$ who is unassigned in $A$ and whose current list is non-empty. As long as this is the case, $r_i$ applies to each hospital $h_j$ at the *head* of her current list (this is the set of one or more hospitals, tied in her current list, which she prefers to all other hospitals in her list). The pair $(r_i, h_j)$ is then added to $A$. If $h_j$ is full or oversubscribed in $A$, we then consider $h_j$'s poorest assignee $r_k$ in $A$ (according to its current list) — if there is more than one, we let $r_k$ be any representative member of this tied group. For each resident $r_l$ inferior to $r_k$ on $h_j$'s current list, we remove $(r_l, h_j)$ from $A$ if it is in that set, and *delete the pair* $(r_l, h_j)$, which refers to the operation of each of $r_l$ and $h_j$ deleting one another from their current lists.

Once the inner while loop terminates (as indeed it must, since every iteration involves a resident becoming assigned, or a pair being deleted), we let $G^z = (R^z, H^z, E^z)$ be the *reduced assignment graph* at phase $z$. To describe this graph, we need some preliminary definitions. By the *tail* of a hospital's list, we mean the set of one or more residents, tied in its current list, to whom it prefers all other residents in its list. A pair $(r_i, h_j) \in A$ is said to be *bound* if $h_j$ is not oversubscribed in $A$, or $r_i$ is not in $h_j$'s tail, or both. The pair $(r_i, h_j)$ is *unbound* if it is not bound. Intuitively, pairs in $A$ that are bound at the termination of the main loop have priority over

## Algorithm 3.3 Algorithm HRT-Strong-Res [364]

**Require:** HRT instance $I$
**Ensure:** return a strongly stable matching $M'$ in $I$ or "no strongly stable matching exists"

1:  $M := \emptyset$;   $A := \emptyset$;   $z := 1$;
2:  **while** a resident $r_i \in R$ is unassigned in $A$ and has a non-empty list **do**
3:     **while** a resident $r_i \in R$ is unassigned in $A$ and has a non-empty list **do**
4:        **for each** hospital $h_j$ at the head of $r_i$'s list **do**
5:           $\{r_i$ applies to $h_j\}$
6:           $A := A \cup \{(r_i, h_j)\}$;
7:           **if** $h_j$ is full or oversubscribed in $A$ **then**
8:              $r_k :=$ worst resident in $A(h_j)$ according to $h_j$'s list;   {any, if > 1}
9:              **for each** strict successor $r_l$ of $r_k$ on $h_j$'s list **do**
10:                $A := A \backslash \{(r_l, h_j)\}$; delete the pair $(r_l, h_j)$;
11:             **end for**
12:          **end if**
13:       **end for**
14:    **end while**
15:    $G^z = (R^z, H^z, E^z) :=$ reduced assignment graph at phase $z$;
16:    **for each** resident $r_i \in R^z$ who is unassigned in $M$ **do**
17:       **if** there exists an alternating path from $r_i$ to a free hospital in $G^z$ **then**
18:          $h_j :=$ free hospital of maximal level reachable
                    from $r_i$ via an alternating path;
19:          $p :=$ alternating path from $r_i$ to $h_j$;
20:          $M := M \oplus p$;
21:       **else**
22:          $Z :=$ set of residents reachable from $r_i$ by alternating paths in $G^z$;
23:          $N(Z) := \{h_j \in H : (r_i, h_j) \in A \wedge r_i \in Z\}$;
24:          **for each** hospital $h_j \in N(Z)$ **do**
25:             **for each** resident $r_k$ at the tail of $h_j$'s list **do**
26:                $A := A \backslash \{(r_k, h_j)\}$; delete the pair $(r_k, h_j)$;
27:             **end for**
28:          **end for**
29:       **end if**
30:       $z$++;
31:    **end for**
32:  **end while**
33:  $M' :=$ feasible matching with respect to $A$;
34:  **if** $M'$ is strongly stable in $I$ **then**
35:     **return** $M'$;
36:  **else**
37:     **return** "no strongly stable matching exists";
38:  **end if**

unbound pairs for inclusion in a potential strongly stable matching, for a bound pair would block a matching if it were not included. A resident $r_i$ is said to be *bound* if $r_i$ is a member of a bound pair of $A$, and *unbound* otherwise. The reduced assignment graph comprises the unbound residents, denoted by $R^z$, the set of hospitals that are collectively assigned in $A$ to residents in $R^z$, denoted by $H^z$, and the edges in $A \cap (R^z \times H^z)$, denoted by $E^z$. In this graph, the capacity of each hospital is $c'_j$, which is defined to be $c_j$ minus the number of bound edges incident to $h_j$ in $A$. In practice, the graph $G^z$ is constructed from $G^{z-1}$, where $G^0$ is the empty graph; see Ref. [364] for further details. Intuitively the algorithm maintains a maximum cardinality matching $M$ in $G^z$, and this is ultimately unioned with the bound pairs in $A$.

For each resident $r_i \in R^z$ who is unassigned in $M$, we search for an alternating path from $r_i$ to a *free* hospital in $G^z$. Such a hospital $h_j$ satisfies $|M(h_j)| < c'_j$. If such a hospital exists, then among all of these, we select $h_j$ to have maximum *level*. Here, the *level* of an edge $(r_k, h_l) \in E^z$ is the minimum $y$ $(1 \le y \le z)$ such that $(r_k, h_l)$ was added to $G^y$ during phase $y$, and the *level* of a hospital $h_l \in H^z$ is the minimum level of the edges incident to $h_l$ in $G^z$. Let $p$ be an alternating path from $r_i$ to $h_j$; we then augment $M$ along $p$. Giving priority to free hospitals with maximum level allows the number of edges traversed upon augmenting path searches to be bounded, when considering the algorithm's complexity analysis.

If an alternating path from $r_i$ to a free hospital in $G^z$ does not exist, then we let $Z$ be the set of residents reachable from $r_i$ by an alternating path in $G^z$, and we let $N(Z)$ be the hospitals that are collectively assigned in $A$ to the residents in $Z$. For each hospital $h_j \in N(Z)$, and for each resident $r_k$ at the tail of $h_j$'s current list, we delete the pair $(r_k, h_j)$ and remove the pair from $A$ if necessary.

Once the outer while loop terminates (as indeed it must, for the same reason as mentioned above for the inner while loop), we let $M'$ be a *feasible matching* with respect to $A$. This comprises $M$ unioned together with all the bound resident–hospital pairs from $A$. If $M'$ is strongly stable in the original instance $I$ then the algorithm outputs $M'$, otherwise it turns out that no other matching is strongly stable in $I$, so the algorithm outputs a message to this effect.

The following result concerning the correctness and complexity of this algorithm, and the optimality of a given matching that it produces, is proved in Refs. [364] and [327, 328].

**Theorem 3.22 ([364, 327, 328]).** *For a given instance I of* HRT, *Algorithm* HRT-Strong-Res *determines in* $O(Cm)$ *time whether or not a strongly stable matching exists, where* $C$ *is the sum of the hospital capacities and* $m$ *is the number of acceptable pairs in* $I$. *If such a matching does exist, all possible executions of the algorithm find one in which every assigned resident is assigned as favourable a hospital as in any strongly stable matching, and every unassigned resident is unassigned in every strongly stable matching.*

An $O(m^2)$ hospital-oriented counterpart of Algorithm HRT-Strong-Res, Algorithm HRT-Strong-Hosp, is described in Ref. [523]. The optimality property of the strongly stable matching it returns (providing one exists) is indicated as follows.

**Theorem 3.23 ([523]).** *For a given instance I of* HRT, *Algorithm* HRT-Strong-Hosp *determines in* $O(m^2)$ *time whether or not a strongly stable matching exists, where* $m$ *is the number of acceptable pairs in* $I$. *If such a matching does exist, all possible executions of the algorithm find one in which every full hospital* $h_j \in H$ *has at least as favourable a set of assignees as it can have in any strongly stable matching, whilst every undersubscribed hospital is assigned a set of residents that constitutes its assignees in every strongly stable matching.*

O'Malley [470] gave an $O(\sqrt{n_1}m)$ algorithm[10] to find a strongly stable matching or report that none exists, given an instance $I$ of HRT such that there is a master list of residents, where $n_1$ is the number of residents and $m$ is the number of acceptable pairs in $I$. He did likewise for the case that $I$ is an instance of HRT with *symmetric preferences* (i.e., $rank(r_i, h_j) = rank(h_j, r_i)$ for any acceptable resident–hospital pair $(r_i, h_j)$).

We have already noted that an instance $I$ of SMTI need not admit a strongly stable matching. One strategy for coping with this could be to find a weakly stable matching in $I$ with the minimum number of blocking pairs of the strong stability type. However this problem is unlikely to be solvable in polynomial time: Abraham *et al.* [27] proved that the problem of finding a weakly stable matching with the fewest number of blocking pairs of the strong stability type is not approximable within $n^{1-\varepsilon}$, for any $\varepsilon > 0$, unless P=NP, where $n$ is the size of $I$.

---

[10]In Ref. [470], the weaker upper bound of $O(\sqrt{C}m)$ was given as the complexity for this algorithm, where $C$ is the total capacity of the hospitals. The improved upper bound follows by the remark in Footnote 6 in Chap. 1.

## 3.4 Super-stability

In this section we consider super-stability in instances of SMT, SMTI and HRT. Super-stability is the most stringent of the three stability criteria that can be defined in the presence of ties. It allows a man and a woman to form a blocking pair in an instance of SMT simply by being indifferent between one another and their partners. This may be seen as somewhat contradictory to human nature, given that (in many situations) people tend towards the status quo unless there is tangible incentive to switch. However, as we will see, super-stable matchings (when they exist) lend themselves to a similar range of structural properties to those that hold for stable matchings in the context of instances of SM and HR. Consequently, a variety of problems associated with the computation of types of super-stable matchings turn out to be solvable in polynomial time, as we will demonstrate in this section. These remarks indicate that super-stability in the context of SMTI and HRT instances is perhaps the stability criterion that is the closest counterpart to classical stability in the context of SMI and HR instances. This discussion motivates the study of super-stable matchings from a theoretical point of view. However a particular practical situation that gives additional motivation for this type of stability is discussed in Sec. 3.5.3.

This section is organised along similar lines to Sec. 3.3. We begin by showing in Sec. 3.4.1 that an SMT instance need not admit a super-stable matching. We also give a necessary and sufficient condition for a matching to be super-stable in an HRT instance in terms of instances of HR obtained by tie-breaking. In Sec. 3.4.2 we show that the Rural Hospitals Theorem holds for super-stable matchings in the context of HRT. We also demonstrate that the existence of a super-stable matching in an HRT instance $I$ yields a Rural Hospitals Theorem for weakly stable matchings in $I$. Section 3.4.3 demonstrates that the set of super-stable matchings in an SMT instance forms a distributive lattice. Then, in Sec. 3.4.4, we describe an $O(m)$ algorithm for finding a super-stable matching or reporting that none exists, given an HRT instance where $m$ is the number of acceptable pairs. Finally in Sec. 3.4.5 we outline some additional results concerning super-stable matchings.

### 3.4.1 *Existence of a super-stable matching*

It is straightforward to construct an instance of SMT with no super-stable matching. For example, an SMT instance containing two men and two women, each of whose preference list is a tie of length 2, admits no super-stable matching. Also the SMT instance shown in Fig. 3.3 admits no strongly stable matching, and hence no super-stable matching. Clearly, in an instance of HRT, super-stability is identical to strong stability when preference lists on one side are strictly ordered.

The following result, stated without proof in Ref. [326], gives an alternative necessary and sufficient condition for a matching to be super-stable in an instance of HRT.

**Lemma 3.24 ([326]).** *Let $I$ be an instance of HRT, and let $M$ be a matching in $I$. Then $M$ is super-stable in $I$ if and only if $M$ is stable in every instance $I'$ of HR derived from $I$ by breaking the ties.*

**Proof.** Let $R = \{r_1, \ldots, r_{n_1}\}$ be the residents in $I$, and let $H = \{h_1, \ldots, h_{n_2}\}$ be the hospitals in $I$. Suppose $M$ is a super-stable matching in $I$, and let $I'$ be an instance of HR obtained from $I$ by breaking the ties in $I$ in some way. Suppose $(r_i, h_j)$ blocks $M$ in $I'$. Then in $I$, either $r_i$ is unassigned in $M$, or $r_i$ prefers $h_j$ to $M(r_i)$, or $r_i$ is indifferent between them. Similarly in $I$, either $h_j$ is undersubscribed, or $h_j$ prefers $r_i$ to at least one member of $M(h_j)$, or $h_j$ is indifferent between them. Hence $(r_i, h_j)$ blocks $M$ in $I$, a contradiction.

Conversely suppose that $M$ is a matching that is super-stable in every instance of HR obtained from $I$ by breaking the ties. Suppose that $M$ is not super-stable in $I$. Then $(r_i, h_j)$ blocks $M$ in $I$. We break the ties in $I$ to form an HR instance $I'$ as follows. If $r_i$ is indifferent between $h_j$ and $M(r_i)$ in $I$, break the tie containing those two hospitals in $r_i$'s list so that $r_i$ prefers $h_j$ to $M(r_i)$ in $I'$. Otherwise, break this tie arbitrarily in $I'$. Similarly if $h_j$ is indifferent between $r_i$ and its worst assignee $r_k$ (or one of its worst assignees, if applicable), then break this tie so that $h_j$ prefers $r_i$ to $r_k$. Otherwise, break this tie arbitrarily in $I'$. Break all other ties in $I$ arbitrarily in $I'$. Then $(r_i, h_j)$ blocks $M$ in $I'$, a contradiction. □

### 3.4.2 *Rural Hospitals Theorem for super-stable matchings in HRT*

We now state a counterpart of Theorem 3.14 for super-stable matchings in an instance of HRT.

**Theorem 3.25 ([326]).** *Let $I$ be a given instance of* HRT. *Then:*

(i) *the same residents are assigned in all super-stable matchings;*

(ii) *each hospital is assigned the same number of residents in all super-stable matchings;*

(iii) *any hospital that is undersubscribed in one super-stable matching is assigned exactly the same set of residents in all super-stable matchings.*

In an HRT instance $I$ that admits a super-stable matching, a counterpart of Theorem 3.25 holds for the set of weakly stable matchings in $I$, as we now prove.

**Theorem 3.26 ([326]).** *Let $I$ be a given instance of* HRT *that admits a super-stable matching. Then:*

(i) *the same residents are assigned in all weakly stable matchings;*

(ii) *each hospital is assigned the same number of residents in all weakly stable matchings;*

(iii) *any hospital that is undersubscribed in one weakly stable matching is assigned exactly the same set of residents in all weakly stable matchings.*

*Proof.* Let $M$ be a super-stable matching in $I$. Then $M$ is weakly stable. Let $M'$ be an arbitrary weakly stable matching in $I$. Then by Lemma 3.1, there is some HR instance $I'$ obtained from $I$ by breaking ties such that $M'$ is stable in $I'$. By Lemma 3.24, $M$ is also stable in $I'$. The result then follows by the Rural Hospitals Theorem (Theorem 1.11) applied to $I'$.  □

### 3.4.3  Super-stable matchings form a lattice

Let $I$ be an instance of HRT. Spieker [548] showed that the set of super-stable matchings in $I$ forms a distributive lattice using the following argument. Let $\mathcal{I}$ denote the set of HR instances that are obtainable from $I$ by breaking the ties in some way. Lemma 3.24 implies that a matching $M$ is super-stable in $I$ if and only if $M$ is stable in every member of $\mathcal{I}$. Hence the set of super-stable matchings in $I$ is precisely $\cap_{I' \in \mathcal{I}} \mathcal{S}_{I'}$, where $\mathcal{S}_{I'}$ denotes the set of stable matchings in the HR instance $I'$. But $\mathcal{S}_{I'}$ forms a distributive lattice (under the dominance partial order as defined in Definition 1.12) [261, Sec. 1.6.5]. Spieker [548] stated that the intersection of these distributive lattices is also a distributive lattice (or is empty), and therefore it follows that the set of super-stable matchings in $I$ forms a distributive lattice.

In fact Spieker stated that *any* intersection of distributive lattices is either a distributive lattice or is empty. However it is not at all clear what is meant by "intersection" in this context. In general, let $P_1 = (S_1, \preceq_1)$ and $P_2 = (S_2, \preceq_2)$ be two partially ordered sets. Define the *intersection* of $P_1$ and $P_2$ to be the pair $P = (S, \preceq)$ where $S = S_1 \cap S_2$ and $\preceq = \preceq_1 \cap \preceq_2$. Clearly $P_1 \cap P_2$ is a partially ordered set. However if $P_1$ and $P_2$ are lattices, it certainly need not be the case in general that $P$ is a lattice. This begs the question as to what definition of "intersection" Spieker had in mind. Moreover, it may be necessary to invoke further assumptions on properties that are satisfied by the lattices that are being intersected before the overall claim is true. These properties may be satisfied by the individual lattices $(S_{I'}, \preceq)$, but again it is unclear as to what assumptions are required.

In Ref. [415], this author gave a brief overview of Spieker's method for showing that super-stable matchings form a distributive lattice without noting the issue discussed in the previous paragraph. However one of the main contributions of Ref. [415] was an alternative proof of Spieker's result which is not based on the intersection of lattices. We feel that our alternative method gives additional insight into the underlying structure of the set of super-stable matchings, and we now outline it here. As in Sec. 3.3.3, for brevity we describe the lattice for SMT, however the structural results of this section may be extended to HRT in a manner analogous to the extension of the lattice results for stable matchings in SM to the HR case [261, Sec. 1.6.5].

We begin with the following result, which demonstrates that if an agent has different partners in two super-stable matchings, then he/she cannot be indifferent between them.

**Lemma 3.27 ([415]).** *Let $I$ be an instance of* SMT, *and let $M, M'$ be two super-stable matchings in $I$. Suppose that, for any agent $p$ in $I$, $(p, q) \in M$ and $(p, q') \in M'$, where $p$ is indifferent between $q$ and $q'$. Then $q = q'$.*

Let $\mathcal{S}$ be the set of super-stable matchings for a given SMT instance $I$, and define the dominance partial order $\preceq$ on $\mathcal{S}$ as in Definition 3.18. The insight into the structure of super-stable matchings in an SMT instance provided by Lemma 3.27 allows us to follow an approach along the lines of that employed in Sec. 3.3.3, in order to show that $(\mathcal{S}, \preceq)$ forms a distributive lattice. We begin with the analogue of Theorem 3.16 for super-stability.

**Theorem 3.28 ([415]).** *Let $I$ be an instance of* SMT, *and let $M, M'$ be two super-stable matchings in $I$. Suppose that $(m_i, w_j) \in M \backslash M'$. Then one of $m_i$, $w_j$ prefers $M$ to $M'$, and the other prefers $M'$ to $M$.*

The following lemmas provide the foundations for the definitions of the "meet" and "join" operations for two super-stable matchings in $I$.

**Lemma 3.29 ([415]).** *Let $I$ be an instance of* SMT, *and let $M, M'$ be two super-stable matchings in $I$. Let $M^*$ be a set of man–woman pairs defined by giving each man the better of his partners in $M$ and $M'$. Then $M^*$ is a super-stable matching in $I$.*

**Lemma 3.30 ([415]).** *Let $I$ be an instance of* SMT, *and let $M, M'$ be two super-stable matchings in $I$. Let $M^*$ be a set of man–woman pairs defined by giving each man the poorer of his partners in $M$ and $M'$. Then $M^*$ is a super-stable matching in $I$.*

We denote by $M \wedge M'$ and $M \vee M'$ the super-stable matchings defined by Lemmas 3.29 and 3.30 respectively. We are now in a position to state our main result of this section.

**Theorem 3.31 ([415]).** *Let $I$ be an instance of* SMT, *and let $S$ be the set of super-stable matchings in $I$. Then $(S, \preceq)$ forms a distributive lattice, with $M \wedge M'$ representing the meet of $M$ and $M'$, and $M \vee M'$ the join, for two super-stable matchings $M, M' \in S$, where $\preceq$ is the dominance partial order on $S$ as defined in Definition 3.18.*

### 3.4.4 Finding a super-stable matching

The problem of finding a super-stable matching, or reporting that none exists, was shown to be solvable in $O(n^2)$ time, given an instance $I$ of SMT of size $n$ [308]. This algorithm was extended to the SMTI case with the same time complexity [414]. In this subsection we present an $O(m)$ algorithm for the analogous problem, given an HRT instance $I$, where $m$ denotes the number of acceptable pairs in $I$. As in Sec. 3.3.4, the algorithm for super-stability is resident-oriented.

For a given instance $I$ of HRT, Algorithm HRT-Super-Res, shown in Algorithm 3.4, finds a super-stable matching or reports that none exists. We will describe informally an execution of this algorithm. This description contains the terms *head*, *tail* and *delete the pair*, whose definitions are unchanged from Sec. 3.3.4.

---

**Algorithm 3.4** Algorithm HRT-Super-Res [326]

**Require:** HRT instance $I$

**Ensure:** return a super-stable matching in $I$, or "no super-stable matching exists"

1: $M := \emptyset$;
2: **while** some resident $r_i \in R$ is unassigned and has a non-empty list **do**
3:     **for each** hospital $h_j$ at the head of $r_i$'s list **do**
4:         $\{r_i$ applies to $h_j\}$
5:         $M := M \cup \{(r_i, h_j)\}$;
6:         **if** $h_j$ is oversubscribed **then**
7:             **for each** resident $r_k$ at the tail of $h_j$'s list **do**
8:                 $M := M\backslash\{(r_k, h_j)\}$;
9:                 delete the pair $(r_k, h_j)$;
10:             **end for**
11:         **end if**
12:         **if** $h_j$ is full **then**
13:             $r_k :=$ worst resident in $M(h_j)$ according to $h_j$'s list;    $\{$any, if $> 1\}$
14:             **for each** strict successor $r_l$ of $r_k$ on $h_j$'s list **do**
15:                 delete the pair $(r_l, h_j)$;
16:             **end for**
17:         **end if**
18:     **end for**
19: **end while**
20: **if** $M$ is a super-stable matching in $I$ **then**
21:     **return** $M$;
22: **else**
23:     **return** "no super-stable matching exists";
24: **end if**

---

Algorithm HRT-Super-Res maintains an assignment $M$ that is initially empty. The algorithm involves a sequence of *apply* operations from the residents to the hospitals, in the spirit of the RGS algorithm for HR. A resident applies simultaneously to *all* hospitals at the head of her list, and all applications are provisionally accepted. If a hospital $h_j$ becomes oversubscribed, it turns out that none of $h_j$'s worst-placed assignees (there must be more than one), nor any residents tied with these assignees in $h_j$'s list, can be a super-stable partner of $h_j$ — the pair $(r_i, h_j)$ is deleted and removed from $M$ if necessary, for any such resident $r_i$. If a hospital $h_j$ is full, then no resident inferior to $h_j$'s worst-placed assignee(s) can be a super-stable partner of $h_j$ — again the pair $(r_i, h_j)$ is deleted for any such resident $r_i$. The sequence of apply operations terminates once every resident either is assigned to a hospital or has an empty list. At this point if $M$ is super-stable in the original instance $I$ then the algorithm outputs $M$, otherwise it turns out that no other matching is super-stable in $I$, so the algorithm outputs a message to this effect.

We now summarise the correctness and complexity of the algorithm.

**Theorem 3.32 ([326]).** *For a given instance $I$ of* HRT, *Algorithm* HRT-Super-Res *determines, in $O(m)$ time, whether or not a super-stable matching exists, where $m$ is the number of acceptable pairs in $I$. If such a matching does exist, all possible executions of the algorithm find one in which every assigned resident has the best partner that she has in any super-stable matching in $I$, and every unassigned resident is unassigned in all super-stable matchings in $I$.*

A hospital-oriented counterpart of Algorithm HRT-Super-Res, Algorithm HRT-Super-Hosp, appears in Ref. [326]. The optimality property of the super-stable matching it returns (providing one exists) is indicated as follows.

**Theorem 3.33 ([326]).** *For a given instance $I$ of* HRT, *Algorithm* HRT-Super-Hosp *determines in $O(m)$ time whether or not a super-stable matching exists, where $m$ is the number of acceptable pairs in $I$. If such a matching does exist, all possible executions of the algorithm find one in which every full hospital $h_j \in H$ is assigned its $c_j$ best super-stable partners, whilst every undersubscribed hospital $h_j$ is assigned a set of residents that constitutes its assignees in every super-stable matching.*

O'Malley [470] gave a simpler $O(m)$ algorithm to find a super-stable matching or report that none exists, given an instance $I$ of HRT such that there is a master list of residents, where $m$ is the number of acceptable pairs in $I$. He did likewise for the case that $I$ is an instance of HRT with symmetric preferences.

### 3.4.5 Optimal super-stable matchings

In Sec. 3.4.3 we described the lattice structure for super-stable matchings in the context of HRT. Given this structure, it is perhaps not surprising that a range of additional problems concerned with computing super-stable matchings in the SMTI context turn out to be solvable in polynomial time, as we demonstrate in this subsection.

Let $I$ be an instance of SMTI. We firstly consider the problem of generating a "succinct certificate" for the unsolvability of $I$. Clearly, if $I$ admits a super-stable matching $M$, $M$ itself is a "certificate" of this fact, and may be used to verify, in $O(m)$ time (by checking for the absence of blocking pairs) that $I$ does indeed admit a super-stable matching, where $m$ is the number of acceptable pairs. However if $I$ does not, then it is less obvious as to what is meant by a "certificate" of this property. Certainly, one can run Algorithm HRT-Super-Res on $I$ and check that it terminates with an assignment $M$ that is not a super-stable matching in $I$. But it is arguable that this is not a "succinct" certificate of the non-existence of a super-stable matching in $I$. Instead, one may use the fact that there is a 1–1 correspondence between the super-stable matchings in $I$ and the satisfying truth assignments of a suitably-constructed 2-SAT instance $J$ [216]. It is well known that a succinct certificate of the unsolvability of a 2-SAT instance $J$ is a cycle in the implication digraph underlying $J$. Given that $J$ can be constructed from $I$ in $O(m)$ time, and $J$ requires $O(m)$ space, it follows that there is a succinct certificate for the unsolvability of $I$ that can be generated in $O(m)$ time, and represented using $O(m)$ space.

Additional polynomial-time solvable problems are as follows. A minimum regret super-stable matching in $I$ can be found in $O(m)$ time, whilst an egalitarian super-stable matching in $I$ can be constructed in $O(m^{1.5})$ time [216]. Here, we are implicitly using the Rural Hospitals Theorem (Theorem 3.25) to discard the unassigned men and women, and then minimising the maximum rank of an agent's partner (in the case of the minimum regret problem) or minimising the sum of the ranks of the agents' partners (in the case of the egalitarian problem) among the assigned men and women that remain.

Furthermore all the super-stable pairs in $I$ can be found in $O(m)$ time, whilst there is an algorithm to list all the super-stable matchings in $I$: the first such matching can be output in $O(m)$ time, and each subsequent super-stable matching can be output in $O(n)$ time [216].

It is likely that the results in this section can be extended to the HRT case, but explicit extensions to this case have yet to be formulated in the literature.

## 3.5    Other results

In this section we outline some additional results involving stable matching problems with indifference.

### 3.5.1 Semi–strong stability

"Semi–strong stability" is a version of stability that, in some sense, lies "in between" weak stability and strong stability. Here, we stipulate which set of agents is allowed to express indifference. For example, in the context of an SMT instance $I$, a matching $M$ is *woman-strongly stable* if there is no man–woman pair $(m_i, w_j)$ such that $m_i$ prefers $w_j$ to $M(m_i)$, and $w_j$ prefers $m_i$ to $M(w_j)$ or is indifferent between them. Clearly *man–strong stability* can be defined similarly. It turns out that, in contrast to the case for strong stability, the problem of deciding whether $I$ admits a woman-strongly stable matching is NP-complete, as we now demonstrate.

**Theorem 3.34.** *The problem of deciding whether an SMT instance admits a woman-strongly stable matching is NP-complete.*

**Proof.** Clearly this problem belongs to NP. To show NP-hardness, we give a reduction from the restriction of COM SMTI in which ties appear on the men's side only, which is NP-complete [419]. Let $I$ be an instance of this problem in which $U = \{m_1, \ldots, m_n\}$ is the set of men and $W = \{w_1, \ldots, w_n\}$ is the set of women. Let $P_i$ denote the preference list of $m_i$, for each $m_i \in U$, and let $Q_j$ denote the preference list of $w_j$, for each $w_j \in W$. We form an instance $I'$ of SMT in which $U \cup \{m_{n+1}\}$ is the set of men, for some new man $m_{n+1}$, and $W \cup \{w_{n+1}\}$ is the set of women, for some new woman $w_{n+1}$. The preference lists of the men and women in $I'$ are as follows:

$$
\begin{aligned}
m_i &: P_i \ w_{n+1} \ [W \backslash P_i] & (1 \leq i \leq n) \\
m_{n+1} &: w_{n+1} \ [W] \\
w_j &: Q_j \ [(U \backslash Q_j) \cup \{m_{n+1}\}] & (1 \leq j \leq n) \\
w_{n+1} &: (U \cup \{m_{n+1}\})
\end{aligned}
$$

In a given agent's preference list, $[S]$ denotes all members of $S$ listed in some arbitrary strict order where the symbol appears. We abuse notation somewhat and use $P_i$ and $Q_j$ to denote the members of the preference lists as well as the preference lists themselves. We claim that $I$ admits a complete weakly stable matching if and only if $I'$ admits a woman-strongly stable matching.

For, suppose that $M$ is a complete weakly stable matching in $I$. It is not difficult to verify that $M \cup \{(m_{n+1}, w_{n+1})\}$ is a woman-strongly stable matching in $I'$.

Conversely suppose that $M'$ is a woman-strongly stable matching in $I$. If $(m_{n+1}, w_{n+1}) \notin M'$ then clearly $(m_{n+1}, w_{n+1})$ blocks $M'$. Moreover if $M'(m_i) \notin P_i$ for some $m_i \in U$ then $(m_i, w_{n+1})$ blocks $M'$ in $I'$, a contradiction. Hence $M = M' \backslash \{(m_{n+1}, w_{n+1})\}$ is a complete matching in $I$. It is straightforward to verify that $M$ is weakly stable in $I$.                $\Box$

Clearly the definition of woman–strong stability may be extended to the SMTI case. If ties belong to the men's side only, clearly woman–strong stability is equivalent to weak stability. Therefore given an SMTI instance $I$, woman-strongly stable matchings can be of different sizes (recall Fig. 3.2), and the problem of finding a maximum woman-strongly stable matching in $I$ is NP-hard (this follows since MAX SMTI is NP-hard even if the ties belong to the men's lists only [419]).

### 3.5.2   Many–many strongly stable matchings

Malhotra [413] studied strongly stable matchings in the multiple partner stable marriage problem with ties, a many–many generalisation of SMTI (see Sec. 5.4). He showed that the set of strongly stable matchings forms a lattice, given an instance of this problem. He also gave a polynomial-time algorithm for finding a strongly stable matching or reporting that none exists, for this problem context. However Chen and Ghosh [139] showed that this algorithm is, in fact, incorrect. They gave a new algorithm which uses the concept of a *critical subgraph* (extending the concept of the *critical set* from Ref. [328]) with complexity $O(m^3 n)$, where $n$ is the number of agents and $m$ is the number of acceptable pairs of agents. The algorithm extends that in Ref. [328] for HRT, but does not use the concept of *level-maximal matchings* from Ref. [364]; thus there could be scope for improving the algorithm's complexity.

### 3.5.3   Partially-ordered preference lists

In most practical applications where preference lists are not strictly ordered, indifference takes the form of ties in the preference lists. However, in some cases, indifference is not expressible in terms of preference lists involving ties. One example context is where there is incomplete information about the preference lists. Suppose that, in a Stable Marriage instance, we wish to find a stable matching (in the classical sense), but for some or all of the agents we have only partial information regarding the true preferences. In general, each preference "list" may be expressible only as a partial order, and the particular linear extension that represents an agent's true preferences is unknown. It is straightforward to show that Lemma 3.24 extends to the HRP case. It follows that a super-stable matching is one that is stable no matter which linear extensions of the various preference posets represent the true preferences.

All the other results from Sec. 3.4 concerning super-stability carry over to HRP, namely the equivalent condition for super-stability (Lemma 3.24), the Rural Hospitals Theorem (Theorem 3.25), the lattice structure for super-stable matchings (Theorem 3.31), and Algorithm HRT-Super-Res (Theorem 3.32). In the case of the statement of Lemma 3.24, we should replace "breaking the ties" with "forming a linear extension of each partial order". Also in the case of Algorithm HRT-Super-Res, the description of the algorithm should be amended so that the *head* of a resident $r_i$'s list is the set of source nodes in the Hasse diagram representing $\prec_{r_i}$, the *tail* of a hospital $h_j$'s list is the set of sink nodes in the Hasse diagram representing $\prec_{h_j}$, and a *worst* resident in $M(h_j)$ is any resident in the tail of $h_j$'s list who is assigned to $h_j$ in $M$.

Concerning strongly stable and weakly stable matchings in HRP, it turns out that the lattice structure that holds for strongly stable matchings in an SMT instance (as established in Sec. 3.3.3) does not carry over to SMP: there is an instance of SMP [415, Sec. 3] that contains no man-optimal strongly stable matching. Moreover the problem of deciding whether an SMP instance admits a strongly stable matching is NP-complete [328]. Clearly the hardness results described in Sec. 3.2 for SMT under weak stability carry over to SMP by restriction.

Fishburn [207] described some further practical situations that give rise to preference structures involving indifference where the structures are not expressible in terms of preference lists involving ties.

Rastegari *et al.* [488] studied instances of SMTI from the point of view of incomplete information. Here, the men and women in a given instance $I$ were termed applicants and employers respectively. Each agent $a_i$ has, in addition their preference list $P(a_i)$ in $I$ (possibly involving ties), a linear order $L(a_i)$, obtained by breaking the ties in $P(a_i)$, representing $a_i$'s true strict underlying preference order over the members of the opposite side. This linear order is initially unknown to $a_i$, and $a_i$ learns information about it through carrying out interviews.

Each interview involving a pair of agents $a_i$ and $a_j$ can be assumed to reveal information for both $a_i$ and $a_j$ about one another — we can think of the outcome as being a unique score assigned by $a_i$ to $a_j$ and vice versa.

Thus if $a_i$ has a tie $T$ of length $k$ in his/her preference list, $k$ interviews are necessary and sufficient in order to produce a strict ranking over the members of $T$.

However, interviews are costly and we wish to minimise their use. Rastegari *et al.* [488] studied the problem of scheduling the minimum number of interviews in $I$ in order to produce an SMTI instance $I'$, obtained from $I$ by breaking some of the ties (as a consequence of the information learned through the interviews), that admits a super-stable matching that is the employer-optimal stable matching no matter which instance $I''$ of SMI, obtained from $I'$ by breaking the remaining ties, is consistent with the true underlying linear orders. (Note that such a schedule always exists: by carrying out all $m$ possible interviews between pairs of agents, where $m$ is the number of acceptable pairs in $I$, we obtain precisely the SMI instance $I''$ that is consistent with the true underlying linear orders, which of course admits a unique employer-optimal stable matching.)

The authors showed that, in general, finding such a minimum schedule of interviews is NP-hard, though solvable in polynomial time if all of the applicants have the same preference list in $I$ initially (though their linear orders need not be the same).

## 3.6   Conclusions and open problems

The quantity of results surveyed in this chapter reflects the steadily expanding body of literature concerning bipartite stable matching problems with indifference. These results indicate that, broadly speaking, whilst a weakly stable matching is guaranteed to exist and can be found in polynomial time, placing any additional constraints on the nature of the weakly stable matching required is likely to yield an NP-hard problem. On the other hand, most problems concerned with finding types of strongly stable and/or super-stable matchings turn out to be solvable in polynomial time, but the drawback here is that each type of matching need not exist. We conclude this chapter with a selection of open problems.

- The approximability results for MAX SMTI and MAX HRT presented in Sec. 3.2.6.1 leave open the question as to whether any improved upper or lower bounds can be found, perhaps for special cases. For example the case of MAX HRT where the hospitals' preference lists are derived from a single master list of residents is a restriction that is particularly relevant in practice, and may turn out to be more amenable to approximation than the general problem.

- As noted in Sec. 3.2.7, it remains open as to whether there is an efficient algorithm for generating all the weakly stable matchings in a given SMT instance.

- As observed in Sec. 3.2.2, an instance of SMTI need not admit a man-optimal weakly stable matching, even if there are no ties in the men's lists. The complexity of the problem of finding a man-optimal weakly stable matching, or reporting that none exists, given an SMTI instance, is open, though we conjecture that the problem is NP-hard. A related decision problem is to determine whether a given weakly stable matching $M$ in $I$ is man-optimal. Gelain et al. [245, 249] gave a polynomial-time algorithm that finds a man-optimal weakly stable matching in $I$, or reports "I don't know" (thus the algorithm does not guarantee to terminate with the correct answer).

- The complexity of the problem of finding a sex-equal weakly stable matching does not appear to have been investigated in the context of SMT where the preference lists of one or both sexes are derived from one or two master lists.

- It is open as to whether the results of Sec. 3.4.5 can be extended to the case of SMTI under strong stability. (Recall that the problem of deciding whether a given SMP instance admits a strongly stable matching is NP-complete.) That is, it is of interest to investigate the complexity of each of the problems of finding a "succinct certificate" for the non-existence of a strongly stable matching, a minimum regret strongly stable matching, an egalitarian strongly stable matching, all strongly stable pairs, and all strongly stable matchings, given an instance of SMT. For the latter problem, partial results concerning the generation of strongly stable matchings within a single equivalence class are contained in Ref. [415].

We note that Feder [201, p.148] conjectures that, for an instance $I$ of SMT, the problem of deciding whether there is a strongly stable matching other than the man-optimal and woman-optimal strongly stable matchings is NP-complete. However given the observations of Sec. 3.3.3, Feder's conjecture would be more appropriately stated as follows: is there a strongly stable matching $M$ in $I$, such that $M \notin [M_a]$ and $M \notin [M_z]$, where $[M_a]$ and $[M_z]$ are the equivalence classes corresponding to a man-optimal and a woman-optimal strongly matching in $I$, respectively? Indeed, it would seem likely that a suitable definition of a rotation in this context could be used to exploit the lattice structure present for strongly stable matchings in $I$ in order to disprove the conjecture.

# Chapter 4

# The Stable Roommates problem

## 4.1 Introduction

As noted in Chap. 1, the Stable Roommates problem is a non-bipartite generalisation of SM, yet prior to 1989 it had been something of a "poor cousin" of its more prominent special case. This changed with the publication of Gusfield and Irving's book [261], in which a whole chapter was devoted to SR, gathering together work in Refs. [306,307,260], in addition to contributing many new key results. As already noted in Chap. 2, Gusfield and Irving concluded with an appendix containing 12 open problems (or more accurately, a range of open problems organised into 12 subsections). Some of these concern SR, and in Sec. 4.2, we update the reader on what is now known, and what is still open, regarding these problems.

Considering more generally the progress that has been made on SR after 1989, a key landmark is the work of Tan (and Hsueh) [556–559] on stable partitions. This structure, which we define formally in Sec. 4.3.1, is present in *every* SR instance, and its existence is strong compensation for the fact that a stable matching need not exist. In Sec. 4.3, we survey the key results relating to stable partitions, including the Tan–Hsueh algorithm for finding a stable partition.

Another important development concerning the structure of stable matchings in the SR context in recent years has been the identification of a new meet semilattice for SR stable matchings. Cheng and Lin [146] established this structure by considering so-called *mirror posets* and *median graphs*, arguing that the new meet semilattice gives rise to a more natural description of SR stable matchings than the earlier structure due to Gusfield and Irving [261]. We describe these results in Sec. 4.4.

Indifference in the context of instances of SMI and HR was surveyed in Chap. 3. Another significant development concerning SR was the study of variants of this problem involving forms of indifference. In some cases, results from SMTI have generalisations to the case of SR with ties that are non-trivial, but perhaps not surprising. However there are some exceptions: for example, whilst the problem of finding a weakly stable matching (of any size) is polynomial-time solvable in the SMTI case, the corresponding problem is NP-hard in the Stable Roommates case. We survey known results corresponding to SR with indifference in Sec. 4.5.

The possible unsolvability of an SR instance motivates methods for finding matchings that are "as stable as possible". Several possible definitions of such matchings are given in Secs. 4.3.4 and 4.3.6; all relate to stable partitions. Another possible interpretation of this concept involves finding matchings with the minimum number of blocking pairs, and is studied in Sec. 4.6.

An important special case of SR arises when there is a global ranking of the edges in the underlying graph, and the agents' preference lists respect this ranking function. We describe results concerning this problem variant in Sec. 4.7.

In Sec. 4.8, we turn our attention to a range of generalisations of SR, including variants where the underlying graph may be capacitated and/or involve parallel edges, some edges may be forbidden, partnerships may be non-integral in size, preference relations may be in the form of choice functions, and agents may form coalitions of arbitrary size.

Finally, in Sec. 4.9 we give some concluding remarks and open problems corresponding to SR and its variants.

## 4.2   Updates to open problems 8–12 from Gusfield & Irving

Problems 8–12 of Gusfield and Irving's appendix [261] specifically relate to SR, and we summarise the work that has been done towards solving each of these problems in this section. In some cases the problem in question has been completely solved, whilst in other cases only partial answers have been given so far.

### 4.2.1   *8: Solvable Roommates Instances*

Let $p_n$ denote the probability that a random instance of SR with $n$ agents (where $n$ is even) is solvable. Open problem 8 focused on (i) whether $p_n$

Table 4.1 Values of $q_{\mathcal{I}_n}$ for various values of $n$

| $n$ | $q_{\mathcal{I}_n}$ | Reference |
|---|---|---|
| 10 | 0.89 | [534] |
| 100 | 0.64 | [534] |
| 500 | 0.45 | [534] |
| 1000 | 0.38 | [448] |
| 5000 | 0.26 | [448] |
| 10000 | 0.23 | [448] |
| 20000 | 0.18 | [448] |

is a strictly decreasing function of $n$, (ii) whether $\lim_{n\to\infty} p_n = 0$, and (iii) finding upper and lower bounds for $p_n$.

Whilst exact values for $p_2$, $p_4$ and $p_6$ can be calculated, namely $p_2 = 1$, $p_4 = \frac{26}{27} \approx 0.963$ [261] and $p_6 = \frac{181431847}{194400000} \approx 0.933$ [448], for each $n > 6$ the number of instances (namely $((n-1)!)^{n-1}$ is too large for all to be generated and tested for solvability systematically. Instead, computational simulations can estimate the value of $p_n$ by randomly generating a (large) sample, say 10000, of SR instances of size $n$ and calculating the proportion that are observed to be solvable.

Empirical results from Refs. [534] and [448] suggest that, as $n$ increases, $p_n$ decreases steeply up to around the $n = 1000$ mark, and then begins to decrease more slowly thereafter. In particular, Sng [534] created a set $\mathcal{I}_n$ of 10000 randomly generated instances of SR, each of size $n$, for various values of $n$ including $n \in \{10, 100, 500\}$. For each such set $\mathcal{I}_n$, he computed $q_{\mathcal{I}_n}$, the proportion of instances of $\mathcal{I}_n$ that were observed to be solvable. (In fact his simulations handled instances of size up to 8000 in intervals of 1000, but for $1000 \le n \le 8000$, the number of generated instances was only 5000.) Additionally, Mertens generated families of 10000 instances $\mathcal{I}_n$ for a range of values of $n$ including $n \in \{1000, 5000, 10000, 20000\}$. Again, he computed $q_{\mathcal{I}_n}$ for each such set. The values of $q_{\mathcal{I}_n}$ for the aforementioned values of $n$ are given in Table 4.1.

Based on his computed values of $q_{\mathcal{I}_n}$, Mertens [448] conjectured that, asymptotically, $p_n \approx e\sqrt{2/\pi}n^{-\frac{1}{4}}$.

Pittel [481] proved that $p_n \ge (1 + o(1))(2e^{\frac{3}{2}})/\sqrt{\pi n}$, whilst Pittel and Irving [482] showed that $\lim_{n\to\infty} p_n \le \sqrt{e}/2 \approx 0.8244$. The empirical evidence therefore suggests that this asymptotic upper bound is not likely to be particularly tight.

Now let $S_n$ be a random variable that denotes the number of stable matchings for a random SR instance with $n$ agents. Pittel [481] also showed that $\lim_{n\to\infty} E(S_n) = \sqrt{e}$.

## 4.2.2   9: Roommates to Marriage

This problem centred around the question of whether there is a polynomial-time reduction from a solvable SR instance $I$ to an SM instance $J$ such that there is a 1–1 correspondence between the sets of stable matchings in $I$ and $J$. This problem is still open (a solution was proposed by Gusfield [258], but was found by Irving to be erroneous).

However Dean and Munshi [162] provided a partial answer, along the following lines. Gusfield and Irving suggested turning each agent in $I$ into two persons (one male and one female) in $J$, whose preference lists are related to that of the original agent in $I$. Following this suggestion, Dean and Munshi transformed each agent $a_i$ in $I$ into a man $m_i$ and a woman $w_i$ in $J$. The preference list of $m_i$ (respectively $w_i$) in $J$ is identical to that of $a_i$ except that each occurrence of $a_j$ on $a_i$'s list is replaced by $w_j$ (respectively $m_j$). Hence in fact $J$ is an instance of SMI; moreover, Dean and Munshi referred to this as a *symmetric* SMI instance. They defined a stable matching $M$ in $J$ to be *symmetric* if $(m_i, w_j) \in M$ if and only if $(m_j, w_i) \in M$. Clearly there is a 1–1 correspondence between the stable matchings in $I$ and the symmetric stable matchings in $J$. The authors gave a polynomial-time algorithm for finding a symmetric stable matching in $J$ by characterising the set of rotations to be eliminated in order to arrive at such a matching. In fact, this reduction holds for the more general Stable Allocation problem (defined in Sec. 4.8.6).

An alternative approach that could still yield some insight into potential structural correspondences between SR and SM is to reduce an *arbitrary* (i.e., not necessarily solvable) instance of SR to an instance of some variant $\Pi$ of SM, where it is not the case that an arbitrary instance of $\Pi$ is guaranteed to admit a stable matching. $\Pi$, for example, could be SMT under super-stability, or even COM SMTI under weak stability. A reduction along these lines, due to Manlove and Abraham (see Ref. [14]), transforms from SR to the problem of deciding whether a complete weakly stable matching exists, given an instance of the variant of SMI in which the preference lists are acyclic relations.

We conclude by remarking that a paper of Hsueh [283] aimed to unify the structures of SR and SM by transforming an SR instance $I$ into an SM instance

$J$ according to Gusfield and Irving's suggestion as described above. The author then inferred structural results about $I$ from those of $J$. Also, for each of SR and SM, Hsueh gave a representation of the stable matchings for a given instance in terms of a so-called *Faigle geometry*. Whilst Hsueh's work contributes some interesting structural correspondences between SR and SM, it does not solve the basic question that this subsection is concerned with.

### 4.2.3   10: Succinct Certificates

Part of this problem focused on whether there is a "succinct certificate" for the unsolvability of a given SR instance. This problem was solved by Tan [557], who demonstrated that an arbitrary instance of SR with $n$ agents admits a structure called a *stable partition*, which can be found in $O(n^2)$ time. From this, it is a straightforward matter to check (in $O(n)$ time) whether $I$ is solvable. Stable partitions are described in more detail in Sec. 4.3.

The stable partition structure provides a natural characterisation of the solvability or otherwise of a given SR instance; the question as to whether such a characterisation could be found was posed in Sec. 10.1 of the Appendix of Ref. [261]. In Sec. 10.2 of the Appendix of Ref. [261], the authors asked whether, given the lower bound results for SM [464] (see Sec. 2.10.10), one can prove an $\Omega(n^2)$ lower bound for the problem of determining whether a given SR instance $I$ with $n$ agents is solvable (note that here we are simply required to decide the solvability or otherwise of $I$, and not to find a stable matching if $I$ is indeed solvable). As far as we are aware, this problem is still open. However Feder [202] showed that the problem of finding a stable matching in $I$ or reporting that none exists is logspace reducible to the problem of deciding whether $I$ admits a stable matching, and vice versa, showing that the two problems have the same parallel complexity.

### 4.2.4   11: Algorithmic Improvements

The part of this problem corresponding to SR concerned whether (i) all rotations for a given SRI instance $I$ can be found in $O(m)$ time, and (ii) the stable matchings in $I$ can be listed in $O(n)$ time per matching (after some polynomial-time initial pre-processing phase), as in the SM case, where $n$ is the number of agents and $m$ is the number of acceptable pairs in $I$.

Results of Feder [202,203] impacted on (i) and (ii) in the following way. For (i), Feder was able to improve the time complexity of the previous algorithm (referred to in Part (ii) Theorem 1.24) from $O(nm \log n)$ to $O(nm)$.

For (ii), Feder was able to answer this question in the affirmative, requiring $O(m)$ pre-processing time. Recall that the previous algorithm (referred to in Part (iii) of Theorem 1.24) listed each stable matching in $O(m)$ time per matching following $O(nm \log n)$ pre-processing time.

In order to describe these results, we begin by extending our definition of the rotation poset as given in Sec. 1.4.4. The so-called *extended rotation poset* $R_I^*$ for $I$ contains all the rotations together with the syntactic duals of the singular rotations, and restricting this structure by excluding these latter elements gives the actual rotation poset. We can find, in $O(m)$ time, a directed graph $R_I$ that represents $R_I^*$, in the sense that the transitive closure of $R_I$ is isomorphic to $R_I^*$ (see Sec. 4.4.1 of Ref. [261]). Digraph $R_I$ is constructed by scanning each preference list in turn, adding a sequence of edges derived from the rotations represented in that list (see Sec. 4.4.1 of Ref. [261]). As a consequence, the *explicit width* of $R_I$ is at most $n$, meaning that we can find a set of at most $n$ vertex–disjoint paths in $R_I$ that cover all the vertices — one such path arises from each preference list.

Digraph $R_I$ turns out to be equivalent to the implication digraph of an instance $J$ of acyclic 2-SAT [261, pp.194–195]. In $J$, each variable and its negation correspond to a rotation and its syntactic dual. The clauses of $J$ are of the form $(\rho \vee \bar{\sigma})$ for any pair of rotations such that $(\rho, \sigma)$ is an edge in $R_I$ (which implies that $\rho$ precedes $\sigma$ in $R_I^*$). Because a singular rotation precedes its syntactic dual in $R_I^*$, the singular rotations are precisely the trivial variables in $J$ — i.e., those that are true in every satisfying truth assignment. Hence the true variables in any satisfying truth assignment for $J$ correspond to a complete closed set of rotations in $I$. The converse is also true, so by Theorem 1.23 there is a 1–1 correspondence between the satisfying truth assignments for $J$ and the stable matchings in $I$. Note that $J$ has $O(m)$ variables and clauses, and can be constructed from $I$ in $O(m)$ time.

The implication digraph $D$ of $J$ has a vertex for each literal and a directed edge $(\sigma, \rho)$ if $(\rho \vee \bar{\sigma})$ is a clause in $J$. So, in fact, $D$ is structurally identical to $R_I$, except that the direction of every edge is reversed.

Feder [203] established that we can construct in $O(nm)$ time a representation of the transitive closure $D^*$ of $D$, which enables us to test in $O(1)$ time whether a given edge is in $D^*$ or not. This allows the singular rotations to be identified, since a rotation $\rho$ is singular if and only if $(\bar{\rho}, \rho) \in D^*$. In turn, this allows the stable pairs of $I$ to be found, since these are precisely the (disjoint) union of the fixed pairs and the pairs that are in some non-singular rotation [261, Lemma 4.4.1]. Furthermore, Feder [203] showed that the satisfying truth assignments of $J$, and hence the stable matchings of $I$, may be listed efficiently. The following result summarises the consequences that arise from the discussion so far.

**Theorem 4.1 ([203]).** *Let $I$ be an instance of* SRI *with $n$ agents, where $m$ is the number of acceptable pairs in $I$, and let $J$ be the instance of 2-*SAT *as described above. Then:*

*(i) the stable pairs for $I$ can be found in $O(nm)$ time;*

*(ii) the rotations for $I$ can be found and determined as singular or non-singular in $O(nm)$ time;*

*(iii) the satisfying truth assignments for $J$, and therefore the stable matchings for $I$, can be listed in $O(n)$ time per solution, after $O(m)$ pre-processing time.*

The time complexities indicated in Parts (i), (ii) and (iii) of Theorem 4.1 therefore improve on the corresponding time complexities given by Theorem 1.24.

Feder [203, p.317] remarked that the problem of computing all stable pairs for a given SRI instance has an inherent dependency on bipartite matching and transitive closure, and hence an $O(m)$ algorithm for this problem may be unlikely to exist. Given that a rotation $\rho = (a_{i_0}, a_{j_0}), \ldots, (a_{i_{r-1}}, a_{j_{r-1}})$ is singular if and only if no $\{a_{i_k}, a_{j_k}\}$ is a stable pair $(0 \leq k \leq r - 1)$ [261, Lemma 4.4.1], Feder's remark also applies to the problem of computing all rotations.

### 4.2.5   12: Optimal Roommates

Gusfield and Irving asked whether the problem of finding a minimum weight stable matching, given an instance of SR, is solvable in polynomial time. Feder [202] gave an answer in the negative: he showed that, in fact, the problem of finding an egalitarian stable matching, given an instance of SR, is NP-hard, and hence finding a minimum weight stable matching is also NP-hard. However the latter problem is approximable within a factor of 2 [263, 202, 203, 565, 162]. Gusfield and Pitt [263] gave a 2-approximation algorithm for the problem of finding a minimum weight stable matching, given an SRI instance $I$, running in $O(m^2)$ time, where $m$ is the number of acceptable pairs in $I$. Feder [202, 203] improved the running time to $O(m \log(n^2/m))$, where $n$ is the number of agents in $I$. Moreover he proved that this problem is approximable within a constant factor of $c$ if and only if the problem of finding a minimum vertex cover in a graph $G$ is approximable within $c$. It is well known that, at present, the best known value of $c$ for the latter problem is 2 — this classical result is due to Gavril [243] (cited in Ref. [241]). Moreover, the approximability of this problem within a constant factor less than 2 would imply the truth of the Unique Games Conjecture [377].

### 4.2.6   *12.1: Linear Programming for Roommates*

Gusfield and Irving asked whether there is an efficiently obtainable linear programming (LP) formulation of SR, similar to that obtainable for SM [576, 520, 510] (see Sec. 2.4).

Abeledo and Rothblum [11] solved this problem by proving that, in the SRI context, the extreme points of the polytope of solutions to the system of linear inequalities formulated by Rothblum [520] for SMI are half-integral and give rise to so-called *stable half-matchings* (see Sec. 4.3.5), and the integral extreme points of this polytope give rise to stable matchings. Abeledo and Rothblum gave some structural results for stable half-matchings, extending some of those obtained by Roth *et al.* [510] for fractional stable matchings in the SMI case (see Sec. 2.4), including a proof for SRI that the *median choice* (see Theorem 1.20) between three stable half-matchings gives rise to a stable half-matching.

Abeledo and Blum [9] showed that, for a given SRI instance $I$, a stable matching, or the non-existence of one, can be determined after solving a series of at most $2m + 1$ linear programs, where $m$ is the number of acceptable pairs in $I$. Again, these linear programs are based on the linear inequalities given by Rothblum [520].

Teo and Sethuraman [565] gave a new LP formulation of SR, which has the property that a feasible solution exists if and only if the corresponding SR instance is solvable. This property is not satisfied by the earlier LP formulation of Abeledo and Rothblum [11]: in that case a given system of linear inequalities could admit a feasible (half-integral) solution even if the corresponding SR instance is unsolvable. Teo and Sethuraman [565] used their LP formulation of SR to derive the structural property established by Theorem 4.16 (see Sec. 4.4). The authors also showed how their LP formulation gives rise to a 2-approximation algorithm for the problem of finding a minimum weight stable matching, given an instance of SR, and in a related paper [564], they showed that their heuristic performs impressively on a range of SR instances of a particular set of sizes.

Given the NP-hardness of finding an egalitarian stable matching in the SR context, as mentioned above, it follows that, in constrast to the case for SMI [576,520,510], the LP formulations of SRI mentioned in this section cannot give rise to an efficient characterisation of minimum weight stable matchings in the context of an SRI instance unless P=NP. However Teo and Sethuraman [566] proposed a further LP formulation for SR which, with the use of a cutting-plane heuristic, is shown to provide solutions that are on average within 6% of the weight of a minimum weight stable matching for a family of randomly generated instances, and always within a factor of 2 from optimal in the worst case for an arbitrary instance.

## 4.3 Stable partitions

### 4.3.1 *Introduction*

As indicated in Sec. 1.4.2, an SR instance may not admit a stable matching. Tan [557] defined an important structure, called a *stable partition*, with a range of useful properties that is, however, present in *every* instance of SRI. For a solvable SRI instance, a stable partition generalises the concept of a stable matching. In Sec. 4.3.2 we define the notion of a stable partition and state some of the properties arising from it. Then in Sec. 4.3.3 we describe an algorithm for constructing a stable partition. In Sec. 4.3.4 we consider the problem of finding, given an SRI instance $I$, a *maximum stable matching* (i.e., a maximum matching in $I$ such that there is no blocking pair involving the assigned agents). Then in Sec. 4.3.5, we explore the concept of a *stable half-matching* [81], which is equivalent to a stable partition. Finally, in Sec. 4.3.6, we describe *P-stable matchings* and *absorbing sets*, which provide additional methods (besides the one given in Sec. 4.3.4) for coping with the possible unsolvability of an SRI instance.

### 4.3.2 *Definition and structure of stable partitions*

We begin by giving a concise definition of a stable partition, due to Pittel and Irving [482]. The definition is valid for SRI instances, so we state it in this more general context, however we require to extend the preference list of agent $a_i$ so that $a_i$ ranks herself last (i.e., after all other agents on her original preference list). We assume henceforth in the remainder of this section that this has implicitly been done, for a given SRI instance.

**Definition 4.2 ([557, 482]).** *Let I be an instance of* SRI *where A is the set of agents. A stable partition is a permutation* $\Pi$ *of A satisfying the following two properties, for each* $a_i \in A$:

*(i) either* $\Pi(a_i) = \Pi^{-1}(a_i)$ *or* $a_i$ *prefers* $\Pi(a_i)$ *to* $\Pi^{-1}(a_i)$;
*(ii) if* $a_i$ *prefers* $a_j$ *to* $\Pi^{-1}(a_i)$ *then* $a_j$ *prefers* $\Pi^{-1}(a_j)$ *to* $a_i$.

*For a given agent* $a_i$, *we define* $\Pi(a_i)$ *and* $\Pi^{-1}(a_i)$ *as the* successor *and* predecessor *of* $a_i$, *respectively, relative to* $\Pi$. *We refer to a cycle in* $\Pi$ *(which could be of length 1) of odd (respectively even) length as an* odd *(respectively* even*) party.*

As stated in Ref. [482], we remark that if $a_i$ is a fixed point of $\Pi$ then $a_i$ is both her own predecessor and successor. Moreover if $(a_i \; a_j)$ forms a transposition of $\Pi$ then $a_i$ is both the predecessor and successor of $a_j$ and vice versa.

The first result of this section indicates that a stable partition is guaranteed to exist in an instance of SRI, and can be found in linear time.

**Theorem 4.3 ([557]).** *Let I be an instance of* SRI. *Then I admits a stable partition, which can be found in* $O(m)$ *time, where m is the number of acceptable pairs of agents in I.*

We will consider algorithms for finding a stable partition in more detail in Sec. 4.3.3.

It is an immediate consequence of Condition (ii) in Definition 4.2 that if every party in a given stable partition $\Pi$ is of length 1 or 2 then the even parties in $\Pi$ give rise to a stable matching $M$, and the odd parties in $\Pi$ correspond to the agents who are unassigned in $M$. It turns out that there is a more general case in which $\Pi$ may give rise to a stable matching, as indicated by the following results.

**Theorem 4.4 ([557]).** *Let I be an* SRI *instance and let* $\Pi$ *be a stable partition in I. Suppose that* $C = (a_1 \; a_2 \; \ldots \; a_{2k})$ *is an even party in* $\Pi$ *for some* $k \geq 2$. *Then* $\Pi' = (\Pi \backslash C) \cup (a_1 \; a_2) (a_3 \; a_4) \ldots (a_{2k-1} \; a_{2k})$ *and* $\Pi'' = (\Pi \backslash C) \cup (a_2 \; a_3) (a_4 \; a_5) \ldots (a_{2k} \; a_1)$ *are also stable partitions in I.*

**Corollary 4.5 ([557]).** *Let I be an* SRI *instance. If* $\Pi$ *is a stable partition in I in which every party has length 1 or 2, then I is solvable. Conversely if I admits a stable matching M then M gives rise to a stable partition* $\Pi$ *in I in which every party has length 1 or 2.*

The next result indicates the effect on the solvability of $I$ if an odd party in a given stable partition in $I$ has length $\geq 3$.

**Theorem 4.6 ([557]).** *Let $I$ be an* SRI *instance and let $\Pi$ be a stable partition in $I$. Suppose that $\Pi$ contains an odd party of length $\geq 3$. Then $I$ is unsolvable.*

The following theorem is key towards proving that Theorem 4.6 gives rise to a characterisation of the unsolvability of an SRI instance.

**Theorem 4.7 ([557]).** *Let $I$ be an* SRI *instance. Then any two stable partitions in $I$ contain exactly the same odd parties.*

An alternative, simpler, proof of Theorem 4.7 appears in [558]. We are now in a position to state Tan's characterisation of the unsolvability of a given SRI instance $I$ in terms of the parties contained in any stable partition of $I$, which is an immediate consequence of Corollary 4.5 and Theorem 4.7.

**Corollary 4.8 ([557]).** *Let $I$ be an* SRI *instance and let $\Pi$ be a stable partition in $I$. Then $I$ is unsolvable if and only if $\Pi$ contains an odd party of length $\geq 3$.*

The existence of an odd party of length $\geq 3$ in a stable partition in $I$ has been referred to as a "succinct certificate" in the literature for the unsolvability of $I$ [557]. This is because, in the absence of the stable partition structure, a human who wishes to verify that $I$ is unsolvable in polynomial time would have to make do with an execution trace of Irving's algorithm ([306]) as applied to $I$. Although this is bound to have $O(m)$ size (where $m$ is the number of acceptable pairs in $I$), such a trace would in general be considerably more complex for a human to verify than a single stable partition $\Pi$ in $I$. Once $\Pi$ has been generated, it is of course a simple matter to check for the existence of an odd party of length $\geq 3$.

As noted by Pittel and Irving [482], the concept of an odd party of length $\geq 3$ in a stable partition, and Corollary 4.8, was implicit in the notion of an "improper rotation" introduced by Irving [307], but the full significance of odd parties was first understood and established by Tan [557].

**Example 4.9 ([482]).** *Consider* SR *instances $I_1$ and $I_2$ shown in Fig. 4.1. In $I_1$, there are five stable partitions, namely $\Pi_1 = (a_1\ a_4\ a_2\ a_6)(a_3\ a_5)$, $\Pi_2 = (a_1\ a_6\ a_3\ a_5)(a_2\ a_4)$, $\Pi_3 = (a_1\ a_4)(a_2\ a_6)(a_3\ a_5)$, $\Pi_4 = (a_1\ a_6)(a_2\ a_4)(a_3\ a_5)$ and $\Pi_5 = (a_1\ a_5)(a_3\ a_6)(a_2\ a_4)$; clearly $\Pi_3$, $\Pi_4$ and*

$$a_1 : a_2 \ a_4 \ a_3 \ a_6 \ a_5 \ a_1 \qquad\qquad a_1 : a_2 \ a_3 \ a_6 \ a_5 \ a_4 \ a_1$$
$$a_2 : a_6 \ a_5 \ a_4 \ a_1 \ a_3 \ a_2 \qquad\qquad a_2 : a_6 \ a_1 \ a_3 \ a_4 \ a_5 \ a_2$$
$$a_3 : a_2 \ a_5 \ a_6 \ a_1 \ a_4 \ a_3 \qquad\qquad a_3 : a_6 \ a_2 \ a_5 \ a_1 \ a_4 \ a_3$$
$$a_4 : a_5 \ a_2 \ a_1 \ a_3 \ a_6 \ a_4 \qquad\qquad a_4 : a_6 \ a_2 \ a_5 \ a_1 \ a_3 \ a_4$$
$$a_5 : a_1 \ a_3 \ a_2 \ a_4 \ a_6 \ a_5 \qquad\qquad a_5 : a_1 \ a_2 \ a_3 \ a_6 \ a_4 \ a_5$$
$$a_6 : a_3 \ a_1 \ a_4 \ a_5 \ a_2 \ a_6 \qquad\qquad a_6 : a_5 \ a_2 \ a_1 \ a_3 \ a_4 \ a_6$$

$$I_1 \qquad\qquad\qquad\qquad I_2$$

Fig. 4.1   Instances $I_1$ and $I_2$ of SR due to Pittel and Irving [482]

$\Pi_5$ *give rise to stable matchings. In* $I_2$ *there is a single stable partition, namely* $\Pi = (a_1 \ a_3 \ a_5)(a_2 \ a_6)(a_4)$, *and therefore* $I_2$ *is unsolvable.*

### 4.3.3   Algorithms for finding a stable partition

As noted in the previous subsection, Tan [557] gave an $O(m)$ algorithm for finding a stable partition in a given SRI instance $I$, where $m$ is the number of acceptable pairs in $I$. This algorithm is an extended and modified version of Irving's algorithm [306] for finding a stable matching in $I$ or reporting that none exists. Later, Tan and Hsueh [559] described an alternative algorithm for constructing a stable partition in $I$ which is conceptually simpler but has $O(n^3)$ complexity, where $n$ is the number of agents in $I$. This algorithm has been referred to as *dynamic* or *incremental* [81] because it essentially assumes that a stable partition $\Pi_r$ is given for an SRI instance $I_r$ with $r$ agents, and shows how to modify $\Pi_r$ to arrive at a stable partition $\Pi_{r+1}$ for an SRI instance $I_{r+1}$, where $I_{r+1}$ is obtained from $I_r$ following the arrival of some additional agent. A nice exposition of the Tan–Hsueh algorithm is given by Pittel and Irving [482], and our description of the algorithm in this section is based on their approach. However unfortunately their pseudocode is not quite correct[1]; we have used the description given by Biró *et al.* [81] in order to correct the error. (The incorrect steps in Pittel and Irving's pseudocode are revealed by Figs. 4 and 5 in Ref. [81]; we use the same example at the end of this subsection to illustrate the problem.)

We begin by describing the Tan–Hsueh algorithm for the offline version of the problem.

Suppose that $I$ is an SRI instance and $A = \{a_1, \dots, a_n\}$ is the set of agents in $I$. Let $I_r$ denote the restriction of $I$ in which the set of agents in $I_r$ is $A_r = \{a_1, \dots, a_r\}$, and the preference list of an agent $a_i \in A_r$ in $I_r$ is

---

[1]It is important to stress that the incorrect steps in the algorithm do not affect the results in Ref. [482], since in the subsequent exposition, it turns out that the affected case in the algorithm in Fig. 1 of that paper (lines 20–22) can never arise.

**Algorithm 4.1** Algorithm Tan–Hsueh [559, 482]

**Require:** $\Pi_r$ is a reduced stable partition of $I_r$
**Ensure:** $\Pi_{r+1}$ is a reduced stable partition of $I_{r+1}$
1: $\Pi_{r+1} := \Pi_r$;
2: $a_p := a_{r+1}$;     {$a_p$ denotes the "proposer"}
3: $Q := \langle\rangle$;
4: *cycling* := false;
5: **loop**
6:     **if** some agent $a_q$ on $a_p$'s list prefers $a_p$ to $a_q$'s predecessor in $\Pi_{r+1}$ **then**
7:         $a_q :=$ most-preferred such agent;     {according to $a_p$'s list in $I_{r+1}$}
8:         append$(Q, \langle a_p, a_q \rangle)$;
9:         **if** *cycling* **and** $a_q = a_x$ **then**
10:            $\langle a_x, a_{i_1}, a_{i_2}, \ldots, a_{i_{2k-1}}, a_p, a_q \rangle :=$suffix$(Q, a_x)$;
11:            $\mathcal{C} := (a_x \ a_{i_1})(a_{i_2} \ a_{i_3}) \ldots (a_{i_{2k-2}} \ a_{i_{2k-1}})$;
12:            $\Pi_{r+1} := (\Pi_{r+1} \setminus \mathcal{C}) \cup (a_p \ a_{i_{2k-1}} \ldots a_{i_2} \ a_{i_1} \ a_x)$;
13:            **return** ;
14:        **else if** $a_q$ is in an odd party $C = (a_q \ a_{i_1} \ a_{i_2} \ \cdots \ a_{i_{2k}})$ of $\Pi_{r+1}$ **then**
15:            $\Pi_{r+1} := (\Pi_{r+1} \backslash C) \cup (a_p \ a_q)(a_{i_1} \ a_{i_2}) \ldots (a_{i_{2k-1}} \ a_{i_{2k}})$;
16:            **return** ;
17:        **else** {$a_q$ is in a transposition of $\Pi_{r+1}$}
18:            $a_t := \Pi_{r+1}(a_q)$;
19:            $\Pi_{r+1} := (\Pi_{r+1} \setminus (a_q \ a_t)) \cup (a_p \ a_q)$;
20:            $a_p := a_t$;     {$a_t$ is the next proposer}
21:            **if** !*cycling* **and** $a_q$ was previously a proposer **then**
22:                *cycling* := true;
23:                $a_x := a_t$;
24:            **end if**
25:        **end if**
26:    **else**
27:        $\Pi_{r+1}(a_p) := a_p$;
28:        **return** ;
29:    **end if**
30: **end loop**

derived from her preference list in $I$ by simply deleting any agent in $A \backslash A_r$. Clearly the unique stable partition of $I_1$ is $\Pi_1 = (a_1)$. Algorithm Tan–Hsueh, described in Algorithm 4.1, constructs a stable partition $\Pi_{r+1}$ in $I_{r+1}$, given a stable partition $\Pi_r$ in $I_r$ ($1 \le r \le n-1$). In fact the algorithm assumes that $\Pi_r$ is a *reduced stable partition* and ensures that $\Pi_{r+1}$ also satisfies this property. A stable partition is *reduced* if every even party is of length 2. By Theorem 4.4, any stable partition can be transformed to a reduced stable partition in linear time; clearly this property is satisfied by $\Pi_2$ in any case.

The algorithm begins by initialising $\Pi_{r+1}$ to be equal to $\Pi_r$. A *proposer* $a_p$ is then identified who will make proposals to certain agents on her preference list in the spirit of the Gale–Shapley algorithm; $a_p$ is initially $a_{r+1}$. Also a queue of agents $Q$ is maintained, which is initially empty. Finally, a boolean *cycling* is maintained, which indicates whether we are travelling round an odd cycle; initially *cycling* is set to false. We then enter a loop which iterates until one of three possible exit conditions loop is reached.

Within a loop iteration, we test whether there is an agent $a_q$ on $a_p$'s list who prefers $a_p$ to $a_q$'s predecessor in $\Pi_{r+1}$. If this is not the case then we simply set $a_p$ to be a fixed point under $\Pi_{r+1}$ and exit. Otherwise we let $a_q$ be the most-preferred agent on $a_p$'s list with the aforementioned property. Implicitly $a_p$ *proposes* to $a_q$ here, denoted by $a_p \to a_q$. We append $a_p$ and $a_q$ to $Q$ in that order.

Assume that *cycling* is still false at this point. If $a_q$ is in an odd party[2] $C = (a_q\ a_{i_1}\ a_{i_2}\ \ldots\ a_{i_{2k}})$ of $\Pi_{r+1}$ for some $k \geq 0$, then we replace $C$ by the transpositions $(a_p\ a_q)\ (a_{i_1}\ a_{i_2})\ \ldots\ (a_{i_{2k-1}}\ a_{i_{2k}})$ in $\Pi_{r+1}$, and exit. Otherwise $a_q$ must be in a transposition $(a_q\ a_t)$ of $\Pi_{r+1}$, since $\Pi_r$ is reduced, and this property is also a loop invariant for $\Pi_{r+1}$. We replace $(a_q\ a_t)$ by $(a_p\ a_q)$ in $\Pi_{r+1}$, and $a_t$ becomes the proposer for the next loop iteration.

If $a_q$ was previously a proposer then *cycling* is set to true and we store $a_t$ in $a_x$. Intuitively, at this point the proposal process has cycled. However a fundamental theorem of Tan and Hsueh [559] states that, if we allow the proposal process to continue with $a_x$, each subsequent proposal $a_p \to a_q$ satisfies the property that $a_q$ belongs to a transposition of $\Pi_{r+1}$, and moreover, the sequence "returns" to the point where $a_q = a_x$. That is, the proposals starting from the very next loop iteration after $a_x$ was initialised are as follows:

$$a_x \to a_{i_1}, a_{i_2} \to a_{i_3}, \ldots, a_{i_{2k-2}} \to a_{i_{2k-1}}, a_{i_{2k}} \to a_q,$$

where $a_q = a_x$ and $\{a_x, a_{i_1}, a_{i_2}, \ldots, a_{i_{2k-1}}, a_{i_{2k}}\}$ are distinct agents. This is recognised by the conditional at line 9 of the algorithm. In that loop iteration, $a_{i_{2k}} = a_p$. This sequence of proposers and proposees is obtained using the method $\text{suffix}(Q, a_x)$, which returns the sub-list of $Q$ starting from the second-last occurrence of $a_x$. Tan and Hsueh's fundamental theorem further states that the transpositions $(a_x\ a_{i_1})\ (a_{i_2}\ a_{i_3})\ \ldots\ (a_{i_{2k-2}}\ a_{i_{2k-1}})$ should be replaced by the odd party $(a_p\ a_{i_{2k-1}}\ a_{i_{2k-2}}\ \ldots\ a_{i_2}\ a_{i_1}\ a_x)$ in $\Pi_{r+1}$. After completing this step, the algorithm terminates. Thus, *cycling* being true signifies that agents are proposing around an odd cycle.

---

[2]In fact this case cannot arise if *cycling* is true, as we later point out.

| Agents' preferences | Proposals $(a_p \to a_q)$ | $a_t$ | $\Pi_7$ |
|---|---|---|---|
| $a_1 : a_3\ a_2$ | 1. $a_7 \to a_6$ | $a_5$ | $(a_1\ a_2)(a_3\ a_4)(a_6\ a_7)$ |
| $a_2 : a_1\ a_4$ | 2. $a_5 \to a_4$ | $a_3$ | $(a_1\ a_2)(a_4\ a_5)(a_6\ a_7)$ |
| $a_3 : a_4\ a_1$ | 3. $a_3 \to a_1$ | $a_2$ | $(a_1\ a_3)(a_4\ a_5)(a_6\ a_7)$ |
| $a_4 : a_2\ a_5\ a_3$ | 4. $a_2 \to a_4$ | $a_5$ | $(a_1\ a_3)(a_2\ a_4)(a_6\ a_7)$ |
| $a_5 : a_6\ a_4\ a_7$ | 5. $a_5 \to a_7$ | $a_6$ | $(a_1\ a_3)(a_2\ a_4)(a_5\ a_7)$ |
| $a_6 : a_7\ a_5$ | 6. $a_6 \to a_5$ | $a_7$ | $(a_1\ a_3)(a_2\ a_4)(a_5\ a_6)$ |
| $a_7 : a_5\ a_6$ | 7. $a_7 \to a_6$ | $a_5$ | $(a_1\ a_3)(a_2\ a_4)(a_5\ a_6\ a_7)$ |

Fig. 4.2 An execution of Algorithm Tan–Hsueh as applied to an SRI instance due to Biró et al. [81].

The time complexity of Algorithm Tan–Hsueh is $O(n^2)$ as noted in Ref. [559]. We summarise the preceding discussion with the following theorem.

**Theorem 4.10 ([559]).** *Let $I$ be an instance of SRI of size $n$. Then, by using Algorithm Tan–Hsueh $n - 2$ times, we can construct a stable partition in $I$ in $O(n^3)$ time.*

For the online version of the problem, let $I_r$ be an SRI instance with agents $A_r = \{a_1, \ldots, a_r\}$. Construct a reduced stable partition $\Pi_r$ in $I_r$ either using Algorithm Tan–Hsueh or otherwise. Now suppose that agent $a_{r+1}$ arrives. Let $I_{r+1}$ denote the SRI instance with agents $A_{r+1} = A_r \cup \{a_{r+1}\}$. The preference lists in $I_{r+1}$ are constructed as follows. We suppose that $a_{r+1}$ ranks in strict order a subset $S$ of the agents in $A_r$ (followed by $a_{r+1}$). Each agent $a_i \in S$ then inserts $a_{r+1}$ into some position $k_i$ her preference list, demoting by one place each agent who was ranked in position $k_i$ or worse in $a_i$'s preference list in $I_r$. Now Algorithm Tan–Hsueh can be used to construct $\Pi_{r+1}$, a reduced stable partition in $I_{r+1}$.

We now give an example to illustrate the operation of Algorithm Tan–Hsueh as applied to the unsolvable SRI instance $I$ with 7 agents, due to Biró et al. [81], as shown in Fig. 4.2. In $I_6$, $\Pi_6 = (a_1\ a_2)(a_3\ a_4)(a_5\ a_6)$ is a stable matching, and hence a reduced stable partition. We now execute Algorithm Tan–Hsueh in order to construct $\Pi_7$ from $\Pi_6$. The proposals, and the values of $a_t$ and $\Pi_7$ at each loop iteration are shown in Fig. 4.2. After proposal 5, $a_q = a_7$ is recognised as previously having been a proposer, so the *cycling* boolean is set to true, and $a_t = a_6$ is recorded as $a_x$. Then, after proposal 7, $a_q = a_x = a_6$, and thus the algorithm recognises that the odd cycle is complete. At this point, suffix$(Q, a_q) = \langle a_6, a_5, a_7, a_6 \rangle$, so the algorithm removes the transposition $(a_5\ a_6)$ from $\Pi_7$, adds the odd cycle $(a_5\ a_6\ a_7)$ to $\Pi_7$, and terminates with $\Pi_7 = (a_1\ a_3)(a_2\ a_4)(a_5\ a_6\ a_7)$ as a reduced stable partition in $I_7$. On the other hand the algorithm of Pittel and Irving [482] tests whether the proposee $a_q$ has previously been a proposer, and if so, it adds a new odd cycle containing all proposees and proposers since $a_q$. Thus in $I_7$, their algorithm terminates after proposal 5 with the cycle $(a_7 \ a_5 \ a_4 \ a_2 \ a_1 \ a_3 \ a_4 \ a_5 \ a_6)$, which is clearly incorrect. The error stems from the fact that the algorithm did not allow the proposal sequence to continue to the "return" in step 7, from where we deduce the true odd party.

### 4.3.4  Maximum stable matchings

To cope with the possible non-existence of a stable matching in a given SRI instance $I$, Tan [556, 558] introduced the notion of a *maximum stable matching* in $I$. This is a matching $M$ of maximum size satisfying the property that there is no blocking pair of $M$ in $I$ involving the agents who are assigned in $M$ (i.e., the assigned agents are said to be *stable within themselves*). This definition should not be ambiguous because, for solvable instances of SRI, all stable matchings have the same cardinality [261, Sec. 4.5.2], and therefore the notion of a stable matching of maximum size is redundant.

More formally, if $A$ is the set of agents in $I$, let $A_M \subseteq A$ denote the set of agents who are assigned in a given matching $M$ in $I$. Also for $S \subseteq A$, let $I \backslash S$ denote the sub-instance of $I$ obtained by deleting each $a_i \in S$ (and by implication, by deleting each such $a_i$ from the preference list of each $a_j \in A \backslash S$). For brevity, we denote $I \backslash (A \backslash A_M)$ by $I_{A_M}$. Then $M$ is a maximum stable matching in $I$ if and only if (i) $M$ is stable in $I_{A_M}$, and (ii) $|M|$ is maximum subject to (i).

Clearly, if there are at least two agents in $I$ then any maximum stable matching in $I$ has size at least 2. For a less trivial example, observe that $M = \{\{a_1, a_3\}, \{a_2, a_6\}\}$ is a maximum stable matching in the SR instance $I_2$ shown in Fig. 4.1.

For the remainder of this section we show how to find a maximum stable matching in $I$. We begin with the following result, proved by Tan [558].

**Proposition 4.11 ([558]).** *Let $I$ be an instance of SRI and let $\Pi$ be a stable partition in $I$. Suppose that $C = (a_{i_1} \ a_{i_2} \ \ldots \ a_{i_{2k+1}})$ is an odd party in $\Pi$ for some $k \geq 1$. Then $\Pi' = (\Pi \backslash C) \cup (a_{i_1} \ a_{i_2})(a_{i_3} \ a_{i_4}) \ldots (a_{i_{2k-1}} \ a_{i_{2k}})$ is a stable partition of $I \backslash \{a_{i_{2k+1}}\}$.*

For a given SRI instance $I$, let $\mathcal{O}(I)$ denote the number of odd parties in any stable partition of $I$ (this number is well-defined by Theorem 4.7). Clearly, in the context of Proposition 4.11, it follows that $\mathcal{O}(I \backslash \{a_{i_{2k+1}}\}) = \mathcal{O}(I) - 1$. It turns out that a greater reduction in the number of odd parties cannot be achieved by deleting an agent from an even party, as the following theorem indicates (in fact the number of odd parties could possibly increase in this case).

**Theorem 4.12 ([558]).** *Let $I$ be an instance of* SRI *and let $A$ be the set of agents in $I$. Then $|\mathcal{O}(I \backslash \{a_i\}) - \mathcal{O}(I)| = 1$ for any $a_i \in A$.*

We now present the main result of this subsection, whose proof indicates how to efficiently find a maximum stable matching in a given SRI instance.

**Theorem 4.13 ([558]).** *Let $I$ be an instance of* SRI. *A maximum stable matching in $I$ can be constructed in $O(m)$ time, where $m$ is the number of acceptable pairs of agents in $I$.*

**Proof.** Construct a stable partition $\Pi$ in $I$; by Theorem 4.3 this can be achieved in $O(m)$ time. Let $k$ be the number of odd parties in $\Pi$. Then $\mathcal{O}(I) = k$, so that, by Theorem 4.12, it is necessary to delete at least $k$ agents from $I$ in order to arrive at an SRI instance $I'$ such that $\mathcal{O}(I') = 0$. Such an instance $I'$ is solvable by Corollary 4.8. Let $I_0 = I$. Following Proposition 4.11, we form $I_1 = I_0 \backslash \{a_i\}$, where $a_i$ is an agent in an odd party in $I_0$, and we let $\Pi_1 = \Pi'$, where $\Pi'$ is as described in Proposition 4.11. We iterate this process, forming $I_2, \ldots, I_k$ and $\Pi_2, \ldots, \Pi_k$. Then $\mathcal{O}(I_j) = \mathcal{O}(I_{j-1}) - 1$ $(2 \leq j \leq k)$ and hence $\mathcal{O}(I_k) = 0$. Thus stable partition $\Pi_k$ contains no odd parties, and thus gives rise to a maximum stable matching in $I$ by Theorem 4.4. Clearly $\Pi_k$ can be constructed from $\Pi$ in $O(m)$ time. $\square$

**Corollary 4.14 ([558]).** *Let $I$ be an instance of* SRI. *Then a maximum stable matching in $I$ has size $(n - \mathcal{O}(I))/2$, where $n$ is the size of $I$.*

Tan [556] also described an alternative $O(m)$ approach for finding a maximum stable matching based on a modified and extended version of Irving's algorithm [306] for SR.

We also remark that the concept of a maximum stable matching is equally applicable when $I$ is solvable. In such a setting, it is tempting to consider the relative sizes of a maximum stable matching and a stable matching in $I$. The following proposition resolves this question.

**Proposition 4.15.** *Let $I$ be a solvable instance of* SRI *and let $M$ be a stable matching in $I$. Then $M$ is a maximum stable matching in $I$.*

**Proof.** Clearly $M$ is stable in $I_{A_M}$. Hence, to show that $M$ is a maximum stable matching, it remains to show that $|M| = |M'|$, where $|M'|$ is a maximum stable matching in $I$. By Corollary 4.14, $|M'| = (n - \mathcal{O}(I))/2$, where $n$ is the size of $I$. But $I$ is solvable, so no odd party in $I$ has length greater than 1. Hence each odd party in $I$ corresponds to an unassigned agent in $M$, i.e., $\mathcal{O}(I) = n - 2|M|$. Thus $|M| = |M'|$.                $\square$

We close this subsection by noting that, as observed by Biró *et al.* [81], the Tan–Hsueh algorithm is equivalent to the Roth–Vande Vate algorithm [516] (see Sec. 2.6) in the bipartite case.

### 4.3.5  *Stable half-matchings*

Biró *et al.* [81] defined the concept of a *stable half-matching* in an instance of SRI, and showed that it is equivalent to a stable partition.[3] Let $I$ be an instance of SRI and let $G = (A, E)$ be the underlying graph of $I$ (ignoring self-loops). A *half-matching* $M_h$ in $I$ is a set of edges in $G$ that can be partitioned into two sets, namely $H$ (*half-weighted edges*) and $M$ (*matching edges*), such that in $G$, each $a_i \in A$ is incident to either (i) one edge of $M$ and no edge of $H$, or (ii) no edge of $M$ and at most two edges of $H$. A half-matching $M_h$ is *stable* if (i), for each $e \in E \backslash M_h$, there exists an endpoint $a_i \in E$ such that either (a) $\{a_i, a_k\} \in M$ for some $a_k$ such that $a_i$ prefers $a_k$ to $a_j$ or (b) $\{\{a_i, a_k\}, \{a_i, a_l\}\} \subseteq H$ for some $a_k$, $a_l$ such that $a_i$ prefers each of $a_k$ and $a_l$ to $a_j$, where $e = \{a_i, a_j\}$, and (ii), for each $e \in H$, there exists an endpoint $a_i \in E$ such that $\{a_i, a_k\} \in H$ for some $a_k$ such that $a_i$ prefers $a_k$ to $a_j$, where $e = \{a_i, a_j\}$.

As described by Biró *et al.* [81], half-matchings can represent a practical situation where agents can create half-time partnerships. For example, if preference lists represent rankings of potential partners for a one-hour tennis match, then an agent could either not play at all, or play for half an hour with one partner, or play for one hour with two different partners (half an hour with each), or play for one hour with one partner. Stability represents the situation in which no two agents would like to improve their assignment by playing together for more time than before (including the

---

[3]In fact the concept was first defined, and its equivalence to a stable partition was first recognised, by Aharoni and Fleiner [33]. However the term *stable half-matching* was used by Biró *et al.* [81], and we base our exposition on their notation and terminology.

case where they were not already assigned to play together), and by possibly rejecting one or both (if applicable) of their existing partnerships in order to do so.

A stable half-matching can also be defined formally as follows. Let $M_h$ be a set of edges in $G$ that is partitioned into two sets, namely $H$ and $M$, as above. Let $x_{M_h} : E \longrightarrow \{0, \frac{1}{2}, 1\}$ be a weight function where, for $e \in E$,

$$x_{M_h}(e) = \begin{cases} 0, & e \notin H \cup M \\ \frac{1}{2}, & e \in H \\ 1 & e \in M. \end{cases}$$

Then $M_h$ is a *half-matching* if, for each $a_i \in A$,

$$\sum \{x_{M_h}(e) : e \in E \wedge a_i \in e\} \leq 1.$$

Now suppose $\{\{a_i, a_j\}, \{a_i, a_k\}\} \subseteq E$. Define $a_j \preceq_{a_i} a_k$ if either $a_i$ prefers $a_j$ to $a_k$ or $a_j = a_k$. A half-matching $M_h$ is *stable* if, for each $e \in E$, there exists $a_i \in e$ such that

$$\sum \{x_{M_h}(e') : e' = \{a_i, a_k\} \in E \wedge a_k \preceq_{a_i} a_j\} = 1,$$

where $e = \{a_i, a_j\}$.

The concept of a stable half-matching is equivalent to that of a stable partition, which may be seen as follows. Let $\Pi$ be a stable partition in $I$. By Theorem 4.4 we may assume that $\Pi$ is reduced. Construct a stable half-matching $M_h$ in $I$ as follows. For each transposition $(a_i \ a_j)$ in $\Pi$, add $\{a_i, a_j\}$ to $M$. If $(a_{i_0}, a_{i_1}, \ldots, a_{i_{2k}})$ is an odd party in $\Pi$ for some $k \geq 1$, add $\{a_{i_r}, a_{i_{r+1}}\}$ to $H$ $(0 \leq r \leq 2k$, where addition is taken modulo $2k + 1)$. Finally let $M_h = M \cup H$. It is straightforward to verify that $M_h$ is a stable half-matching in $I$. Conversely suppose that $M_h$ is a stable half-matching in $I$. Form a stable partition $\Pi$ as follows. For each edge $\{a_i, a_j\} \in M$, add the transposition $(a_i \ a_j)$ to $\Pi$. The stability condition implies that the edges in $H$ form a collection of cycles in $G$, each of length at least 3; add each such cycle to $\Pi$. It may be verified that $\Pi$ is a stable partition in $I$.

Aharoni and Fleiner [33] used a game-theoretic result of Scarf [521] to independently prove the existence of a stable half-matching for an instance of SRI.

### 4.3.6  P-stable matchings and absorbing sets

In Sec. 4.3.4, we showed that a stable partition can be used to construct a matching that is "stable" in a weaker sense than classical stability, as a means of coping with the possible non-existence of a stable matching. In this subsection we survey two alternative structures for achieving a similar outcome, both of which can be derived from a stable partition. The first such structure, defined by Inarra et el. [303], is called a *P-stable matching*, and can be defined as follows.

Let $I$ be an SRI instance and let $\Pi$ be a stable partition in $I$. Then a *P-stable matching* is a matching $M$ in $I$ such that, for each party $C = (a_{i_1} \ a_{i_2} \ \ldots \ a_{i_r})$ in $\Pi$, and for each $j$ $(1 \leq j \leq r)$, $M(a_{i_j}) \in \{\Pi^{-1}(a_{i_j}), \Pi(a_{i_j})\}$, apart from a unique $k$ such that $a_k$ is unassigned in $M$ in the case that $r$ is odd. Clearly (as a consequence of Theorem 4.3), $I$ admits at least one P-stable matching, and if $I$ is solvable then the set of P-stable matchings, taken over all stable partitions, coincides with the set of stable matchings. Moreover, the P-stable matchings are precisely the maximum stable matchings as defined in Sec. 4.3.4 (to verify this, the reader is referred to the proof of Theorem 4.13).

Inarra *et al.* [303] gave an algorithm for finding a P-stable matching based on the operation of *satisfying a blocking pair*. This operation can be described as follows: if $M$ and $M'$ are two matchings in $I$, then $M'$ is obtained from $M$ by *satisfying a blocking pair* $\{a_i, a_j\}$ of $M$ if (i) $\{a_i, a_j\} \in M'$, (ii) $M(a_i)$ (if it exists) is unassigned in $M'$, (iii) $M(a_j)$ (if it exists) is unassigned in $M'$, and (iv) every pair of $M$ not including $a_i$ or $a_j$ is in $M'$. The algorithm of Inarra *et al.* proceeds along the following lines: starting from an arbitrary matching $M_0$, a sequence of matchings $M_1, M_2, \ldots, M_k$ is constructed such that $M_k$ is a P-stable matching, and $M_i$ is obtained from $M_{i-1}$ by satisfying a blocking pair $(1 \leq i \leq k)$.

This algorithm generalises that of Diamantoudi *et al.* [167], who provided an algorithm along similar lines for *solvable* instances of SRI. In particular, Diamantoudi *et al.* showed that a similar sequence exists for solvable SRI instances with the outcome that the final matching $M_k$ is stable. Their algorithm in turn generalises the Roth–Vande Vate algorithm [516] for SMI (see Sec. 2.6).

Chung [153] had already given worked towards generalising Roth and Vande Vate's results to the SRI case: he showed that, for SRI instances satisfying the so-called "no odd ring" condition (see Ref. [153] for further details), the Roth–Vande Vate algorithm [516] converges to a stable matching with probability 1. This condition is not equivalent to the property that a stable partition contains no odd party. Moreover, whilst the "no odd ring" condition is sufficient for the existence of a stable matching, it is not a necessary condition, as pointed out by Diamantoudi *et al.* [167]. Thus Chung's algorithm is not applicable to the wider class of all solvable SRI instances. Also Lebedev *et al.* [403] showed that, for SRI instances with so-called *acyclic* preferences, which arise in the context of P2P networks (see Sec. 4.7.2), a stable matching can always be reached from an initial matching by iteratively satisfying blocking pairs. See also Refs. [429–431].

Biró and Norman [95] greatly simplified the landscape by showing that the Tan–Hsueh algorithm (see Sec. 4.3.3) can be used in order to construct a sequence of matchings, each one formed from its predecessor by satisfying a blocking pair, that ultimately yields a P-stable matching. The proof of correctness of this approach is simpler than that of Inarra *et al.* [303]. For solvable instances of SRI, the sequence ultimately produces a stable matching, and thus the result also implies that of Diamantoudi *et al.* [167] (and again, Biró and Norman have a shorter correctness proof for their approach). Moreover Biró and Norman showed that the sequence terminates after satisfying at most $O(nm)$ blocking pairs (here $n = |A|$ and $m = |E|$, where $G = (A, E)$ is the underlying graph of $I$); note that Inarra *et al.* [303] and Diamantoudi *et al.* [167] could only guarantee that their sequence was finite in each case, rather than terminating in a polynomial number of steps.

When defining the notion of a P-stable matching, a key aim of Inarra *et al.* [303] was to formulate a structure that (i) is always present for an abitrary SRI instance $I$, (ii) coincides with a stable matching if $I$ is solvable, and (iii) satisfies a weaker notion of stability if $I$ is unsolvable. Besides P-stable matchings, in another paper, Inarra *et al.* [304] define a further structure (although inherently related to P-stable matchings) that also satisfies these three properties, namely so-called *absorbing sets*.

Let $I$ be an SRI instance and let $\mathcal{M}$ be the set of all matchings in $I$. Absorbing sets are defined in terms of a dominance concept. In this context, a matching $M' \in \mathcal{M}$ is said to *dominate* a matching $M \in \mathcal{M}$ if there is a finite sequence of matchings $M = M_0, M_1, \ldots, M_k = M'$ in $\mathcal{M}$ such that $M_i$ can be obtained from $M_{i-1}$ by satisfying a blocking pair $(1 \le i \le k)$. A set of matchings $\mathcal{A} \subseteq \mathcal{M}$ is an *absorbing set* if (i) every pair of distinct matchings in $\mathcal{A}$ dominate one another, and (ii) no matching in $\mathcal{A}$ is dominated by a matching in $\mathcal{M} \backslash \mathcal{A}$. Thus, $A$ comprises matchings that are reachable from one another by successively satisfying blocking pairs, and such that no matching in $\mathcal{M}\backslash\mathcal{A}$ can be reached from a matching in $\mathcal{A}$ by successively satisfying blocking pairs.

Inarra *et al.* [304] proved that, if $I$ is solvable then an absorbing set is a singleton consisting of a stable matching, and conversely, a singleton set consisting of a stable matching is an absorbing set. More generally, for an arbitrary SRI instance $I$, and for an absorbing set $\mathcal{A}$ in $I$, the authors proved that $\mathcal{A}$ can be obtained from some stable partition $\Pi$ in $I$ by taking the union of the P-stable matchings relative to $\Pi$ together with the matchings that dominate them. Furthermore, since not all stable partitions in $I$ induce absorbing sets in this way, the authors characterised those stable partitions that *do* give rise to an absorbing set; such a stable partition is bound to exist in $I$.

## 4.4   Mirror posets and median graphs

Cheng and Lin [146] established important structural relationships between the set of stable matchings $S$ in a given SR instance $I$ and so-called *mirror posets* and *median graphs*. Their observations help to provide a meet-semilattice structure for $S$ that the authors consider to be more "natural" than the semilattice proposed by Gusfield and Irving [261] and summarised by Theorem 1.21. The structural correspondence also sheds light on the observed "local/global median phenomenon" of stable matchings in $I$ (see Sec. 2.7, and Theorem 2.19 in particular). In particular, as we will see in this section, an analogue of this phenomenon, and of Theorem 2.19 in particular, holds in the SR context. In this section we give an overview of these structural results.

We begin by considering generalised median stable matchings in the SR context. Given an SR instance $I$ and a set of stable matchings $\mathcal{T}$ in $I$, it is straightforward to extend the definitions of $P_{\mathcal{T}}(a_i)$ and $P_{j,\mathcal{T}}(a_i)$ given on Page 90 to $I$, where $a_i$ is an agent in $I$, $1 \leq j \leq t$ and $t = |\mathcal{T}|$. Teo and Sethuraman [565] proved the following version of Theorem 2.9 for SR:

**Theorem 4.16 ([565]).** *Let $I$ be an SR instance, let $\mathcal{T}$ be a set of stable matchings in $I$ and let $t = |\mathcal{T}|$. If $t$ is odd then the set of pairs $M$ in which $M(a_i) = p_{(t+1)/2,\mathcal{T}}(a_i)$ for each agent $a_i$ is a stable matching in $I$. If $t$ is even then there is a stable matching $M$ in $I$ such that $M(a_i) \in \{p_{t/2,\mathcal{T}}(a_i), p_{t/2+1,\mathcal{T}}(a_i)\}$ for each agent $a_i$.*

We refer to the stable matching $M$ whose existence is guaranteed by Theorem 4.16 as a *median stable matching* in $I$. Klaus and Klijn [391] independently proved Theorem 4.16, showing that it holds even if $\mathcal{T}$ is a multiset containing (not necessarily distinct) stable matchings.

We now turn to the structural correspondence between mirror posets and median graphs, where the latter concept has a close relationship with median stable matchings. Cheng and Lin [146] defined a poset $\mathcal{P} = (P, \blacktriangleleft)$ to be a *mirror poset* if the elements of $P$ can be partitioned into dual pairs, where the dual of $\rho \in P$ is denoted by $\bar{\rho}$, such that (i) $\rho \not\blacktriangleleft \bar{\rho}$, and (ii) $\sigma \prec \rho$ if and only if $\bar{\rho} \prec \bar{\sigma}$ for any $\rho$, $\sigma$ in $P$. Given an SR instance $I$, let $\mathcal{S}$ denote the set of stable matchings in $I$, let $R = R(I)$ denote the set of rotations in $I$, and let $\lhd$ be the partial order on rotations defined in Sec. 1.4.4. Let $R'$ denote the non-singular rotations of $I$. Then Lemma 4.3.7 of Ref. [261] implies that $(R', \lhd)$ is a mirror poset.

A graph $G = (V, E)$ is defined to be a *median graph* if, for any three vertices $u, v, w$ in $V$, there is a unique vertex $x \in V$ such that $x$ lies on a shortest path from $u$ to $v$, from $u$ to $w$ and from $v$ to $w$ in $G$. A vertex $v \in V$ is a *median* of $G$ if $\sum_{w \in V} d(v, w)$ is minimum, taken over all vertices in $V$, where $d(v, w)$ is the length of a shortest path from $v$ to $w$ in $G$. Median graphs were first studied by Avann [51], and independently introduced by Nebeský [460], Mulder [458], and Mulder and Schrijver [457].

Recall from Sec. 1.4.4 the definitions of *closed* and *complete* subsets of rotations in $I$. It is straightforward to generalise these definitions to the case of an arbitrary mirror poset $\mathcal{P} = (P, \blacktriangleleft)$. That is, a set $S \subseteq P$ is *closed* if, for each $\rho \in S$ and $\sigma \in P$, $\sigma \prec \rho$ implies that $\sigma \in S$. Also $S$ is *complete* if $S$ contains exactly one of each pair of dual elements of $P$.

We are now in a position to summarise Cheng and Lin's structural results via the following theorem.

**Theorem 4.17 ([146]).** *The structural relationships between mirror posets, median graphs and stable matchings in an SR instance, as shown in Fig. 4.3, hold. In particular:*

*(i) Given an SR instance $I$, $(R', \lhd)$ is a mirror poset, where $R'$ is the set of non-singular rotations in $I$ and $\lhd$ is the partial order on rotations defined in Sec. 1.4.4. By Theorem 1.23, the complete closed subsets of $R'$ are in 1–1 correspondence with the stable matchings in $I$.*

*(ii) Given a mirror poset $\mathcal{P} = (P, \blacktriangleleft)$, let $\mathcal{S}_{\mathcal{P}}$ be the set of all closed complete subsets of $P$. Define $G(\mathcal{S}_{\mathcal{P}})$ to be the graph with vertex set $\mathcal{S}_{\mathcal{P}}$, where two vertices $R_1$ and $R_2$ are adjacent if and only if they differ in one*

*dual element, i.e., there is an element $\rho \in P$ such that $\rho \in R_1$, $\bar{\rho} \in R_2$, and $R_1 \backslash \{\rho\} = R_2 \backslash \{\bar{\rho}\}$. Then $G(\mathcal{S}_P)$ is a median graph.*

(iii) *Given an* SR *instance $I$, there is a corresponding median graph. This relationship, which follows by (i) and (ii), can be described as follows. Let $\mathcal{S}$ be the set of stable matchings and let $R'$ be the set of non-singular rotations in $I$. By Theorem 1.23, the complete closed subsets of $R'$ are in 1-1 correspondence with the matchings in $\mathcal{S}$. Given any $M \in \mathcal{S}$, let $R_M$ denote the associated subset of $R'$ according to this 1-1 correspondence. Define $G(\mathcal{S})$ to be the graph with vertex set $\mathcal{S}$, where two vertices $M$ and $M'$ are adjacent if and only if $R_M$ and $R_{M'}$ differ by one rotation. (Since $R_M$ and $R_{M'}$ are complete subsets of $R'$, this is equivalent to the condition that there exists $\rho \in R'$ such that $\rho \in R_M$, $\bar{\rho} \in R_{M'}$, and $R_M \backslash \{\rho\} = R_{M'} \backslash \{\bar{\rho}\}$.) Clearly $G(\mathcal{S})$ is isomorphic to $G(\mathcal{S}_{R'})$, as defined in (ii), and hence by (ii), $G(\mathcal{S})$ is a median graph.*

(iv) *Let $\mathcal{P}$ be a mirror poset with $2n$ elements. Then there is an* SR *instance $I$ with $O(n^2)$ agents such that $R'$ is isomorphic to $\mathcal{P}$, where $R'$ is the set of non-singular rotations in $I$. Moreover, when the dual of each element in $\mathcal{P}$ is given, $I$ can be constructed in $O(n^2)$ time.*

(v) *Let $G = (V, E)$ be a median graph. Then there is an* SR *instance $I$ such that $G(\mathcal{S})$ (as defined in (iii)) is isomorphic to $G$, where $\mathcal{S}$ is the set of stable matchings in $I$.*

(vi) *Given a median graph, there is a corresponding mirror poset (this follows by (v) and (i); alternatively, see Lemma 14 of Ref. [146] for a direct construction of a mirror poset from a median graph).*

Cheng and Lin argued that, for a given SR instance $I$, $G(\mathcal{S})$ (defined by Part (iii) of Theorem 4.17) is the "natural representation" of the set of stable matchings $\mathcal{S}$ in $G$. This is because (as established by Part (iii) of Theorem 4.17) $G(\mathcal{S})$ is a median graph, and thus can be viewed as a median semilattice. Also, as Cheng and Lin assert, this median semilattice is isomorphic to the meet-semilattice for $\mathcal{S}$ due to Gusfield and Irving [261]. Further evidence that $G(\mathcal{S})$ is the "right way" to generalise the distributive lattice structure for SM stable matchings to the SR case is given by Fig. 4.4. The purpose of this figure is to illustrate that, in the SM case, dualities hold involving distributive lattices, posets and SM stable matchings that are analogous to the correspondences labelled (i)–(vi) in Fig. 4.3, in terms of median graphs, mirror posets and SR stable matchings. We explain each of the correspondences from Fig. 4.4 in the SM case as follows:

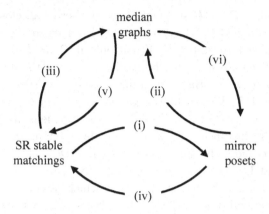

Fig. 4.3  Structural relationships for SR established by Cheng and Lin [146]

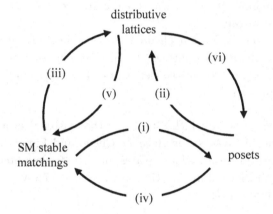

Fig. 4.4  Structural relationships for SM generalised by Cheng and Lin [146] to SR

(i) For a given SM instance $I$, the corresponding poset is given by $(R(I), \lhd)$, where $R(I)$ is the set of rotations in $I$ and $\lhd$ is the partial order on rotations defined inSec. 1.3.4.3. By Theorem 1.16, there is a 1–1 correspondence between the stable matchings in $I$ and the closed subsets of $R(I)$.

(ii) For a given poset, a corresponding distributive lattice may be constructed as follows. Let $\mathcal{D}$ be a distributive lattice. An element $\rho \in \mathcal{D}$ is *join-irreducible* if $\rho$ is not the join of a finite set of other elements of $\mathcal{D}$. Equivalently, $\rho$ is join-irreducible if it is neither the bottom element of $\mathcal{D}$ nor the join of any two smaller elements. Now suppose we are given a poset $\mathcal{P} = (P, \blacktriangleleft)$. Let $\mathcal{S}_{\mathcal{P}}$ be the set of all closed subsets of $P$. Define $G'(\mathcal{S}_{\mathcal{P}})$ to be the digraph with vertex set $\mathcal{S}_{\mathcal{P}}$, where vertex $R_1$ is adjacent to vertex $R_2$ if and only if $R_2 = R_1 \cup \{\rho\}$ for some $\rho \in P$. Then $\mathcal{S}_{\mathcal{P}}$ is a distributive lattice under the operation of set containment, whose Hasse diagram is given by $G'(\mathcal{S}_{\mathcal{P}})$. Moreover the sub-poset of $\mathcal{S}_{\mathcal{P}}$ formed by the join-irreducible elements of $\mathcal{S}_{\mathcal{P}} \backslash \{M_0\}$ is isomorphic to $\mathcal{P}$, where $M_0$ denotes the top element of $\mathcal{S}_{\mathcal{P}}$.

(iii) Given an SM instance $I$, the set $\mathcal{S}$ of stable matchings forms a distributive lattice under the dominance relation of Definition 1.12.

(iv) Irving and Leather [319] proved that, given any poset $\mathcal{P} = (P, \blacktriangleleft)$, there is a corresponding SM instance $I$, which can be constructed from $\mathcal{P}$ in polynomial time, such that the subsets of $P$ that are closed under $\blacktriangleleft$ are in 1–1 correspondence with the stable matchings in $I$.

(v) Gusfield *et al.* [262] proved that any distributive lattice is the set of stable matchings (together with the dominance relation over them) for a small SM instance.

(vi) Birkhoff's Representation Theorem for distributive lattices [76] asserts that, for a distributive lattice $\mathcal{D}$, the closed subsets of the poset induced by its join-irreducible elements form a distributive lattice that is isomorphic to $\mathcal{D}$.

Recall Theorem 4.16, which establishes the existence of a median stable matching for a given SR instance $I$. Cheng and Lin [146] proved the following structural result, which relates median stable matchings in $I$ to medians of $G(\mathcal{S})$ (defined in Part (iii) of Theorem 4.17), where $\mathcal{S}$ is the set of stable matchings in $I$.

**Corollary 4.18 ([146]).** *Let $I$ be an* SR *instance and let $\mathcal{S}$ be the set of stable matchings in $I$. A stable matching $M \in \mathcal{S}$ is a median stable matching in $I$ if and only if $M$ is a median of the graph $G(\mathcal{S})$ as defined in Part (iii) of Theorem 4.17.*

Corollary 4.18 implies that the local/global phenomenon for median stable matchings in the SM context (as observed in Sec. 2.7) also carries over to SR. This result provides further evidence that $G(\mathcal{S})$ is the "natural representation" of the set of stable matchings $\mathcal{S}$ in $I$.

We also remark that, as part of their construction, Cheng and Lin also deduced Theorem 4.16 even in the case that $\mathcal{T}$ is a multiset of stable matchings (i.e., the matchings in $\mathcal{T}$ need not be distinct).

## 4.5 Indifference

### 4.5.1 *Introduction*

In this section we study structural and algorithmic results for variants of SR involving indifference. Recall from Sec. 1.4.5 the definitions of weak stability, strong stability and super-stability that were given for SRPI. We consider results for each of these criteria in Sections 4.5.2, 4.5.3 and 4.5.4 respectively.

### 4.5.2 *Weakly stable matchings*

We firstly remark that it is straightforward to generalise Lemma 3.1 to the SRTI case [321]. That is, if $I$ is an instance of SRTI and $M$ is a matching in $I$, then $M$ is weakly stable in $I$ if and only if $M$ is stable in some instance $I'$ of SRI derived from $I$ by breaking the ties.

In contrast to the case for SR, Ronn [493] proved that WEAK SRT, the problem of deciding whether a weakly stable matching exists, given an instance of SRT, is NP-complete. He showed that NP-completeness holds even if each preference list is either strictly ordered or contains a tie of length 2 at the head. Alternative (shorter) proofs can be found in Refs. [261, Sec. 4.5.3] and [321]. In fact, starting from the NP-complete problem COM SMTI (see Sec. 3.2.4), there is a very short reduction showing the NP-completeness of WEAK SRT, as we now demonstrate.

**Theorem 4.19 ([321]).** WEAK SRT *is NP-complete.*

*Proof.* Clearly WEAK SRT is in NP. To show NP-hardness, we transform from COM SMTI, which is NP-complete by Theorem 3.6. Hence let $I$ be an instance of SMTI in which $U = \{u_1, \ldots, u_n\}$ is the set of men and $W = \{w_1, \ldots, w_n\}$ is the set of women. For notational convenience, we collectively denote the set of men and women in $I$ by $A = \{a_1, \ldots, a_{2n}\}$. Given any agent $a_i \in A$, let $P(a_i)$ denote the preference list of $a_i$ in $I$.

We construct an instance $I'$ of SRT as follows. The set of agents in $I'$ is $A \cup A' \cup A''$, where $A' = \{a_i' : a_i \in A\}$ and $A'' = \{a_i'' : a_i \in A\}$. Given any agent $a_i \in A$, the preference lists of $a_i$, $a_i'$ and $a_i''$ in $I'$ are as follows:

$$a_i : P(a_i) \ a_i' \ a_i'' \ \ldots$$
$$a_i' : a_i'' \ a_i \ \ldots$$
$$a_i'' : a_i \ a_i' \ \ldots$$

In a given agent's preference list in $I'$, the symbol "..." denotes all remaining agents in $I'$ in arbitrary strict order.

Given a complete weakly stable matching $M$ in $I$, we form a complete matching $M'$ in $I'$ by adding the pairs $\{a_i', a_i''\}$ to $M$, for each $a_i \in A$; clearly $M'$ is weakly stable in $I'$. Conversely, given a weakly stable matching $M'$ in $I'$, it is straightforward to verify that $\{a_i', a_i''\} \in M'$, and $a_i$ is assigned to some agent in $P(a_i)$, for each $a_i \in A$. Hence $M = M' \setminus \{\{a_i', a_i''\} : a_i \in A\}$ is a complete weakly stable matching in $I$. $\qquad\square$

As observed following Corollary 3.7, COM SMTI is NP-complete even if each person's list is strictly ordered or is a tie of length 2 [419]. It follows by inspection of the construction given in the proof of Theorem 4.19 that this reduction yields NP-completeness for WEAK-SRT under the same conditions that held under Ronn's reduction.

Given an instance $I$ of SRTI, Chung [153] defined a sufficient condition for the existence of a weakly stable matching based on the absence of an *odd ring*. As mentioned in Sec. 4.3.6, this structure is similar to the notion of an odd party in a stable partition, though they are not quite the same concept. Most importantly, the non-existence of an odd ring is not a necessary condition for the existence of a weakly stable matching. Chung showed that, in the absence of an odd ring, a natural extension of the Roth–Vande Vate algorithm [516] to the SRTI context converges to a weakly stable matching in $I$ with probability 1. Chung also identified various types of preference profiles that give rise to solvable instances of SRTI.

When preference lists are incomplete, certain results from the SMTI context (see Sec. 3.2.3 and Corollary 3.7) imply that, given an SRTI instance $I$, weakly stable matchings in $I$ can have different sizes (should they exist), and the problem of finding a maximum weakly stable matching is NP-hard. It is also straightforward to generalise the proof of Theorem 3.4 to the SRTI case, in order to deduce that an SRTI instance $I$ that is solvable under weakly stability admits a weakly stable matching of size $k$, for each $s^-(I) \leq k \leq s^+(I)$, where $s^-(I)$ (respectively $s + (I)$) denotes the minimum (respectively maximum) size of a weakly stable matching in $I$. Also, Footnote 1 on Page 136 holds in the context of a general graph, and therefore it follows that $s^+(I) \leq 2s^-(I)$.

### 4.5.3 Strongly stable matchings

Scott [523] studied strongly stable matchings in the context of SRTI. He showed that, given an instance $I$ of SRTI that is solvable under strong stability, the same set of agents are unassigned in all strongly stable matchings in $I$ [523, Corollary 3.2.3]. Moreover he presented the following result, answering an open problem in Ref. [321]:

**Theorem 4.20 ([523]).** *Let $I$ be an instance of* SRTI. *There is an $O(m^2)$ algorithm for finding a strongly stable matching in $I$ or reporting that none exists, where $m$ is the number of acceptable pairs of agents in $I$.*

This algorithm is based on a combination of Irving's algorithms for SR [306] and for SMT under strong stability [308]. Scott [523] conjectured that an $O(nm)$ algorithm should be possible, where $n$ is the size of $I$, using similar techniques that Kavitha *et al.* [364] employed in order to improve the $O(m^2)$ algorithm of Irving *et al.* for HRT under strong stability [328] to achieve a running time of $O(nm)$.

O'Malley [470] gave an $O(\sqrt{n}m)$ algorithm to find a strongly stable matching or report that none exists, given an instance $I$ of SRTI such that there is a master list of agents. He did likewise for the case that $I$ is an instance of SRTI with *symmetric preferences* (i.e., $rank(a_i, a_j) = rank(a_j, a_i)$ for any acceptable pair of agents $\{a_i, a_j\}$).

### 4.5.4 Super-stable matchings

We firstly observe that it is straightforward to adapt the proof of Lemma 3.24 in order to obtain the following result, as noted in Ref. [321].

**Lemma 4.21 ([321]).** *Let $I$ be an instance of* SRTI, *and let $M$ be a matching in $I$. Then $M$ is super-stable in $I$ if and only if $M$ is stable in every instance $I'$ of* SRI *derived from $I$ by breaking the ties.*

It follows immediately from Lemma 4.21 and from Theorem 4.5.2 of Ref. [261] that, as in the strong stability case, the same set of agents are unassigned in all super-stable matchings in $I$ (as observed in Theorem 5.1 of Ref. [321]). For, if $M$ and $M'$ are two super-stable matchings in $I$, then $M$ and $M'$ are both stable in some instance $I'$ of SRI obtained from $I$ by breaking the ties. But Theorem 4.5.2 of Ref. [261] implies that, when $M$ and $M'$ are viewed as stable matchings in $I'$, the same set of agents are assigned in $M$ and $M'$.

Irving and Manlove [321] proved a counterpart of Theorem 4.20 for super-stability, as follows:

**Theorem 4.22 ([321]).** *Let $I$ be an instance of* SRTI. *There is an $O(m)$ algorithm for finding a super-stable matching in $I$ or reporting that none exists, where $m$ is the number of acceptable pairs of agents in $I$.*

This algorithm is based on a combination of Irving's algorithms for SR [306] and for SMT under super-stability [308]. Irving and Manlove [321] also showed how to extend their algorithm for SRTI under super-stability to the SRPI case; indeed the structural results described in Lemma 4.21 and its subsequent paragraph also hold true in such a context (in Lemma 4.21, "breaking the ties" should be replaced by "forming a linear extension of each partial order").

O'Malley [470] gave a simpler $O(m)$ algorithm to find a super-stable matching or report that none exists, given an instance $I$ of SRTI such that there is a master list of agents, where $m$ is the number of acceptable pairs of agents in $I$. He did likewise for the case that $I$ is an instance of SRTI with symmetric preferences.

Using a polynomial-time reduction from SRTI under super-stability to 2-SAT, we can deduce additional algorithmic results for problems involving super-stable matchings in the context of SRTI, including computing all super-stable pairs, all super-stable matchings and a minimum regret super-stable matching — see Sec. 4.8.2 for more details.

## 4.6    "Almost stable" matchings

### 4.6.1    *Introduction*

As mentioned in Sec. 4.2.1, the probability that a random instance of SR with $n$ agents is solvable would seem to decrease fairly steeply as $n$ increases up to the value of 1000, and continues to decrease beyond this value, albeit less dramatically. In particular, given a random instance with 10,000 agents, experiments estimate that the solvability probability is less then 25% [448].

Therefore, as $n$ grows large, these results suggest that an arbitrary matching in a random SR instance with $n$ agents is likely to admit at least one blocking pair. However, as discussed in Sec. 2.8, a blocking pair need not necessarily lead to a given matching being undermined in practice. Hence a limited number of blocking pairs might be tolerated as a means of

$$\left.\begin{array}{l} a_{4i+1} : a_{4i+2} \quad a_{4i+3} \quad a_{4i+4} \\ a_{4i+2} : a_{4i+3} \quad a_{4i+1} \quad a_{4i+4} \\ a_{4i+3} : a_{4i+1} \quad a_{4i+2} \quad a_{4i+4} \\ a_{4i+4} : a_{4i+1} \quad a_{4i+2} \quad a_{4i+3} \end{array}\right\} 0 \le i \le r - 1$$

$$M_r^1 = \{\{a_{4i+1}, a_{4i+2}\} : 0 \le i \le r - 1\}$$
$$M_r^2 = M_r^1 \cup \{\{a_{4i+3}, a_{4i+4}\} : 0 \le i \le r - 1\}$$

Fig. 4.5    Instance $I_r$ of SR and two matchings $M_r^1, M_r^2$ in $I_r$

coping with the possible non-existence of a stable matching. This motivates the problem of finding, given an unsolvable SRI instance $I$, a matching in $I$ that admits the minimum number of blocking pairs [466, 199, 195]. As in Sec. 2.8, such a matching is, in the sense described here, "as stable as possible".

Let $I$ be an instance of SRI and let $\mathcal{M}$ denote the set of matchings in $I$. Given $M \in \mathcal{M}$, it is straightforward to extend the definitions of $bp(I, M)$ and $ba(I, M)$ from Sec. 2.8 to the SRI context. Also we define $bp(I)$ and $ba(I)$ as follows:

$$bp(I) = \min\{|bp(I, M)| : M \in \mathcal{M}\}$$

$$ba(I) = \min\{|ba(I, M)| : M \in \mathcal{M}\}.$$

Finding a matching $M$ in $I$ such that $|bp(M)| = bp(I)$ is just one method of trying to cope with the potential unsolvability of $I$. Alternative approaches have already been described in Secs. 4.3.4 and 4.3.6, where the former section focuses on finding a maximum stable matching (that is, a matching of maximum size such that the assigned agents are stable within themselves). However such a matching may only be half the size of a maximum (cardinality) matching in $I$. Yet in many applications we seek to assign as many agents as possible, and as discussed above, in order to satisfy this property, in many cases a certain number of blocking pairs may be sanctioned.

For example, suppose that $r \ge 1$ and consider the SR instance $I_r$ and matchings $M_r^1, M_r^2$ as shown in Fig. 4.5. Since $I_r$ is built up from $r$ copies of unsolvable SR instances with 4 agents, Tan's algorithm is bound to construct a matching $M$ in $I_r$ of size $r$ (such as $M_r^1$). Any such matching $M$ satisfies $|bp(I_r, M)| \ge 2r$. However $M_r^2$ is a matching in $I$ such that $|M_r^2| = 2r$ and $|bp(I_r, M_r^2)| = r$. In particular $M_r^1$ is half the size of $M_r^2$ and admits twice as many blocking pairs in $I_r$.

In this section we review complexity and approximability results for problems involving finding matchings with the minimum number of blocking pairs, given instances of SR and its variants. In Sec. 4.6.2 we present hardness results for the minimisation problem, whilst in Sec. 4.6.3 we show that the problem of finding a matching with exactly $K$ blocking pairs is fixed, if $K$ is a constant. In Sec. 4.6.4 we consider preference lists of bounded length, and finally in Sec. 4.6.5 we discuss some open problems.

### 4.6.2    *Hardness results*

Define MIN BP SRI to be the problem of finding, given an SRI instance $I$, a matching $M$ in $I$ such that $|bp(M)| = bp(I)$ (MIN BP SR and MIN BP SRT are defined analogously; note that, if $I$ is an SR instance, clearly $M$ must assign $n - \left( \lceil \frac{n}{2} \rceil - \lfloor \frac{n}{2} \rfloor \right)$ agents, where $n$ is the size of $I$). Also, define EXACT BP SRI to be the problem of deciding, given an SRI instance $I$ and an integer $K$, whether $I$ admits a matching $M$ such that $|bp(M)| = K$ (EXACT BP SR and EXACT BP SRT are defined analogously). Our first result, proved by Abraham *et al.*[16], indicates that MIN BP SR is NP-hard and very difficult to approximate.

**Theorem 4.23 ([16]).** MIN BP SR *is not approximable within* $n^{\frac{1}{2}-\varepsilon}$, *for any* $\varepsilon > 0$, *unless P=NP, where* $n$ *is the number of agents in a given instance.*

It also turns out that EXACT BP SR is a hard problem, as the next result reveals.

**Theorem 4.24 ([16]).** EXACT BP SR *is NP-complete.*

We now turn to the case that ties may be present in the preference lists. It turns out that, in this context, MIN BP SRT may be even harder to approximate than MIN BP SR.

**Theorem 4.25 ([16]).** MIN BP SRT *is not approximable within* $n^{1-\varepsilon}$, *for any* $\varepsilon > 0$, *unless P=NP, where* $n$ *is the number of agents in a given instance. The result holds even if all preference lists are complete, there is at most one tie per list, and each tie is of length 2.*

We now remark on the format of the inapproximability results presented above for MIN BP SR and MIN BP SRT. We implicitly assume that a given instance $I$ of MIN BP SR is unsolvable, so that $bp(I) \geq 1$. Recall that the solvability or otherwise of $I$ can be determined in $O(m)$ time [306, 261], where $m$ is the number of acceptable pairs of agents in $I$. Hence $bp(I)$ can be regarded as the objective function for measuring performance guarantee. On the other hand, given an instance $I$ of MIN BP SRT, we do not assume that $I$ is unsolvable, since the problem of deciding whether this is the case is NP-complete [493, 321]. Hence possibly $bp(I) = 0$, and therefore we use $opt(I)$ to measure performance guarantee, where $opt(I) = 1 + bp(I)$. In fact the proof of Theorem 4.25 in Ref. [16] shows that, given any $\varepsilon > 0$, it is NP-hard to distinguish between the cases that $I$ admits a stable matching, and $bp(I) \geq n^{1-\varepsilon}$.

### 4.6.3    Matchings with a constant number of blocking pairs

In this subsection we consider EXACT BP SRI. Let $I$ be an SRI instance and let $K \geq 1$ be a fixed constant. We give an $O(m^{K+1})$ algorithm, due to Abraham *et al.* [16], that finds a matching $M$ in $I$ such that $|bp(I, M)| = K$, or reports that no such matching exists, where $m$ is the number of acceptable pairs of agents in $I$. Later, we show how to modify this algorithm if we require that $|bp(I, M)| \leq K$.

Let $G = (A, E)$ be the underlying graph of $I$. The algorithm is based on generating subsets $B$ of edges of $G$, where $|B| = K$ — these edges will form the blocking pairs with respect to a matching to be constructed in a subgraph of $G$. Given such a set $B$, we form a subgraph $G_B = (A, E_B)$ of $G$ as follows. For each agent $a_i$ incident to an edge $e = \{a_i, a_j\} \in B$, if $e$ is to be a blocking pair of a matching $M$, it follows that $\{a_i, a_j\} \notin M$ and $a_i$ cannot be assigned in $M$ to an agent whom she prefers to $a_j$ in $I$. Hence we delete $\{a_i, a_j\}$ from $E_B$, and also we delete $\{a_i, a_k\}$ from $E_B$ for any $a_k$ such that $a_i$ prefers $a_k$ to $a_j$ in $I$. If any such edge $\{a_i, a_k\}$ is not in $B$, then we require that $\{a_i, a_k\}$ is not a blocking pair of a constructed matching $M$. This can only be achieved if $a_k$ is assigned in $M$ to an agent whom she prefers to $a_i$ in $I$. Hence we invoke $truncate_{a_k}(a_i)$, which represents the operation of deleting $\{a_k, a_l\}$ from $E_B$, for any $a_l$ such that $a_k$ prefers $a_i$ to $a_l$ in $I$. Additionally we add $a_k$ to a set $P$ to subsequently check that $a_k$ is assigned in $M$.

Having completed the construction of $G_B$, we denote by $I_B$ the SRI instance with underlying graph $G_B$ and preference lists obtained by restricting the preferences in $I$ to $E_B$. By construction of $G_B$, it is immediate that any matching $M$ in $G_B$ satisfies $B \subseteq bp(I, M)$. To avoid any additional blocking pairs in $I$, we seek a stable matching in $I_B$ in which all agents in

---

**Algorithm 4.2** Algorithm $K$-BP-SR [16]

---

**Require:** an SRI instance $I$ and an integer $K$
**Ensure:** return a matching $M$ such $|bp(M)| = K$ or "no matching with $K$ blocking pairs exists"

1: **for each** $B \subseteq E$ such that $|B| = K$ **do**
2:    $E_B := E$; $\{G_B = (A, E_B)$ is a subgraph of $G\}$
3:    $P := \emptyset$;
4:    **for each** agent $a_i$ incident to some $\{a_i, a_j\} \in B$ **do**
5:       delete $\{a_i, a_j\}$ from $E_B$;
6:       **for each** agent $a_k$ such that $a_i$ prefers $a_k$ to $a_j$ in $I$ **do**
7:          delete $\{a_i, a_k\}$ from $E_B$;
8:          **if** $\{a_i, a_k\} \notin B$ **then**
9:             $truncate_{a_k}(a_i)$;
10:             $P := P \cup \{a_k\}$;
11:          **end if**
12:       **end for**
13:    **end for**
14:    **if** there is a stable matching $M$ in $I_B$ **then**
15:       **if** every agent in $P$ is assigned in $M$ **then**
16:          **return** $M$;
17:       **end if**
18:    **end if**
19: **end for**
20: **return** "no matching with $K$ blocking pairs exists";

---

$P$ are assigned. We apply Irving's algorithm for SRI [261] to $I_B$ — suppose it finds a stable matching $M$ in $I_B$. If all agents in $P$ are matched then $bp(I, M) = B$, and hence $|bp(I, M)| = K$ — thus we may return $M$ and halt. If some agents in $P$ are unassigned in $M$ then we need not consider any other stable matching in $I_B$, since Theorem 4.5.2 of Ref. [261] asserts that the same agents are assigned in all stable matchings in $I_B$. Hence (and also in the case that no stable matching in $I_B$ is found), we may consider the next subset $B$. If we complete the generation of all subsets $B$ without having returned a matching $M$, we report that no matching with the desired property exists. The algorithm is displayed as Algorithm $K$-BP-SR in Algorithm 4.2. The following theorem establishes its correctness and complexity.

**Theorem 4.26 ([16]).** *Given an SRI instance $I$ and a fixed constant $K$, Algorithm $K$-BP-SR finds a matching with exactly $K$ blocking pairs, or reports that no such matching exists, in $O(m^{K+1})$ time, where $m$ is the number of acceptable pairs of agents in $I$.*

Note that it is straightforward to modify Algorithm $K$-BP-SR so that it outputs the largest stable matching taken over all subsets $B$ — we may then find a matching $M$ such that (i) $|bp(I, M)| = K$, and (ii) $M$ is of maximum cardinality with respect to (i). This extension uses the fact that all stable matchings in $I_B$ have the same size [261, Theorem 4.5.2], so that the choice of stable matching constructed by the algorithm is not of significance for Condition (ii).

Finally we remark that Algorithm $K$-BP-SR may easily be modified in order to find a matching $M$ such that $bp(I, M) \leq K$: the outermost loop iterates over all subsets $B$ of $E$ such that $|B| \leq K$. Again, one can find a maximum such matching if required. The time complexity of the algorithm remains unchanged.

The above algorithm does not extend to the SRT case unless P=NP, as the next result shows.

**Theorem 4.27 ([16]).** EXACT BP SRT *is NP-complete for each fixed integer* $K \geq 0$.

### 4.6.4 Bounded length preference lists

We now consider the case where the lengths of the preference lists in a given SRI instance are bounded above by some constant $d \geq 2$. Define MIN BP $d$-SRI to be the restriction of MIN BP SRI to such an instance. Biró *et al.* [92] presented complexity and approximability results for MIN BP $d$-SRI. We begin with the following result, which concerns preference lists of length at most 2. It turns out that this restriction is easily solvable in polynomial time by constructing a stable partition.

**Theorem 4.28 ([92]).** MIN BP *2*-SRI *is solvable in* $O(m)$ *time, where* $m$ *is the number of acceptable pairs of agents in a given instance* $I$. *Moreover* $bp(I)$ *is equal to the number of odd parties of size* $\geq 3$ *in a stable partition in* $I$.

On the other hand, when preference lists are of length at most 3, MIN BP $d$-SRI is NP-hard and not approximable within some $c > 1$ unless P=NP.

**Theorem 4.29 ([92]).** MIN BP *3*-SRI *is not approximable within* $\frac{1017}{1016} - \delta$, *for any* $\delta > 0$, *unless P=NP*.

We now consider upper bounds for the approximability of MIN BP $d$-SRI for $d \geq 3$. A straightforward $2d - 2$ approximation algorithm follows from upper and lower bounds for $bp(I)$, which were established in Ref. [16]. In order to describe these bounds, we require the following notation. Let $\mathcal{C}$ denote the set of odd parties of length $\geq 3$ in a stable partition $\Pi$ in $I$. Recall that $\mathcal{C}$ is a property of $I$ and is independent of the particular choice of $\Pi$ by Theorem 4.7. Given $C_i \in \mathcal{C}$, let $d_i = \min_{a_j \in C_i} d_G(a_j)$, where $d_G(a_j)$ denotes the degree of vertex $a_j$ in the underlying graph $G$ of $I$. Then the following bounds for $bp(I)$ hold:

**Theorem 4.30 ([16]).** *Let $I$ be an* SRI *instance. Then*

$$\left\lceil \frac{|\mathcal{C}|}{2} \right\rceil \leq bp(I) \leq \sum_{C_i \in \mathcal{C}} (d_i - 1).$$

**Corollary 4.31 ([92]).** *For each $d \geq 3$,* MIN BP $d$-SRI *is approximable within $2d - 2$.*

**Proof.** Let $I$ be an instance of MIN BP $d$-SRI. A simple approximation algorithm achieving this performance guarantee is based on constructing a stable partition $\Pi$ in $I$, and modifying $\Pi$ as follows. Let $\mathcal{C}$ be the set of odd parties in $\Pi$ of length $\geq 3$. For each $C_i \in \mathcal{C}$, delete from $\Pi$ a vertex of minimum degree in the underlying graph of $I$. Having done this, all parties in $\Pi$ are either of length 1 or of even length, and hence $\Pi$ can be decomposed into a reduced stable partition $\Pi'$ by Theorem 4.4. $\Pi'$ gives rise to a matching $M$ in $I$ that implies the upper bound for $bp(I)$ given in Theorem 4.30 (see the proof of Lemma 3 in Ref. [16] for more details), and hence $|bp(I, M)| \leq (d-1)|\mathcal{C}|$. The lower bound given by Theorem 4.30 gives $bp(I) \geq |\mathcal{C}|/2$, which implies the result.                                    □

By making a more careful choice of vertex to delete from each odd party in $\Pi$, an improved approximation algorithm with performance guarantee $2d - 3$ can be arrived at. This further improves to $2d - 4$ in the absence of an *elitist odd party*, which we now define.

**Definition 4.32 ([92]).** *Let $\Pi$ be a stable partition for an* SRI *instance $I$. An* elitist odd party *is an odd party $P = (a_{i_0}, a_{i_1}, \ldots, a_{i_k})$ in $\Pi$ with $k \geq 2$ such that $\Pi(a_{i_j})$ and $\Pi^{-1}(a_{i_j})$ are the first and second entries, respectively, of $a_{i_j}$'s preference list for $0 \leq j \leq k$.*

**Theorem 4.33 ([92]).** *For each fixed* $d \geq 3$, MIN BP $d$-SRI *is approximable within* $2d - 3$. *If the instance contains no elitist odd party, this bound improves to* $2d - 4$.

### 4.6.5 Open problems

Analogues of the problems studied in this section can be obtained if we replace the requirement to minimise the number of blocking pairs with the goal to minimise the number of blocking agents. It remains open to determine which of the results in this section carry over to the case that $ba(I, M)$ should be minimised for a given matching $M$, rather than $bp(I, M)$, for a given instance $I$ of SR and its variants.

## 4.7 Globally-ranked pairs

### 4.7.1 Definitions and motivation for the SRTI-GRP model

Let $I$ be an instance of SRTI and let $G = (A, E)$ be the underlying graph of $I$. Then $I$ is an instance of the *Stable Roommates problem with Globally Ranked Pairs* (SRTI-GRP) if there is a ranking function $rank : E \longrightarrow \mathbb{R}$ such that, if $e = \{a_i, a_j\} \in E$ and $e' = \{a_i, a_k\} \in E$, then $a_i$ prefers $a_j$ to $a_k$ if and only if $rank(e) < rank(e')$, and $a_i$ is indifferent between $a_j$ to $a_k$ if and only if $rank(e) = rank(e')$.

A global ranking of the edges in the graph underlying an instance of SRTI can be an appropriate model for a number of the applications described in Sec. 1.4.6. For example, with reference to kidney exchange, when two patient–donor pairs are matched with each other (in order to swap donors), we are not certain if the swap can occur until expensive last-minute compatibility tests are performed on the donors and patients. If an incompatibility involving one of the donors and her recipient patient in the exchange is identified, the swap is cancelled and the two patients must wait for the next matching run. Since doctors can rank potential swaps by their chance of success, and patients prefer swaps with better chances of success, this is exactly the preference model of SRTI-GRP.

Regarding the P2P file-sharing network example, the presence of a master list of peers (according to properties such as download / upload bandwidth, latency and storage capacity) gives rise to a global ranking function as follows: an instance of SRTI-GRP can be obtained by defining $rank(\{a_i, a_j\})$ to be $rank(a_i) + rank(a_j)$, where $\{a_i, a_j\}$ is any edge and $rank(a_k)$ is the rank of agent $a_k$ in the master list.

The remainder of this section is organised as follows. In Sec. 4.7.2, we consider a problem model that is closely related to SRTI-GRP. Then in Sec. 4.7.3 we give algorithmic results for problems involving computing weakly stable and strongly stable matchings in instances of SRTI-GRP. Finally, in Sec. 4.7.4 we discuss related work.

### 4.7.2   Globally acyclic preferences

The generality of the SRTI-GRP model can be demonstrated by defining a second restriction of SRTI called the *Stable Roommates problem with Globally Acyclic Preferences* (SRTI-GAP) [27]. Instances of SRTI-GAP satisfy the following characterisation test: given an arbitrary instance $I$ of SRTI with underlying graph $G = (A, E)$, construct a digraph $P(G)$, containing one vertex $e$ for each edge $e \in E$, and an arc from $e = \{a_i, a_j\} \in E$ to $e' = \{a_i, a_k\} \in E$ if $a_i$ prefers $a_k$ to $a_j$. For each $e = \{a_i, a_j\}$ and $e' = \{a_i, a_k\}$ in $E$, if $a_i$ is indifferent between $a_j$ and $a_k$, merge vertices $e$ and $e'$. Note that a merged vertex may contain several original edge–vertices and have self-loops. Then $I$ is an instance of SRTI-GAP if and only if $P(G)$ is acyclic.

Instances of SRTI-GRP satisfy the SRTI-GAP test, since any directed path in $P(G)$ consists of arcs with monotonically improving ranks, and so no cycles are possible. In the reverse direction, given any instance of SRTI-GAP, we can derive a suitable rank function from a reverse topological sort on $P(G)$, i.e., $rank(e) < rank(e')$ if and only if $e$ appears before $e'$. The following proposition is clear:

**Proposition 4.34 ([27]).** *Let $I$ be an instance of* SRTI. *Then $I$ is an instance of* SRTI-GRP *if and only if $I$ is an instance of* SRTI-GAP.

SRTI-GAP is a model that has practical applications in the context of P2P networks — see Refs. [232, 429, 431] for more details.

### 4.7.3   Weakly and strongly stable matchings in SRTI-GRP

In contrast to the general SRTI case, it turns out that necessary and sufficient conditions for the existence of weakly stable and strongly stable matchings can be given in terms of the ranks of the edges in the underlying graph. Let $I$ be an instance of SRTI-GRP and let $G = (A, E)$ be the underlying graph of $I$. Let $n = |A|$ and $m = |E|$. Without loss of generality we assume that $rank : E \longrightarrow \{1, 2, \ldots m\}$. For each $i$ $(1 \leq i \leq m)$, we define $E_i$ to be the set of edges with rank $i$, and $E_{\leq i}$ to be the set $E_1 \cup E_2 \cup \ldots \cup E_i$.

As well as modelling real-world problems, SRTI-GRP is an important theoretical restriction of SRTI. For, as described in Secs. 1.4.3 and 4.5.2, it is well known that an instance $I$ of SRT need not admit a weakly stable matching (even if $I$ is an instance of SR) and moreover the problem of deciding whether $I$ admits a weakly stable matching is NP-complete. It turns out that SRTI-GRP has neither of these undesirable properties, as demonstrated by the following lemma.

**Lemma 4.35 ([27]).** *Let $I$ be an instance of SRTI-GRP and let $G = (V, E)$ be the underlying graph of $I$. Then $M$ is a weakly stable matching in $I$ if and only if $M \cap E_{\leq i}$ is a maximal matching of $E_{\leq i}$, for all $i$ ($1 \leq i \leq m$), where $m = |E|$ and $E_{\leq i}$ are as defined above ($1 \leq i \leq m$).*

Hence we can construct a weakly stable matching in $O(n + m)$ time by finding a maximal matching on rank–1 edges, removing the matched vertices, finding a maximal matching on rank–2 edges, and so on.

Strongly stable matchings are also easy to characterise in SRTI-GRP.

**Lemma 4.36 ([27]).** *Let $I$ be an instance of SRTI-GRP and let $G = (V, E)$ be the underlying graph of $I$. Then $M$ is a strongly stable matching in $I$ if and only if, for each $i$ ($1 \leq i \leq m$), $M \cap E_i$ is a perfect matching of*

$$\{e \in E_i : e \text{ is not adjacent to any } e' \in M \cap (E_{\leq i} \backslash E_i)\},$$

*where $m = |E|$, and $E_i$, $E_{\leq i}$ are as defined above ($1 \leq i \leq m$).*

Of course, even $E_1$ may not admit a perfect matching, and so strongly stable matchings may not exist. However, we can find a strongly stable matching, or prove that no such matching exists in $O(\sqrt{n}m)$ time by using the maximum matching algorithm of Micali and Vazirani for non-bipartite graphs [451,577]. This improves on the best known running time of $O(m^2)$ for general SRTI (see Sec. 4.5.3).

Lemmas 4.35 and 4.36 indicate that SRTI-GRP can be "more tractable" than SR. However, the possible non-existence of a strongly stable matching motivates the search for weakly stable matchings with desirable properties. A *rank-maximal* matching [318,371] includes the maximum possible number of rank–1 edges, and subject to this, the maximum possible number of rank–2 edges, and so on[4]. More formally, for this subsection only, define

---

[4]Note that this definition of rank-maximality in the context of SRTI-GRP is based on edge ranks, and is slightly different from the notion of rank-maximality that we define in a more general SRTI instance in Sec. 8.2.5.

the *profile* of a matching $M$ as $\langle p_1, p_2, \ldots, p_m \rangle$, where $p_i$ is the number of rank-$i$ edges in $M$. Then a matching is rank-maximal if and only if its profile is lexicographically maximum, taken over all matchings.

Recall from Lemma 4.36 that a strongly stable matching is perfect on rank-1 edges, and subject to removing the matched vertices, perfect on rank-2 edges, and so on. It is clear that a rank-maximal matching is strongly stable, when strong stability is possible. If no strongly stable matching exists, then a rank-maximal matching, which by Lemma 4.35 is always weakly stable, seems a natural substitute. Irving *et al.* [318] gave a polynomial-time algorithm for the problem of finding a rank-maximal matching in an instance of HAT. Abraham *et al.* [27] generalised this algorithm to the SRTI-GRP case, giving rise to the following result.

**Theorem 4.37 ([27]).** *Let $I$ be an instance of* SRTI-GRP *with underlying graph $G = (A, E)$. Then a rank-maximal matching in $I$ can be found in time $O(\min(n + r^*, r^* \sqrt{n})m)$, where $n = |A|$, $m = |E|$ and $r^*$ is the rank of the worst-ranked edge in a rank-maximal matching in $I$.*

### 4.7.4 Related work

Several papers have focused on instances of SRTI-GAP that arise from P2P networks. In particular, Lebedev *et al.* [403] independently proved Lemma 4.35 by showing that every instance of SRTI-GAP (and hence SRTI-GRP by Proposition 4.34) admits a weakly stable matching. Gai *et al.* [232] showed that every instance of SRI with a master list is an instance of SRTI-GAP, but the converse need not be true. They also considered instances of SRTI with *symmetric preferences* (see Sec. 3.2.7). See also Refs. [233, 429, 430].

Arkin *et al.* [50] defined GEOMETRIC SRTI, which is a restriction of SRTI-GRP in which the agents are points in $\mathbb{R}^d$ (for some fixed $d \geq 1$), all agents are mutually acceptable, and the ranking function maps a pair of agents $a_i$, $a_j$ to the Euclidean distance $||a_i - a_j||_d$ between them. The case in which $d = 1$ (so-called *single-peaked preferences*) and there are no ties had already been studied by Bartholdi and Trick [68], who showed that, for a given instance of this problem, a unique stable matching exists and can be found in $O(n)$ time; here $n$ is the number of agents in a given instance. For the case that $d \geq 2$, Arkin *et al.* proved that (i) there is an $O(n \log n)$ algorithm for finding a super-stable matching or reporting that none exists; (ii) there is an $O(n^{1.5})$ algorithm for finding a strongly stable matching or reporting that none exists (the time complexity is $O(n^{1.19})$ if $d = 2$), and

(iii) a weakly stable matching always exists and can be found in $O(n \log n)$ time.

## 4.8 Other extensions of SR

### 4.8.1 *Introduction*

In this section we survey algorithmic results for a range of problems, each of which extend the basic SR problem model in some way. The problems that we consider include the so-called Stable Roommates problem with Forbidden Pairs (Sec. 4.8.2), Stable Crews problem (Sec. 4.8.3), Stable Fixtures problem (4.8.4), Stable Multiple Activities problem (Sec. 4.8.5), Stable Allocation problem (4.8.6), Stable Roommates problem with Choice Functions (4.8.7) and Coalition Formation Games (Sec. 4.8.8).

### 4.8.2 *Stable Roommates problem with Forbidden Pairs*

The *Stable Roommates problem with Forbidden pairs* (SRF) [216] is an extension of SRI in which an instance $I$ additionally involves a set $F \subseteq E$ of *forbidden pairs*, where $G = (A, E)$ is the underlying graph of $I$. A matching $M$ in $I$ is *stable* if (i) $M$ is stable in the underlying SRI instance obtained from $I$ by deleting the forbidden pairs, and (ii) $M \cap F = \emptyset$. Hence although a stable matching in $I$ must contain no forbidden pair, it is nevertheless possible for a forbidden pair to be a blocking pair of a given matching.

It turns out that SRTI under super-stability and SRF are closely related: they are polynomial-time reducible to one another. We will describe the reduction from the former problem to the latter; before doing so, we require the following lemma, proved by Fleiner *et al.* [216].

**Lemma 4.38 ([216]).** *Let $I$ be an instance of SRTI and let $A$ be the set of agents in $I$. Let $I'$ be any instance of SRI obtained from $I$ by breaking the ties. For any $a_i \in A$, let $S_i$ be the set of stable partners of $a_i$ in $I'$. Now suppose that $\{a_j, a_k\} \subseteq S_i$, where $a_i$ is indifferent between $a_j$ and $a_k$ in $I$, and $a_i$ prefers $a_j$ to $a_k$ in $I'$. Then $a_j$ is not a super-stable partner of $a_i$ in $I$.*

**Proof.** Suppose that $M$ is a super-stable matching in $I$ containing $\{a_i, a_j\}$. Then $M$ is stable in $I'$ by Lemma 4.21. Also by assumption there exists a matching $M'$ containing $\{a_i, a_k\}$ that is stable in $I'$. As $a_i$ prefers $a_j$ to $a_k$ in $I'$, it follows by Lemma 4.3.9 of [261] that $a_k$ prefers

$M'(a_k) = a_i$ to $M(a_k)$. Hence $a_k$ either prefers $a_i$ to $M(a_k)$ in $I$ or is indifferent between them. Also by assumption $a_i$ is indifferent between $a_j$ and $a_k$ in $I$, so $\{a_i, a_k\}$ blocks $M$ in $I$, a contradiction. □

We now describe the reduction from SRTI under super-stability to SRF, due to Fleiner *et al.* [216].

**Theorem 4.39 ([216]).** *Let $I$ be an instance of* SRTI. *Then we may construct in polynomial time an instance $J$ of* SRF *such that there is a 1–1 correspondence between the super-stable matchings in $I$ and the stable matchings in $J$.*

*Proof.* Let $G = (A, E)$ be the underlying graph of $I$, where $n = |A|$ and $m = |E|$. We may as well assume that $I$ admits some super-stable matching $M$, for if not then we can let $J$ be any unsolvable instance of SR and the result trivially holds. Let $I'$ be an instance of SRI obtained from $I$ by breaking the ties arbitrarily. For each agent $a_i \in A$, let $S_i$ be the set of stable partners of $a_i$ in $I'$ (by Theorem 4.1, all such sets $S_i$ can be constructed in $O(nm)$ overall time).

We construct a set $F$ of forbidden pairs and a second instance $I''$ of SRI, obtained from $I$ by breaking the ties, as follows. Let $a_i \in A$, and let $T$ be a tie in $a_i$'s preference list in $I$. If $T$ contains no agent from $S_i$, break $T$ arbitrarily in $I''$. Otherwise, let $a_j$ be the worst agent in $S_i$, according to $a_i$'s preferences in $I'$, that belongs to $T$ (by Lemma 4.38, $a_j$ is the only possible super-stable partner of $a_i$ in $I$ from $T$). Break $T$ in $I''$ such that $a_i$ prefers $a_k$ to $a_j$, for all $a_k \in T \setminus \{a_j\}$. Additionally, add $\{a_i, a_k\}$ to $F$ for each such $a_k$. Having repeated this for each agent $a_r \in A$, and for each tie in $a_r$'s preference list in $I$, we define $J$ by taking $I''$ together with the pairs in $F$. We claim that a matching $M$ is super-stable in $I$ if and only if $M$ is stable in $J$.

To see this, suppose that $M$ is super-stable in $I$. Then $M \cap F = \emptyset$ by Lemma 4.38. Hence, and by Lemma 4.21, $M$ is stable in $J$. Conversely let $M$ be stable in $J$, where $M \cap F = \emptyset$. Suppose that $\{a_i, a_j\}$ blocks $M$ in $I$. If $a_i$ is indifferent between $a_j$ and $M(a_i)$ in $I$, then by construction $a_i$ prefers $a_j$ to $M(a_i)$ in $J$, since $\{a_i, M(a_i)\} \notin F$. A similar remark holds for $a_j$, and hence $\{a_i, a_j\}$ blocks $M$ in $J$, a contradiction. □

We now consider the reduction in the opposite direction to that described by Theorem 4.39. Fleiner *et al.* [216] modified a reduction of Cechlárová and Fleiner [122] to show that, given an SRF instance $I$, we can construct in $O(m)$ time an instance $J$ of SRT such that a stable matching in $I$ can be derived in $O(m)$ time from a super-stable matching in $J$, and vice-versa, where $m = |E|$. Given that we can find a super-stable matching in $J$, or report that none exists, in $O(m)$ time (see Theorem 4.22), it follows that we can find a stable matching in $I$, or report that none exists, in the same time complexity.

A more powerful reduction [216] transforms an instance $I$ of the more general *Stable Roommates problem with Preference posets and Forbidden pairs* (SRPF) to an instance $J$ of 2-SAT (this reduction was mentioned in Sec. 3.4.5 for the special case that $I$ is an instance of SMTI under super-stability). In the SRPF case, $J$ can be constructed from $I$ in $O(nm)$ time, and as $J$ requires $O(m)$ space, it follows that there is a succinct certificate for the unsolvability of $I$ that can be generated in $O(nm)$ time, and represented using $O(m)$ space, where $n = |A|$. Moreover the reduction implies an $O(nm)$ algorithm for finding a super-stable matching in $I$ or reporting that none exists.

Other consequences of the reduction are as follows. A minimum regret super-stable matching in $I$ can be found in $O(nm)$ time. Also all the super-stable pairs in $I$ can be found in $O(nm)$ time, whilst there is an algorithm to list all the super-stable matchings in $I$: the first such matching can be output in $O(nm)$ time, and each subsequent super-stable matching can be output in $O(n)$ time [216].

Tighter complexity bounds for the above problems, plus an algorithm for finding an egalitarian super-stable matching, exist in the case that $I$ is an instance of the *Stable Marriage problem with Preference posets and Forbidden pairs* (SMPF) — see Ref. [216] for more details.

A direct algorithm for finding a super-stable matching or reporting that none exists, given an instance $I$ of SRPF, is described in Ref. [217]. This algorithm has $O(m(n + m))$ complexity and avoids the need to transform $I$ to a 2-SAT instance.

We close this subsection by mentioning a problem that is in some sense "opposite" to SRF, namely the *Stable Roommates problem with FRee pairs* (SRFR) [123]. An instance of this problem is an SRI instance $I$ that additionally involves a set $F \subseteq E$ of *free pairs*, where $G = (A, E)$ is the underlying graph of $I$. If a pair of agents is free then this pair can belong to a stable matching (though is not obliged to) but cannot be a blocking pair. The intuition behind this definition is that, in some practical applications it may be the case that certain pairs can be identified that cannot lead to a given matching being undermined in practice, since the agents involved only have partial information about one another. Thus the set of potential blocking pairs is reduced, which increases the likelihood of a stable matching existing (see also Sec. 2.10.9). Nevertheless, Cechlárová and Fleiner [123] showed that, in contrast to the case for SRF, the problem of determining whether an instance of SRFR admits a stable matching is NP-complete.

### 4.8.3  Stable Crews problem

The *Stable Crews problem* (SC) [121] is a generalisation of SRI in which each agent $a_i$ has one of two possible roles in a given matching $M$ for a given instance $I$. If $a_i$ has role $k$ ($k \in \{1, 2\}$) in $M$ then this is denoted by $a_i^k$; note that the notion of role is dependent on $M$ and not $I$. For example role 1 might be "driver" and role 2 might be "navigator" for a car journey. A matching $M$ is a set of pairs of mutually acceptable agents with roles such that each agent appears in at most one pair, and each pair contains exactly one agent in each role, i.e., if $\{a_i^k, a_j^l\} \in M$ then $k = 3 - l$. Each agent has a strictly-ordered preference list over the others according to the roles that they may have in a given matching, for example $a_i$ may prefer $a_j$ as a navigator to $a_j$ as a driver.

Based on this extended notion of preference, Cechlárová and Ferková [121] defined an appropriate notion of stability. They showed that SR is essentially a special case of SC in the sense that, given an instance $I$ of SR, one may construct in polynomial time an instance $J$ of SC such that the stable matchings of $I$ and $J$ are in 1–1 correspondence. The main result of the paper, however, is an $O(n^2)$ algorithm for finding a stable matching or reporting that none exists, given an instance of SC, where $n$ is the number of agents. This algorithm is an extension of Irving's algorithm for SR [306] that uses an extended notion of a rotation, together with an extra step involving the elimination of so-called *double favourites*. Finally, the authors show that a result analogous to Theorem 1.19 holds for SC: that is, for a solvable instance $I$, the same agents are unassigned in all stable matchings in $I$.

### 4.8.4  Stable Fixtures problem

The *Stable Fixtures problem* (SF) [330] is an many–many extension of SRI in which each agent $a_i$ has a capacity $c_i \geq 1$. This problem can also be regarded as a non-bipartite extension of HR. Indeed, it is straightforward to adapt the notation and terminology defined for HR in Sec. 1.3.2 to the SF

$$a_1 : a_2 \quad a_3 \quad a_4 \qquad c_1 = 2$$
$$a_2 : a_1 \quad a_3 \quad a_4 \qquad c_2 = 2$$
$$a_3 : a_1 \quad a_2 \quad a_4 \qquad c_3 = 2$$
$$a_4 : a_1 \quad a_2 \quad a_3 \qquad c_4 = 2$$

Fig. 4.6   An instance of SF

context, obtaining in particular definitions of a matching and stability in an SF instance. The problem name stems from a practical situation where the agents are players (or teams) who are to play against one another in a tournament. Each player ranks their potential opponents in order of preference, and the task is to construct a set of *fixtures*, consisting of distinct matches (each involving two players, where the number of matches that a player is involved in may be more than one but cannot exceed its capacity), that is stable. Another application for SF comes from P2P networking [430, 253, 254].

One curious property of SF is the following. Suppose that $I$ is an instance $I$ of SF with $n = 2k$ players (for some $k \geq 1$) where all preference lists are complete and each player has a capacity of $c$, for some $c$ $(1 \leq c \leq 2k - 1)$. If $c = 1$, clearly $I$ is an instance of SR and any stable matching has size $k$. For general $c > 1$ we might expect any stable matching to be of size $ck$. However this is not the case, as the example SF instance in Fig. 4.6 illustrates: here $k = 2$, $c = 2$ and any stable matching has size 3.

Irving and Scott [330] described an $O(m)$ algorithm for finding a stable matching or reporting that none exists, given an SF instance $I$ where $m$ is the number of acceptable pairs of agents. Their algorithm is again an extension of Irving's algorithm for SR [306] and involves the elimination of rotations, given a rotation definition that is adapted appropriately to the SF context. They also showed that an analogue of sorts of Theorem 1.11 holds in the SF context. In particular, for a given instance $I$ of SF, (i) every player has the same number of assignees in all stable matchings in $I$, (ii) any player that is undersubscribed in one stable matching has exactly the same set of assignees in all stable matchings in $I$, and (iii) all stable matchings in $I$ have the same size.

Scott [523] studied the extension of SF where preference lists may include ties (referred to as SFT — the Stable Fixtures problem with Ties). He formulated three notions of stability that extend the definitions given in the SR context (see Sec. 4.5). The problem of deciding whether a weakly stable matching exists, given an SFT instance, is NP-complete — this follows by restriction, given that the corresponding problem is NP-complete for SRT (see Sec. 4.5.2). By constrast, Scott [523] described an $O(m)$ algorithm for finding a super-stable matching or reporting that none exists, given an instance of SFT. It is an open question as to whether the corresponding problem for strong stability is polynomial-time solvable or NP-hard.

Note that SF is a very general stable matching problem. Clearly SR is the special case of SF in which each agent has capacity 1. Alternatively, we may consider restrictions of SF in which the underlying graph $G$ is bipartite. This gives rise to many–many bipartite stable matching problems (see Sec. 5.4). Its restriction in which the capacity of each vertex on one side of $G$ is 1 is HR (Sec. 1.3).

Motivated by P2P networking, Georgiadis and Papatriantafilou [253, 254] studied a method for coping with SF instances that do not admit a stable matching. They defined a notion of *satisfaction* for each agent $a_i$ with respect to a given matching $M$ — this takes into account $a_i$'s capacity, the number of assignees of $a_i$ in $M$, the ranks of these assignees in $a_i$'s preference list, and the length of $a_i$'s list. The authors then considered MAX SF SAT, the problem of finding a matching that maximises the overall satisfaction of the agents in a given SF instance relative to this definition. They presented distributed approximation algorithms for MAX SF SAT in a static and dynamic setting. Note that MAX SF SAT is not known to be NP-hard.

### 4.8.5   Stable Multiple Activities problem

The *Stable Multiple Activities problem* (SMA) [122] is a generalisation of SF in which the underlying graph may have parallel edges. This represents a practical situation in which players may form multiple partnerships according to different sports activities: for example, $a_i$ might play a match with $a_j$ for each of tennis, chess and badminton. A solution to an instance of this problem is referred to as a *stable b-matching* [122]. SMA has also been referred to in the literature as the *stable b-matching problem* [212,104].

Cechlárová and Fleiner [122] showed that there is a polynomial-time reduction from an SMA instance $I$ to an SR instance $J$ such that a stable $b$-matching in $I$ corresponds to a stable matching in $J$, and vice versa, however the mapping is not injective in general. The authors described an $O(m^2)$ algorithm for finding a stable $b$-matching or reporting that none exists, given an SMA instance, where $m$ is the number of edges in the under-

lying multigraph. Again, this algorithm is an extension of Irving's algorithm
for SR [306]. However, unlike Irving's two-phase approach, the algorithm of
Cechlárová and Fleiner has only a single phase.

A two-phase algorithm for SMA was described by Borbel'ová and
Cechlárová [104]. The first phase involves the deletion of certain edges
from the underlying multigraph so as to obtain an instance that satisfies
the so-called *first–last property*. This property is a generalisation of the
notion of a *stable table* that is the outcome of phase 1 of Irving's algorithm
for a solvable instance of SR [306] (see also Chapter 4 of Ref. [261]). The
second phase of the algorithm of Borbel'ová and Cechlárová involves the
elimination of rotations in the SMA context. The concept of a rotation here
extends the definition in the SR context (see Sec. 1.4.4) and were first defined
in the SMA setting by Cechlárová and Fleiner [122]. The main advantage
of the two-phase approach for SMA is that the algorithm's complexity is
$O(m)$, improving on the previous algorithm in Ref. [122]. Borbel'ová and
Cechlárová [104] also proved that each player has the same number of assignees in every stable $b$-matching for a given SMA instance. Note that a
two-phase $O(m)$ algorithm for SMA was independently formulated by Ando
and Kanemaru [45].

In subsequent work, Borbel'ová and Cechlárová [106] studied rotations
in the SMA context in greater depth. They proved an analogue of Theorem
1.23 in the context of an SMA instance $I$ and used it in order to derive
efficient algorithms for finding all stable pairs, a minimum regret stable
$b$-matching, and an egalitarian stable $b$-matching in the case that the underlying graph of $I$ is bipartite.

### 4.8.6  Stable Allocation problem

The *Stable Allocation problem* (SA) (or *Ordinal Transportation problem*)
was introduced by Baïou and Balinski [59]. In its most general form, as
defined by Biró and Fleiner [82], an instance involves a multigraph $G =
(V, E)$, where each vertex $v \in V$ has a linear order $\prec_v$ over its adjacent
edges. Also there are functions $b : V \longrightarrow \mathbb{R}^+$, representing the *bounds* of
the vertices, and $c : E \longrightarrow \mathbb{R}^+$, representing the *capacities* of the edges. A
function $x : E \longrightarrow \mathbb{R}^+$ is an *allocation* if $x(v) \equiv \sum_{v \in e} x(e) \le b(v)$ for all
$v \in V$ and $0 \le x(e) \le c(e)$ for all $e \in E$. A vertex $v \in V$ is *saturated* if
$x(v) = b(v)$ and similarly an edge $e \in E$ is *saturated* if $x(e) = c(e)$. An edge
$e \in E$ is *dominated* at a vertex $v \in E$ if $\sum_{e \preceq_v f} x(f) = b(v)$. Allocation $x$ is
*stable* if, given an unsaturated edge $e \in E$, there exists a saturated vertex

$v \in e$ such that $e$ is dominated at $v$. Baïou and Balinski [59] defined the restriction of SA in which $G$ is bipartite.

SA is a very general stable matching problem: as observed by Biró and Fleiner [82], its various restrictions encompass many of the stable matching problems already introduced in this book. We begin by considering the *Integral Stable Allocation problem* (ISA), in which $x(e) \in \mathbb{Z}$ for all $e \in E$. The special case of ISA in which $c(e) = 1$ for all $e \in E$ is SMA (Sec. 4.8.5). If, in addition, $G$ is simple, we obtain SF (Sec. 4.8.4). Restrictions of SF (e.g., where $b(v) = 1$ for all $v \in V$, and/or $G$ is bipartite, have already been noted in Sec. 4.8.4. Finally, the restriction of SMA in which $b(v) = 1$ for all $v \in V$ and each pair of vertices has at most two parallel edges is SC (Sec. 4.8.3).

Baïou and Balinski [59] described an algorithm for SA which has exponential complexity in the worst case. They also described an inductive algorithm for the special case of SA in which $G$ is bipartite.

Dean and Munshi [162] showed that Baïou and Balinski's algorithm for SA in bipartite graphs can be implemented to run in $O(nm)$ time without the use of sophisticated data structures, where $n = |V|$ and $m = |E|$. With the aid of such structures, Dean and Munshi proved that this algorithm can be implemented to run in $O(m \log n)$ time. They also showed how to solve general SA (in non-bipartite graphs) in the same time complexity, with high probability. For ISA in the case of non-bipartite graphs, Biró and Fleiner [82] gave a weakly polynomial-time algorithm that runs in $O(m^3 \log B)$ time, where $B = \max\{b(v) : v \in V\}$. Further, Dean and Munshi [162] proved that a weighted version of SA in bipartite graphs can be solved in polynomial time, whilst there is a 2-approximation algorithm for the corresponding problem in the non-bipartite case, which is NP-hard. Finally, Dean and Swar [163] considered a variant of SA involving edge multipliers.

### 4.8.7   Stable Roommates problem with Choice Functions

The *Stable Roommates problem with Choice Functions* (SRCF) [215] is a many–many extension of SRI that can be defined as follows. An instance comprises a graph $G = (A, E)$, together with a function $Ch_{a_i} : \mathbb{P}(N(a_i)) \longrightarrow \mathbb{P}(N(a_i))$, for each agent $a_i \in A$, where $N(a_i)$ denotes the vertices adjacent to $a_i$ in $G$ and $\mathbb{P}(X)$ denotes the power set of $X$, for any $X \subseteq V$.

The function $Ch_{a_i}$ is called the *choice function* for agent $a_i$ and intuitively maps any subset $X$ of $N(a_i)$ to the set of agents that $a_i$ chooses from $X$. In an instance of SRI, each agent $a_i \in A$ has a linear order $\prec_{a_i}$ over the agents in $N(a_i)$, and thus for any $X \subseteq N(a_i)$, $Ch_{a_i}(X)$ is equal to the minimal element of $X$ according to $\prec_{a_i}$. In this case $Ch_{a_i}$ is referred to as a *linear choice function*.

In general, a natural and commonly-studied property of choice functions is *substitutability* [374,514]. For any $a_i \in A$, $Ch_{a_i}$ is said to be *substitutable* if, for any set $X \subseteq N(a_i)$ and for any two distinct agents $a_j, a_k$ in $X$, $a_j \in Ch_{a_i}(X)$ implies $a_j \in Ch_{a_i}(X \backslash \{a_k\})$. Roughly speaking, this means that if $a_i$ chooses $a_j$ from $X$ then $a_i$ continues to choose $a_j$ even if some other options in $X$ become unavailable.

Now let $M \subseteq E$. We say that $M$ is *individually rational* if, for each $a_i \in A$, $Ch_{a_i}(M(a_i)) = M(a_i)$, where $M(a_i)$ denotes the set of neighbours of $a_i$ in $M$ ($M(a_i)$ is empty if there are no edges of $M$ incident to $a_i$). Intuitively, the individual rationality of $M$ guarantees that no agent $a_i$ would prefer to lose any member of $M(a_i)$. A pair $\{a_i, a_j\} \in E$ is *blocking* if $a_j \in Ch_{a_i}(M(a_i) \cup \{a_j\})$ and $a_i \in Ch_{a_j}(M(a_j) \cup \{a_i\})$. Intuitively, a blocking pair signifies a situation in which both $a_i$ and $a_j$ would prefer to augment their existing partnerships by choosing each other. We say that $M$ is a *stable partnership* if $M$ is individually rational and admits no blocking pair. In the case where $Ch_{a_i}$ is a linear choice function, for each $a_i \in A$, the definition of a stable partnership coincides with that of a stable matching, and thus SRI is a special case of SRCF.

Fleiner [215] defined a choice function to be *increasing* if, for any agent $a_i \in A$ and for any subsets $X$ and $Y$ of $N(a_i)$, we have $Y \subseteq X$ implies that $|Ch_{a_i}(Y)| \leq |Ch_{a_i}(X)|$. Intuitively this means that, relative to $Y$, if some additional options become available, then the set that $a_i$ chooses (from the superset $X$ of $Y$) will be at least as large as the set that $a_i$ would choose from $Y$. Fleiner showed that, for SRCF instances with increasing and substitutable choice functions, there is an efficient algorithm for finding a stable partnership or reporting that none exists. (In fact, Fleiner's algorithm can be used to construct a *stable half-partnership*, which is a generalisation of a stable half-matching (see Sec. 4.3.5), given an arbitrary instance of SRCF.) By contrast, for general substitutable choice functions, he showed that the problem of deciding whether a stable partnership exists is NP-complete.

The bipartite version of SRCF with substitutable preferences is solvable in polynomial time — see Sec. 5.4.4 for more details.

### 4.8.8   Coalition Formation Games

The *Coalition Formation Game* (CFG) [171, 66, 264] is an extension of SRI in which the set of agents $A$ is to be partitioned into *coalitions* (i.e., sets of arbitrary size) subject to a stability requirement. A distinction between CFG and SRI is that, in the latter case, coalitions must be of size at most 2. CFG has also been referred to as a *Hedonic Game* or as the *Stable Partition problem* in the literature (see Refs. [264, 120] for more details). However we note that the latter terminology conflicts with the concept of a stable partition, as defined in Sec. 4.3.

The different types of preference structures for agents are categorised into two main groups in Ref. [120]: (i) each agent $a_i$ ranks in (not necessarily strict) order of preference a list of coalitions (to which $a_i$ would potentially belong), and (ii) each agent ranks in order of preference a list of other agents (again, ties are permitted in the preference lists). Problems in model (i) in particular have been referred to as Hedonic Games [102, 65]. A drawback with this model from a computational point of view is that, as discussed in Ref. [126], a representation of the preference structure inherent in an instance of a Hedonic Game would, in general, have a space requirement that is super-polynomial in the number of agents.

Turning to model (ii), problems in this category have been most commonly referred to in the literature as Coalition Formation Games or Stable Partition problems. In order for an agent to decide between two coalitions that she may be involved in, it is necessary to extend the notion of preference over individual agents to preference over coalitions. Various methods for doing this have been defined in the literature, including so-called $B$-*preferences*, $W$-*preferences*, $BW$- and $WB$-*preferences* [125, 131, 126–128]. Other methods are surveyed in Refs. [264, 120].

Suppose that there is a suitable definition of an agent's preferences over coalitions. Subject to this, and given a partition $M$ of $A$, we define a set of agents $A'$ to be *blocking* if each agent $a_i \in A'$ prefers $A'$ to $M(a_i)$, the coalition containing $a_i$ in $M$. Similarly, $A'$ is *weakly blocking* if no agent $a_i \in A'$ prefers $M(a_i)$ to $A'$, and some agent $a_j \in A'$ prefers $A'$ to $M(a_j)$. The relevant stability concepts can now be defined as follows. A partition $M$ is defined to be a *core partition* if $M$ admits no blocking coalition, whilst $M$ is a *strong core partition* if $M$ admits no weakly blocking coalition.

The various computational problems that arise in this setting involve determining whether a given partition is in the core or strong core, deciding whether a core or strong core partition exists, and finding one if so, and describing the structure of core or strong core partitions. Variants of the problems arise when the size of the coalitions is bounded [129], and/or when the order that the agents appear in a given coalition is important [124, 80, 129]. These restrictions are relevant in the context of kidney exchange (see Sec. 1.4.6): here every coalition corresponds to a cycle $(a_{i_0}, a_{i_1}, \ldots, a_{i_{r-1}})$, where each $a_{i_j}$ corresponds to a patient–donor pair $(p_{i_j}, d_{i_j})$, and the cycle represents the transplantation of a kidney from donor $d_{i_j}$ to patient $p_{i_{j+1}}$ ($0 \leq j \leq r - 1$ and addition is taken modulo $r$). Since all operations in a cycle must take place simultaneously, cycles should be as short as possible for logistical reasons. Further details regarding what is known about these questions for a range of preference models are given in Refs. [264, 120].

We close this section by remarking that a number of desirable properties involving partitions in a CFG instance such as *Nash-stability*, *individual stability* and *contractual individual stability* (see Ref. [264] for definitions of these terms) have recently been studied in the context of SMTI and SRTI instances by Aziz [54] from an algorithmic point of view.

## 4.9 Conclusions and open problems

The wealth of results for SR and its variants surveyed in this chapter indicates that over the last 25 years this problem has, in some sense, been "tamed" — this is in stark contrast to the situation during the 1960s and 1970s when the problem had something of a mysterious nature. During this time, the only tangible observation was due to Gale and Shapley [235], who gave an example to show that an instance may be unsolvable, and this prompted Knuth [394] to ask whether the problem of deciding whether a stable matching exists, given an SR instance, might be NP-complete. This was of course later proved not to be the case by Irving [306].

Nevertheless, despite the substantial advances that have been made since Irving's paper, a number of intriguing open questions remain. These have been variously stated at different points throughout the chapter, but we gather some of the most significant ones together as follows:

- Can a tighter asymptotic upper bound for the solvability probability of a random SR instance be found? As described in Sec. 4.2.1, the best current upper bound is about 0.82 [482], but empirical evidence suggests that this bound is not likely to be very tight.
- Is there a polynomial-time reduction from a solvable instance $I$ of SR to an SM instance $J$ such that there is a 1–1 correspondence between

the stable matchings in $I$ and $J$? As mentioned in Sec. 4.2.2, Dean and Munshi [162] have given a partial answer to this question, but their reduction is not in the "true spirit" of what Gusfield and Irving had intended.

- Is there an $\Omega(n^2)$ lower bound for the problem of determining whether a given SR instance with $n$ agents is solvable? As mentioned in Sec. 4.2.3, only an answer to the decision problem is required, and not an actual stable matching if one does exist.

- The concepts of rank-maximality and stability can be combined to define the notion of a *rank-maximal stable matching*, given an instance of SRI. (That is, given a solvable SRI instance $I$, find a stable matching in which the maximum number of agents obtain their first choice, and subject to this, the maximum number of agents obtain their second choice, and so on, among all stable matchings in $I$.) A rank-maximal stable matching is formally defined in Sec. 8.2.4 in the SMI context and it is straightforward to adapt that definition to the SRI case (using the definition of rank-maximality from Sec. 8.2.5). In Sec. 8.2.4, we note that the problem of finding a rank-maximal stable matching in an instance of SMI is solvable in polynomial time. Is the same true in the SRI context?

- In Sec. 2.10.3 we considered the problem RATIONALIZABILITY in the SM context. One can also consider this problem in the SR case. That is, given a set $\mathcal{M}$ of matchings involving $n$ agents, the problem is to determine whether there is an SR instance $I$ that admits a set of stable matchings $\mathcal{S}$ such that $\mathcal{M} \subseteq \mathcal{S}$. To the best of the author's knowledge, it is unknown as to whether this problem is solvable in polynomial time or NP-complete.

- An instance $I$ of SRF may not admit a stable matching, yet the underlying SR instance $J$, obtained by omitting the forbidden pairs in $I$, may be solvable (see Sec. 4.8.2). In such a setting, what is the complexity of finding a stable matching in $J$ that has the minimum number of forbidden pairs in $I$?

- What is the complexity of the problem of finding a strongly stable matching or reporting that none exists, given an instance of SFT (see Sec. 4.8.4)?

Chapter 5

# Further stable matching problems

## 5.1 Introduction

In this chapter we focus on matching problems with preference lists, not already considered in the preceding chapters, where the solution criterion in each case involves some stability property. Satisfying such a property invariably involves establishing the absence of a blocking pair (or coalition of agents of size two or greater) who would prefer to be assigned to one another (or, in some cases, to one another's partners) than remain with their partners in the given matching. In some cases the notion of a blocking pair or coalition is closely based on classical (Gale–Shapley) stability, with the definition suitably modified and/or extended to fit the problem context. However in other cases the definition is rather different (this is perhaps best illustrated by *exchange-stability*, as defined below), though the satisfaction of such a blocking pair or coalition still represents an overall improvement for the agents concerned.

Many of the problems that we consider are extensions of HR motivated by practical applications, with suitably amended stability criteria. We also study three-dimensional extensions of SM and SR that have received attention in the literature. Furthermore, we describe *exchange-stability*, which is in some sense orthogonal to classical stability, in the SM and SR contexts — this criterion involves the absence of coalitions of agents who envy one another's partners. Again, our emphasis is on algorithmic results for the problems covered in this chapter, though structural results are also important as they often impact on a problem's algorithmic behaviour.

The problems that we consider are as follows. In Sec. 5.2, we focus on variants of HR involving lower and common quotas. A lower quota for a hospital indicates the minimum number of assignees that it requires in a matching in order to be viable. This is a counterpart to a hospital's upper quota, which we have previously referred to as its capacity. We describe two models for HR with lower quotas that have been considered in the literature. Common quotas arise when a group of hospitals has a common upper bound on the number of residents that may be collectively assigned to them. We also study a variant of HR in which each hospital classifies its acceptable residents into (possibly overlapping) categories, with each category having a lower and upper quota.

An extension of HR that has important practical applications occurs when couples supply joint preferences over pairs of hospitals (that are typically geographically close to one another) so that any matching respects these joint lists in a precise sense. Unfortunately a stable matching need not exist in this context, and furthermore the problem of deciding whether such a matching exists is NP-complete even in some highly restricted cases. Nevertheless some restrictions of the general problem do turn out to be solvable in polynomial time. We discuss results for HR with couples in Sec. 5.3.

We have seen that HR is a many–one extension of SM. It is natural, then, to consider the many–many extension of HR — in the literature this many–many bipartite stable matching problem has been presented in the context of matching workers to firms, where each agent can have multiple assignees (up to some fixed capacity). Two basic models have been considered in the literature: firstly when agents have preferences over individual agents from the other set, and secondly when agents rank in order of preferences subsets of agents that constitute their entire set of potential assignees in a given matching. The latter case has an obvious drawback, namely that the length of a given agent's preference list need not be polynomially-bounded in the number of agents. Sec. 5.4 discusses these two formulations of the many–many bipartite stable matching problem in greater detail. (Note that many–many generalisations of SR were considered in Sec. 4.8.4, Sec. 4.8.5 and Sec. 4.8.6.)

Another generalisation of HR is motivated by the problem of allocating students to projects in a university department. Each project is offered by some lecturer, and both projects and lecturers have capacities indicating their maximum number of assignees. Students rank projects in order of preference, whilst lecturers may rank in preference order either their projects, or the students who find their projects acceptable, or both. We thus obtain different models for this variant of HR, with different structural and algorithmic properties. These models are described in Sec. 5.5.

Three-dimensional versions of SM and SR have been considered in the literature. In the three-dimensional extension of SM we have three disjoint sets of agents rather than two (Knuth [394] suggested that men, women and dogs could be the categories of agents involved) and the task is to form triples of agents (each containing a member from each of the three disjoint sets) that are stable in a particular sense. The three-dimensional version of SR involves a set of agents that is some multiple of 3 in cardinality, and the task is to form either unordered or ordered triples of these agents so as to again satisfy some stability property. For many of these three-dimensional stable matching problems, a stable matching need not exist and the problem of deciding whether one does exist is NP-complete. We survey results for three-dimensional stable matching problems in Sec. 5.6.

In Sec. 5.7 we study variants of SM and SR where the solution criterion for the particular matching that we seek involves *exchange-stability* rather than classical stability. Informally, if a matching is exchange-stable then no two agents can improve by swapping their partners. The concept can be generalised to the case where we seek a matching in which no coalition of agents can improve by swapping their partners in a cyclic fashion. It is even possible to combine exchange-stability with classical stability, with some interesting algorithmic consequences.

Two other stable matching problems are considered in Sec. 5.8. In Sec. 5.8.1, we study the problem of finding a matching in a given SM or SR instance that is stable with respect to both the original preference lists and the reverse of these. In Sec. 5.8.2 we focus on a variant of SR in which a matching must be robust against *blocking cycles* of length 2 or more. The stability concept here is subtly different from exchange-stability, and has applications to kidney exchange.

Finally, in Sec. 5.9 we outline a selection of open questions that are collected together from the problems considered in this chapter.

## 5.2  HR **with lower and common quotas**

### 5.2.1  *Introduction*

In the classical Hospitals / Residents problem [235, 261, 514], each hospital has a capacity, or *upper quota*, limiting its number of assignees from above. However in many practical applications, a hospital might have a *lower quota* [84, 291, 275, 218], indicating that the training programme that it offers is not viable unless it has a given minimum level of participation. This leads to the *Hospitals / Residents problem with Lower Quotas* (HR-LQ). We consider two formulations of HR-LQ, denoted HR-LQ-1 and HR-LQ-2, that arise in the presence of lower quotas.

The first model, HR-LQ-1, studied by Biró *et al.* [84], allows a given hospital to be either *open* (meaning that its number of assignees lies between the lower and upper quotas) or *closed* (implying that it has no assignees). We define this problem formally and discuss some associated complexity results in Sec. 5.2.2. In the second model, considered by Hamada *et al.* [275], no hospital is allowed to be closed, and in a feasible matching every hospital must meet its lower quota (with respect to its number of assignees). Of course this implies in general that a feasible matching need not exist, though Hamada *et al.* place some restrictions on a given instance of the problem to guarantee the existence of such a matching. We define the second model (HR-LQ-2) formally and present some algorithmic results for it in Sec. 5.2.3.

Another generalisation of HR arises when groups of hospitals impose *common quotas*, limiting (from above) the number of residents that may be collectively assigned to them. We then obtain the *Hospitals / Residents problem with Common Quotas* (HR-CQ), as studied by Biró *et al.* [84].

Motivation for this variant comes, for example, from the Japan Residency Matching Program (JRMP), the matching scheme for allocating medical students to hospitals in Japan [596, 354, 355]. Here, in common with medical matching programmes in many other countries, the Japanese government was concerned about the geographical distribution of residents assigned to hospitals, and in particular the shortage of assignees to rural hospitals [354, 355]. In an attempt to combat this, *regional caps* were introduced, which limit the number of residents collectively assigned to hospitals in a given prefecture, the intention being that more residents will then dissipate to rural areas.

We define HR-CQ formally in Sec. 5.2.4. As is the case with the introduction of lower quotas, the presence of common quotas leads to NP-hardness for the problem of finding a stable matching (or reporting that none exists). However as we will illustrate, if the sets of hospitals which collectively have common quotas satisfy a certain natural structural condition, a stable matching always exists and can be found in polynomial time. Moreover in these circumstances the stable matchings have some nice structural properties.

Another extension of HR that involves lower and common quotas (albeit with a different manifestation) is the *Classified Stable Matching* problem, as introduced by Huang [291]. This is a generalisation of HR in which residents are classified by the hospitals (e.g., according to their area of medical expertise), and each class comes with upper and lower quotas that indicate the maximum and minimum number of residents belonging to that particular class that must be assigned to a given hospital in a feasible matching. Huang's motivation for the problem came from academic hiring, where residents correspond to applicants, hospitals correspond to academic institutions, and each institution classifies its applicants according to various criteria, which may include their research area, for example. The classification of applicants for a given institution need not give rise to disjoint classes, reflecting the possibility that an applicant may be affiliated with more than one research area. We give a formal definition of the Classified Stable Matching problem, and discuss some algorithmic and structural results for the problem in Sec. 5.2.5. Whilst the two models of HR with lower quotas studied in Secs. 5.2.2 and 5.2.3 lead to NP-completeness for the problem of deciding whether a stable matching exists, somewhat surprisingly the presence of lower quotas in the Classified Stable Matching problem does not necessarily lead to NP-completeness. As we will show, under certain conditions on the resident classifications, the problem of finding a stable matching or reporting that none exists becomes solvable in polynomial time.

Further motivation for variants of HR with lower and common quotas comes from the Hungarian higher education matching scheme [84]. The "residents" (i.e., applicants) rank in order of preference the courses of study that they would like to follow (together with an indication as to whether their chosen programme of study is to be financed privately or by the state). The "hospitals" (i.e., colleges and universities) rank in order of preference their applicants on the basis of academic performance (note that the evaluation criteria may vary from one institution to the next, so it need not be the case that there is a single master list of all applicants). The matching scheme has run since 1985, and typically over 100,000 students take part annually (140,953 students in 2011).

What makes the scheme relevant in this context is the presence of lower and common quotas. A course at a given college or university may have a lower quota, indicating the minimum number of applicants that must be assigned to it in order for it to run (or else the course is suspended for that particular year). Biró *et al.* [84] also reported that, since 2007, there have been common quotas for each field of study at each academic institution, in addition to a common quota for state-financed places for each field of study nationally.

## 5.2.2 HR *with lower quotas (model 1)*

Recall that in an instance $I$ of HR, each hospital $h_j \in H$ has a positive integral capacity denoted by $c_j$, which indicates the maximum number of residents that may be assigned to $h_j$. Suppose in addition that $h_j$ has a non-negative *lower quota* $l_j$, indicating the minimum number of residents that must be assigned to $h_j$ before this hospital can open. For consistency, we also denote $c_j$ by $u_j$ in this section, and we refer to $u_j$ as the *upper quota* of $h_j$. We assume that $l_j \leq u_j$ for each $h_j \in H$. We let HR-LQ-1 denote the extension of HR in which hospitals have lower quotas as well as upper quotas.

A *matching* $M$ in this context requires that every hospital $h_j \in H$ satisfies $|M(h_j)| = 0$ or $l_j \leq |M(h_j)| \leq u_j$. We say that $h_j$ is *closed* if $|M(h_j)| = 0$, and *open* otherwise. We now present the stability concept for HR-LQ-1, as defined by Biró *et al.* [84].

**Definition 5.1 ([84]).** *Let $I$ be an instance of* HR-LQ-*1 and let $M$ be a matching in $I$. Then $M$ is* stable *if the following two conditions are satisfied:*

*(1)* *(no blocking pair) there is no acceptable resident–hospital pair $(r_i, h_j)$ such that (i) $r_i$ is either unassigned or prefers $h_j$ to $M(r_i)$, (ii) $h_j$ is open, and (iii) $h_j$ is either undersubscribed or prefers $r_i$ to a member of $M(h_j)$;*

*(2)* *(no blocking coalition) there is no closed hospital $h_j$ (blocking hospital) and set of $l_j$ residents, each of whom is either unassigned (and finds $h_j$ acceptable) or prefers $h_j$ to her assigned hospital.*

Under certain circumstances, an HR-LQ-1 instance $I$ is guaranteed to admit a stable matching. For, let $J$ be the HR instance obtained from $I$ by disregarding the lower quotas. Suppose that $|M_a(h_j)| \geq l_j$ for all $h_j \in H$, where $M_a$ is the resident-optimal stable matching in $J$. Then no hospital is closed, so clearly $M_a$ is stable in $I$. In fact, by the Rural Hospitals Theorem (Theorem 1.11), any stable matching $M$ in $J$ also satisfies $|M(h_j)| \geq l_j$ for all $h_j \in H$, which implies that $M$ is also stable in $I$. On the other hand if some hospital in $J$ does not achieve its lower quota in $M_a$, then $I$ may not admit a stable matching, as we now illustrate.

Residents' preferences        Hospitals' preferences
$r_1 : h_1 \ h_2$              $h_1 : 2 : 2 : r_1 \ r_2$
$r_2 : h_2 \ h_1$              $h_2 : 1 : 1 : r_1 \ r_2$

Fig. 5.1  An instance of HR-LQ-1 with no stable matching due to Biró *et al.* [84]

Residents' preferences        Hospitals' preferences
$r_1 : h_1 \ h_2$              $h_1 : 2 : 2 : r_1 \ r_2$
$r_2 : h_2 \ h_1$              $h_2 : 3 : 3 : r_1 \ r_2 \ r_3$
$r_3 : h_2$

Fig. 5.2  An instance of HR-LQ-1 with stable matchings of different sizes

Consider the HR-LQ-1 instance $I$, due to Biró *et al.* [84], shown in Fig. 5.1. Next to each hospital $h_j$ we show $l_j$, then $u_j$, followed by $h_j$'s preference list. Let $M$ be a matching in $I$. As the sum of the lower bounds is 3 and there are only two residents, some hospital must be closed in $M$. Suppose firstly that $h_1$ is closed. If $M(h_2) = \{r_1\}$ then $\{r_1, r_2\}$ forms a blocking coalition with $h_1$. If $M(h_2) = \{r_2\}$ then $(r_1, h_2)$ forms a blocking pair. Now suppose that $h_2$ is closed. Then $r_2$ forms a blocking coalition with $h_2$. Hence $I$ admits no stable matching.

Even when an HR-LQ-1 instance admits a stable matching, there could be stable matchings of different sizes. To see this, consider the HR-LQ-1 instance $I$, adapted from Example 2 in Ref. [83], shown in Fig. 5.2. It is easy to verify that each of $M_1 = \{(r_1, h_1), (r_2, h_1)\}$ and $M_2 = \{(r_1, h_2), (r_2, h_2), (r_3, h_2)\}$ is stable in $I$.

It turns out that, in contrast to the case for HR, the problem of deciding whether a stable matching exists, given an HR-LQ-1 instance, is NP-complete, as the following result indicates.

**Theorem 5.2 ([84]).** *The problem of deciding whether a given* HR-LQ-1 *instance admits a stable matching is NP-complete, even if each upper and lower quota is equal to 3.*

The complexity of the decision problem is open in the case that each lower quota is at most 2.

One way of coping with an unsolvable HR-LQ-1 instance is to find a matching (with some hospitals potentially closed) that meets only Condition 1 of Definition 5.1. Such a matching is referred to as a *pairwise stable matching*. A pairwise stable matching always exists (which can be seen by simply closing all the hospitals). However one may seek to maximise the number of assigned residents, which motivates MAX PS HR-LQ-1, the problem of finding a maximum cardinality pairwise stable matching in an HR-LQ-1 instance. Alternatively, when deciding which hospitals to close, one may try to give priority to hospitals $h_j$ such that $|M_a(h_j)| \geq l_j$, where $M_a$ is the resident-optimal stable matching in the HR instance obtained by ignoring the lower quotas. Such a hospital is called *popular*. In an HR-LQ-1 instance, a pairwise stable matching in which each popular hospital is open is called a *popular pairwise stable matching*. Again, such a matching is bound to exist (this can be seen by starting from $M_a$ and then closing all of the unpopular hospitals), but as before one may wish to maximise the number of assigned residents. This gives MAX POP PS HR-LQ-1, the problem of finding a maximum size popular pairwise stable matching in a given HR-LQ-1 instance. Each of MAX PS HR-LQ-1 and MAX POP PS HR-LQ-1 is NP-hard, even if each lower and upper quota is equal to 3 [83]. Again, the complexity of both problems is open if no lower quota exceeds 2.

### 5.2.3  HR *with lower quotas (model 2)*

Hamada *et al.* [275] considered an alternative version of HR-LQ-1 in which hospitals are not permitted to be closed, and a feasible matching must satisfy the lower and upper quotas for *each* hospital. To ensure that such a matching exists, the authors make the following assumptions, namely that the number of residents is at least the sum of the lower quotas (which seems reasonable) and that each resident finds acceptable all hospitals that have a positive lower quota (a strong assumption, which is less likely to be satisfied in practice). We refer to both of these assumptions collectively as Assumption A.

We define HR-LQ-2 as follows. An instance of this problem is the same as for HR-LQ-1, subject to the condition that Assumption A holds. In the HR-LQ-2 context, a *matching* $M$ now requires that $l_j \leq |M(h_j)| \leq u_j$ for each hospital $h_j \in H$. The notion of stability for $M$ just corresponds to classical Gale–Shapley stability. The usage of the HR-LQ-2 notation is thus intended to emphasise that the matching and stability concepts here are distinct from those defined for HR-LQ-1.

Although a matching is bound to exist in an HR-LQ-2 instance as noted above, a stable matching need not. To see this, consider the simple HR-LQ-2 instance $I$ shown in Fig. 5.3 (again each hospital is followed by its lower quota, upper quota and preference list). The unique matching in $I$ is $M =$

Resident's preferences     Hospitals' preferences

$r_1 : h_1 \; h_2$                       $h_1 : 0 : 1 : r_1$

                                          $h_2 : 1 : 1 : r_1$

Fig. 5.3    An instance of HR-LQ-2 with no stable matching

$\{(r_1, h_2)\}$ due to $h_2$'s lower quota, but clearly $M$ is not stable. It is of course straightforward to determine whether an instance $I$ of HR-LQ-2 admits a stable matching: just find the resident-oriented stable matching $M_a$ in the underlying HR instance obtained by ignoring the lower quotas. If $|M_a(h_j)| \geq l_j$ for each $h_j \in H$ then clearly $M_a$ is a stable matching in $I$, otherwise by Theorem 1.11, $I$ admits no stable matching.

In order to cope with the possible non-existence of a stable matching, the authors considered matchings that minimise either the number of blocking pairs, or the number of *blocking residents* (i.e., the number of residents who are involved in a blocking pair) — this approach was also taken in Secs. 2.8 and 4.6 for SMI and SRI respectively.

Let MIN BP HR-LQ-2 (MIN BR HR-LQ-2) denote respectively the problems of finding a matching with the minimum number of blocking pairs (residents), given an instance of HR-LQ-2. Hamada *et al.* proved the following results.

**Theorem 5.3 ([275]).** *Let $n_1$ and $n_2$ denote respectively the numbers of residents and hospitals in a given HR-LQ-2 instance. The following results hold for* MIN BP HR-LQ-*2 and* MIN BR HR-LQ-*2:*

(i) MIN BP HR-LQ-*2 is not approximable within* $(n_1 + n_2)^{1-\varepsilon}$*, for any* $\varepsilon > 0$*, unless P=NP. The result holds even if (i) all the preference lists are complete, (ii) all hospitals have the same preference list, and (iii) $u_j = 1$ for each $h_j \in H$.*

(ii) MIN BP HR-LQ-*2 is approximable within* $n_1 + n_2$*.*

(iii) MIN BP HR-LQ-*2 is solvable in* $O(b^2(n_2(n_1 + b))^{b+1})$ *time, where $b$ is the number of blocking pairs in an optimal solution, for the restriction that $u_j = 1$ for each $h_j \in H$.*

(iv) MIN BR HR-LQ-*2 is NP-hard. The result holds even if each (i) hospital's preference list is obtained from a "master list" of residents (see Sec. 1.3.6), and (ii) $u_j = 1$ for each $h_j \in H$.*

(v) MIN BR HR-LQ-*2 is approximable within* $\sqrt{n_1}$*.*

(vi) *If* MIN BR HR-LQ-*2 is approximable within $c$, for some constant $c > 1$,* *then the Dense k-Subgraph problem*[1] *is approximable within* $(1 + \varepsilon)c^4$, *for any* $\varepsilon > 0$.

It is of interest to consider the approximability of MIN BR HR-LQ-2 in the absence of Assumption A. Of course, in such a setting a matching need not exist, but one can decide in polynomial time whether a matching does exist (by solving the Degree Constrained Subgraph problem [226]). We can thus define HR-LQ-3 to be the variant of HR-LQ-2 in which (i) Assumption A is not necessarily satisfied, (ii) the concepts of a matching and stability are defined as in the HR-LQ-2 case, and (iii) a matching exists. We then define MIN BR HR-LQ-3 as the problem of finding a matching with the minimum number of blocking residents, given an instance of HR-LQ-3. Hamada *et al.* [275] proved that MIN BR HR-LQ-3 is not approximable within $n^{1-\varepsilon}$, for any $\varepsilon > 0$, unless P=NP. MIN BP HR-LQ-3 (the counterpart of MIN BP HR-LQ-3 where we seek to minimise the number of blocking pairs rather than the number of blocking residents) is also worthy of investigation, although it is likely that this problem is as hard to approximate as MIN BR HR-LQ-3.

Fragiadakis *et al.* [221] also considered HR-LQ-2, but they strengthened Assumption A and insisted that each resident ranks *all* hospitals and vice versa. Even so, a stable matching still need not exist in these circumstances (again Fig. 5.3 illustrates this). The authors proposed strategy-proof mechanisms for finding matchings that satisfy criteria that are weaker than stability, including Pareto optimality, and the elimination of so-called *strong justified envy*.

### 5.2.4 HR *with common quotas*

In this subsection we consider HR with common quotas, denoted by HR-CQ, as studied by Biró *et al.* [84]. Although lower quotas are no longer present, sets of hospitals may have common upper quotas that limit the number of residents that may be collectively assigned to them. Formally, an instance $I$ of HR-CQ is an HR instance together with a set $\mathcal{H} \subseteq \mathbb{P}(H)$, which we refer to as a *set system of hospitals*, comprising the so-called *bounded sets* of hospitals. Each bounded set of hospitals $H_k \in \mathcal{H}$ has a *common quota* denoted by $U_k$. Here, the upper case '$U$' is intended to distinguish the upper quota of a bounded set of hospitals from the upper quota $u_j$ of an individual hospital $h_j \in H$.

---

[1] An instance of this problem is a graph $G$ and an integer $k$, and the problem is to find an induced subgraph of $G$ with $k$ vertices that contains as many edges as possible. This problem is not currently known to have a constant-factor approximation algorithm.

Moreover, in $I$ each $H_k \in \mathcal{H}$ has a preference list over a subset of $R$ that constitutes a *master list* for the hospitals in $H_k$. That is, the preference lists in $I$ satisfy the following three conditions:

(1) for each $H_k \in \mathcal{H}$, $A(H_k) = \bigcup_{h_j \in H_k} A(h_j)$, where $A(h_j)$ (respectively $A(H_k)$) denotes the set of residents that $h_j$ ($H_k$) finds acceptable;
(2) for each $H_k \in \mathcal{H}$, and for each $h_j \in H_k$, $h_j$'s preference list is obtained from that of $H_k$ by deleting the residents in $A(H_k)\backslash A(h_j)$;
(3) for any two bounded sets $\{H_k, H_l\} \subseteq \mathcal{H}$, let $H' = H_k \cap H_l$ and let $A' = \bigcup_{h_r \in H'} A(h_r)$. Then for any $\{r_i, r_j\} \subseteq A'$, $r_i$ precedes $r_j$ on $H_k$'s list if and only if $r_i$ precedes $r_j$ on $H_l$'s list.[2]

A *matching* $M$ in $I$ is a matching in the underlying HR instance (obtained by ignoring the common quotas) such that $|M(H_k)| \leq U_k$ for each $H_k \in \mathcal{H}$, where $M(H_k) = \bigcup_{h_j \in H_k} M(h_j)$. We may as well assume that $U_k < \sum_{h_j \in H_k} u_j$, for otherwise the common quota of $H_k$ is redundant as far as the definition of a matching is concerned, and $H_k$ can be deleted from the set system. The definitions of *undersubscribed, full* and *oversubscribed* for a hospital $h_j \in H$ carry over in a natural way to a bounded set $H_k \in \mathcal{H}$. We now present the concept of stability, as defined by Biró *et al.* [84].

**Definition 5.4 ([84]).** *Let $I$ be an instance of* HR-CQ *and let $M$ be a matching in $I$. A blocking pair of $M$ is an acceptable resident–hospital pair $(r_i, h_j) \in (R \times H)\backslash M$ such that:*

*(1) either $r_i$ is unassigned in $M$ or prefers $h_j$ to $M(r_i)$;*
*(2) either $h_j$ is undersubscribed in $M$ or prefers $r_i$ to the worst resident in $M(h_j)$;*
*(3) for each $H_k \in \mathcal{H}$ such that $h_j \in H_k$, either $H_k$ is undersubscribed or prefers $r_i$ to the worst resident in $M(H_k)$.*

*M is* stable *if $M$ admits no blocking pair.*

---

[2]Note that in Ref. [84], the authors erroneously claimed that a condition similar to the third condition in the above list followed as a consequence of the first two. This observation was then used later in the paper (in the final paragraph of Sec. 4.1 of Ref. [84]) when considering nested set systems (defined below). Thus, in order for certain results in Ref. [84] to hold true, the third condition must be added as a property to be satisfied by an instance of HR-CQ. The property states that the master lists of two bounded sets $H_k$ and $H_l$ must agree with respect to the relative order of the applicants who are collectively found acceptable by the hospitals in the intersection of $H_k$ and $H_l$. Since the condition is arguably a natural one, it is not unreasonable to insist that it be satisfied by an instance of HR-CQ.

Residents' preferences        Hospitals' preferences
$r_1 : h_1 \ h_4$                     $h_1 : 1 : r_1$
$r_2 : h_2$                          $h_2 : 1 : r_2$
$r_3 : h_4 \ h_3$                    $h_3 : 1 : r_3$
                                      $h_4 : 1 : r_1 \ r_3$
                                      $H_1 = \{h_1, h_2\} : 1 : r_2 \ r_1$
                                      $H_2 = \{h_2, h_3\} : 1 : r_3 \ r_2$

Fig. 5.4   An instance of HR-CQ with no stable matching due to Biró *et al.* [84]

Residents' preferences        Hospitals' preferences
$r_1 : h_4 \ h_1$                     $h_1 : 1 : r_1$
$r_2 : h_2 \ h_4$                     $h_2 : 1 : r_2$
$r_3 : h_3$                          $h_3 : 1 : r_3$
                                      $h_4 : 1 : r_2 \ r_1$
                                      $H_1 = \{h_1, h_2\} : 1 : r_1 \ r_2$
                                      $H_2 = \{h_2, h_3\} : 1 : r_2 \ r_3$

Fig. 5.5   An instance of HR-CQ with stable matchings of different sizes

An instance of HR-CQ need not admit a stable matching. To see this, consider the HR-CQ instance $I$, due to Biró *et al.* [84], shown in Fig. 5.4 (next to each hospital $h_j$ we show $u_j$, followed by $h_j$'s preference list, and we do likewise for each bounded set $H_k$). Suppose for a contradiction that $M$ is a stable matching in $I$. If $r_1$ is unassigned in $M$ then $(r_1, h_4)$ blocks $M$. Otherwise, if $M(r_1) = h_1$ then $(r_3, h_4) \in M$, for otherwise $(r_3, h_4)$ blocks $M$. But then $(r_2, h_2)$ blocks $M$. Hence $M(r_1) = h_4$, which implies that $(r_3, h_3) \in M$, for otherwise $(r_3, h_3)$ blocks $M$. But then $(r_2, h_2) \notin M$, so $(r_1, h_1)$ blocks $M$, a contradiction.

Even when an HR-CQ instance does admit a stable matching, it turns out that the stable matchings can have different sizes. To illustrate this, consider the HR-CQ instance $I$ shown in Fig. 5.5. It may be verified that each of $M_1 = \{(r_1, h_4), (r_2, h_2)\}$ and $M_2 = \{(r_1, h_1), (r_2, h_4), (r_3, h_3)\}$ is stable in $I$.

Even determining whether a stable matching exists in a given HR-CQ instance is hard, as the next result indicates.

**Theorem 5.5 ([84]).**   *The problem of deciding whether a given HR-CQ instance admits a stable matching is NP-complete, even if (i) every hospital and every bounded set has upper quota 1, (ii) each bounded set contains two hospitals, and (iii) each hospital appears in at most two bounded sets.*

What makes HR-CQ hard is the existence of pairs of bounded sets of hospitals that have a non-empty intersection, where neither bounded set is contained in the other. In the absence of such bounded sets, efficient algorithms and elegant structural results can be derived for this problem. In a given HR-CQ instance, define the set system of hospitals $\mathcal{H}$ to be *nested*[3] if, for any bounded sets $H_k, H_l \in \mathcal{H}$ such that $H_k \cap H_l \neq \emptyset$, either $H_k \subseteq H_l$ or $H_l \subseteq H_k$. Let HR-CQ-NSS denote the restriction of HR-CQ in which the set system of hospitals is nested.

Biró *et al.* [84] presented two algorithms for finding stable matchings in a given instance of HR-CQ-NSS. These algorithms are resident-oriented and hospital-oriented in that they find stable matchings that are resident-optimal and hospital-optimal in precise senses, respectively. As a by-product of establishing the correctness of these algorithms, the authors deduced that every instance of HR-CQ-NSS admits at least one stable matching. The following theorems summarise the discussion in this paragraph, indicating the complexity of the algorithms and the precise optimality properties that are satisfied in each case.

**Theorem 5.6 ([84]).** *Given an instance of* HR-CQ-NSS, *the resident-oriented algorithm finds a stable matching $M$ that is resident-optimal in the following sense. In $M$, each assigned resident is assigned to the best hospital that she could obtain in any stable matching, and each unassigned resident is unassigned in every stable matching. The complexity of this algorithm is $O(km + pn_1)$, where $k$ is the maximum level of nesting (i.e., the maximum integer $k$ such that there exist bounded sets $H_{i_1} \subset H_{i_2} \subset \cdots \subset H_{i_k}$), $m$ is the number of acceptable resident–hospital pairs, $n_1$ is the number of residents and $p = |\mathcal{H}|$ is the number of bounded sets.*

**Theorem 5.7 ([84]).** *Given an instance of* HR-CQ-NSS, *the hospital-oriented algorithm finds a stable matching $M$ that is hospital-optimal in the following sense. For any hospital $h_j \in H$, there is no stable matching in which $h_j$ is assigned a resident $r_i \in R \backslash M(h_j)$ whom $h_j$ prefers to some member of $M(h_j)$. Also in $M$, each assigned resident is assigned to the worst hospital that she could obtain in any stable matching, and each unassigned resident is unassigned in any stable matching. The complexity of this algorithm is $O(n_2 m)$, where $n_2$ is the number of hospitals and $m$ is the number of acceptable resident–hospital pairs.*

---

[3] A nested set system is also referred to in the literature as a *laminar family* [290, 218].

A consequence of each of Theorems 5.6 and 5.7 is that all stable matchings for a given HR-CQ-NSS instance have the same cardinality. We also note that, although we refer to the matching $M$ output by the hospital-oriented algorithm as being *hospital-optimal*, it is nevertheless possible that a hospital $h_j \in H$ could obtain a set of assignees in another stable matching $M'$ where $M'(h_j) \supset M(h_j)$. However Theorem 5.7 implies that any resident in $M'(h_j) \backslash M(h_j)$ is worse than $h_j$'s worst assignee in $M$.

Using a matroid-theoretic approach, Biró *et al.* [84] established the following structural results that generalise the Rural Hospitals Theorem (Theorem 1.11):

**Theorem 5.8 ([84]).** *Let $I$ be an instance of* HR-CQ-NSS[4] *and let $M$ be a stable matching in $I$. The following properties hold:*

 *(i) the same residents are assigned in all stable matchings;*

 *(ii) for a given bounded set $H_k \in \mathcal{H}$, if there is no bounded set $H_l \in \mathcal{H}$ such that $H_k \subset H_l$ and $H_l$ is full in $M$, then $H_k$ has the same number of assignees in all stable matchings;*

 *(iii) for a given bounded set $H_k \in \mathcal{H}$, if there are no bounded sets $H_j, H_l \in \mathcal{H}$ such that $H_j \subseteq H_k \subset H_l$ and each of $H_j$ and $H_l$ is full in $M$, then $H_k$ has the same set of assignees in all stable matchings.*

Note that Theorem 5.8 does indeed generalise the corresponding result for HR (Theorem 1.11), since, an HR instance $I$ can be considered as an HR-CQ-NSS instance $J$ by letting each hospital in $h_j$ in $I$ become a bounded set $\{h_j\}$ in $J$ such that $U_j = u_j$. Then each of Parts (ii) and (iii) of Theorem 1.11 can be deduced from the corresponding part of Theorem 5.8 by considering the bounded set $\{h_j\}$ for any hospital $h_j$.

We further remark that Biró *et al.* [84] also established a counterpart of Theorem 2.9 for HR-CQ-NSS.

We close this subsection by noting that Kamada and Kojima [354, 355] considered an alternative model for HR-CQ motivated by regional caps in the JRMP, as described in Sec. 5.2.1. In their model, there is no master list of residents for a given bounded set of hospitals, and hence there is no requirement for the hospitals in a given bounded set to have preference lists that are consistent with one another relative to a master list. The stability definition given by the authors is very different to that defined by Biró *et al.*

---

[4]Parts (ii) and (iii) of this result are derived from Theorem 17 in Ref. [84], which is erroneously stated in terms of HR-CQ rather than HR-CQ-NSS. However the proof of this theorem does require the set system to be nested.

[84] for HR-CQ. Kamada and Kojima [354, 355] defined a feasible *matching* (i.e., a matching in the underlying HR instance that satisfies the common upper quotas specified by the regional caps) to be *stable* if the only blocking pairs that it admits (where *blocking pair* is defined relative to classical Gale–Shapley stability) involve a resident $r_i$ and a hospital $h_j$ such that (i) the number of residents assigned to $h_j$'s region is equal to the regional cap, (ii) $h_j$ prefers all of its assignees to $r_i$, and (iii) either $r_i$ is not already assigned within $h_j$'s region, or the movement of $r_i$ to $h_j$ "equalises" the excesses over certain *target capacities* (see Ref. [354, 355] for more details). Thus, certain blocking pairs are tolerated in theory with respect to this definition, but in practice the government would prevent these blocking pairs from leading to any disruption of the matching by imposing harsh penalties on agents who deviate from their assignment. Kamada and Kojima [354, 355] presented an algorithm that always finds a matching that is stable relative to their definition, though they do not discuss the computational complexity of their algorithm.

### 5.2.5 *Classified stable matching*

In Sec. 5.2.1, we described informally the *Classified Stable Matching* problem as a further generalisation of HR involving lower and common quotas (this time the two types of quota are both present). Huang [291] studied this problem in the context of assigning applicants to academic institutions. However, for consistency with the preceding subsections, we will describe the problem in terms of residents and hospitals, and refer to it as the *Hospitals / Residents problem with Classified Residents* (HR-CR).

Formally, an instance $I$ of HR-CR comprises an HR instance such that each hospital $h_j \in H$ has a family of sets of residents, or *classes*, denoted by $\mathcal{R}_j$, defined as follows:

$$\mathcal{R}_j = \{R_j^k : R_j^k \subseteq A(h_j) \wedge 1 \leq k \leq |\mathcal{R}_j|\},$$

where $A(h_j)$ denotes the set of residents that are acceptable to $h_j$. The family of sets $\mathcal{R}_j$ is referred to as $h_j$'s *classification* of its acceptable residents. As noted in Sec. 5.2.1, two classes in $\mathcal{R}_j$ need not be disjoint. Each class $R_j^k \in \mathcal{R}_j$ has an *upper quota* $u_j^k \in \mathbb{Z}^+$ and a *lower quota* $l_j^k \in \mathbb{Z}_0^+$. We lose no generality in assuming that $\bigcup_{k=1}^s R_j^k = A(h_j)$, where $s = |\mathcal{R}_j|$, for if some resident $r_i \in A(h_j)$ does not belong to any class in $h_j$'s classification, we can always add a new class $R_j^{s+1} = \{r_i\}$ to $\mathcal{R}_j$ where $u_j^{s+1} = 1$ and $l_j^{s+1} = 0$.

A *matching* in $I$ is a matching $M$ in the underlying HR instance (obtained by ignoring the hospitals' classifications) such that, for each hospital $h_j \in H$ and for each class $R_j^k \in \mathcal{R}_j$, $l_j \leq |M(h_j) \cap R_j^k| \leq u_j$. That is, the number of residents assigned to $h_j$ in $M$ that belong to the class $R_j^k$ must respect the lower and upper quotas of that class. We remark that, by adding a special class $R_j^0 = A(h_j)$ to $\mathcal{R}_j$, where $u_j^0 = u_j$ and $l_j^0 = 0$, we need not explicitly ask that $|M(h_j)| \leq u_j$ in the definition of a matching, since this is enforced by $R_j^0$. However at least for our purposes here, there is no real benefit in making this addition to $\mathcal{R}_j$. We now present the stability concept for HR-CR, as defined by Huang [291].

**Definition 5.9 ([291]).** *Let $I$ be an instance of* HR-CR *and let $M$ be a matching in $I$. A* blocking coalition *of $M$ comprises a hospital $h_j \in H$ and a set of residents $R' \subseteq A(h_j)$ such that:*

*(1) $|M(h_j)| \leq |R'| \leq u_j$;*

*(2) for each $R_j^k \in \mathcal{R}_j$, $l_j^k \leq |R' \cap R_j^k| \leq u_j^k$;*

*(3) for each $r_i' \in R'$, either $r_i'$ is unassigned in $M$ or $r_i'$ prefers $h_j$ to $M(r_i')$, or $M(r_i') = h_j$;* [5]

*(4) there exists some $r_i' \in R'$ such that either $r_i'$ is unassigned in $M$ or $r_i'$ prefers $h_j$ to $M(r_i')$;*

*(5) for each $i$ ($1 \leq i \leq t$), either $h_j$ prefers $r_i'$ to $r_i$ or $r_i' = r_i$, where $M(h_j) = \{r_1, \ldots, r_t\}$ and $R' = \{r_1', \ldots, r_{t'}'\}$, and without loss of generality the elements of each set are listed in decreasing order of preference according to $h_j$'s preference list;*

*(6) either $t' > t$, or there exists some $i$ ($1 \leq i \leq t$) such that $h_j$ prefers $r_i'$ to $r_i$.*

$M$ *is* stable *if $M$ admits no blocking coalition.*

Conditions 1 and 2 of the blocking coalition definition state that the residents $R'$ involved must respect the quotas of the hospital $h_j$ and those of the classes in $h_j$'s classification. Moreover, since $R'$ would be $h_j$'s new set of assignees, the cardinality of $R'$ must be at least as great as the number of $h_j$'s existing assignees. Conditions 3 and 4 state that no resident in $R'$ should be worse off by moving to $h_j$, whilst at least one resident in $R'$ should be better off. Similarly, Condition 5 states that $h_j$ should not

---

[5]The definition of a *blocking group* in Ref. [291], on which we base our definition of a blocking coalition, is erroneous in that it allows a resident to become worse off after switching to $h_j$. We correct this error in our definition.

obtain a worse set of assignees, whilst Condition 6 states that $h_j$ should either obtain more assignees or at least one better assignee.

It is straightforward to come up with an HR-CR instance that has a matching, but no stable matching. For, consider the HR-LQ-2 instance shown in Fig. 5.3. By ignoring the hospitals' lower quotas, and by letting $\mathcal{R}_1 = \emptyset$ and $\mathcal{R}_2 = \{R_2^1\}$, where $R_2^1 = \{r_1\}$ and $l_2^1 = u_2^1 = 1$, we obtain an instance of HR-CR which clearly admits no stable matching. Huang [289, Fig. 3] gave an instance of HR-CR which admits stable matchings of different sizes.

Huang [291] proved the following algorithmic results concerning HR-CR. In what follows, HR-CR-NSS denotes the restriction of HR-CR in which each hospital $h_j$'s classification is a nested set system.

**Theorem 5.10 ([291]).** *The problem of deciding whether a given instance of* HR-CR *admits a stable matching is NP-complete. The result holds even if each hospital's classification contains only two classes, and each of these classes has lower quota 0.*

**Theorem 5.11 ([291]).** *Let $I$ be an instance of* HR-CR-NSS. *There is an $O(m^2)$ algorithm to find a stable matching in $I$ or report that none exists, where $m$ is the number of acceptable resident–hospital pairs in $I$. For each hospital $h_j$, if each class in $h_j$'s classification has lower quota 0, then a stable matching always exists. In general, if $I$ does admit a stable matching, then the algorithm constructs the resident-optimal stable matching $M$ in $I$, which satisfies the following properties:*

*(i) in $M$, each assigned resident is assigned to the best hospital that she could obtain in any stable matching, and each unassigned resident is unassigned in every stable matching;*

*(ii) each hospital $h_j \in H$ has the worst possible set of assignees in $M$: that is, suppose that $M'$ is a stable matching in $I$ such that $M(h_j) \neq M'(h_j)$, and let $M(h_j) = \{r_1, \ldots, r_t\}$ and $M'(h_j) = \{r_1', \ldots, r_t'\}$ [6], where without loss of generality the elements of each set are listed in decreasing order of preference according to $h_j$'s preference list. Then there exists some $s$ ($1 \leq s \leq t$) such that $r_i = r_i'$ ($1 \leq i < s$) and $h_j$ prefers $r_s'$ to $r_s$.*

---

[6]Theorem 5.12 states that every hospital has the same number of assignees in every stable matching in $I$.

In fact, Huang [291] proved a dichotomy result concerning the complexity of HR-CR. For each hospital $h_j \in H$ he defined a *class inclusion poset*, which is the inclusion poset on the set of non-empty intersections of pairs of (not necessarily distinct) classes in $h_j$'s classification $\mathcal{R}_j$. A poset is defined to be a *downward forest* if any two successors of a given element are comparable. Huang's dichotomy result is as follows: if every class inclusion poset is isomorphic to a downward forest, then the problem of finding a stable matching or reporting that none exists is solvable in polynomial time. Otherwise, the problem of deciding whether a stable matching exists is NP-complete. As Huang remarks, if every class inclusion poset is isomorphic to a downward forest then each hospital's classification is a nested set system, and thus we obtain an instance of HR-CR-NSS and can apply Theorem 5.11.

In the HR-CR-NSS context, Huang [291] established the following structural results that generalise the Rural Hospitals Theorem (Theorem 1.11):

**Theorem 5.12 ([291]).** *Let $I$ be an instance of HR-CR-NSS that admits a stable matching, and let $M$ be a stable matching in $I$. For any hospital $h_j \in H$, let $\mathcal{R}'_j$ be the set of maximal (under set inclusion) classes $R^k_j \in \mathcal{R}_j$ such that $|M(h_j) \cap R^k_j| = u^k_j$. Let $\mathcal{R}''_j$ be the set of classes $R^k_j \in \mathcal{R}_j$ such that $R^k_j \cap R^l_j = \emptyset$ for all $R^l_j \in \mathcal{R}'_j$. Then the following properties hold:*

*(i) the same residents are assigned in all stable matchings (and hence all stable matchings in $I$ have the same size);*

*(ii) $h_j$ has the same number of assignees in all stable matchings;*

*(iii) each class in $\mathcal{R}'_j \cup \mathcal{R}''_j$ has the same number of assignees in all stable matchings;*

*(iv) each class in $\mathcal{R}''_j$ has the same set of assignees in all stable matchings.*

Note that Theorem 5.12 is a true generalisation of Theorem 1.11, for given an HR instance $I$ we can form an HR-CR-NSS instance $J$ from $I$, such that for each $h_j \in H$, $\mathcal{R}_j = \{R^1_j\}$, where $R^1_j = A(h_j)$, $l^1_j = 0$ and $u^1_j = u_j$. Given a stable matching $M$ in $I$, if some hospital $h_j$ is undersubscribed in $M$ then $R^1_j \in \mathcal{R}''_j$, and hence Part (iv) of Theorem 5.12 implies Part (iii) of Theorem 1.11.

Denote by HR-CR-NSS-0 the special case of HR-CR-NSS in which the lower bound of every class in a given hospital's classification is 0. For this problem, Huang proved that stability as defined by Definition 5.9 is equivalent to establishing the absence of a blocking pair involving a single resident and a hospital. Further, Huang [291] gave a polyhedral characterisation for HR-CR-NSS-0. In particular, he proved that the extreme points of a polytope that characterises fractional stable matchings are integral. He gave a

separation oracle based on dynamic programming that enables a solution
to the LP constraints to be found in polynomial time. Moreover, a suitable
choice of objective function allows optimal stable matchings (for various
optimality criteria) to be found efficiently. Huang also proved an analogue
of Theorem 2.9 for HR-CR-NSS-0. He left open whether his polyhedral characterisation could be extended to more general instances of HR-CR-NSS in
which classes in the classifications of hospitals could have a positive lower
quota.

Fleiner and Kamiyama [218] answered this question in the affirmative,
not just for HR-CR-NSS, but in fact for a more general many–many bipartite
stable matching problem where agents on each side can classify agents on
the other side, and again the classifications form nested set systems. Using
a matroid-theoretic approach, and by defining a natural stability concept
(that turns out to be equivalent to Huang's definition in the HR-CR-NSS
case), they gave an $O(m^3)$ algorithm to find a stable matching or report
that none exists, where $m$ is the number of acceptable pairs of agents in an
instance. Moreover, they gave a polyhedral characterisation of the set of
stable matchings for an instance of this problem, leading to a polynomial-
time algorithm for finding an optimal stable matching (relative to a linear
cost function), thus solving Huang's open problem. Furthermore, they
showed that the set of stable matchings for a given problem instance $I$
forms a lattice, and proved that, if $I$ admits a stable matching, then a
matching that is analogous to the man-optimal stable matching in the SM
case can be found in $O(m^3)$ time.

## 5.3  HR with couples

### 5.3.1  *Introduction*

The existence of *couples* who wish to be located at the same hospital, or
at hospitals geographically close to one another, gives rise to an important
variant of HR called the *Hospitals / Residents problem with Couples* (HRC)
[498,261,493,514,91]. Couples rank acceptable pairs of hospitals in order of
preference, where each pair represents a simultaneous assignment of both
residents involved in the couple to two (not necessarily distinct) hospitals.
The study of HRC was motivated by the fact that, in the mid-1970s, participation in the NRMP was observed to decrease, as a result of the original
algorithm being unable to cope satisfactorily with the complicated preference structure of couples [498]. The NRMP algorithm was redesigned in
1983 and in the mid-1990s to address this issue, among others [498,507].

Two crucial distinctions between HR and HRC are that (i) an instance of HRC need not admit a stable matching, and (ii) the problem of deciding whether an HRC instance admits a stable matching is NP-complete. In this section we will elaborate on these results, as well as discussing additional algorithmic results for HRC. The section is organised as follows. In Sec. 5.3.2, we define HRC formally and present some preliminary results that mainly illustrate the lack of structure in a given problem instance $I$ (as compared to HR, say). This includes the observation that $I$ may not admit a stable matching. In Sec. 5.3.3, we survey algorithmic results for HRC.

We then look at two restrictions of HRC which involve additional assumptions about the structure of the couples' preference lists. The first, called the *Hospitals / Residents problem with Consistent Couples* (HRCC) and studied in Sec. 5.3.4, relates to the case where the couples' lists are consistent in a precise sense with derived preference lists over individual hospitals involving the residents in the couples. The second, called the *Hospitals / Residents problem with Inseparable Couples* (HRIC) and studied in Sec. 5.3.5, is the special case of HRCC in which both members of a given couple must either be assigned to the same hospital or not at all.

We remark that there is an excellent, comprehensive survey of HRC due to Biró and Klijn [91]. Our purpose here is to focus largely on algorithmic results. Whilst this inevitably leads to a small degree of overlap with Ref. [91], the reader is strongly encouraged to refer to Ref. [91] for additional background and results relating to HRC that we do not cover here.

## 5.3.2  *Problem definition and preliminary results*

The *Hospitals / Residents problem with Couples* (HRC) may be defined formally as follows. An instance $I$ of HRC involves a set $R = \{r_1, \ldots, r_{n_1}\}$ of *residents* and a set $H = \{h_1, \ldots, h_{n_2}\}$ of *hospitals*. A subset $R'$ of the residents in $R$ is partitioned into ordered pairs, each called a *couple*, such that each resident in $R'$ belongs to exactly one couple. The set of couples is denoted by $R_C$. We denote by $R_S = R \backslash R'$ the *single residents*, i.e., the residents who do not belong to a couple in $R_C$.

Each single resident $r_i \in R_S$ has a strict preference list over acceptable hospitals. Each couple $(r_i, r_j) \in R_C$ submits a joint (strict) preference list over acceptable pairs of hospitals. Each entry in this list is an ordered pair $(h_k, h_l) \in H \times H$ of (not necessarily distinct) hospitals representing the assignment of $r_i$ to $h_k$ and of $r_j$ to $h_l$. The fact that $r_i$ precedes $r_j$ in a given ordered pair does not suggest that $r_i$ has any kind of precedence over $r_j$ in a given matching mechanism, but the notation merely serves to disambiguate which hospital each member of the couple will be assigned to. We denote by $A(r_i, r_j)$ the set of pairs of hospitals that $(r_i, r_j)$ find acceptable. Hence, $r_i$ finds acceptable the set of hospitals $A(r_i) = \{h_k : (h_k, h_l) \in A(r_i, r_j)\}$, and similarly $r_j$ finds acceptable the set of hospitals $A(r_j) = \{h_l : (h_k, h_l) \in A(r_i, r_j)\}$.[7,8] Finally, each hospital $h_j \in H$ has a *capacity* $c_j \in \mathbb{Z}^+$, and ranks those residents in $R$ (whether single or a member of a couple) who find $h_j$ acceptable in strict order of preference.

A *matching* $M$ in $I$ is, as in the HR case, a set of mutually acceptable resident–hospital pairs such that (i) each resident appears in at most one pair, (ii) each hospital $h_j$ appears in at most $c_j$ pairs, and additionally (iii) for each $(r_i, r_j) \in R_C$, $(M(r_i), M(r_j)) \in A(r_i, r_j)$. That is, the hospitals that $r_i$ and $r_j$ are jointly assigned to in $M$ are compatible for them according to their joint preference list.

Stability definitions for a matching in $I$ have been given previously in the literature [498, 261, 493, 178, 115, 387, 390, 392]. However it appears that none of the stability definitions in these sources adequately takes account of the possibility that, given a couple $(r_i, r_j)$ who prefer a pair $(h_k, h_l)$ to $(M(r_i), M(r_j))$, it may be that $h_k = h_l$ (that is, the two residents concerned wish to move to the same hospital rather than to remain with their current hospitals). Biró and Klijn [91] gave a detailed discussion of precisely this issue, and remarked that there appear to be four sources in the literature that do take account of this possibility when defining stability in HRC, namely Refs. [444,87,428,396] (see also the comprehensive rationale given in Ref. [87] for the authors' stability definition). Biró and Klijn gave examples that indicate that the stability definitions in Refs. [444, 87, 396] are all distinct from one another. The definitions of Refs. [444] and [428] are very similar apart from one missing case in the definition of Ref. [428][9].

Throughout this section we will adopt the following definition of McDermid and Manlove [444] as our stability definition for HRC.

---

[7]A couple may wish to indicate in their preference list the possibility that one member is unassigned, whilst the other is assigned to a hospital. For example, $(r_i, r_j)$ might prefer $(h_{k_1}, *)$ to $(h_{k_2}, h_{k_3})$, where "$*$" represents a resident being unassigned. To facilitate this, we can simply introduce a new "dummy" hospital $h_{n_2+1}$, where $c_{n_2+1} = n_1$, and the assignment of a resident $r_j$ to $h_{n_2+1}$ represents $r_j$ being unassigned in practice.

[8]Of course it need not follow that $h_k \in A(r_i)$ and $h_l \in A(r_j)$ implies $(h_k, h_l) \in A(r_i, r_j)$.

[9]The authors have overlooked the possibility that a couple $(r_i, r_j)$ prefers $(h_k, h_k)$ to their assignment $(h_k, h_l)$ in a given matching $M$, where $h_k$ and $h_l$ are distinct, $c_k = 2$, $h_k$ prefers $r_i$ to $r_j$ to $r_s$, and $r_s$ is a single resident in $M(h_k)$. It seems reasonable that $(r_i, r_j)$ should block $M$ with $h_k$ in a stability definition, though this is not possible according to the definition of Marx and Schlotter [428].

**Definition 5.13 ([444]).** *Let I be an instance of* HRC *and let M be a matching in I. A* blocking pair *relative to M satisfies at least one of the following properties:*

*(1) it involves a single resident $r_i \in R_S$ and a hospital $h_j \in H$ such that $(r_i, h_j)$ blocks M as per Definition 1.8;*

*(2) it involves a couple $(r_i, r_j) \in R_C$ and a hospital $h_k \in H$ such that either*

    *(a) $(r_i, r_j)$ prefers $(h_k, M(r_j))$ to $(M(r_i), M(r_j))$, and $h_k$ is either undersubscribed in M or prefers $r_i$ to some member of $M(h_k)\backslash\{r_j\}$; or*

    *(b) $(r_i, r_j)$ prefers $(M(r_i), h_k)$ to $(M(r_i), M(r_j))$, and $h_k$ is either undersubscribed in M or prefers $r_j$ to some member of $M(h_k)\backslash\{r_i\}$;*

*(3) it involves a couple $(r_i, r_j) \in R_C$ and a pair of (not necessarily distinct) hospitals $h_k, h_l \in H$ such that $h_k \neq M(r_i)$, $h_l \neq M(r_j)$, $(r_i, r_j)$ prefers $(h_k, h_l)$ to $(M(r_i), M(r_j))$, and either*

    *(a) $h_k \neq h_l$, and $h_k$ (respectively $h_l$) is either undersubscribed in M or prefers $r_i$ (respectively $r_j$) to at least one of its assigned residents in M; or*

    *(b) $h_k = h_l$, and $c_k - |M(h_k)| \geq 2$; or*

    *(c) $h_k = h_l$, and $c_k - |M(h_k)| = 1$, and $h_k$ prefers at least one of $r_i, r_j$ to some member of $M(h_k)$); or*

    *(d) $h_k = h_l$, $h_k$ is full in M, $h_k$ prefers $r_i$ to some $r_s \in M(h_k)$, and $h_k$ prefers $r_j$ to some $r_t \in M(h_k)\backslash\{r_s\}$.*

*M is* stable *if it admits no blocking pair.*

Note that when all hospitals have capacity 1, Conditions 3(b), 3(c) and 3(d) in Definition 5.13 cannot occur, and the disparity between the stability definitions in the literature caused by a blocking pair involving a couple, both members of whom wish to move to the same hospital, is eliminated. One of the issues with stability definitions for HRC concerns whether *scorelimits* [90] are preserved by blocking pairs — see Ref. [87] for a more detailed discussion of this point.

Biró and Klijn [91] gave an instance $I$ of HRC that admits no stable matching, which is shown in Fig. 5.6. Here, $(r_1, r_2)$ is a couple and $r_3$ is a single resident, whilst adjacent to each hospital we show its capacity followed by its preference list. There are three possible matchings for this instance: $M_1 = \{(r_1, h_1), (r_2, h_2)\}$, $M_2 = \{(r_3, h_1)\}$ and $M_3 = \{(r_3, h_2)\}$. $M_1$ is blocked by $(r_3, h_2)$, $M_2$ is blocked by $((r_1, r_2), (h_1, h_2))$, whilst $M_3$ is

Residents' preferences     Hospitals' preferences

$(r_1, r_2) : (h_1, h_2)$         $h_1 : 1 : r_1 \ \ r_3$

$\qquad r_3 : h_1 \ \ h_2$         $h_2 : 1 : r_3 \ \ r_2$

Fig. 5.6    An instance of HRC with no stable matching due to Biró and Klijn [91]

Residents' preferences     Hospitals' preferences

$(r_1, r_2) : (h_1, h_1)$        $h_1 : 2 : r_1 \ \ r_3 \ \ r_2$

$\qquad r_3 : h_1$

Fig. 5.7    A second instance of HRC with no stable matching

blocked by $(r_3, h_1)$. If we allow hospitals to have non-unitary capacity, an even smaller HRC instance $I'$, shown in Fig. 5.7, demonstrates the possible non-existence of a stable matching. A similar argument to that employed for $I$ can be used to show that $I'$ admits no stable matching. Biró and Klijn remarked that Roth [498] was the first to demonstrate that an HRC instance need not admit a stable matching.

Biró and Klijn [91] also gave example HRC instances with the following properties:

(i) there is no stable matching that is optimal for either the residents or the hospitals, as originally shown by Aldershof and Carducci [37];

(ii) the stable matchings have different sizes, as originally shown by Aldershof and Carducci [37];

(iii) there exists a matching $M$ from which we cannot reach a stable matching by iteratively satisfying blocking pairs (see Sec. 2.6), as originally shown by Klaus and Klijn [390];

(iv) there exists no strategy-proof mechanism that produces a stable matching whenever one exists (this was shown by illustrating that, with respect to an instance with a unique stable matching $M_1$, a single resident $r$ can misrepresent her preferences in order to obtain an instance with a unique stable matching $M_2$, where $r$ is assigned in $M_2$ and unassigned in $M_1$).

### 5.3.3   *Algorithmic results*

The main algorithmic result for HRC is due to Ronn [492, 493], who showed that, in contrast to HR, the problem of finding a stable matching if one exists is unlikely to be solvable in polynomial time. Moreover the result holds for a restricted case of HRC, as follows.

**Theorem 5.14** ([492, 493]).   *Given an instance of* HRC, *the problem of deciding whether a stable matching exists is NP-complete. The result holds even if each hospital has capacity 1 and there are no single residents.*

The possible non-existence of a stable matching in a given HRC instance in $I$ motivates the problem of finding a matching in $I$ with the minimum number of blocking pairs. Theorem 5.14 implies that this problem is NP-hard, though its approximability is open. Also open is the complexity of the problem of finding a maximum matching in $I$ with the minimum number of blocking pairs. It is likely that this problem is NP-hard, though again its approximability is of interest.

Theorem 5.14 was independently proved by Ng and Hirschberg [463], who actually established NP-completeness for a slightly more restricted case.

**Theorem 5.15** ([463]).   *Given an instance of* HRC, *the problem of deciding whether a stable matching exists is NP-complete. The result holds even if each hospital has capacity 1, there are no single residents, and each couple finds acceptable every distinct pair of hospitals.*

Ng and Hirschberg also considered the variant of HRC in which $H$ is partitioned into two sets $H_1$ and $H_2$, where $H_1 = \bigcup_{(r_i, r_j) \in R_C} A(r_i)$ and $H_2 = \bigcup_{(r_i, r_j) \in R_C} A(r_j)$. That is, if we regard the set of all first (respectively second) members of each couple to be collectively the men (respectively women), then $H_1$ (respectively $H_2$) is the set of hospitals that the men (respectively women) collectively find acceptable. Since $H$ is partitioned into $H_1$ and $H_2$, this implies that the job market for the men is disjoint from the job market for the women. We denote this restriction of HRC by HRC-DUAL-MARKET. Ng and Hirschberg proved that NP-completeness also holds for HRC-DUAL-MARKET.

**Theorem 5.16** ([463]).   *Given an instance of* HRC-DUAL-MARKET, *the problem of deciding whether a stable matching exists is NP-complete. The result holds even if each hospital has capacity 1 and there are no single residents.*

Biró et al. [87] established NP-completeness for a special case of HRC in which the preference lists of hospitals are derived from a single master list of residents. This case is important from the standpoint of practical applications (for example, in the SFAS context, there is a single ranking of residents according to academic "scores").

**Theorem 5.17 ([87]).** *Given an instance of* HRC, *the problem of deciding whether a stable matching exists is NP-complete. The result holds even if each hospital has capacity 1, and the preference list of each single resident, couple and hospital is derived from a strictly-ordered master list of hospitals, pairs of hospitals, and residents, respectively.*[10]

The parameterized complexity of the problem of finding a stable matching in an HRC instance has been considered in Refs. [87] and [428], as indicated by the following results.

**Theorem 5.18 ([428]).** *Given an instance I of* HRC, *the problem of deciding whether a stable matching exists is W[1]-hard relative to parameterization* $|R_C|$, *where* $R_C$ *is the set of couples in I. The result holds even if each hospital has capacity 1.*

**Theorem 5.19 ([87]).** *Given an instance I of* HRC, *the problem of finding a stable matching or reporting that none exists belongs to FPT if (i) the hospitals' lists are derived from a strictly-ordered master list of residents, and (ii) the problem is parameterized by* $|R_C|$, *where* $R_C$ *is the set of couples in I.*

Marx and Schlotter [428] also considered the parameterized complexity of the problem of finding a maximum cardinality stable matching (or reporting that none exists), given an instance of HRC. They proved that if W[1]$\neq$FPT, there is no FPT local search algorithm for this problem if it is parameterized by $l$, the size of the neighbourhood to be searched, even if each hospital has capacity 1. However if the problem is parameterized by both $l$ and $|R_C|$, the number of couples, then there is an FPT local search algorithm (with no assumption on the hospital capacities).

Biró *et al.* [87] developed a range of heuristics for the problem of finding a stable matching or reporting that none exists in a given HRC instance,

---

[10]In fact the conditions under which NP-completeness is established are a little stronger: in the HRC instance constructed, each resident (whether single or a member of a couple) has a preference list derived from a master list of individual hospitals. The preference list of each couple $c = (r_i, r_j)$ is *responsive* to the individual lists of each member of the couple [387] — that is, if $r_i$ prefers $h_k$ to $h_l$ and $\{(h_k, h_t), (h_l, h_t)\} \subseteq A(c)$, then $c$ prefers $(h_k, h_t)$ to $(h_l, h_t)$; similarly if $r_j$ prefers $h_k$ to $h_l$ and $\{(h_t, h_k), (h_t, h_l)\} \subseteq A(c)$, then $c$ prefers $(h_t, h_k)$ to $(h_t, h_l)$.

and subjected them to a detailed empirical evaluation based on randomly-generated data. Klaus and Klijn [387] (see also Klaus *et al.* [392]) showed that, for instances of HRC with so-called *weakly responsive preferences*, a stable matching always exists and can be found in polynomial time. They also showed that, for preferences satisfying this property, one can always arrive at a stable matching, starting from an arbitrary matching, by satisfying a sequence of blocking pairs [390].

Cantala [115] and Sethuraman *et al.* [526] suggested a special case of HRC involving *tiered preferences* that arise from geographical constraints. This is based on the assumption that both members of a given couple wish to be assigned to hospitals geographically close to one another. Each couple partitions the hospitals into *regions* (both members of the couple must agree on this partition, although the partitions need not be the same for all couples). Each couple ranks in strict order of preference the regions that form their partition (so, for example, the couple may decide that they prefer to be assigned to hospitals in the West of Scotland rather than in Northern Scotland). Again, both members of the couple must agree on this ranking. This is the point of departure for the models in the two papers.

In the model of Cantala [115], the couple then rank in order of preference their pairs of acceptable hospitals that belong to each region. When interpreting the couple's preference structure, their preference over regions has first priority, followed by their preference over pairs of hospitals. Cantala showed that, even if preference lists in a given HRC instance are structured this way, an instance may not admit a stable matching.

In the model of Sethuraman *et al.* [526], each member of the couple has a strict preference list over their acceptable hospitals in each region. These two individual preference lists need not agree. So, for example, the preference list of a resident $r_i$ who belongs to a couple $(r_i, r_j)$ might have the following form:

$$h_{1,1} \ \ldots \ h_{1,n_1} \ h_{2,1} \ \ldots \ h_{2,n_2} \ \ldots \ h_{k,1} \ \ldots h_{k,n_k}$$

where the couple have identified the regions as $R_1, \ldots, R_k$ (preferring $R_p$ to $R_q$ whenever $p < q$), and $\{h_{p,1}, \ldots, h_{p,n_p}\}$ is the set of hospitals that $r_i$ finds acceptable in region $p$ ($1 \le p \le k$).

The stability definition in this context is classical (Gale–Shapley) stability. Thus we have a standard HR instance with an additional constraint that both members of each couple should either be assigned to the same region, or both should be unassigned. Sethuraman *et al.* [526] devised an

LP-based method to find a stable matching for an instance of this problem, or report that none exists. However Cheng *et al.* [148] observed that the additional constraint satisfies the *identification property* as defined in Sec. 2.10.8 since the members of a couple agree on a common ranking of the regions (see Ref. [148] for further explanation). Cheng *et al.* showed that the problem can be solved in $O(n_1 n_2)$ time using their framework, where $n_1$ and $n_2$ are the numbers of residents and hospitals respectively.

We close this section by remarking that Aldershof and Carducci [38] described a genetic algorithm for HRC, and by giving three further citations relevant to HRC that were not referred to by Biró *et al.* [91], namely Refs. [74, 579, 580].

### 5.3.4 *Consistent couples*

A natural special case of HRC arises when the preference list of each couple is *consistent* in a precise sense [444]. That is, suppose that $c_k = (r_i, r_j) \in R_C$. Their preference list $\prec_{c_k}$ over a subset of $H \times H$ is *consistent* if there exist individual preference lists $\prec_{r_i}$ and $\prec_{r_j}$ for $r_i$ and $r_j$ respectively, each over a subset of $H$, such that, if $(h_{p_1}, h_{q_1})$ and $(h_{p_2}, h_{q_2})$ are two distinct pairs in $H \times H$, then $(r_i, r_j)$ prefers $(h_{p_1}, h_{q_1})$ to $(h_{p_2}, h_{q_2})$ relative to $\prec_{c_k}$ only if (i) either $r_i$ prefers $h_{p_1}$ to $h_{p_2}$ relative to $\prec_{r_i}$, or $h_{p_1} = h_{p_2}$, and (ii) either $r_j$ prefers $h_{q_1}$ to $h_{q_2}$ relative to $\prec_{r_j}$, or $h_{q_1} = h_{q_2}$.[11] Here a matching is defined as in the HRC case, and stability is defined as in Definition 5.13. We refer to this restriction of HRC as the *Hospitals / Residents problem with Consistent Couples* (HRCC).

We first note that an instance of HRCC need not admit a stable matching. For, consider the HRC instance $I$ shown in Fig. 5.6. Clearly $I$ is an instance of HRCC, and we have already observed that $I$ does not admit a stable matching. Moreover consistent couples do not make it any easier to decide whether a stable matching exists, as indicated by the following result.

---

[11] In fact consistent preference lists need not be responsive (see Footnote 10 for a definition of responsive preferences). To see this, suppose that the preference list of a couple $c = (r_1, r_2)$ includes the pairs $(h_1, h_3)$, $(h_2, h_3)$, $(h_1, h_4)$ and $(h_2, h_4)$ in that order. Then this list is responsive to the individual preference lists for $r_1$ and $r_2$ where $r_1$ prefers $h_1$ to $h_2$ and $r_2$ prefers $h_3$ to $h_4$. However $c$'s list is not consistent with the individual lists of $r_1$ and $r_2$. Also we remark that consistent preferences need not be *weakly responsive* as defined by Klaus and Klijn [387], and hence their polynomial-time algorithm for HRC with weakly responsive preferences does not automatically apply to HRCC. This is due to the fact that weakly responsive preferences must allow one member of a couple the possibility of being unassigned, which need not be the case with consistent preferences.

**Theorem 5.20 ([444]).** *Let I be an instance of* HRCC. *The problem of deciding whether I admits a stable matching (where stability is defined relative to Definition 5.13) is NP-complete. The result holds even if the preference list of each single resident and the joint preference list of each couple is of length at most 3, the preference list of each hospital is of length at most 6, and each hospital has capacity at most 2.*

We remark that, by inspection of the preferences lists constructed in the proof of Theorem 5.17 (see Ref. [87]), it turns out that this theorem for HRC also applies to HRCC, and hence we can also deduce NP-completeness for the restriction of HRCC in which the preference lists are derived from master lists, and each hospital has capacity 1.

We can instead define the stability of a matching in terms of classical (Gale–Shapley) stability, relative to the preference lists of the individual residents and hospitals (as in Definition 1.8). (Of course, in a given matching, an assigned couple must still obtain a pair of hospitals that they find mutually acceptable according to their joint list). Note that if the joint preference list of a couple is consistent, then the derived individual preference lists for the two residents in the couple are unique, and therefore the notion of stability is not dependent on any particular choice of these individual lists. A matching that satisfies classical stability is stable relative to Definition 5.13, but the converse is not true in general. It turns out that, in contrast to Theorem 5.20, finding a stable matching or reporting that none exists can be accomplished in polynomial time for classical stability, given an HRCC instance.

**Theorem 5.21 ([444]).** *Let I be an instance of* HRCC. *There is an $O(m)$ algorithm that finds a stable matching or reports that none exists (where the stability definition is classical stability defined relative to the preference lists of the individual residents and hospitals), where $m$ is the sum of the lengths of the individual residents' lists in I.*

### 5.3.5 *Inseparable couples*

A natural restriction of HRCC arises when the joint preference list of each couple $(r_i, r_j)$ has a very specific structure: that is, each element is of the form $(h_k, h_k)$ for some hospital $h_k \in H$ [444]. Thus $r_i$ and $r_j$ wish to be either assigned to the same hospital, or both be unassigned. We refer to this restriction of HRCC as the *Hospitals / Residents problem with Inseparable Couples* (HRIC).

As is the case for HRC and HRCC, an instance of HRIC need not admit a stable matching (where stability is as defined in Definition 5.13). To see this, observe that the HRC instance shown in Fig. 5.7, which admits no stable matching, is an instance of HRIC.

Moreover, even the restricted nature of HRIC is unlikely to lead to a polynomial-time algorithm for finding a stable matching (or reporting that none exists), as Theorem 5.20 also holds for instances of HRIC [444].

Given the structure of a couple $c_k = (r_i, r_j)$'s preference list in an HRIC instance, it is natural to replace $(r_i, r_j)$ by a single entity $c_k$ whose preference list is obtained from that of $(r_i, r_j)$ by replacing each occurrence of $(h_p, h_p)$ by $h_p$. Each single resident occupies one post at a given hospital, whilst each couple occupies two posts. This formulation of HRIC has a natural generalisation to the case where each resident $r_i \in R$ has a *size* $s_i \in \mathbb{Z}^+$, indicating the number of posts that $r_i$ occupies at any hospital. Hospitals rank residents of any size as a single entity. We refer to this variant of HRIC as the *Hospitals / Residents problem with Sizes* (HRS) [444].

Let $I$ be an HRS instance and let $M$ be an assignment in $I$. Given a hospital $h_j \in H$, we denote $\sum \{s_i : r_i \in M(h_j)\}$ by $O_j^M$ and refer to this as the *occupancy* of $h_j$ in $M$. We say that $h_j$ is *undersubscribed* if $O_j^M < c_j$. A *matching* is an assignment $M$ such that $|M(r_i)| \leq 1$ for each $r_i \in R$, and $O_j^M \leq c_j$ for each $h_j \in H$. In other words, each resident is assigned to at most one hospital, and the sum of the sizes of the residents assigned to a hospital does not exceed its capacity. A pair $(r_i, h_j) \in R \times H$ *blocks* a matching $M$, or is a *blocking pair* for $M$, if

(1) $r_i$ is unmatched, or $r_i$ prefers $h_j$ to $M(r_i)$, and
(2) $O_j^M + s_i \leq c_j$, or $h_j$ prefers $r_i$ to each of residents $r_{k_1}, \ldots r_{k_t} \in M(h_j)$, such that

$$O_j^M + s_i - \sum_{p=1}^{t} s_{k_p} \leq c_j.$$

A matching in $I$ is *stable* if it admits no blocking pair.

The definition implies that $h_j$ could participate in a blocking pair with $r_i$ if either (i) $h_j$ currently has room for $r_i$, or (ii) $h_j$ can make room for $r_i$ by rejecting a set of residents, each of whom it finds worse than $r_i$. A matching is *stable* if it admits no blocking pair.

We assume without loss of generality that, for each $r_i \in R$ and for each hospital $h_j$ on $r_i$'s preference list, $s_i \leq c_j$, for otherwise $(r_i, h_j)$ could never belong to a stable matching, nor could $(r_i, h_j)$ form a blocking pair.

Residents' preferences     Hospitals' preferences

$r_1 : 1 : h_2 \ h_1$          $h_1 : 2 : r_1 \ r_3 \ r_2$

$r_2 : 1 : h_1 \ h_2$          $h_2 : 1 : r_2 \ r_1$

$r_3 : 2 : h_1$

Fig. 5.8    An instance of HRIC with no stable matching [444]

Figure 5.8 illustrates an HRS instance $I$, due to McDermid and Manlove [444], that admits no stable matching. In this figure, next to each resident $r_i$ we show $s_i$ and then $r_i$'s preference list. As before, next to each hospital $h_j$ we show $c_j$ followed by $h_j$'s preference list. Suppose that $M$ is a stable matching in $I$. If $(r_3, h_1) \in M$, then $(r_1, h_2) \in M$, for otherwise $(r_1, h_1)$ blocks $M$. But then $(r_2, h_2)$ blocks $M$. Hence $r_3$ is unassigned in $M$. Thus $(r_2, h_1) \in M$, for otherwise $(r_2, h_1)$ blocks $M$. Hence $(r_1, h_2) \in M$, for otherwise $(r_1, h_2)$ blocks $M$. But then $(r_3, h_1)$ blocks $M$, a contradiction.

McDermid and Manlove [444] proved the following algorithmic results for HRS.

**Theorem 5.22 ([444]).** *Given an instance $I$ of HRS, the problem of deciding whether $I$ admits a stable matching is NP-complete. The result holds even if the size of each resident and the capacity of each hospital is at most 2, and the length of each preference list is at most 3. However, if each hospital's list is of length at most 2 (and the sizes of the residents, the capacities of the hospitals and the lengths of the residents' lists are unrestricted), a stable matching always exists and can be found in $O(m)$ time, where $m$ is the number of acceptable resident–hospital pairs.*

The stability definition for HRS assumes that if a resident $r_i$ prefers a hospital $h_j$ to $M(r_i)$, and $h_j$ prefers $r_i$ to a set of residents of total size that is sufficient to free up enough space for $r_i$, then $(r_i, h_j)$ forms a blocking pair. This includes the case, for example, where $r_i$ has size 1, and $h_j$ is assigned a single resident of size 10. Thus, the occupancy of $h_j$ will decrease by 9 if the blocking pair is satisfied. An alternative definition would allow a hospital only to participate in a blocking pair if its occupancy would not decrease as a consequence of satisfying the blocking pair. The algorithmic complexity of finding a stable matching (or reporting that none exists) in HRS is open for this alternative definition of stability.

A version of HRS, called the *Unsplittable Stable Marriage problem*, was studied previously by Dean et al. [161]; their problem differs from HRS in that they permit a hospital $h_j$'s capacity to be exceeded by the assignment of a resident to $h_j$. They provided a polynomial-time algorithm that finds a stable matching in which the capacity of each oversubscribed hospital is exceeded by at most the size of its largest acceptable resident.

## 5.4 Many–many stable matching

### 5.4.1 Introduction

As noted in Sec. 1.3.6.2, many–many extensions of SM (and by implication HR) have been considered in the literature [544, 57, 212, 426, 182, 67, 398, 189]. These extensions have variously been referred to as the many–many (bipartite) stable matching (or marriage) problem [543, 189], the multiple partner stable marriage problem [67] and the bipartite stable $b$-matching problem [212]. What is, however, common to these sources is that the problems are most often referred to in terms of assigning *workers* to *firms*, where each agent can be multiply assigned (up to a given capacity). We will therefore refer to the general many–many bipartite stable matching problem as the *Workers / Firms problem* (WF).

Two variations of the basic WF model have been considered in the literature. The first version, which we denote by WF-1, involves each worker ranking in strict order of preference a set of individual acceptable firms, and vice versa for each firm. In the second version, denoted by WF-2, each worker ranks in strict order of preference all possible subsets of firms, and vice versa for the each firm. Bansal *et al.* [67] noted that WF-1 has generally been studied mainly by the computer science community, whilst the economics community has mainly focused on WF-2. One reason for this is that WF-2 suffers from the drawback that the length of an agent's preference list is in general exponential the number of agents. A consequence of this is that the practical applicability of any algorithm for WF-2 would be severely limited. On the other hand, this problem does not arise with WF-1.

This section is organised as follows. In Sec. 5.4.2 we define formally the basic WF model. We then define WF-1, and discuss structural and algorithmic results for this problem, in Sec. 5.4.3. We do likewise for WF-2 in Sec. 5.4.4.

### 5.4.2 Definition of the basic WF model

An instance $I$ of WF involves a set $W = \{w_1, \ldots, w_{n_1}\}$ of *workers* and a set $F = \{f_1, \ldots, f_{n_2}\}$ of *firms*. The *agents* are the members of $W \cup F$. Each agent $a_k \in W \cup F$ has a positive integral *capacity* denoted by $c(a_k)$. Also there is a set $E \subseteq W \times F$ of *acceptable* worker–firm pairs. Each worker $w_i \in W$ has an *acceptable* set of firms $A(w_i)$, where

$$A(w_i) = \{f_j \in F : (w_i, f_j) \in E\}.$$

Similarly each firm $f_j \in H$ has an acceptable set of workers $A(f_j)$, where

$$A(f_j) = \{w_i \in W : (w_i, f_j) \in E\}.$$

An *assignment* $M$ is a subset of $E$. If $(w_i, f_j) \in M$, $w_i$ is said to be *assigned* to $f_j$, and $f_j$ is *assigned* to $w_i$. For each $a_k \in W \cup F$, the set of assignees of $a_k$ in $M$ is denoted by $M(a_k)$. An agent $a_k \in W \cup F$ is *undersubscribed*, *full* or *oversubscribed* according as $|M(a_k)|$ is less than, equal to, or greater than $c(a_k)$, respectively.

The definition of a stable matching in $I$ depends on the notion of preference, and in particular whether agents have preferences over individual agents or over subsets of agents. The former case leads to WF-1, whilst the latter leads to WF-2. The concept of a stable matching in each of these cases is defined in the following subsections.

### 5.4.3   WF-*1: preferences over individuals*

The variant WF-1 of WF is obtained when each agent $a_k \in W \cup F$ has a strictly-ordered preference list over $A(a_k)$. In an instance $I$ of WF-1, a *matching* is an assignment $M$ in $I$ such that $|M(a_k)| \leq c(a_k)$ for each $a_k \in W \cup F$. A pair $(w_i, f_j) \in E \backslash M$ *blocks* a matching $M$, or is a *blocking pair* for $M$, if (i) either $w_i$ is undersubscribed or prefers $f_j$ to some member of $M(w_i)$, and (ii) either $f_j$ is undersubscribed or prefers $w_i$ to some member of $M(f_j)$. $M$ is said to be *stable* if it admits no blocking pair.

Baïou and Balinski [57] modelled WF-1 in terms of the marriage graph defined in Sec. 2.10.5, and they used this representation to prove a number of structural and algorithmic results. Firstly, they showed that a stable matching always exists in $I$ and can be found in $O(n^2)$ time, where $n = \max\{n_1, n_2\}$. They then showed, given an agent $a_k \in W \cup F$ and given two stable matchings $M, M'$ in $I$, that $|M(a_k)| = |M'(a_k)|$, and if $|M(a_k)| < c(a_k)$ then $M(a_k) = M'(a_k)$ (thus generalising Theorem 1.11 for HR).

In order for agents to compare their partners in two stable matchings, the authors introduced the *max–min criterion* which states that an agent $a_k \in W \cup F$ *prefers* one stable matching $M$ to another $M'$ if (i) $|M(a_k)| \geq |M'(a_k)|$ and (ii) $a_k$ prefers the worst assignee in $M(a_k)$ to that in $M'(a_k)$. On this basis, the authors showed that *worker-optimal* and *firm-optimal* stable matchings (generalising resident-optimal and hospital-optimal stable matchings in the HR context) exist and can be found in $O(n^2)$ time. Moreover by ordering stable matchings according to the max-min criterion, Baïou and Balinski proved that the set of stable matchings in $I$ forms a lattice.

Bansal *et al.* [67] also studied WF-1. As noted in Sec. 2.10.8, they defined a *meta-rotation* in a given WF-1 instance $I$, which generalises of the concept of a rotation in SMI (see Sec. 1.3.4.3). They proved that in $I$, the closed subsets of the meta-rotation poset are in 1–1 correspondence with the stable matchings. On the basis of this result, the authors gave an $O(n^6)$ algorithm for finding an *egalitarian stable matching* in $I$ — in such a stable matching, which generalises the corresponding notion for SM (see Sec. 1.3.4.1), the sum of the ranks of the agents' partners is minimised. An $O(n^3(\log n)^2)$ algorithm for the same problem was given by Eirinakis *et al.* [189]. This latter algorithm can be extended (with unchanged time complexity) to the case where preference lists are weighted, and we seek a *minimum weight stable matching* — in this matching, which again generalises the same concept for SM (see Sec. 1.3.4.1), the sum of the weights of the agents' partners is minimised. In another paper, Eirinakis *et al.* [190] gave an $O(n^2)$ algorithm to find a *minimum total regret stable matching* — this is a stable matching that minimises the maximum *total regret* over all agents, where the *total regret* for any agent $a_k \in W \cup F$ is the sum of the ranks of $a_k$'s partners in $M$.

Eirinakis *et al.* [189] also gave an $O(n^2)$ algorithm to find all the *stable pairs* in $I$ — these are the worker–firm pairs that appear in some stable matching in $I$. Let $S$ denote the set of stable matchings in $I$. The authors also gave an $O(n^2 + n|S|)$ algorithm to list all members of $S$.

As noted in Sec. 2.4, Fleiner [212] gave an LP-based characterisation of stable matchings in an instance of WF-1. Also, as mentioned in Sec. 2.5, Eirinakis *et al.* [189] gave a CSP encoding of a WF-1 instance. Further, we mention that, given an instance $I$ of WF-1 with ties, one can generalise the definitions of weakly stable, strongly stable and super-stable matchings in HRT (see Sec. 1.3.5) to $I$. We refer the reader to Sec. 3.5.2 for details of an algorithm for finding a strongly stable matching or reporting that none exists in $I$. We close this section by remarking that, in the same sense that SR is a non-bipartite version of SM, SF (see Sec. 4.8.4) is a non-bipartite version of WF-1.

### 5.4.4  WF-*2*: *group preferences*

WF-2 is the variant of WF where agents have *group preferences*. That is, each worker $w_i \in W$ has a strictly-ordered preference list over a subset $\mathcal{S}(w_i)$ ($w_i$'s *acceptable sets* of partners) of $\mathbb{P}(F)\backslash\{\emptyset\}$, where $\bigcup_{S\in\mathcal{S}(w_i)} S = A(w_i)$ (hence $w_i$'s acceptable partners are precisely the agents that collectively belong to $w_i$'s acceptable sets of partners). The preference list for a firm $f_j \in F$ is defined similarly. In an instance $I$ of WF-2, a *matching* is an assignment $M$ such that, for each $a_k \in W \cup F$, either $M(a_k) = \emptyset$ or $M(a_k) \in \mathcal{S}(a_k)$ .

In an instance of WF-2, it is no longer necessary to explicitly retain capacity information for the agents. For, in view of the definition of a matching, for any agent $a_k \in W \cup F$, if $S \in \mathcal{S}(a_k)$ and $|S(a_k)| > c(a_k)$, then $a_k$ can simply declare $S$ as unacceptable.

Given a worker $w_i \in W$ and a set $S \in \mathbb{P}(F)\backslash\{\emptyset\}$, define $Ch(w_i, S)$ to be $w_i$'s most-preferred subset of $S$ in $\mathcal{S}(w_i)$. (That is, $Ch(w_i, S)$ is intuitively the set of partners from among $S$ that $w_i$ desires the most.) $Ch(f_j, S)$ is defined similarly for a firm $f_j \in F$ and a set $S \in \mathbb{P}(W)\backslash\{\emptyset\}$. A matching $M$ is *individually rational* if, for any $a \in W \cup F$, $M(a) = Ch(a, M(a))$. Thus a matching would fail to be individually rational if some agent would prefer to reject one or more of its assignees.

A matching $M$ in $I$ is *pairwise stable* [499] if it is individually rational and $M$ admits no *blocking pair*, which is a worker–firm pair $(w_i, f_j) \in (W \times F)\backslash M$ such that $f_j \in Ch(w_i, M(w_i)\cup\{f_j\})$ and $w_i \in Ch(f_j, M(f_j)\cup\{w_i\})$. In a blocking pair, $w_i$ and $f_j$ would prefer to be assigned to sets of partners that include one another than to remain with their current assignees.

A WF-2 instance need not admit a pairwise stable matching — see Example 2.7 in Ref. [514] for more details. However it turns out that the existence of such a matching is guaranteed if the agents' preferences satisfy a property known as *substitutability* [374]. The preference list of a worker $w_i \in W$ is *substitutable* [374] if, for any set $S \in \mathcal{S}(w_i)$ containing distinct firms $f_j, f_k$, if $f_j \in Ch(w_i, S)$ then $f_j \in Ch(w_i, S\backslash\{f_k\})$. (That is, $w_i$'s desire for $f_j$ does not depend on the presence of $f_k$.) Substitutability for a firm $f_j \in F$ is defined similarly.

A stronger condition than substitutability that has been considered in the literature is *responsiveness* [500]. The preference list of a worker $w_i \in W$ is *responsive* [500] if (i) for any $S \in \mathcal{S}(w_i)$ and for any $f_j \in A(w_i)\backslash S$, $w_i$ prefers $S \cup \{f_j\}$ to $S$, and (ii) for any two subsets $S, S' \in \mathcal{S}(w_i)$ such that $S' = (S\backslash\{f_j\}) \cup \{f_k\}$ for two distinct firms $f_j \in S$ and $f_k \in S'$, $w_i$

prefers $S'$ to $S$ if and only if $w_i$ prefers $\{f_k\}$ to $\{f_j\}$. (That is, if one set of assignees is obtained from another by adding an acceptable firm, then the worker prefers the larger set of assignees, and if two sets of assignees differ by replacing one firm $f_j$ by another, $f_k$, then the worker prefers the set with the most-preferred firm from among $f_j$ and $f_k$).Responsiveness for a firm $f_j \in F$ is defined similarly. Responsive preferences are clearly substitutable.

Roth [498] proved that, given an instance of WF-2 where every agent's preference list is substitutable, a pairwise stable matching always exists. Moreover he gave an algorithm for finding one, and proved the existence of worker-optimal and firm-optimal stable matchings. Martínez *et al.* [426] gave an algorithm to compute the entire set of pairwise stable matchings in this context. Blair [97] proved that the set of pairwise stable matchings in an instance of WF-2, where again every agent's preference list is substitutable, forms a lattice. As mentiond in Sec. 2.6.4, Kojima and Ünver [398] showed that, given an instance of WF-2, and starting from an arbitrary matching, one can arrive at a pairwise stable matching by satisfying a sequence of blocking pairs, as long as the agents on one side have responsive preferences, whilst the agents on the other side have substitutable preferences.

A drawback of the pairwise stability concept is that a pairwise stable matching is only robust against two agents who might act together to undermine it. A more powerful stability definition, called *setwise stability* was formulated by Sotomayor [544], which prevents a matching from being undermined by a coalition of agents who could improve relative to it. Intuitively, a matching $M$ is *setwise stable* [544] if there is no coalition of agents who, by forming new partnerships only among themselves, possibly dissolving some partnerships of $M$ and possibly keeping other ones, can all obtain a strictly preferred set of partners. Formally, following Echenique and Oviedo [182], $M$ is *setwise stable* if it is individually rational and there is no triple $(W', F', M')$, where $W' \subseteq W$, $F' \subseteq F$ and $M'$ is a matching, such that (i) $W' \cup F' \neq \emptyset$, and for all $a_k \in W' \cup F'$, (ii) $M'(a_k) \in \mathcal{S}(a_k)$, (iii) $M'(a_k) \backslash M(a_k) \subseteq W' \cup F'$, (iv) $a_k$ prefers $M'(a_k)$ to $M(a_k)$, and (v) $M'(a_k) = Ch(a_k, M'(a_k))$. Clearly a setwise stable matching is pairwise stable.

Under certain conditions, setwise stable matchings are bound to exist. The preference list of an agent $a_k \in W \cup F$ is *strongly substitutable* [182] if, for any $S, S' \in \mathcal{S}(a_k)$ such that $a_k$ prefers $S$ to $S'$, and for any $a_l \in A(a_k)$, if $a_l \in Ch(a_k, S' \cup \{a_l\})$ then $a_l \in Ch(a_k, S \cup \{a_l\})$. Echenique and Oviedo [182] proved that, given a WF-2 instance where the preferences on one side are substitutable and those on the other side are strongly substitutable, the sets of pairwise and setwise stable matchings coincide. Results above then indicate that in such a case, there is an algorithm to find a setwise stable matching [499] and the set of such matchings forms a lattice [97].

Note that a non-bipartite version of WF-2 with substitutable preferences was considered in Sec. 4.8.7. Further sources that are relevant in the WF-2 context are Refs. [501, 41–43, 398, 117, 189].

## 5.5   The Student–Project Allocation Problem

### 5.5.1   *Introduction*

In many university departments, students seek to undertake a project in a given field of speciality as part of the upper level of their degree programme. Typically a wide range of available projects is offered, and usually the total number of project places exceeds the number of students, to provide something of a choice. Also, typically each lecturer will offer a variety of projects, but does not necessarily expect that all will be taken up.

Each student has preferences over the available projects that she finds acceptable, whilst a lecturer will normally have some form of preference list over the projects she offers and/or the students who find them acceptable. There may also be upper bounds on the number of students that can be assigned to a particular project, and the number of students that a given lecturer is willing to supervise. In this section we consider the problem of allocating students to projects based on these preference lists and capacity constraints – the so-called *Student–Project Allocation problem* (SPA).

Variants of SPA arise according to the nature of the preference lists that lecturers provide. In the case of some centralised matching schemes that assign students to projects, lecturer preferences are not permitted [485, 563, 47]. However at the time of writing, the Department of Computer Science at the University of York permits lecturer preferences over students in its centralised student–project allocation process [179, 372, 568]. This leads to our first variant of SPA, namely the *Student–Project Allocation problem with lecturer preferences over Students* (SPA-S) [23], in which each lecturer $l$ ranks in order of preference the students who find acceptable at least one project that $l$ offers. Such a preference list may reflect $l$'s assessment of the students' academic suitability for her projects.

An alternative variant of SPA is the *Student–Project Allocation problem with lecturer preferences over Projects* (SPA-P) [422, 346], in which each lecturer $l$ ranks in order of preference the projects that she offers. This preference list may reflect the possibility that $l$ prefers to supervise projects that are closely connected with her research, whilst the remaining projects that $l$ offers (perhaps only proposed to ensure that the students have adequate choice) have a lesser priority. The final variant is a hybrid version of SPA-S and SPA-P. In the *Student–Project Allocation problem with lecturer preferences over Student–Project pairs* (SPA-(S,P)), each lecturer $l$ has a preference list that depends on not just the students who find acceptable a project that $l$ offers, but also the particular projects of $l$'s that these students would undertake.

Although the SPA problem model and its variants are introduced and motivated in the context of Student–Project allocation, they are equally valid in other scenarios, for example where applicants seek posts at large organisations, each split into several departments.

In this section we define formally each of these variants of SPA. We then describe structural and algorithmic results for these problem models. The section is organised so that SPA-S, SPA-P and SPA-(S,P) are dealt with in Secs. 5.5.2, 5.5.3 and 5.5.4 respectively.

### 5.5.2 *Lecturer preferences over students:* SPA-S

In this section we consider SPA-S. We define an instance $I$ of the problem formally and discuss related work in Sec. 5.5.2.1. We give an algorithm for finding a stable matching in $I$ in Sec. 5.5.2.2, and in Sec. 5.5.2.3 we describe properties of stable matchings in $I$. The algorithm given in Sec. 5.5.2.2 is student-oriented; in Sec. 5.5.2.4 we briefly discuss a counterpart that is lecturer-oriented. Finally, some open problems for SPA-S are given in Sec. 5.5.2.5.

#### 5.5.2.1 *Introduction*

We begin with a formal definition of SPA-S. An instance of SPA-S comprises a set $S = \{s_1, s_2, \ldots, s_{n_1}\}$ of *students*, a set $P = \{p_1, p_2, \ldots, p_{n_2}\}$ of *projects*, and a set $L = \{l_1, l_2, \ldots, l_{n_3}\}$ of *lecturers*. Each student $s_i$ supplies a preference list, ranking a subset of $P$ in strict order. If project $p_j$ appears on $s_i$'s preference list, we say that $s_i$ finds $p_j$ *acceptable*. Denote by $A(s_i)$ the set of projects that $s_i$ finds acceptable.

Each lecturer $l_k$ *offers* a non-empty set of projects $P_k$, where $P_1, P_2, \ldots, P_{n_3}$ partitions $P$. Lecturer $l_k$ supplies a preference list, denoted by $\mathcal{L}_k$, ranking in strict order of preference those students who find at least one project in $P_k$ acceptable. Also, $l_k$ has a capacity constraint $d_k$, indicating the maximum number of students that she is willing to supervise. Similarly, each project $p_j$ carries a capacity constraint $c_j$, indicating the maximum number of students that could be assigned to $p_j$. We assume that $\max\{c_j : p_j \in P_k\} \leq d_k \leq \sum\{c_j : p_j \in P_k\}$.

For any $p_j \in P_k$, we denote by $\mathcal{L}_k^j$ the *projected preference list of $l_k$ for $p_j$* — this is obtained from $\mathcal{L}_k$ by deleting those students who do not find $p_j$ acceptable. In this way, the ranking of $\mathcal{L}_k^j$ is inherited from $\mathcal{L}_k$.

An *assignment* $M$ is a subset of $S \times P$ such that:

(1) $(s_i, p_j) \in M$ implies that $p_j \in A(s_i)$ (i.e., $s_i$ finds $p_j$ acceptable).
(2) For each student $s_i \in S$, $|\{(s_i, p_j) \in M : p_j \in P\}| \leq 1$.

A number of definitions for SPA-S follow by a straightforward analogy from the definitions of the corresponding terms in the HR context (see Sec. 1.3.2). Firstly, if $(s_i, p_j) \in M$, we say that $s_i$ is *assigned* to $p_j$, and $p_j$ is *assigned* $s_i$. For notational convenience, if $s_i$ is assigned in $M$ to $p_j$, we may also say that $s_i$ is *assigned* to $l_k$, and $l_k$ is *assigned* $s_i$, where $p_j \in P_k$.

For any student $s_i \in S$, if $s_i$ is assigned in $M$ to some project $p_j$, we let $M(s_i)$ denote $p_j$; otherwise we say that $s_i$ is *unassigned* in $M$. For any project $p_j \in P$, we denote by $M(p_j)$ the set of students assigned to $p_j$ in $M$. Project $p_j$ is *undersubscribed, full* or *oversubscribed* according as $|M(p_j)|$ is less than, equal to, or greater than $c_j$, respectively. Similarly, for any lecturer $l_k \in L$, we denote by $M(l_k)$ the set of students assigned to $l_k$ in $M$. Lecturer $l_k$ is *undersubscribed, full* or *oversubscribed* according as $|M(l_k)|$ is less than, equal to, or greater than $d_k$ respectively.

A *matching* $M$ is an assignment such that:

(3) For each project $p_j \in P$, $|M(p_j)| \leq c_j$.
(4) For each lecturer $l_k \in L$, $|M(l_k)| \leq d_k$.

A student–project pair $(s_i, p_j) \in (S \times P) \backslash M$ *blocks* a matching $M$, or is a *blocking pair* of $M$, if:

(1) $p_j \in A(s_i)$ (i.e., $s_i$ finds $p_j$ acceptable);
(2) either $s_i$ is unassigned in $M$, or $s_i$ prefers $p_j$ to $M(s_i)$;
(3) either

   (a) $p_j$ is undersubscribed and $l_k$ is undersubscribed, or
   (b) $p_j$ is undersubscribed, $l_k$ is full, and either $s_i \in M(l_k)$ or $l_k$ prefers $s_i$ to the worst student in $M(l_k)$, or

Student preferences     Lecturer preferences

| | | | | | | |
|---|---|---|---|---|---|---|

$s_1 : p_1 \ p_7$

$s_2 : p_1 \ p_2 \ p_3 \ p_4 \ p_5 \ p_6$

$s_3 : p_2 \ p_1 \ p_4$

$s_4 : p_2$

$s_5 : p_1 \ p_2 \ p_3 \ p_4$

$s_6 : p_2 \ p_3 \ p_4 \ p_5 \ p_6$

$s_7 : p_5 \ p_3 \ p_8$

$l_1 : s_7 \ s_4 \ s_1 \ s_3 \ s_2 \ s_5 \ s_6$     $l_1$ offers $p_1, p_2, p_3$

$l_2 : s_3 \ s_2 \ s_6 \ s_7 \ s_5$     $l_2$ offers $p_4, p_5, p_6$

$l_3 : s_1 \ s_7$     $l_3$ offers $p_7, p_8$

Project capacities: $c_1 = 2$, $c_i = 1$ ($2 \leq i \leq 8$)

Lecturer capacities: $d_1 = 3$, $d_2 = 2$, $d_3 = 2$

Fig. 5.9   An instance of SPA-S due to Abraham *et al.* [23]

(c) $p_j$ is full and $l_k$ prefers $s_i$ to the worst student in $M(p_j)$,

where $l_k$ is the lecturer who offers $p_j$.

A matching is *stable* if it admits no blocking pair.

The blocking pair definition attempts to capture the various practical scenarios in which $s_i$ and $l_k$ could both simultaneously improve relative to $M$ by permitting an assignment between $s_i$ and $p_j$. For a more detailed discussion of the different cases in the blocking pair definition, see Ref. [23].

An example SPA-S instance $I$ from Ref. [23] is shown in Fig. 5.9. It turns out that

$$M = \{(s_1, p_1), (s_2, p_5), (s_3, p_4), (s_4, p_2), (s_7, p_3)\}$$

is the unique stable matching in $I$.

Clearly HR is a special case of SPA-S in which $n_2 = n_3$, $c_j = d_j$ and $P_j = \{p_j\}$ ($1 \leq j \leq n_2$). Essentially the projects and lecturers are indistinguishable in this case. We have already seen that an HR instance is reducible to an SMI instance using the technique of "cloning" hospitals (see Sec. 3.2.6.5). However there is no straightforward reduction involving cloning from a SPA-S instance to an HR instance, due to the projects and lecturers being distinct entities, each having capacity constraints.

We remark that SPA-S is a special case of HR-CQ-NSS (see Sec. 5.2.4). An instance $I$ of SPA-S corresponds to an instance $J$ of HR-CQ in which the students correspond to residents, the projects correspond to hospitals, and the lecturers are represented by bounded sets of hospitals. Since every pair of distinct bounded sets is disjoint, it follows that $J$ is in fact an instance of HR-CQ-NSS.

We now describe other related work. Fleiner [209, 211] developed a matroid-theoretic characterisation of stable matchings in bipartite matching models. This is based on imposing two ordered partition matroids,

$\mathcal{M}_A$ and $\mathcal{M}_B$, one on each side of a bipartite graph $G$. A matching is an independent set that is common to both $\mathcal{M}_A$ and $\mathcal{M}_B$. Moreover a stable matching corresponds to an $\mathcal{M}_A\mathcal{M}_B$-*kernel*, and it is shown that such a structure is bound to exist [209, 211]. Fleiner [213] noted that the SPA-S problem model may be included in this characterisation by imposing a student matroid as a partition matroid, and a lecturer matroid as the truncation of a direct sum of uniform matroids (thus ensuring that all project and lecturer capacities are satisfied). Here the vertices on one side of $G$ correspond to students, the vertices on the other side correspond to lecturers, and the edges correspond to acceptable student–project pairs (so that $G$ is in general a multigraph).

Also Eguchi *et al.* [185] (see also Ref. [555]) formulated a model for two-sided matching problems in which preferences are based on $M^\natural$-*concave functions*, which arise in discrete convex analysis. They gave an algorithm for finding a stable matching in such a context, however the algorithm does not, in general, run in polynomial time for an arbitrary $M^\natural$-concave function. Their model includes the possibility of capacities and multiple partners; moreover since linear orders gives rise to $M^\natural$-concave functions, it follows that the model of Eguchi *et al.* [185] includes SPA-S as a special case.

### 5.5.2.2   *Overview of Algorithm* SPA-S-*student*

We now present a linear-time algorithm for finding a stable matching in a given SPA-S instance $I$. This algorithm is *student-oriented* in that it involves a series of iterations, during each of which a student *applies* to a project — this is an analogous operation to a man proposing to a woman in the context of the man-oriented Gale–Shapley algorithm. Moreover, as we will see, the stable matching $M$ returned by the algorithm is *student-optimal*, in the sense that each student obtains the best project that she could obtain in any stable matching. The algorithm is a generalisation of the resident–oriented Gale–Shapley algorithm for HR [261, Section 1.6.3].

Throughout the course of the algorithm's execution, apply operations lead to provisional assignments between students, projects and lecturers; such assignments can subsequently be broken during the algorithm's execution. Also, throughout the execution, entries are possibly deleted from the preference lists of students, and from the projected preference lists of lecturers. We use the abbreviation *delete* $(s_i, p_j)$ to denote the operation of deleting $p_j$ from the preference list of $s_i$, and deleting $s_i$ from $\mathcal{L}_k^j$, where $l_k$ is the lecturer who offers $p_j$.

Initially the matching $M$ constructed by the algorithm is empty. As long as there is some student $s_i$ who is unassigned and who has a non-empty list, $s_i$ applies to the first project $p_j$ on her list. We let $l_k$ be the lecturer who offers $p_j$. Immediately, $(s_i, p_j)$ is added to $M$.

If $p_j$ is oversubscribed, then $l_k$ rejects the worst student $s_r$ assigned to $p_j$. The pair $(s_r, p_j)$ will be deleted by the subsequent conditional that tests for $p_j$ being full. Similarly, if $l_k$ is oversubscribed, then $l_k$ rejects her worst assigned student $s_r$. The pair $(s_r, p_t)$ will be deleted by either of the two subsequent conditionals, where $p_t$ is $s_r$'s assigned project in $M$.

Regardless of whether any rejections occurred as a result of the two situations described in the previous paragraph, we have two further (possibly non-disjoint) cases in which deletions may occur. If $p_j$ is full, we let $s_r$ be the worst student assigned to $p_j$ (according to $\mathcal{L}_k^j$) and delete $(s_t, p_j)$ for each successor $s_t$ of $s_r$ on $\mathcal{L}_k^j$. Similarly if $l_k$ is full, we let $s_r$ be the worst student assigned to $l_k$, and delete $(s_t, p_u)$ for each successor $s_t$ of $s_r$ on $\mathcal{L}_k$, and for each project $p_u$ offered by $l_k$ that $s_t$ finds acceptable.

The algorithm is described in pseudocode form in Algorithm 5.1 as Algorithm SPA-S-student. The following result, proved in Ref. [23], establishes the correctness and complexity of the algorithm.

**Theorem 5.23 ([23]).** *Every SPA-S instance $I$ admits a stable matching. Moreover, any execution of Algorithm SPA-S-student applied to $I$ constructs the unique stable matching in which each assigned student is assigned to the best project that she could obtain in any stable matching, whilst each unassigned student is unassigned in every stable matching. The algorithm runs in $O(m)$ time and $O(n_1 n_2)$ space, where $m$ is the number of acceptable student–project pairs, $n_1$ is the number of students and $n_2$ is the number of projects.*

Given the optimality property established by Theorem 5.23, the stable matching returned by Algorithm SPA-S-student is defined to be the *student-optimal stable matching*.

### 5.5.2.3 *Properties of stable matchings in an instance of SPA-S*

In this subsection we explore which properties of Theorem 1.11 (the "Rural Hospitals Theorem" for HR) have analogues in the context of SPA-S. Whilst some properties do carry over, certain others, with natural analogues for SPA-S, perhaps surprisingly do not hold. The following result indicates which properties of Theorem 1.11 do hold in the SPA-S setting.

---

**Algorithm 5.1** Algorithm SPA-S-student [23]

---

**Require:** SPA-S instance $I$
**Ensure:** return the student-optimal stable matching $M$ in $I$
1: $M := \emptyset$;
2: **while** some student $s_i$ is unassigned **and** $s_i$ has a non-empty list **do**
3:      $p_j :=$ first project on $s_i$'s list;
4:      $l_k :=$ lecturer who offers $p_j$;
5:      $\{s_i$ applies to $p_j\}$
6:      $M := M \cup \{(s_i, p_j)\}$;
7:      **if** $p_j$ is oversubscribed **then**
8:          $s_r :=$ worst student assigned to $p_j$;          $\{$according to $\mathcal{L}_k^j\}$
9:          $M := M \backslash \{(s_r, p_j)\}$;
10:      **else if** $l_k$ is oversubscribed **then**
11:          $s_r :=$ worst student assigned to $l_k$;          $\{$according to $\mathcal{L}_k\}$
12:          $p_t := M(s_r)$;
13:          $M := M \backslash \{(s_r, p_t)\}$;
14:      **end if**
15:      **if** $p_j$ is full **then**
16:          $s_r :=$ worst student assigned to $p_j$;          $\{$according to $\mathcal{L}_k^j\}$
17:          **for each** successor $s_t$ of $s_r$ on $\mathcal{L}_k^j$ **do**
18:              delete $(s_t, p_j)$;
19:          **end for**
20:      **end if**
21:      **if** $l_k$ is full **then**
22:          $s_r :=$ worst student assigned to $l_k$;          $\{$according to $\mathcal{L}_k\}$
23:          **for each** successor $s_t$ of $s_r$ on $\mathcal{L}_k$ **do**
24:              **for each** project $p_u \in P_k \cap A(s_t)$ **do**
25:                  delete $(s_t, p_u)$;
26:              **end for**
27:          **end for**
28:      **end if**
29: **end while**
30: **return** $M$;

---

**Theorem 5.24 ([23]).** *For a given SPA-S instance, the following statements hold:*

*(i) each lecturer has the same number of students in all stable matchings;*
*(ii) exactly the same students are unassigned in all stable matchings;*
*(iii) a project offered by an undersubscribed lecturer has the same number of students in all stable matchings.*

Two key properties of Theorem 1.11 have no analogue for SPA-S. Firstly, Part (ii) of the theorem states that in HR, each hospital is assigned the same number of residents in any stable matching. However, in a SPA-S instance, a project offered by a lecturer who is full in one stable matching need not be assigned the same number of students in all stable matchings (see Fig. 4 in Ref. [23] for more details).

Secondly, Part (iii) of the theorem states that in HR, any hospital that is undersubscribed in one stable matching is assigned the same set of residents in all stable matchings. However, in a SPA-S instance, a lecturer who is undersubscribed in one stable matching need not be assigned the same set of students in all stable matchings (see Fig. 3 in Ref. [23] for more details).

#### 5.5.2.4 *Lecturer-oriented algorithm*

A lecturer-oriented counterpart of Algorithm SPA-S-student, namely Algorithm SPA-S-lecturer, is presented in Ref. [23]. Algorithm SPA-S-lecturer produces the *lecturer-optimal stable matching*, in which each lecturer obtains the best set of students that she could obtain in any stable matching. However the definition of "best" needs some care in the SPA-S context. To obtain a precise definition of lecturer-optimality, we require to define the *prefers* relation on pairs of stable matchings for a given lecturer.

Let $M$ and $M'$ be two stable matchings for a given instance of SPA-S. By Theorem 5.24, we know that $|M| = |M'|$ and $|M(l_k)| = |M'(l_k)|$. For a given lecturer $l_k$ who is assigned different sets of students in $M$ and $M'$, suppose that

$$M(l_k) \setminus M'(l_k) = \{s_1, \ldots, s_r\}$$

and

$$M'(l_k) \setminus M(l_k) = \{s'_1, \ldots, s'_r\},$$

where, in each case, the students are enumerated in the order in which they appear in $\mathcal{L}_k$. If $l_k$ prefers $s_i$ to $s'_i$ for all $i$ $(1 \leq i \leq r)$ we say that $l_k$ *prefers* $M$ to $M'$. Alternatively, and equivalently, $l_k$ *prefers* $M$ to $M'$ if there is a one-to-one mapping $f$ from $M'(l_k) \setminus M(l_k)$ to $M(l_k) \setminus M'(l_k)$ with the property that $l_k$ prefers $f(s'_i)$ to $s'_i$ for all $s'_i \in M'(l_k) \setminus M(l_k)$.

The following theorem summarises the key properties of the stable matching resulting from any execution of Algorithm SPA-S-lecturer, and the complexity of that algorithm.

**Theorem 5.25.** *For a given instance of* SPA-S, *any execution of Algorithm* SPA-S-lecturer *constructs the unique stable matching $M$ satisfying the following two properties:*

*(i) each student is unassigned or is assigned to the worst project she has in any stable matching;*

*(ii) each lecturer prefers M to any stable matching in which she has a different set of assigned students.*

The algorithm runs in $O(m)$ time and $O(n_1 n_2)$ space, where $m$ is the number of acceptable student–project pairs, $n_1$ is the number of students and $n_2$ is the number of projects.

### 5.5.2.5 *Open problems*

There are two interesting generalistions of SPA-S that give rise to potential avenues for future research. Firstly, if we allow ties in the preference lists of students and lecturers, different stability definitions are possible. These can be obtained by extending the three stability definitions that have been applied to HRT (see Sec. 1.3.5). Under the analogue of weak stability, every instance of SPA-S with ties admits a weakly stable matching (this follows by breaking the ties arbitrarily and applying Algorithm SPA-S-student to the resulting instance of SPA-S, for example). However such matchings could be of different sizes for a given SPA-S instance with ties, and the problem of finding a maximum weakly stable matching is NP-hard (this follows by Corollary 3.7). Under the analogues of the two stronger stability criteria, namely strong stability and super-stability, an instance of SPA-S with ties need not admit a matching satisfying either criterion (again this follows by considering the analogous observations for SMT — see Secs. 3.3.1 and 3.4.1). However it remains open to construct algorithms for finding such a matching in each case, or reporting that none exists, for a given instance of SPA-S with ties.

A second direction is to consider the case that each project $p_j$ carries a lower bound, indicating the minimum number of assignees that $p_j$ must obtain in a given matching in order for the project to run. As in the HR case (see Secs. 5.2.2 and 5.2.3), there are two possibilities according to whether a project is allowed to be closed or not in a given matching. If a project can be closed, one can extend the stability definition from Sec. 5.2.2 to the SPA-S case. As HR is a special case of SPA-S, Theorem 5.2 then implies that the problem of deciding whether, given a SPA-S instance $I$ with project lower bounds where projects can be closed, $I$ admits a stable matching is NP-complete. If no project can be closed, then the stability definition for SPA-S without project lower bounds is unchanged. However, in contrast to the case for HR-LQ-2, the problem of deciding whether a stable matching exists need not be quite so straightforward. This is due to the fact that a given project need not obtain the same number of assignees in all stable matchings (see Sec. 5.5.2.3). It is open as to whether there exists a polynomial-time algorithm for finding a stable matching or reporting that none exists, given an instance of SPA-S with lower bounds for the projects, where no project can be closed.

### 5.5.3 *Lecturer preferences over projects*

We now consider the case that lecturers rank in order of preference the projects that they offer, rather than the students who find their projects acceptable. Formally, the definition of an instance $I$ of SPA-P is identical to that of SPA-S, except that the preference list of each lecturer $l_k \in L$ comprises a strict ranking of the projects in $P_k$. The definitions of an assignment and a matching in $I$ are unchanged from the SPA-S case, though the stability definition is rather different.

A student–project pair $(s_i, p_j) \in (S \times P) \backslash M$ *blocks* a matching $M$, or is a *blocking pair* of $M$, if the following conditions are satisfied relative to $M$ in $I$:

(1) $p_j \in A(s_i)$ (i.e., $s_i$ finds $p_j$ acceptable);
(2) either $s_i$ is unassigned or $s_i$ prefers $p_j$ to $M(s_i)$;
(3) $p_j$ is undersubscribed and either

(a) $s_i \in M(l_k)$ and $l_k$ prefers $p_j$ to $M(s_i)$, or
(b) $s_i \notin M(l_k)$ and $l_k$ is undersubscribed, or
(c) $s_i \notin M(l_k)$ and $l_k$ is full and $l_k$ prefers $p_j$ to her worst project $p_r$ satisfying $M(p_r) \neq \emptyset$,

where $l_k$ is the lecturer who offers $p_j$.

In SPA-P, it turns out that a matching can be undermined not just by a student and lecturer acting together, but also by a group of students colluding. To this end, a *blocking coalition* with respect to a matching $M$ is defined to be a set of students $\{s_{i_0}, \ldots, s_{i_{r-1}}\}$, for some $r \geq 2$, each of whom is assigned in $M$, such that $s_{i_j}$ prefers $M(s_{i_{j+1}})$ to $M(s_{i_j})$ $(0 \leq j \leq r - 1$, where addition is taken modulo $r$).

A matching is defined to be *stable* if it admits no blocking pair and no blocking coalition. Some intuition for the stability definition is given in Ref. [422]. The following result indicates that a stable matching is guaranteed to exist and can be found in linear time.

Student preferences        Lecturer preferences
$s_1 : p_1 \ p_2$          $l_1 : p_1$                      $c_1 = d_1 = 1$
$s_2 : p_1$                $l_2 : p_2$                      $c_2 = d_2 = 1$

Fig. 5.10   A SPA-P instance due to Manlove and O'Malley [422]

**Theorem 5.26 ([422]).** *Let $I$ be an instance of* SPA-P. *Then $I$ admits a stable matching, which can be found in $O(m)$ time, where $m$ is the total length of the students' preference lists in $I$.*

It is not difficult to find a SPA-P instance where stable matchings might have different sizes. Consider the SPA-P instance $I$ shown in Fig. 5.10. It may be verified that each of the matchings $M_1 = \{(s_1, p_1)\}$ and $M_2 = \{(s_1, p_2), (s_2, p_1)\}$ is stable in $I$. In practical situations, often a key priority is to match as many students to acceptable projects as possible, so this naturally leads one to consider the complexity of MAX SPA-P, the problem of finding a maximum stable matching, given a SPA-P instance.

Manlove and O'Malley [422] showed that MAX SPA-P is NP-hard and not approximable within $\delta$, for some constant $\delta > 1$, unless P=NP. However the constant $\delta$ was very close to 1; Iwama *et al.* [346] gave a tighter lower bound on the approximability of MAX SPA-P, as indicated by the following result.

**Theorem 5.27 ([346]).** MAX SPA-P *is not approximable within $21/19 - \varepsilon$, for any $\varepsilon > 0$, unless P=NP.*

Iwama *et al.* [346] also showed that it is UGC-hard[12] to approximate MAX SPA-P within $4/3 - \varepsilon$, for any $\varepsilon > 0$, unless P=NP.

Regarding upper bounds for the approximability of MAX SPA-P, Manlove and O'Malley gave a straightforward 2-approximation algorithm for this problem. In fact the algorithm simply constructed an arbitrary stable matching; it was then shown that any two stable matchings differ in size by at most a factor of 2. Iwama *et al.* [346] modified this algorithm by applying a technique of Király (see Sec. 3.2.6.2), which amounts to giving rejected students a second chance to apply to projects on their preference list. This resulted in the following improved upper bound.

**Theorem 5.28 ([346]).** MAX SPA-P *is approximable within 3/2.*

---

[12]See Footnote 2 on Page 138.

O'Malley [470] considered a stronger form of stability for SPA-P than the version defined in this section for SPA-P. He referred to this stronger version of stability as *strong stability*. He showed that, for a given instance of SPA-P, a strongly stable matching need not exist. Moreover, he gave a linear-time algorithm to find a strongly stable matching or report that none exists.

### 5.5.4 *Lecturer preferences over student–project pairs*

Recall that SPA-(S,P) is the generalisation of each of SPA-S and SPA-P in which lecturers have preferences over student–project pairs. This means that a lecturer is able to express the fact that her preferences over students may depend on the particular project that they would undertake (e.g., they may feel that student $s_i$ is better-suited to project $p_{j_1}$ than project $p_{j_2}$). The study of SPA-(S,P) was first suggested by Abraham *et al.* [23], and Abu El-Atta and Moussa [29] were the first to define the problem formally.

To give a formal definition of SPA-(S,P), we require to define a set $B(l_k)$ of acceptable student–project pairs for each lecturer $l_k \in L$ as follows:

$$B(l_k) = \{(s_i, p_j) \in S \times P : p_j \in A(s_i) \cap P_k\}.$$

That is, $B(l_k)$ contains those student–project pairs $(s_i, p_j)$ such that $s_i$ finds acceptable a project $p_j$ that $l_k$ offers. With this definition, an instance of SPA-(S,P) is then defined in the same way as an instance of SPA-S: that is, each lecturer ranks $B(l_k)$ in strict order of preference.

The definitions of an assignment and a matching in $I$ are unchanged from the SPA-S case, and it turns out that the blocking pair definition is somewhat similar, except that a little more care is necessary to deal with the possibility that a student is trying to switch projects offered by the same lecturer. Given a matching $M$, in the SPA-(S,P) context we first need to redefine $M(l_k)$ as follows:

$$M(l_k) = \{(s_i, p_j) \in S \times P : s_i \in M(p_j) \wedge p_j \in P_k\}.$$

That is, $M(l_k)$ contains those student–project pairs $(s_i, p_j)$ such that $s_i$ is assigned to $p_j$ in $M$ and $l_k$ offers $p_j$. Abu El-Atta and Moussa [29] defined a student–project pair $(s_i, p_j) \in (S \times P) \backslash M$ to *block* a matching $M$, or to be a *blocking pair* of $M$, if:

(1) $p_j \in A(s_i)$ (i.e., $s_i$ finds $p_j$ acceptable);
(2) either $s_i$ is unassigned in $M$, or $s_i$ prefers $p_j$ to $M(s_i)$;
(3) either

(3.1) $p_j$ is undersubscribed, and *either*

(a) $s_i$ is assigned in $M$ and $M(s_i) \in P_k$, and $l_k$ prefers $(s_i, p_j)$ to $(s_i, M(s_i))$, *or*

(b) $s_i$ is unassigned in $M$ or $M(s_i) \notin P_k$, and $l_k$ is undersubscribed, *or*

(c) $s_i$ is unassigned in $M$ or $M(s_i) \notin P_k$, $l_k$ is full and $l_k$ prefers $(s_i, p_j)$ to some pair $(s_r, p_t) \in M(l_k)$;

(3.2) $p_j$ is full, and $l_k$ prefers $(s_i, p_j)$ to some pair $(s_r, p_j) \in M(l_k)$, and *either*

(a) $s_i$ is assigned in $M$ and $M(s_i) \in P_k$, and $l_k$ prefers $(s_i, p_j)$ to $(s_i, M(s_i))$, *or*

(b) $s_i$ is unassigned in $M$ or $M(s_i) \notin P_k$;

where $l_k$ is the lecturer who offers $p_j$.

A matching is *stable* if it admits no blocking pair.

A discussion of the stability definition now follows. Parts (1) and (2) are as before. Now consider Part 3.1, which corresponds to the case that $p_j$ is undersubscribed. In Part 3.1(a), $s_i$ is trying to change from one project $p_t$ to another project $p_j$, where both are offered by $l_k$. Although $p_j$ is undersubscribed, $l_k$ would only agree to the switch if she thinks that $s_i$ is better-suited to $p_j$ than $p_t$. The number of students assigned to $l_k$ would not change as a result. In Part 3.1(b), $s_i$ was not previously assigned to one of $l_k$'s projects, and since both $p_j$ and $l_k$ are undersubscribed, $l_k$ would accept $s_i$ to do $p_j$. In Part 3.1(c), again $s_i$ was not previously assigned to one of $l_k$'s projects, but this time $l_k$ is full, in which case $l_k$ would only agree to the switch if she improves by rejecting $s_r$ from $p_t$, for some student–project pair $(s_r, p_t)$ to which she prefers $(s_i, p_j)$.

We now consider Part 3.2, which corresponds to the case that $p_j$ is full. Lecturer $l_k$ would only allow $s_i$ to move to $p_j$ if $l_k$ thinks that $s_i$ is better-suited to project $p_j$ than some other student $s_r$ already assigned to $p_j$. In Part 3.2(a), $s_i$ is trying to switch projects offered by $l_k$, and as before $l_k$ would only allow that if she believes that $s_i$ is better-suited to $p_j$ than to $M(s_i)$. However in Part 3.2(b), $s_i$ was not previously assigned to one of $l_k$'s projects, and therefore $l_k$ would accept $s_i$ to do $p_j$.

By way of illustration, consider the SPA-(S,P) instance $I$ shown in Fig. 5.11. It may be verified that the matching $\{(s_i, p_i) : 1 \leq i \leq 4\}$ is stable in $I$.

Student preferences      Lecturer preferences

$s_1 : p_2 \; p_1$                 $l_1 : (s_1, p_1) \; (s_2, p_2) \; (s_1, p_2) \; (s_3, p_1) \; (s_3, p_2)$

$s_2 : p_2 \; p_3$                 $l_2 : (s_2, p_3) \; (s_3, p_3) \; (s_4, p_3) \; (s_4, p_4)$

$s_3 : p_1 \; p_2 \; p_3$

$s_4 : p_3 \; p_4$              $l_1$ offers $p_1, p_2; c_1 = 1, c_2 = 2, d_1 = 2$

                               $l_2$ offers $p_3, p_4; c_3 = 1, c_4 = 1, d_2 = 2$

Fig. 5.11    A SPA-(S,P) instance

Abu El-Atta and Moussa extended the student-oriented algorithm for SPA-S to the SPA-(S,P) setting, establishing the following result.

**Theorem 5.29 ([29]).** *Every instance of* SPA-(S,P) *admits a stable matching, and moreover such a matching can be found in* $O(m)$ *time, where* $m$ *is the total length of the students' preference lists.*

Data structures for representing SPA-(S,P) instances (together with the changes that may occur to these structures during an execution of the student-oriented algorithm in Ref. [29]) were discussed by Moussa and Abu El-Atta [456]. The authors also described a Java implementation that gives a visualisation of these data structures during an algorithm execution.

## 5.6  3D stable matching problems

One of Knuth's research problems [394] concerned whether it is possible to generalise SM to three sets of elements, which he referred to as men, women and dogs. This leads to three-dimensional, or tripartite, variants of SM, which we collectively refer to as Three-Dimensional Stable Marriage (3DSM) problems, and consider in Sec. 5.6.1. In these problems, the goal is to partition the sets of agents into triples (each triple consisting of a man, a woman and a dog) such that the set of triples is stable in some sense.

Different problem models can be obtained depending on the nature of the agents' preference lists. For example, each agent might rank in order of preference the pairs of other agents that they are prepared to form triples with. We consider this preference structure in Sec. 5.6.1.1, and the generalisation where ties are permitted in Sec. 5.6.1.2. In the presence of preference lists over pairs, the agents' lists might be based on a lexicographic ordering over the pairs – we consider this case in Sec. 5.6.1.3. Another possibility is that the agents' preference lists involve only individual agents (e.g., if men rank only women in order of preference, women's lists contain only dogs, and dogs rank only men). We consider this variant of 3DSM in Sec. 5.6.1.4.

Another direction is to consider three-dimensional generalisations of SR, which we collectively refer to as 3DSR problems, and study in Sec. 5.6.2. This time the aim is to partition the agents into disjoint unordered sets of size 3 (intuitively to share 3-bed rooms). Preference models can be defined similar to those indicated in the previous paragraph for 3DSM problems. Agents might rank unordered pairs in order of preference, indicating their rankings over potential roommates. Preferences of this nature are considered in Sec. 5.6.2.1, and the extension where ties are permitted is studied in Sec. 5.6.2.2. Alternatively, agents might rank only individual agents in order of preference — variants of 3DSR focusing on this model of preference are described in Sec. 5.6.2.3. In Sec. 5.6.2.4 we consider the case that the triples (in both a given matching and a potential blocking triple) are ordered, which gives rise to a variant of 3DSR that has applications to kidney exchange. Finally Sec. 5.6.2.5 focuses on a geometric variant of 3DSR.

### 5.6.1   *3D variants of* SM

#### 5.6.1.1   *Strictly-ordered preferences over pairs*

Ng and Hirschberg [465] defined the Three-Gender Stable Marriage Problem (3GSM) as follows. An instance $I$ comprises a set $U = \{m_1, \ldots, m_n\}$ of *men*, a set $W = \{w_1, \ldots, w_n\}$ of *women* and a set $D = \{d_1, \ldots, d_n\}$ of *dogs*. Define the *size* of $I$ to be $n$. Each man, woman and dog has a strict preference list over the pairs in $W \times D$, $U \times D$ and $U \times W$ respectively.

A *matching* $M$ in $I$ is a disjoint set of $n$ triples (i.e., each man, woman and dog appears in exactly one triple in $M$). Given an agent $a \in U \cup W \cup D$, we define $M[a]$ to be the triple containing $a$, and $M(a)$ to be the pair formed by removing $a$ from $M[a]$. Similarly, given a triple $t \in U \times W \times D$ and an agent $a$ in $t$, define $t(a)$ to be the pair formed by removing $a$ from $t$. A *blocking triple* is a man–woman–dog triple $t \in U \times W \times D$ such that each member $a$ of $t$ prefers $t(a)$ to $M(a)$. A matching is *stable* if it admits no blocking triple.

Ng and Hirschberg [465] gave an example 3GSM instance of size 2, illustrated in Fig. 5.12, that admits no stable matching: each of the four possible matchings admits a blocking triple, as shown in Table 5.1.

Earlier, Alkan [40] had also given an example to show that there are instances of 3GSM for which no stable matching exists. However in his

$$m_1 : \quad (w_1, d_2) \quad (w_1, d_1) \quad (w_2, d_2) \quad (w_2, d_1)$$
$$m_2 : \quad (w_2, d_2) \quad (w_1, d_1) \quad (w_2, d_1) \quad (w_1, d_2)$$
$$w_1 : \quad (m_2, d_1) \quad (m_1, d_2) \quad (m_1, d_1) \quad (m_2, d_2)$$
$$w_2 : \quad (m_2, d_1) \quad (m_1, d_1) \quad (m_2, d_2) \quad (m_1, d_2)$$
$$d_1 : \quad (m_1, w_2) \quad (m_1, w_1) \quad (m_2, w_1) \quad (m_2, w_2)$$
$$d_2 : \quad (m_1, w_1) \quad (m_2, w_2) \quad (m_1, w_2) \quad (m_2, w_1)$$

Fig. 5.12  A 3GSM instance with no stable matching due to Ng and Hirschberg [465]

Table 5.1  A blocking triple for each matching in the 3GSM instance shown in Fig. 5.12 due to Ng and Hirschberg [465]

| Matching | Blocking triple |
|---|---|
| $\{(m_1, w_1, d_1), (m_2, w_2, d_2)\}$ | $(m_1, w_1, d_2)$ |
| $\{(m_1, w_1, d_2), (m_2, w_2, d_1)\}$ | $(m_2, w_1, d_1)$ |
| $\{(m_1, w_2, d_1), (m_2, w_1, d_2)\}$ | $(m_1, w_1, d_2)$ |
| $\{(m_1, w_2, d_2), (m_2, w_1, d_1)\}$ | $(m_2, w_2, d_2)$ |

example, the agents' preference lists were incomplete (i.e., they did not contain all possible pairs of agents from the other two sets) and the instance was of size 3.

Ng and Hirschberg proved the following result, which indicates that when we move from a two-dimensional version of the stable marriage problem to a three-dimensional version, the problem changes from being polynomial-time solvable to NP-complete. In complexity theory, when a parameter inherently associated with a given problem changes from 2 to 3, the transition from polynomial-time solvability to NP-completeness is very prevalent.

**Theorem 5.30 ([465]).**  *The problem of deciding whether a given 3GSM instance has a stable matching is NP-complete.*

As noted in Sec. 2.3.1, Subramanian [551] independently proved Theorem 5.30.

Following comments by a referee of their paper, Ng and Hirschberg [465] observed that the preference lists in the 3GSM instance as constructed by the reduction in their proof of Theorem 5.30 are *inconsistent* in general. An example of an inconsistent preference list is that of dog $d_1$ in the 3GSM instance shown in Fig. 5.12. Here, $d_1$ prefers $w_2$ to $w_1$ when paired with $m_1$, but prefers $w_1$ to $w_2$ when paired with $m_2$. Thus dog $d_1$ does not have a consistent preference of $w_1$ over $w_2$ (or vice versa). Huang [287] observed that the preference lists in the 3GSM instance constructed by Subramanian's reduction [551] are not consistent in general either. Ng and Hirschberg [465] asked whether NP-completeness for 3GSM would still hold if preference lists did not contain such inconsistencies.

Formally, Huang [287] defined the notion of a *consistent* preference list for a man as follows. Suppose that man $m_i \in U$ has strictly-ordered preference lists (linear orders) $\prec_{m_i}^W$ and $\prec_{m_i}^D$ over all women and dogs respectively. Define $\prec_{m_i}$ to be the *product order* on $W \times D$, i.e., $(w_{j_1}, d_{k_1}) \prec_{m_i} (w_{j_2}, d_{k_2})$ if and only if either (i) $w_{j_1} \prec_{m_i}^W w_{j_2}$ and $d_{k_1} = d_{k_2}$, or (ii) $w_{j_1} = w_{j_2}$ and $d_{k_1} \prec_{m_i}^D d_{k_2}$, or (iii) $w_{j_1} \prec_{m_i}^W w_{j_2}$ and $d_{k_1} \prec_{m_i}^D d_{k_2}$. Then $m_i$'s overall preference list (i.e., his linear order over all pairs in $W \times D$) is *consistent* if it is a linear extension of the partial order $\prec_{m_i}$. The definitions of consistency for the overall preference lists of women and dogs are analogous.

Huang [287, 288] answered Ng and Hirschberg's open problem with the following result.

**Theorem 5.31 ([287, 288]).** *Given a 3GSM instance with consistent preference lists, the problem of deciding whether a stable matching exists is NP-complete.*

### 5.6.1.2  *Preferences over pairs with ties*

Huang [287] also defined a notion of consistency for preference lists over pairs that possibly involve ties. Suppose again that man $m_i \in U$ has strictly-ordered preference lists (linear orders) $\prec_{m_i}^W$ and $\prec_{m_i}^D$ over the women and dogs respectively. As in the previous subsection, let $\prec_{m_i}$ be the product order on $(W, \prec_{m_i}^W)$ and $(D, \prec_{m_i}^D)$. Then $m_i$'s overall preference list $\prec_{m_i}'$ over all pairs in $W \times D$ (now possibly involving ties) is *consistent* if it is a *relaxed linear extension* of $\prec_{m_i}$: that is, $(w_{j_1}, d_{k_1}) \prec_{m_i} (w_{j_2}, d_{k_2})$ implies that $m_i$ prefers $(w_{j_1}, d_{k_1})$ to $(w_{j_2}, d_{k_2})$ in $\prec_{m_i}'$, and two elements are tied in $\prec_{m_i}'$ only if they are incomparable in $\prec_{m_i}$. The definitions of consistency for the overall preference lists of women and dogs are analogous.

With ties in the agents' overall preference lists, Huang [287] defined four levels of stability, along the lines of Irving's definitions of weak stability, strong stability and super-stability for SMT [308] (see Sec. 1.3.5). As usual, these stability definitions are based on the absence of a blocking structure, in this case a blocking triple. In what follows, define the *degree* of a triple $t$ with respect to a matching $M$, denoted by $deg_M(t)$, to be the number of agents $a$ in $t$ who prefer $t(a)$ to $M(a)$. In each case, $t$ is a *blocking triple* if each agent $a$ in $t$ either prefers $t(a)$ to $M(a)$ or is indifferent between them, and in addition, a further constraint may be present on $deg_M(t)$, depending on the level of stability, as follows:

- *weakly stable matching*: $deg_M(t) \geq 3$;
- *strongly stable matching*: $deg_M(t) \geq 2$;
- *super-stable matching*: $deg_M(t) \geq 1$;
- *ultra-stable matching*: $deg_M(t) \geq 0$ (i.e., no constraint on $deg_M(t)$.

Notice that if the agents' overall preference lists are strictly ordered, then weak stability coincides with stability as defined by Ng and Hirschberg [465].

One way of generating consistent preference lists with ties, starting from preference lists for each agent over individual agents of the other two types, was suggested by Huang [287] as follows. Given a man $m_i \in U$ and two distinct pairs $(w_{j_1}, d_{k_1})$ and $(w_{j_2}, d_{k_2})$ in $W \times D$, if

$$rank(m_i, w_{j_1}) + rank(m_i, d_{k_1}) < rank(m_i, w_{j_2}) + rank(m_i, d_{k_2})$$

then $m_i$ prefers $(w_{j_1}, d_{k_1})$ to $(w_{j_2}, d_{k_2})$, and if

$$rank(m_i, w_{j_1}) + rank(m_i, d_{k_1}) = rank(m_i, w_{j_2}) + rank(m_i, d_{k_2})$$

then $m_i$ is indifferent between $(w_{j_1}, d_{k_1})$ and $(w_{j_2}, d_{k_2})$ (here it is assumed that $rank(m_i, w_{j_r})$ is defined relative to $\prec_{m_i}^W$ for $r \in \{1, 2\}$, and $rank(m_i, d_{j_r})$ is defined relative to $\prec_{m_i}^D$ for $r \in \{1, 2\}$). The construction of consistent preference lists with ties for women and dogs is analogous. Huang refers to instances of 3GSM built in this way as satisfying the *PON (Preference by Ordinal Number) scheme*. We will use 3GSM-PON to denote such instances.

Huang [287, 288] established the following complexity results for problems related to finding matchings satisfying the above-mentioned levels of stability in instances of 3GSM-PON.

**Theorem 5.32 ([287, 288]).** *Given an instance of 3GSM-PON, each of the problems of deciding whether there exists a matching that is strongly stable, super-stable or ultra-stable is NP-complete.*

Recall from Theorem 5.31 that the problem of deciding whether a weakly stable matching exists is NP-complete even if we are given a 3GSM instance with consistent and strictly-ordered preference lists.

### 5.6.1.3    *Lexicographic preferences over pairs of agents*

Danilov [160] discussed a very special case of 3GSM with strictly-ordered preferences over pairs in which men care primarily about women (and then dogs), women primarily about men (and then dogs), whilst the dogs' overall preference lists over pairs are not constrained in any way. More specifically, if a man $m_i \in U$ prefers $(w_j, d_{r_1})$ to $(w_k, d_{r_2})$, then $m_i$ prefers $(w_j, d_{r_3})$ to $(w_k, d_{r_4})$ for any dogs $d_{r_3}$ and $d_{r_4}$. This is equivalent to assuming that $m_i$ has a strict preference list $\prec_{m_i}^{W}$ over the individual women, with his overall list over pairs satisfying the property that, for two distinct women $w_j$ and $w_k$ in $W$, $m_i$ prefers $(w_j, d_{r_1})$ to $(w_k, d_{r_2})$ if and only if $w_j \prec_{m_i}^{W} w_k$, for any two dogs $d_{r_1}$ and $d_{r_2}$ in $D$. The constraints on the women's overall preference lists over pairs are defined analogously. Eriksson *et al.* [198] used the term *lexicographically acylic preferences* to describe this preference structure.

Danilov [160] showed that, for this restriction of 3GSM, a stable matching always exists and can be found in linear time using two applications of the Gale–Shapley algorithm. This indicates that there is at least one non-trivial variant of 3DSM that is solvable in polynomial time.

**Theorem 5.33 ([160]).**  *Let $I$ be an instance of 3GSM with lexicographically acyclic preferences, and let $n$ be the size of $I$. Then $I$ admits a stable matching, which can be found in $O(n^2)$ time.*

A natural follow-on from Theorem 5.33 is to consider what happens in the case of *lexicographically cyclic preferences*. Here, we are given a 3GSM instance in which men care primarily about women (and then dogs), women care primarily about dogs (and then men), and dogs care primarily about men (and then women). Boros *et al.* defined this problem and gave an example 3GSM instance of size 3 with lexicographically cyclic preferences having no stable matching. They did however show that every 3GSM instance of size 2 with lexicographically cyclic preferences admits a stable matching.

For 3GSM instances of size at least 3 having lexicographically cyclic preferences, it remains open as to whether the problem of finding a stable matching (or reporting that none exists) is solvable in polynomial time or NP-hard.

### 5.6.1.4    *Preferences over individual agents*

An intriguing variant of 3DSM, called the *Three-Dimensional Stable Marriage problem with Cyclic Preferences*(3DSM-CYC) was mentioned by Ng and Hirschberg [465], who attributed the problem to Knuth. An instance $I$ of this problem is as for 3GSM, but the difference lies in the fact that each agent ranks only individual agents from a certain set in order of preference. Specifically, men care only about women, women care only about dogs, and dogs care only about men. More formally, each man $m_i \in U$ ranks all the women in $W$ in strict order of preference, but he is indifferent as to which dog he is assigned with. (If we were to envisage $m_i$'s preference list as a ranking of woman–dog pairs, then it would effectively consist of $n$ ties, each of length $n$: all the pairs containing the same woman are tied.) The preference lists of each woman and dog are constructed analogously.

Given a triple $t$ and a man $m_i$ in $t$ in the 3DSM-CYC context, define $t(m_i)$ to be the woman $w_j \in W$ such that $t = (m_i, w_j, d_k)$ for some $d_k \in D$. Define $t(w_j)$ and $t(d_k)$ for a woman $w_j \in W$ and a dog $d_k \in D$ similarly. Given a matching $M$ and an agent $a \in U \cup W \cup D$, define $M(a)$ to be $t(a)$, where $t$ is the unique triple in $M$ containing $a$.

A *strongly blocking triple* relative to $M$ is a triple $t \in U \times W \times D$ such that each agent $a_i$ in $t$ prefers $t(a_i)$ to $M(a_i)$. A matching is *weakly stable* (also referred to as *stable* in the literature [109,198]) if it admits no strongly blocking triple.

A *weakly blocking triple* relative to $M$ is a triple $t \in U \times W \times D$ such that at least two agents $a_i$ in $t$ prefer $t(a_i)$ to $M(a_i)$, and the remaining agent $a_i$ in $t$ either prefers $t(a_i)$ to $M(a_i)$ or is indifferent between them (where preference lists are strictly ordered, the case of indifference is equivalent to the case that $t(a_i) = M(a_i)$). A matching is *strongly stable* if it admits no weakly blocking triple.

Boros *et al.* [109] proved that every instance of 3DSM-CYC of size at most 3 admits a weakly stable matching. Eriksson *et al.* [198] extended this result to instances of size 4, and conjectured that every instance of 3DSM-CYC admits a weakly stable matching. This problem is still open, though Biró and McDermid [94] showed that if preference lists are allowed to be incomplete, then a weakly stable matching need not exist.

Formally, define 3DSMI-CYC to be the variant of 3DSM-CYC in which agents may have unacceptable partners (i.e., each man ranks in order of preference his acceptable set of women, and the women and dogs do likewise with respect to the dogs and men respectively). In a 3DSMI-CYC instance $I$, a *matching* $M$ is a disjoint set of triples such that $M(a)$ is acceptable for $a$, for each agent $a$ who is assigned in $M$. The definitions of a strongly and a weakly blocking triple are extended from the 3DSM-CYC case by noting that each agent prefers to be assigned (to an acceptable partner) than to

$$m_1 : w_1 \ w_3 \ w_2 \qquad w_1 : d_1 \ d_3 \ d_2 \qquad d_1 : m_2 \ m_3 \ m_1$$
$$m_2 : w_2 \ w_1 \ w_3 \qquad w_2 : d_2 \ d_1 \ d_3 \qquad d_2 : m_1 \ m_2 \ m_3$$
$$m_3 : w_1 \ w_2 \ w_3 \qquad w_3 : d_2 \ d_1 \ d_3 \qquad d_3 : m_1 \ m_2 \ m_3$$

Fig. 5.13   A 3DSM-CYC instance with no strongly stable matching due to Irving [311]

Table 5.2   A blocking triple for a matching containing each possible triple involving $m_1$ in the 3DSM-CYC instance shown in Fig. 5.13

| Matching triple | Blocking triple | |
| --- | --- | --- |
| $(m_1, w_1, d_1)$ | $(m_3, w_1, d_1)$ | |
| $(m_1, w_1, d_2)$ | $(m_1, w_1, d_3)$ | |
| $(m_1, w_1, d_3)$ | $\begin{cases} (m_2, w_2, d_2) \\ (m_3, w_1, d_1) \end{cases}$ | if this is not a matching triple<br>otherwise |
| $(m_1, w_2, d_1)$ | $(m_1, w_1, d_1)$ | |
| $(m_1, w_2, d_2)$ | $(m_1, w_3, d_2)$ | |
| $(m_1, w_2, d_3)$ | $(m_1, w_2, d_2)$ | |
| $(m_1, w_3, d_1)$ | $(m_1, w_3, d_2)$ | |
| $(m_1, w_3, d_2)$ | $\begin{cases} (m_2, w_2, d_1) \\ (m_1, w_1, d_3) \end{cases}$ | if this is not a matching triple<br>otherwise |
| $(m_1, w_3, d_3)$ | $(m_1, w_3, d_2)$ | |

remain unassigned. Biró and McDermid [94] gave a 3DSMI-CYC instance of size 6 with no weakly stable matching. They also proved the following complexity result concerning weakly stable matchings in 3DSMI-CYC.

**Theorem 5.34 ([94]).** *The problem of deciding whether a given instance of 3DSMI-CYC admits a weakly stable matching is NP-complete.*

On the other hand, a strongly stable matching need not exist in a given instance of 3DSM-CYC. Irving [311] showed this via a 3DSM-CYC instance of size 3, illustrated in Fig. 5.13. For each possible triple $t$ containing $m_1$, a blocking triple for a matching containing $t$ is indicated in Table 5.2.

Biró and McDermid [94] gave the following NP-completeness result for strongly stable matchings in 3DSM-CYC.

**Theorem 5.35 ([94]).** *The problem of deciding whether a given instance of 3DSM-CYC admits a strongly stable matching is NP-complete.*

Huang [290] considered the case that individual preference lists in a given instance of 3DSM-CYC may involve ties — we denote by 3DSMT-CYC this generalisation of the problem. Recall the definition of the *degree* of a blocking triple from Sec. 5.6.1.2. This definition can also be applied to the 3DSMT-CYC context, and by using it we can also obtain the notions of a *weakly stable*, *strongly stable* and *super-stable* matching in a 3DSMT-CYC instance, as previously defined in Sec. 5.6.1.2. Huang proved the following result concerning the algorithmic complexity of computing super-stable matchings.

**Theorem 5.36 ([290]).** *The problem of deciding whether a given instance of* 3DSMT-CYC *admits a super-stable matching is NP-complete, even if each tie is of length at most 3 and occurs at the head of some agent's list.*

Huang [290] also proved a number of other results concerning weakly and strongly stable matchings in a given 3DSMT-CYC instance $I$, namely (i) the set of strongly stable matchings in $I$ is a union of distributive lattices, (ii) the numbers of weakly and strongly stable matchings in $I$ can be exponential in the size of $I$, and (iii) the problem of counting the number of strongly stable matchings in $I$ is #P-complete.

Cui and Jia [157] generalised the study of 3DSM-CYC to the case in which men, women and dogs can be multiply assigned up to some given capacity.

### 5.6.2  3D variants of SR

#### 5.6.2.1  Strictly-ordered preferences over pairs

Ng and Hirschberg [465] defined the Three-Person Stable Assignment Problem (3PSA) as a three-dimensional extension of SR. A 3PSA instance $I$ comprises a set $A = \{a_1, \ldots, a_n\}$ of *agents*, where $n = 3k$ for some $k \geq 1$. Define the *size* of $I$ to be $n$, and for some $S \subseteq A$ where $|S| \geq 2$, let $\mathbb{P}_2(S)$ denote the set of subsets of $S$ of size 2. Each agent $a_i \in A$ has a strict preference list over $\mathbb{P}_2(A \backslash \{a_i\})$.

A *matching* $M$ in $I$ is a partition of $A$ into $k$ triples[13]. Given an agent $a_i \in A$, we define $M[a_i]$ and $M(a_i)$ as in Sec. 5.6.1.1, except that these are now unordered sets rather than tuples. Similarly, given a triple $t \subseteq A$ and an agent $a_i \in t$, we define $t(a_i)$ as in Sec. 5.6.1.1, except that the pair is now unordered. A *blocking triple* relative to $M$ is a triple $t \subseteq A$ such that each $a_i \in t$ prefers $t(a_i)$ to $M(a_i)$. A matching is *stable* if it admits no blocking triple.

---

[13]For convenience, in this subsection we refer to a set of size 3 as a *triple*, even though the elements are unordered.

Ng and Hirschberg gave a straightforward reduction from 3GSM to 3PSA, showing that the problem of deciding whether a stable matching exists, given an instance of the latter problem, is NP-complete.

**Theorem 5.37 ([465]).** *The problem of deciding whether a given 3PSA instance has a stable matching is NP-complete.*

As in Sec. 5.6.1.1, the preference lists in the 3PSA instance as constructed by the reduction in Ng and Hirschberg's proof of Theorem 5.30 are inconsistent in general. Huang [287] established the hardness of 3PSA for consistent preference lists. Here, the definition of consistency is similar to that for the 3GSM case.

Formally, suppose that each agent $a_i \in A$ has a strictly-ordered preference list (linear order) $\prec_{a_i}$ over all agents in $A \backslash \{a_i\}$. Define the partial order $\prec'_{a_i}$ on $\mathbb{P}_2(A \backslash \{a_i\})$ as follows. Given $\{a_p, a_q\} \in \mathbb{P}_2(A \backslash \{a_i\})$ and $\{a_r, a_s\} \in \mathbb{P}_2(A \backslash \{a_i\})$, where, without loss of generality, $a_p \prec_{a_i} a_q$ and $a_r \prec_{a_i} a_s$, define $\{a_p, a_q\} \prec'_{a_i} \{a_r, a_s\}$ if and only if either (i) $a_p \prec_{a_i} a_r$ and $a_q = a_s$, or (ii) $a_p = a_r$ and $a_q \prec_{a_i} a_s$, or (iii) $a_p \prec_{a_i} a_r$ and $a_q \prec_{a_i} a_s$. Then $a_i$'s overall preference list (i.e., her linear order over $\mathbb{P}_2(A \backslash \{a_i\})$) is *consistent* if it is a linear extension of $\prec'_{a_i}$.

Huang [287, 288] proved the following counterpart of Theorem 5.31 for 3PSA with consistent preference lists.

**Theorem 5.38 ([287, 288]).** *Given a 3PSA instance with consistent preference lists, the problem of deciding whether a stable matching exists is NP-complete.*

### 5.6.2.2 *Preference lists over pairs with ties*

As in the 3GSM case, Huang [287] also defined a notion of consistency for preference lists over pairs that possibly involve ties in the case of 3PSA. Formally, suppose again that each agent $a_i \in A$ has a strictly-ordered preference list (linear order) $\prec_{a_i}$ over all agents in $A \backslash \{a_i\}$. Define the partial order $\prec'_{a_i}$ on $\mathbb{P}_2(A \backslash \{a_i\})$ as in the previous subsection. Then $a_i$'s overall preference list (i.e., her linear order $\prec''_{a_i}$ over $\mathbb{P}_2(A \backslash \{a_i\})$) is *consistent* if it is a *relaxed linear extension* of $\prec'_{a_i}$: that is, if $\{a_p, a_q\} \in \mathbb{P}_2(A \backslash \{a_i\})$ and $\{a_r, a_s\} \in \mathbb{P}_2(A \backslash \{a_i\})$, where, without loss of generality, $a_p \prec_{a_i} a_q$ and $a_r \prec_{a_i} a_s$, then $\{a_p, a_q\} \prec'_{a_i} \{a_r, a_s\}$ implies that $a_i$ prefers $\{a_p, a_q\}$ to $\{a_r, a_s\}$ in $\prec''_{a_i}$, and two elements are tied in $\prec''_{a_i}$ only if they are incomparable in $\prec'_{a_i}$.

As in Sec. 5.6.1.2, we can define four levels of stability (weak, strong, super- and ultra-stability) in the presence of ties in the agents' overall preference lists. Again, if the agents' overall preference lists are strictly ordered, then weak stability coincides with stability as defined by Ng and Hirschberg [465].

For each agent $a_i \in A$, starting from $\prec_{a_i}$, we can arrive at a consistent preference list with ties for $a_i$ over $\mathbb{P}_2(A\backslash\{a_i\})$ as in Sec. 5.6.1.2. That is, given $\{a_p, a_q\} \in \mathbb{P}_2(A\backslash\{a_i\})$ and $\{a_r, a_s\} \in \mathbb{P}_2(A\backslash\{a_i\})$, where, without loss of generality, $a_p \prec_{a_i} a_q$ and $a_r \prec_{a_i} a_s$, if

$$rank(a_i, a_p) + rank(a_i, a_q) < rank(a_i, a_r) + rank(a_i, a_s)$$

then $a_i$ prefers $\{a_p, a_q\}$ to $\{a_r, a_s\}$, and if

$$rank(a_i, a_p) + rank(a_i, a_q) = rank(a_i, a_r) + rank(a_i, a_s)$$

then $a_i$ is indifferent between $\{a_p, a_q\}$ and $\{a_r, a_s\}$ (where $rank(a_i, a_t)$ is defined relative to $\prec_{a_i}$ for $a_t \in A$). As in Sec. 5.6.1.2, we will use 3PSA-PON to denote 3PSA instances constructed in this way.

Huang [287, 288] established the following counterpart of Theorem 5.32.

**Theorem 5.39 ([287, 288]).** *Given an instance of 3PSA-PON, each of the problems of deciding whether there exists a matching that is strongly stable, super-stable or ultra-stable is NP-complete.*

Theorem 5.38 has already established NP-completeness for the problem of deciding whether a weakly stable matching exists is NP-complete in a given 3PSA instance with consistent and strictly-ordered preference lists.

### 5.6.2.3 *Preferences over individual agents*

Iwama *et al.* [337] defined another variant of 3DSR, called the *Stable Roommates problem with Triple Rooms* (SR-TR). An instance $I$ of this problem is as for 3PSA, except that each agent $a_i \in A$ has a strict preference list $\prec_{a_i}$ over all the individual agents in $A\backslash\{a_i\}$. Let $\preceq_{a_i}$ denote the reflexive closure of $\prec_{a_i}$. Given $\{a_p, a_q\} \in \mathbb{P}_2(A\backslash\{a_i\})$, where without loss of generality $a_i$ prefers $a_p$ to $a_q$, let $f_i(\{a_p, a_q\}) = a_p$ and let $s_i(\{a_p, a_q\}) = a_q$.

Given a matching $M$ in $I$, a triple $t \subseteq A$ is a *blocking triple* of $M$ if $t \notin M$, and for each each $a_i \in t$, $f_i(t(a_i)) \preceq_{a_i} f_i(M(a_i))$ and $s_i(t(a_i)) \preceq_{a_i} s_i(M(a_i))$. A matching is *stable* if it admits no blocking triple.

Iwama *et al.* proved the following result regarding the complexity of determining whether a SR-TR instance admits a stable matching.

$$a_1 : a_5 \quad a_6 \quad a_2 \quad a_3 \quad a_4$$
$$a_2 : a_3 \quad a_6 \quad a_4 \quad a_1 \quad a_5$$
$$a_3 : a_2 \quad a_6 \quad a_4 \quad a_1 \quad a_5$$
$$a_4 : a_2 \quad a_3 \quad a_1 \quad a_5 \quad a_6$$
$$a_5 : a_1 \quad a_6 \quad a_3 \quad a_4 \quad a_2$$
$$a_6 : a_1 \quad a_2 \quad a_3 \quad a_4 \quad a_5$$

Fig. 5.14   An instance of SR-TR

**Theorem 5.40 ([337]).** *Given an instance of* SR-TR, *the problem of deciding whether a stable matching exists is NP-complete.*

In Ref. [338], a maximisation variant of SR-TR is shown to be APX-hard.

It is tempting to believe that the problem of finding a stable matching in the context of SR-TR is equivalent to the problem of finding a weakly stable matching, given an instance of 3PSA-PON. However there is a subtle difference between the two problems, which can be illustrated by the SR-TR instance $I$ shown in Fig. 5.14. The preference lists over individual agents indicate that, in the corresponding 3PSA-PON instance $I'$, $a_6$ prefers $\{a_2, a_3\}$ to $\{a_1, a_5\}$, for example. With this observation in mind, it can be verified that $M = \{\{a_1, a_5, a_6\}, \{a_2, a_3, a_4\}\}$ is stable in $I$ but not weakly stable in $I'$, because $\{a_2, a_3, a_6\}$ is a blocking triple of $M$ in $I'$.

#### 5.6.2.4   *Three-Way Kidney Transplant*

Huang [290] considered a variant of 3DSR in which the agents correspond to patient–donor pairs[14] in a kidney exchange setting (see Sec. 1.4.6), and the triples (that either belong to a matching or constitute a blocking triple) are ordered. Moreover, a triple $(a_{i_0}, a_{i_1}, a_{i_2})$ represents the donation of a kidney from the donor in pair $a_{i_{r+1}}$ to the patient in pair $a_{i_r}$ (where $0 \leq r \leq 2$ and addition is taken modulo 3).

Formally, an instance $I$ of the *Three-Way Kidney Transplant* problem (3WKT) is the same as for SR-TR as defined in the previous subsection, except that triples in a matching and in a blocking triple are ordered. Let $A$ be the set of agents in $I$ and assume that $n = |A| = 3k$ for some $k \geq 1$. A *matching* $M$ in $I$ is a set of $k$ triples such that each agent appears in exactly one triple in $M$. Given a triple $t = (a_{i_0}, a_{i_1}, a_{i_2})$, we define $t(a_{i_r}) = a_{i_{r+1}}$ for each $r$ (where $0 \leq r \leq 2$), and addition is taken modulo 3), and we refer to $t(a_{i_r})$ as the *successor* of $a_{i_r}$. If $t \in M$, we let $M(a_{i_r})$ denote $t(a_{i_r})$.

---

[14]Note that in this subsection, usage of the term *pair* signifies a single agent, representing a patient–donor pair.

As already discussed, in matching $M$ the donor in pair $M(a_{i_r})$ donates a kidney to the patient in pair $a_{i_r}$. The preferences of $a_{i_r}$ over individual agents correspond to the patient in $a_{i_r}$'s compatibility with the donors in the other pairs. For example, following Huang [290], let $I$ be the instance of 3WKT involving three agents $a_0, a_1, a_2$, where $a_i$ prefers $a_{i-1}$ to $a_{i+1}$ ($0 \le i \le 2$ and arithmetic is taken modulo 3). Then, for example, the patient in $a_0$ prefers to receive a kidney from the donor in $a_2$ rather than the donor in pair $a_1$.

The concept of the *degree* of a blocking triple as defined in Sec. 5.6.1.2 can also be applied to the 3WKT context, and by using it we can also obtain the notions of a *weakly stable, strongly stable* and *super-stable* matching in a 3WKT instance, as previously defined in Sec. 5.6.1.2.

A curious phenomenon that arises as a result of the 3WKT problem definition is that it is possible for a matching $M$ in a given instance to be blocked by a triple of agents who already belong to a triple in $M$ (albeit in a different order). For example in the 3WKT instance $I$ defined above, the matching $M = \{(a_0, a_1, a_2)\}$ is blocked by the triple $(a_0, a_2, a_1)$ with respect to weak stability, strong stability or super-stability.

Huang proved the following results concerning the algorithmic complexity of computing stable matchings with respect to the three aforementioned stability criteria.

**Theorem 5.41 ([290]).** *Each of the problems of deciding whether a given 3WKT instance admits a weakly stable, strongly stable or super-stable matching is NP-complete.*

Huang [290] also proved that, given a 3WKT instance $I$, the numbers of weakly and strongly stable matchings in $I$ can be exponential in the size of $I$, and the problem of counting the number of strongly stable matchings in $I$ is #P-complete.

Biró and McDermid [94] defined the *b-way stable l-way exchange problem* ($b \ge 2$ and $l \ge 2$) as a generalisation of 3WKT in which a given instance is defined in the same way as for 3WKT, but now a matching can involve tuples of length $r$, where $2 \le r \le l$, and a blocking tuple can be of length $s$, where $2 \le s \le b$. In such a problem instance, Biró and McDermid [94] defined a matching to be *strongly stable* (corresponding to super-stability as defined in Sec. 5.6.1.2) if there is no blocking tuple $t$ (of length at most $b$) in which at least one agent $a_i$ in $t$ prefers $t(a_i)$ to $M(a_i)$, whilst no agent $a_j$ in $t$ prefers $M(a_j)$ to $t(a_j)$. A matching can also be defined to be *weakly stable* (referred to as *stable* in Ref. [94]) if there is no blocking tuple $t$ (of length at most $b$) in which each agent $a_i$ in $t$ prefers $t(a_i)$ to $M(a_i)$.

If $l = \infty$, the problem of finding a weakly stable matching is solvable in linear time using the Top Trading Cycles (TTC) algorithm [527]. Roth and Postlewaite [508] showed that in fact the TTC algorithm (see Sec. 6.2.1.2 and Sec. 6.2.1.4) yields a strongly stable matching.

If $l = 2$ and $b = 2$, we obtain SR (see Chap. 4). If $l = 2$ and $b > 2$, we obtain the *Cycle Stable Roommates problem* (see Sec. 5.8.2). In this setting the problem of determining whether a weakly stable matching exists is NP-complete in each of the cases that $b = \infty$ or $b = 3$ (see Sec. 5.8.2 for more details).

If $l = 3$ and $b = 3$, we return to 3WKT. As observed by Biró and McDermid, in this context, Theorem 5.34 implies that the problem of determining whether a weakly stable matching exists is NP-complete even for tripartite directed graphs (where the list of agents is partitioned into men, women and dogs). Also Theorem 5.35 implies that the same is true for strong stability. Both of these observations implicitly assume that we extend the problem definition to allow preference lists to be incomplete. Biró and McDermid noted that the complexity of determining whether a weakly stable or strongly stable matching exists is open in the case that $l = 3$ and $b = 2$.

### 5.6.2.5    *Geometric* 3DSR

Arkin *et al.* [50] considered a geometric variant of 3PSA, which could also be described as a three-dimensional variant of GEOMETRIC SR (see Sec. 4.7.4). An instance $I$ of GEOMETRIC 3DSR comprises a set $A$ of agents, where $n = |A| = 3k$ for some $k \geq 1$, and where the agents in $A$ correspond to points in $\mathbb{R}^d$, for some given $d \geq 1$. It is assumed that all agents are mutually acceptable, and we also define the notation $M(a_i)$, $t(a_i)$ and $\mathbb{P}_2(S)$, and the terms *triple* and *matching*, as in Sec. 5.6.2.1. We remark that the "three-dimensional" aspect of GEOMETRIC 3DSR refers to the size of the triples, rather than to the particular value of $d$.

Given an agent $a_i \in A$ and two sets $\{a_p, a_q\}$ and $\{a_r, a_s\}$ in $\mathbb{P}_2(A \backslash \{a_i\})$, we say that $a_i$ *prefers* $\{a_p, a_q\}$ and $\{a_r, a_s\}$ if

$$||a_i - a_p||_d + ||a_i - a_q||_d < ||a_i - a_r||_d + ||a_i - a_s||_d$$

and we say that $a_i$ is *indifferent between* the two pairs if equality holds. A matching $M$ is *weakly stable* if there is no triple of agents $t$ such that $a_i$ prefers $t(a_i)$ to $M(a_i)$ for each $a_i \in t$. Arkin *et al.* [50] gave an example GEOMETRIC 3DSR instance with 12 agents that does not admit a weakly stable matching. They also remarked that, in general, the complexity of the problem of deciding whether a weakly stable matching exists, given a GEOMETRIC 3DSR instance, is open. Arkin *et al.* [50] also generalised the concept of an $\alpha$-stable matching (see Sec. 2.10.10) to the GEOMETRIC 3DSR case, presenting some associated algorithmic results.

## 5.7 Exchange-stable matching problems

In this section we consider criteria for matchings that are based on the absence of pairs or coalitions of agents who can swap partners so as to improve relative to their existing assignees. We study *exchange-stability* in Section 5.7.1, which corresponds to the absence of a pair of agents who envy each other's partners. Section 5.7.2 deals with the case where the coalitions of agents who can swap partners so as to improve may be of size greater than two. Finally in Section 5.7.3 we outline results for the case that the matching must be stable (in the classical sense) in addition to being resistant against coalitions of agents who envy one another's partners.

### 5.7.1 *Exchange-stability as a solution concept*

Alcalde [35] defined an alternative notion of stability, so-called *exchange-stability*, in the context of an instance $I$ of SR. He defined a matching $M$ in $I$ to be *exchange-stable* if $M$ admits no *exchange-blocking pair*, which is a pair of agents $\{a_i, a_j\}$, each of whom prefers the other's partner in $M$ to their own. That is, $a_i$ prefers $M(a_j)$ to $M(a_i)$, and $a_j$ prefers $M(a_i)$ to $M(a_j)$, so $a_i$ and $a_j$ would prefer to swap partners than remain with their existing ones. As in the case of classical stability, the preferences of $M(a_i)$ and $M(a_j)$ are not considered as far as the potential swap is concerned.

Alcalde showed that exchange-stability and classical stability are independent notions, i.e., neither criterion implies the other. Indeed, he constructed an SR instance $I_1$, shown in Fig. 5.15, that admits a stable matching (namely $\{\{a_1, a_2\}, \{a_3, a_4\}\}$) but no exchange-stable matching, and an SR instance $I_2$, shown in Fig. 5.16, that admits an exchange-stable matching (namely $\{\{a_1, a_3\}, \{a_2, a_4\}\}$) but no stable matching.

Alcalde argued that, in situations when participants have "property rights", exchange-stability could be more appropriate than classical stability. For example, in the context of assigning $2n$ students to $n$ two-bed rooms, an individual's property would be the bed that she occupies. A

$$a_1 : a_2 \ a_3 \ a_4$$
$$a_2 : a_4 \ a_1 \ a_3$$
$$a_3 : a_1 \ a_4 \ a_2$$
$$a_4 : a_3 \ a_2 \ a_1$$

$$a_1 : a_2 \ a_3 \ a_4$$
$$a_2 : a_3 \ a_1 \ a_4$$
$$a_3 : a_1 \ a_2 \ a_4$$
$$a_4 : a_1 \ a_2 \ a_3$$

Fig. 5.15 Instance $I_1$ of SR due to Alcalde [35]

Fig. 5.16 Instance $I_2$ of SR due to Alcalde [35]

Men's preferences
$$m_1 : w_1 \ w_2$$
$$m_2 : w_2 \ w_1$$

Women's preferences
$$w_1 : m_2 \ m_1$$
$$w_2 : m_1 \ m_2$$

Fig. 5.17   An instance of SM with no exchange-stable matching due to Cechlárová [118]

blocking pair $\{a_i, a_j\}$ in the classical sense could not lead to any disruption of the matching in practice, since there is no extra room for $a_i$ and $a_j$ to occupy, and moreover each of the partners of $a_i$ and $a_j$ in the matching would exercise their property rights by refusing to give up their bed in order to make a room available. However an exchange-blocking pair $\{a_i, a_j\}$ would in practice lead $a_i$ and $a_j$ to simply swap beds.

The problem of deciding whether an SR instance admits an exchange-stable matching was shown to be NP-complete by Cechlárová [118], however in the problem instances she constructed, the preference lists were both incomplete and *inconsistent* (i.e., there were agents $a_i, a_j$ such that $a_i$ found $a_j$ acceptable but not vice versa). Later, Cechlárová and Manlove [130] showed that NP-completeness holds for instances of SR (with complete preference lists).

**Theorem 5.42 ([130]).** *The problem of determining whether a given* SR *instance admits an exchange-stable matching is NP-complete.*

The concept of exchange-stability may be applied to an instance $I$ of SM: a matching $M$ in $I$ is *exchange-stable* if there are no two agents of the same sex, both assigned in $M$, each of whom prefers the other's partner to his/her own partner in $M$. Cechlárová [118] gave an example SM instance, shown in Fig. 5.17, that admits no exchange-stable matching. Cechlárová and Manlove [130] proved the following result.

**Theorem 5.43 ([130]).** *The problem of determining whether a given* SM *instance admits an exchange-stable matching is NP-complete.*

Ref. [101] considers exchange-stable matchings in a matching market where agents express preferences via utility functions rather than via traditional ordinal preferences.

### 5.7.2  Exchange-blocking coalitions

In the SM context, it is also possible to consider a weaker form of exchange-stability, where exchange-blocking pairs are only permitted to contain two men (or analogously, two women). A matching $M$ in an SM instance $I$ is defined to be *man-exchange-stable* if there is no exchange-blocking pair involving two men (a *woman-exchange-stable* matching may be defined similarly).

A group of agents who wish to swap partners so as to improve their allocation may involve more than just a pair of agents, of course. More generally, an *exchange-blocking coalition* is a sequence of agents $\langle a_0, a_1, \ldots, a_{k-1} \rangle$, for some $k \geq 2$, such that, for each $i$ $(0 \leq i \leq k-1)$, $a_i$ prefers $M(a_{i+1})$ to $M(a_i)$, where addition is taken modulo $k$. An exchange-blocking coalition involving only men (respectively women) is a *man-exchange-blocking coalition* (*woman-exchange-blocking coalition*). A matching is *coalition-exchange-stable* if it admits no exchange-blocking coalition. The definitions of *man-coalition-exchange-stable* and *woman-coalition-exchange-stable* are analogous. Indeed, a matching is man-coalition-exchange stable if and only if it is Pareto optimal for the men.

The notion of exchange-stability for a given matching $M$ in an SMI instance $I$ is well-defined if we require that $M$ is a perfect matching in the underlying graph of $I$; if a perfect matching does not exist then we simply say that $I$ admits no exchange-stable matching. Similar remarks apply to the other variants of exchange-stability introduced in the preceding paragraph.

Cechlárová and Manlove [130] established the following results regarding exchange-blocking coalitions in SMI instances.

**Theorem 5.44 ([130]).** *Let $I$ be an instance of* SMI, *where $n$ is the size of $I$ and $m$ is the number of acceptable pairs in $I$. Then*

    *(i) a matching can be tested for coalition-exchange-stability in $O(m)$ time, and the same bound holds for man/woman-coalition-exchange-stability;*

    *(ii) we can find a man-coalition-exchange-stable matching in $I$ or report that none exists in $O(\sqrt{n}m)$ time;*

|  Men's preferences  |  Women's preferences  |
|---|---|

$$m_1 : w_1 \quad w_4 \quad w_2 \quad w_3 \qquad\qquad w_1 : m_4 \quad m_1 \quad m_2 \quad m_3$$
$$m_2 : w_3 \quad w_2 \quad w_4 \quad w_1 \qquad\qquad w_2 : m_1 \quad m_2 \quad m_4 \quad m_3$$
$$m_3 : w_2 \quad w_3 \quad w_4 \quad w_1 \qquad\qquad w_3 : m_3 \quad m_2 \quad m_1 \quad m_4$$
$$m_4 : w_3 \quad w_2 \quad w_4 \quad w_1 \qquad\qquad w_4 : m_2 \quad m_4 \quad m_3 \quad m_1$$

Fig. 5.18    An instance of SM with no stable matching that is man-exchange-stable due to Irving [316].

(iii) *if $I$ is an instance of* SM, *then $I$ admits a man-coalition-exchange-stable matching, and such a matching can be found in $O(n^2)$ time;*

(iv) *if preference lists in $I$ are allowed to be inconsistent, the problem of deciding whether a man-exchange-stable matching in $I$ exists is NP-complete.*

With respect to the above theorem, Part (i) may be established with the aid of the *envy graph*. This digraph $D_M$ is defined relative to a given matching $M$, and contains a vertex for each agent who is assigned in $M$, and an arc from agent $a_i$ to agent $a_j$ if $a_i$ prefers $M(a_j)$ to $M(a_i)$. Clearly $M$ is coalition-exchange-stable if and only if $D_M$ is acyclic, and $M$ is man-coalition-exchange-stable (respectively woman-coalition-exchange-stable) if and only if $D_M$ admits no cycle involving only men (women). Part (ii) follows by using the algorithm for constructing a maximum Pareto optimal matching in the HA instance obtained from $I$ by ignoring the women's preferences. Also the algorithm for Part (iii) is the Random Serial Dictatorship Mechanism. See Chapter 6 for more details about both of these algorithms.

Cechlárová and Manlove reported that the complexity of the problem of finding a coalition-exchange-stable matching or reporting that none exists, for a given SM instance, is open, though conjecture that the problem is NP-hard.

### 5.7.3    Stable matchings that are exchange-stable

Cechlárová and Manlove [130], Irving [316] and McDermid *et al.* [440] also considered matchings that are both stable and coalition-exchange-stable (implicitly the definition of an exchange-blocking coalition is extended in this case so that it involves only agents who are assigned relative to the matching in question). It turns out that an SM instance need not admit a stable matching that is even man-exchange-stable. Consider the SM instance $I$, due to Irving [316], illustrated in Fig. 5.18. There are two stable matchings in $I$, namely the man-optimal and woman optimal stable matchings $M_a$ and $M_z$ respectively, where

$$M_a = \{(m_1, w_1), (m_2, w_2), (m_3, w_3), (m_4, w_4)\}$$
$$M_z = \{(m_1, w_2), (m_2, w_4), (m_3, w_3), (m_4, w_1)\}$$

$M_a$ admits the man-exchange-blocking pair $\{m_2, m_3\}$, whilst $M_z$ admits the man-exchange-blocking pairs $\{m_1, m_2\}$ and $\{m_1, m_4\}$.

Cechlárová and Manlove [130] observed that if a stable matching $M$ is to be man-coalition-exchange-stable then it must be the man-optimal stable matching. For, if $M$ is not the man-optimal stable matching then there is some stable matching $M'$ and rotation $\rho = (m_0, w_0), (m_1, w_1), \ldots, (m_{r-1}, w_{r-1})$ exposed in $M'$ such that $M = M'/\rho$. But then $\langle m_{r-1}, \ldots, m_1, m_0 \rangle$ is a man-exchange-blocking coalition of $M$, a contradiction. A similar observation holds for woman-coalition-exchange-stability, and hence a necessary condition for the existence of a stable matching $M$ that is coalition-exchange-stable is that $M$ is the unique stable matching. We have already seen that the envy graph can be used to test $M$ for coalition-exchange-stability.

In the SRI case, it turns out that there is also a strong necessary condition for the existence of a stable matching that is coalition-exchange-stable. Consider an execution of Irving's algorithm [306] as applied to a given solvable SRI instance $I$. Suppose that Phase 2 of the algorithm is executed, terminating with stable matching $M$. This phase involves the elimination of one or more rotations. Let $\rho$ be the final rotation to be eliminated. Then $\rho = (x_0, y_0), (x_1, y_1), \ldots, (x_{r-1}, y_{r-1})$ for some $r \geq 2$. Hence $\langle x_{r-1}, \ldots, x_1, x_0 \rangle$ is an exchange-blocking coalition of $M$. It follows that a necessary condition for $I$ to admit a stable matching that is coalition-exchange-stable is that Phase 1 of Irving's algorithm terminates with a stable matching $M$ (which is therefore unique). To check that $M$ is coalition-exchange-stable, the envy graph $D_M$ for $M$ (whose definition is analogous to that given above for the SMI case) may again be used.

The following result summarises the discussion in the preceding two paragraphs.

**Theorem 5.45 ([130]).** *Let $I$ be an instance of* SRI. *We can find a stable matching that is coalition-exchange-stable or report that none exists in* $O(m)$ *time, where $m$ is the number of acceptable pairs of agents. The same is true if $I$ is an instance of* SMI, *and also if the stable matching is required to be man/woman-coalition-exchange-stable.*

Irving [316] considered the case that stable matchings are required to be man-exchange-stable (i.e., there must be no man-exchange-blocking coalitions of size 2). This problem was motivated by the fact that, in a previous run of SFAS (the Scottish medical matching scheme for allocating junior doctors to hospital posts), two participants discovered that, were they to exchange their allocated hospitals, they would both be better off [310]. Of course, the stability of the matching ensured that the hospitals would not agree to the switch, since they would each be worse off, were the residents to swap. However this incident nevertheless led to some feelings of dissatisfaction among the participants involved, and raised the question as to whether it would be possible to find an efficient algorithm to construct a stable matching that admits no man-exchange-blocking pair, or report that no such matching exists. Irving showed that this is unlikely in general, though there is hope if the men's preference lists are short. Irving also showed that the problem is hard in general if the stable matching is required to be exchange-stable.

**Theorem 5.46 ([316]).**

*(i) The problem of deciding whether a given* SM *instance admits a stable matching that is man-exchange-stable is NP-complete. NP-completeness also holds if the stable matching is instead required to be exchange-stable.*

*(ii) Given an instance of* SMI *of size n in which the men's preference lists are of length at most 3, there is an $O(n)$ algorithm that finds a stable matching that is man-exchange-stable, or reports that no such matching exists.*

There is a straightforward reduction from an SM instance $I$ to an SR instance $J$ such that the stable matchings in $I$ are in 1–1 correspondence with the stable matchings in $J$ [261, Lemma 4.1.1]. The same reduction yields a similar correspondence for exchange-stable matchings [130, Lemma 3.1]. Putting these two observations together with Theorem 5.46, we obtain the following corollary.

**Corollary 5.47 ([316]).** *The problem of deciding whether a given* SR *instance admits a stable matching that is exchange-stable is NP-complete.*

Irving conjectured that the problem of deciding whether an SMI instance admits a stable matching that is man-exchange-stable is NP-complete, even if the men's lists are of length at most 4. This was shown to be true by McDermid *et al.*, using an adaptation of Irving's reduction from Ref. [316], with NP-completeness holding even when the women's lists are also of bounded length.

**Theorem 5.48 ([440]).** *Let $k$ and $k'$ be two integers where $k \geq 4$ and $k' \geq 5$. Let $I$ be an SMI instance where the men's lists are of length at most $k$ and the women's lists are of length at most $k'$. Then the problem of deciding whether $I$ admits a stable matching that is man-exchange-stable is NP-complete.*

McDermid *et al.* remarked that if the preference lists in their constructed SMI instances are allowed to be inconsistent, then the lower bound for $k'$ in the context of Theorem 5.48 improves to 3.

For SMI instances that do not admit a stable matching that is man-exchange-stable, a natural alternative is to seek a stable matching that has the smallest number of man-exchange-blocking pairs. Unfortunately, however, McDermid *et al.* proved that this problem is also NP-hard even in a highly restricted setting.

**Theorem 5.49 ([440]).** *Let $k$ and $k'$ be two (fixed) integers where $k \geq 3$ and $k' \geq 3$. Let $I$ be an SMI instance where the men's lists are of length at most $k$ and the women's lists are of length at most $k'$. Then the problem of deciding whether $I$ admits a stable matching that admits at most $K$ man-exchange-blocking pairs, for some (non-fixed) integer $K \geq 0$, is NP-complete.*

Again, McDermid *et al.* remarked that if the preference lists in their constructed SMI instances are allowed to be inconsistent, then the lower bound for $k'$ in the context of Theorem 5.49 improves to 2.

## 5.8 Two additional stable matching problems

### 5.8.1 *Bistable matching problems*

Given an instance $I$ of SM or SR, let $\hat{I}$ denote the instance obtained by reversing each agent's preference list in $I$. A matching in $I$ is *bistable* if it is stable in both $I$ and $\hat{I}$. The notion of bistability was introduced by Weems in SM and SR [585]. The concept is interesting from a theoretical standpoint, but the practical motivation for the definition is less clear!

Men's preferences        Women's preferences
$m_1 : w_1 \ \ w_2$                $w_1 : m_1 \ \ m_2$
$m_2 : w_2 \ \ w_1$                $w_2 : m_2 \ \ m_1$

Fig. 5.19    An instance of SM with no bistable matching

We firstly remark that an SM instance need not admit a bistable matching, as illustrated by the instance in Fig. 5.19. On the other hand, Weems [585] observed that the SM instance of size 4 shown in Fig. 2.1 satisfies the property that each of the 10 stable matchings is bistable. This follows from the fact that for each $i$ ($1 \leq i \leq 4$), the reverse of $m_i$'s preference list is precisely $w_i$'s preference list (with $w$ replaced by $m$ for each element of $m_i$'s preference list).

Weems [585] described a simple extension of the MEGS algorithm for finding a man-optimal bistable matching in a given SM instance $I$. Since he did not present the algorithm explicitly in pseudocode form, we give such a description in the form of Algorithm Bistable shown in Algorithm 5.2. In what follows we assume that $U$ and $W$ are the sets of men and women in $I$ respectively, and we define a *reverse blocking pair* of a matching $M$ in $I$ to be a blocking pair of $M$ in $\hat{I}$. Clearly $M$ is bistable if and only if it admits no blocking pair and no reverse blocking pair in $I$.

Algorithm Bistable deletes entries from the preference lists as per the MEGS algorithm for SM [261, Sec. 1.2.4]: if a man $m_i$ proposes to a woman $w_j$ then, as usual, we delete each pair $(m_k, w_j)$ such that $w_j$ prefers $m_i$ to $m_k$ (recall that *delete the pair* $(m_k, w_j)$ means deleting $m_k$ from $w_j$'s list and vice versa). The new addition is that, for each such man $m_k$, we now delete each pair $(m_k, w_l)$ such that $m_k$ prefers $w_l$ to $w_j$. The reasoning is that such a pair $(m_k, w_l)$ could never belong to a bistable matching $M'$. For, suppose otherwise. In any bistable matching, $w_j$ must obtain a partner who is no worse than $m_i$. Thus $(m_k, w_j)$ is a reverse blocking pair of $M'$.

It follows that the algorithm never deletes a *bistable pair* (i.e., a pair that belongs to some bistable matching), and hence the algorithm correctly reports that no bistable matching exists if it reaches line 15. On the other hand, if the algorithm returns some matching $M$ then it follows from the correctness of the MEGS algorithm that $M$ admits no blocking pair. Moreover $M$ admits no reverse blocking pair $(m_i, w_j)$, for if $w_j$ prefers $M(w_j)$ to $m_i$, then the pair $(m_i, w_j)$ is deleted along with each pair $(m_i, w_k)$ such that $m_i$ prefers $w_k$ to $w_j$, and hence $m_i$ prefers $w_j$ to $M(m_i)$.

We summarise this discussion with the following theorem.

---

**Algorithm 5.2** Algorithm Bistable [585]

**Require:** SM instnance $I$
**Ensure:** return the man-optimal bistable matching $M$ in $I$, or "no bistable matching exists"

1: $M := \emptyset$;
2: **while** some man $m_i \in U$ is unassigned in $M$ **do**
3:     $w_j :=$ most-preferred woman on $m_i$'s list;     $\{m_i$ proposes to $w_j\}$
4:     **if** $w_j$ is assigned in $M$ **then**
5:        $M := M\backslash\{(M(w_j), w_j)\}$;
6:     **end if**
7:     $M := M \cup \{(m_i, w_j)\}$;
8:     **for each** successor $m_k$ of $m_i$ on $w_j$'s list **do**
9:        delete the pair $(m_k, w_j)$;
10:       **for each** predecessor $w_l$ of $w_j$ on $m_k$'s list **do**
11:          delete the pair $(m_k, w_l)$;
12:       **end for**
13:     **end for**
14:     **if** some preference list is empty **then**
15:        **return** "no bistable matching exists";
16:     **end if**
17: **end while**
18: **return** $M$;

---

**Theorem 5.50 ([585]).** *Given an* SM *instance $I$ of size $n$, Algorithm* Bistable *returns the unique man-optimal bistable matching in $I$, or else reports that no bistable matching in $I$ exists. The complexity of Algorithm* Bistable *is $O(n^2)$.*

Weems [585] also extended the LP-based formulation of SM as given by Vande Vate (see Sec. 2.4) to the bistability case. Further, he showed that the set of bistable matchings in a given SM instance $I$ forms a distributive lattice. Finally, he obtained a characterisation of bistable matchings based on closed subsets of the *bistable permutation order*: here, each bistable permutation is the composition of a sequence of rotations.

Sethuraman and Teo [525] also considered bistability in the context of SM. They gave a simple equivalent definition of bistability, namely, a matching $M$ is bistable if and only if, for each pair $(m_i, w_j) \in (U \times W)\backslash M$, exactly one of $m_i$ and $w_j$ prefers the other to their partner in $M$. They gave an LP-based characterisation of bistable matchings in a given SM instance $I$, proving that its polytope is the convex hull of the set of bistable matchings in $I$. The authors also proved a direct analogue of Theorem 2.9, concerning generalised median stable matchings, for the case of bistable matchings.

We now turn to bistable matchings in the case of SR. Weems [584] described a characterisation of bistable matchings in a given SR instance $I$ in terms of SAT clauses — these clauses essentially extend the 2-SAT characterisation of SR [261, pp.194–195] that encodes the rotation poset in $I$ for classical stability (see Sec. 4.2.4). Further clauses are present that encode so-called *backward rotations*, which must also be considered in addition to classical (*forward*) rotations in order to enforce bistability. The difficulty is that some clauses in Weems' characterisation ended up being of length 3, which led him to speculate as to whether the problem of finding a bistable matching in $I$ is NP-hard.

Sethuraman and Teo [525] showed that this is in fact not the case. They gave an LP-based characterisation of bistable matchings in a given SR instance $I$, extending their earlier LP formulation for stable matchings in $I$ [565] (see Sec. 4.2.6). They also proved a direct analogue of Theorem 4.16 (concerning generalised median stable matchings in SR) for bistable matchings.[15]

### 5.8.2   *The Cycle Stable Roommates problem*

Irving [312] considered a variant of SRI in which we seek a matching $M$ that admits no *blocking cycle*. This is a coalition of agents $\langle a_0, a_1, \ldots, a_{k-1} \rangle$, for some $k \geq 2$, such that, for each $i$ ($0 \leq i \leq k-1$), either (i) $a_i$ is unassigned in $M$ and finds $a_{i+1}$ acceptable, or (ii) $a_i$ prefers $a_{i+1}$ to $M(a_i)$, where addition is taken modulo $k$. He defined a matching to be *cycle stable* if it admits no blocking cycle. Since a blocking pair corresponds to a blocking cycle of length 2, a cycle stable matching is stable in the classical sense.

The motivation for considering this problem comes from kidney exchange. Recall from Sec. 1.4.6 that the problem of constructing kidney exchanges between patients with willing but incompatible donors can be modelled via SRI. In practice, pairwise kidney exchanges (involving two patient–donor pairs) are the most likely type of exchange to proceed, however longer cycles (in which more than two patient–donor pairs exchange kidneys) are possible. In general, cycles need to be as short as possible as all operations must take place simultaneously.

Hence one could ask, for example, whether there is a stable matching in a given SRI instance that admits no short blocking cycle, say of length

---

[15]In fact, Theorems 3.2 and 3.3 in Ref. [525], which are intended to establish the analogue of Theorem 4.16 for bistable matchings, are erroneously stated in terms of stability rather than bistability. However the results do indeed hold in the case of bistability [524].

$$a_1 : a_3 \ a_5 \ a_2 \ a_6 \ a_4 \qquad a_1 : a_3 \ a_2 \ a_5 \ a_4 \ a_6$$
$$a_2 : a_1 \ a_4 \ a_6 \ a_3 \ a_5 \qquad a_2 : a_1 \ a_4 \ a_5 \ a_6 \ a_3$$
$$a_3 : a_6 \ a_2 \ a_5 \ a_1 \ a_4 \qquad a_3 : a_5 \ a_4 \ a_2 \ a_6 \ a_1$$
$$a_4 : a_2 \ a_6 \ a_5 \ a_3 \ a_1 \qquad a_4 : a_3 \ a_2 \ a_6 \ a_1 \ a_5$$
$$a_5 : a_6 \ a_3 \ a_1 \ a_4 \ a_2 \qquad a_5 : a_1 \ a_6 \ a_4 \ a_3 \ a_2$$
$$a_6 : a_1 \ a_4 \ a_3 \ a_2 \ a_5 \qquad a_6 : a_5 \ a_4 \ a_2 \ a_1 \ a_3$$

Fig. 5.20  Instances $I_1$ and $I_2$ of SR due to Irving [312].

at most 3. This is equivalent to asking for a matching, which corresponds to a set of pairwise exchanges, that admits no blocking cycle of length at most 3. The fact that a solution itself can contain only pairs and not 3-cycles, and yet must be resistant to triples of agents who could improve by swapping kidneys among themselves, could be seen as a shortcoming of the model insofar as the kidney exchange application is concerned. (Were the model to allow triples in a solution, we would be in the realm of a coalition formation game — see Sec. 4.8.8 for more details.) Nevertheless it is arguable that the problem is interesting in its own right.

**Example 5.51 ([312]).** *Consider the* SR *instances $I_1$ and $I_2$, shown in Fig. 5.20, due to Irving [312]. Matching $M_1 = \{\{a_1, a_2\}, \{a_3, a_5\}, \{a_4, a_6\}\}$ is stable (in the classical sense) but not cycle stable because of the blocking cycle $\langle a_1, a_5, a_6 \rangle$. However matching $M_2 = \{\{a_1, a_5\}, \{a_2, a_4\}, \{a_3, a_6\}\}$ is cycle stable in $I_1$. Instance $I_2$ admits a stable matching $M_3 = \{a_1, a_2\}, \{a_3, a_4\}, \{a_5, a_6\}\}$ which is the unique stable matching in $I_2$. However $M_3$ admits the blocking cycle $\langle a_1, a_3, a_5 \rangle$, and hence $I_2$ has no cycle stable matching.*

Clearly a matching $M$ in a given SRI instance may be tested for cycle stability with the aid of the following digraph $D'_M$ (a modification of the envy graph defined in Sec. 5.7.2). The vertex set of $D'_M$ is the set of agents in $I$, and an arc $(a_i, a_j)$ is in $D'_M$ if either (i) $a_i$ is unassigned in $M$ and finds $a_j$ acceptable, or (ii) $a_i$ prefers $a_j$ to $M(a_i)$. It follows easily that $M$ is cycle stable if and only if $D'_M$ is acyclic.

The main result of Irving's paper [312] is the following.

**Theorem 5.52 ([312]).** *The problem of deciding whether a given* SR *instance admits a cycle stable matching is NP-complete. The result also holds even if we insist that the length of a blocking cycle must be at most 3.*

Thus Irving's NP-completeness result holds for the restricted version of the problem that is motivated by the kidney exchange application. An obvious open problem is whether NP-completeness still holds for SRI with bounded preference lists, either for the case that blocking cycles may be of unbounded length, or where their length is again bounded, say by 3.

## 5.9    Conclusions and open problems

In this chapter we have considered a diverse range of matching problems involving various forms of stability criteria. Many algorithmic results are known and have been described here, but some intriguing open problems remain. Here we list a selection of these.

(1) As discussed in Sec. 5.2.2, the problem of deciding whether an instance of HR-LQ-1 admits a stable matching is NP-complete. The result holds even if each hospital's upper and lower quota is equal to 3, but the complexity of the decision problem is open if each lower quota is at most 2.

(2) To cope with the possible non-existence of a stable matching in a given HRC instance in $I$, we might try to find a matching that is "almost stable" in a particular sense. That is, given the widespread practical applications of HRC, there is strong motivation for considering the problem of finding a matching in $I$ with the minimum number of blocking pairs. Theorem 5.14 implies that this problem is NP-hard in general, though its approximability is open.

(3) In Sec. 5.4.3 we described a model for many–many bipartite stable matching, namely WF-1, where the size of the instance is polynomially-bounded in the number of agents. In an instance of WF-1, each worker and firm has a strictly-ordered preference list over individual agents from the other set. Variants of WF-1 where these preference lists can involve ties (or other forms of indifference) can also be considered. It is possible to formulate analogues of weak stability, strong stability and super-stability in WF-1 with ties (see Sec. 1.3.5 for definitions of these criteria in the HR case). Clearly negative results for HRT also hold for WF-1 with ties, but it remains open to extend exact and approximation algorithms for HRT to the many–many case. A result along these lines has already been obtained in the case of strong stability (see Sec. 3.5.2).

(4) Sec. 5.5.2.2 described an efficient algorithm for finding a student-optimal stable matching in a given instance of SPA-S. As mentioned

in Sec. 5.5.2.5, it remains open to consider, from an algorithmic point of view, variants of SPA-S with ties, and also the case where project lower bounds may be present and no project may be closed.

(5) Perhaps the most intriguing open problem in this list, at least in view of the number of authors that have mentioned it, concerns 3DSM-CYC (defined in Sec. 5.6.1.4), and in particular the question of whether every instance $I$ of this problem admits a weakly stable matching. Moreover, it is open as to whether there is a polynomial-time algorithm for finding a weakly stable matching in $I$ (or reporting that none exists, if it turns out that $I$ need not admit such a matching).

# PART 2
# Other Optimal Matching Problems

# Chapter 6

# Pareto optimal matchings

## 6.1 Introduction

In Sec. 1.5.3 we defined the notion of a Pareto optimal matching in a given instance of HA. The concept can equally be defined in instances of HAT, CHA, SMI, HR and SR. In this chapter we will describe structural and algorithmic results for Pareto optimal matchings in instances of all of these problems.

We have already seen (via Fig. 1.2) that, in an HA instance, a Pareto optimal matching could be half the size of a maximum cardinality matching (clearly the instance in Fig. 1.2 can be replicated as many times as necessary to produce an arbitrarily large HA instance with this property). This motivates the problem of finding a Pareto optimal matching of maximum size, which we refer to as a *maximum Pareto optimal matching*. This problem will form a major focus of this chapter in particular.

Pareto optimal matchings (and even maximum Pareto optimal matchings) can have a relatively poor profile. The following example illustrates this.

**Example 6.1.** *Consider the HA instance I illustrated by Fig. 6.1, where n, the number of applicants, is some integer $\geq 2$. Consider the following*

$$a_1 : h_1 \ h_n$$
$$a_2 : h_1 \ h_2$$
$$a_3 : h_2 \ h_1 \ h_3$$
$$\dots$$
$$a_n : h_{n-1} \ h_1 \ \dots \ h_{n-2} \ h_n$$

Fig. 6.1 An instance of HA with a "bad" Pareto optimal matching

*maximum Pareto optimal matchings in* $I$:

$$M_1 = \{(a_i, h_i) : 1 \leq i \leq n\}$$
$$M_2 = \{(a_1, h_n)\} \cup \{(a_i, h_{i-1}) : 2 \leq i \leq n\}$$

*Then applicant* $a_i$ *has her ith-choice house in* $M_1$ *(*$1 \leq i \leq n$*), whilst* $p(M_2) = \langle n - 1, 1 \rangle$.

Despite the possible shortcomings of Pareto optimal matchings in terms of size, profile and indeed weight, they are regarded by economists as a fundamental solution concept and a minimum requirement for any "reasonable" solution to a cooperative game. Moreover, mechanisms for producing Pareto optimal matchings are often strategy-proof, whilst the same need not be true of mechanisms that produce matchings that are optimal with respect to size, weight or profile.

Pareto optimal matchings can be constructed using a classical algorithm called the *Serial Dictatorship Mechanism* (see e.g., Ref. [5]), which we subsequently refer to as Algorithm SDM. This is a straightforward greedy algorithm that takes each applicant in turn and assigns her to the most-preferred available house on her preference list. The order in which the applicants are processed will, in general, affect the outcome. If a lottery is used in order to determine the applicant ordering, then we obtain the *Random Serial Dictatorship Mechanism* [5]. Alternatively, the applicants might be prioritised in some objective way. Roth and Sotomayor [514, Example 4.3] remark that when the United States Naval Academy matches graduating students to their first posts as Naval Officers using an approach based on Algorithm SDM, students are considered in non-decreasing order of graduation results.

Clearly Algorithm SDM may be implemented to execute in $O(m)$ time, where $m$ is the number of acceptable applicant–house pairs. In addition, the mechanism is *group strategy-proof* (i.e., no coalition of applicants can jointly misrepresent their true preferences in order for at least one member of the coalition to improve, whilst no other coalition member is worse off; see e.g., Ref. [553]). However despite these desirable properties, an arbitrary execution of Algorithm SDM need not produce a Pareto optimal matching that is optimal with respect to either size, weight or profile. For example, with respect to Example 6.1, Algorithm SDM produces $M_1$ by considering the agents in increasing indicial order, whilst $M_2$ is produced if the algorithm starts with $a_2$. It is thus of interest to consider algorithms for computing Pareto optimal matchings with additional properties.

In this chapter we study Pareto optimal matchings in a range of problem domains, including HA, HAT, CHA, HR and SR. We present structural characterisations of Pareto optimal matchings, leading to efficient algorithms for checking a matching for Pareto optimality. We also focus on the algorithmic complexity of problems associated with computing particular types of Pareto optimal matchings, including maximum Pareto optimal matchings. We further consider matchings in the *core* (a stronger notion compared to Pareto optimality) for associated housing markets.

The remaining sections are organised as follows: Sec. 6.2 concerns the House Allocation problem, with the strict preference case (HA) considered in Sec. 6.2.1 and preference lists with ties (HAT) dealt with in Sec. 6.2.2. In each of these subsections of Sec. 6.2 we also discuss matchings in the core of associated housing market problems with and without ties. Subsequently, results for Pareto optimal matchings are described in instances of CHA, the many–one extension of HA, in Sec. 6.3; in HR, the variant of CHA in which both sets of agents have preference lists, in Sec. 6.4; and in SRI, the non-bipartite generalisation of HA, in Sec. 6.5. Finally in Sec. 6.6 we present some concluding remarks and open problems.

## 6.2 House Allocation problem

### 6.2.1 *Strictly-ordered preferences*

#### 6.2.1.1 *Testing for Pareto optimality*

Abraham *et al.* [18] gave a characterisation of Pareto optimal matchings in a given HA instance $I$ that leads to an $O(m)$ algorithm for testing for Pareto optimality, where $m$ is the number of acceptable applicant–house pairs. To describe this algorithm, we require some initial definitions. In what follows, we recall definitions from Sec. 1.5.2.

We say that a matching $M \in \mathcal{M}$ is *trade-in-free* if there is no applicant–house pair $(a_i, h_j)$ such that $a_i$ is assigned in $M$, $h_j$ is unassigned in $M$, and $a_i$ prefers $h_j$ to $M(a_i)$. Also $M$ is *cyclic coalition-free* if $M$ admits no *cyclic coalition*, which is a sequence of applicants $C = \langle a_{i_0}, a_{i_1}, \ldots, a_{i_{r-1}} \rangle$, for some $r \geq 2$, all assigned in $M$, such that $a_{i_j}$ prefers $M(a_{i_{j+1}})$ to $M(a_{i_j})$ $(0 \leq j \leq r - 1)$ (all subscripts are taken modulo $r$ when reasoning about cyclic coalitions). The matching

$$M' = (M \setminus \{(a_{i_j}, M(a_{i_j})) : 0 \leq j \leq r-1\}) \cup \{(a_{i_j}, M(a_{i_{j+1}})) : 0 \leq j \leq r-1\}$$

is defined to be the matching obtained from $M$ by *satisfying* $C$.

The following proposition gives necessary and sufficient conditions for a matching to be Pareto optimal.

**Proposition 6.2 ([18]).** *Let $M$ be a matching in a given instance of* HA. *Then $M$ is Pareto optimal if and only if $M$ is maximal, trade-in-free and cyclic coalition-free.*

For a given matching $M$, we can trivially check whether $M$ satisfies the maximality and trade-in-free properties in $O(m)$ time. To check for the absence of cyclic coalitions, we construct the *envy graph* $D'_M$ of $M$. This is similar to the envy graph $D_M$ as defined in Sec. 5.7.2, but with one subtle distinction, namely the vertex set of $D'_M$ comprises only those applicants who are assigned in $M$. As in the case of $D_M$, $D'_M$ has an arc $(a_i, a_j)$ whenever $a_i$ prefers $M(a_j)$ to $M(a_i)$. It is clear that $M$ is cyclic coalition-free if and only if $D'_M$ is acyclic. We can perform this last check in $O(m)$ time using depth-first search in $D'_M$. Putting these observations together, we have the following result.

**Proposition 6.3 ([18]).** *Let $M$ be a matching in a given instance of* HA. *We may check whether $M$ is Pareto optimal in $O(m)$ time, where $m$ is the number of acceptable applicant–house pairs.*

It is straightforward to verify that a matching constructed by Algorithm SDM as described in Sec. 6.1 is Pareto optimal. We therefore have the following result.

**Proposition 6.4.** *Let $I$ be an instance of* HA. *Then we may find a Pareto optimal matching in $I$ in $O(m)$ time using Algorithm* SDM, *where $m$ is the number of acceptable applicant–house pairs.*

We have already seen that Pareto optimal matchings can have different sizes (see Sec. 1.5.3). Indeed, relative to Fig. 1.2, Algorithm SDM will produce a Pareto optimal matching of size $i$ if applicant $a_i$ is processed first, for $i \in \{1, 2\}$. Thus we require an alternative approach if a maximum Pareto optimal matching is required.

### 6.2.1.2 *Maximum Pareto optimal matchings*

In this section we describe an $O(\sqrt{n_1}m)$ algorithm, due to Abraham *et al.* [18], for finding a maximum Pareto optimal matching in a given HA instance $I$, where $n_1$ is the number of applicants and $m$ is the number of acceptable applicant–house pairs. The algorithm operates in three phases, with each phase enforcing one of the conditions for Pareto optimality given in Proposition 6.2. We remark that, in Sec. 6.3 we present an algorithm, having the same time complexity, for finding a maximum Pareto optimal matching in the more general CHA case. However it is instructive to begin by describing the algorithm for the 1–1 case, not just because it aids understanding of the algorithm for CHA, but also because it illustrates how to obtain an $O(m)$ implementation of the classical Top Trading Cycles algorithm.

**Phase 1 of the algorithm.** Phase 1 involves using the Hopcroft–Karp algorithm [281] to compute a maximum matching $M$ in the underlying graph $G$ of $I$ as defined in Sec. 1.5.2. This phase guarantees that $M$ is maximal, takes $O(\sqrt{n_1} m)$ time, and dominates the overall runtime of the algorithm.

**Phase 2 of the algorithm.** In this phase, we transform $M$ into a trade-in-free matching by repeatedly identifying and promoting applicants who prefer an unassigned house to their existing assignment. Each promotion breaks an existing assignment, thereby freeing a house, which itself may be a preferred assignment for a different applicant. With the aid of suitable data structures, we can ensure that the next applicant to be identified for promotion can be found efficiently.

For each house $h_j \in H$, we maintain a linked list $L_j$ of pairs $(a_i, r)$, where $a_i \in A$ is an assigned applicant who finds $h_j$ acceptable, and $r = rank(a_i, h_j)$. Initially the pairs in $L_j$ involve only those assigned applicants $a_i$ who prefer $h_j$ to $M(a_i)$, though subsequently the pairs in $L_j$ may contain applicants $a_i$ who prefer $M(a_i)$ to $h_j$. The initialisation of these lists can be carried out using one traversal of the applicant preference lists, which we assume are represented as doubly linked lists or arrays, in $O(m)$ time.

For each assigned applicant $a_i$, we also use this traversal to initialise a variable, denoted by $curr_i$, which stores $rank(a_i, M(a_i))$. This variable is maintained during the execution of the algorithm. One final initialisation remains: construct a stack $S$ of all unassigned houses $h_j$ where $L_j$ is non-empty. We now enter the loop described by Algorithm Phase 2 in Algorithm 6.1.

During each loop iteration we pop an unassigned house $h_j$ from $S$ and remove the first pair $(a_i, r)$ from the list $L_j$ (which must be non-empty). If $a_i$ prefers $h_j$ to $M(a_i)$ (i.e., $r < curr_i$) then $a_i$ is promoted from $h_k = M(a_i)$ to $h_j$, also $M$ and $curr_i$ are updated, and finally $h_k$, which is now

---

**Algorithm 6.1** Algorithm Phase 2 for HA [18]

---

**Require:** HA instance $I$ and a maximal matching $M$
**Ensure:** $M$ is a maximal and trade-in-free matching in $I$
1: **while** $S \neq \emptyset$ **do**
2:    $h_j := S.\text{pop}()$;
3:    $(a_i, r) := L_j.\text{removeHead}()$;
4:    **if** $r < curr_i$ **then**
5:       $\{h_j$ is unassigned in $M$, $a_i$ is assigned in $M$ and prefers $h_j$ to $M(a_i)\}$
6:       $h_k := M(a_i)$;
7:       $M := (M \backslash \{(a_i, h_k)\}) \cup \{(a_i, h_j)\}$;
8:       $curr_i := r$;
9:       $h_j := h_k$;
10:   **end if**
11:   **if** $L_j \neq \emptyset$ **then**
12:     $S.\text{push}(h_j)$;
13:   **end if**
14: **end while**

---

unassigned, is pushed onto $S$ if $L_k$ is non-empty. If $a_i$ prefers $M(a_i)$ to $h_j$ then $h_j$ is pushed back onto $S$ if $L_j$ is still non-empty.

Each iteration of the loop removes a pair from a list $L_j$. Since applicant preference lists are finite and no new pair is added to a list $L_j$ during a loop iteration, the while loop must eventually terminate with $S$ empty. At this point no assigned applicant $a_i$ would trade $M(a_i)$ for an unassigned house, and so $M$ is trade-in-free. Additionally, $M$ remains a maximum matching, since any applicant assigned at the end of Phase 1 is also assigned at the end of Phase 2. Finally, it is clear that this phase runs in $O(m)$ time given the data structures described above.

**Phase 3 of the algorithm.** In this phase, we transform $M$ into a cyclic coalition-free matching. Recall that cyclic coalitions in $M$ correspond to cycles in the envy graph $D'_M$. So a natural algorithm involves repeatedly finding and satisfying cyclic coalitions in $D'_M$ until no more cyclic coalitions remain. This algorithm has a runtime of $O(m^2)$, since there are $O(m)$ cyclic coalitions, and cycle-detection takes $O(m)$ time.

A better starting point for an efficient algorithm is Gale's Top Trading Cycles (TTC) algorithm [527], which has been the focus of much attention, particularly in the game theory and economics literature [527,508,497,592, 6]. This method is also based on repeatedly finding and satisfying cyclic coalitions, however the number of iterations is reduced by the following observation: no applicant assigned to her first choice can be in a cyclic coalition. We remove such applicants from consideration, and since the houses assigned to them are no longer exchangeable, they can be deleted from the preference lists of the remaining applicants. This observation can now be recursively applied to the reduced preference lists. At some point, either no applicants remain, in which case the matching is cyclic coalition-free, or no applicant is assigned to her *reduced first choice* (i.e., the first choice on her reduced preference list).

In this last case, it turns out that there must be a cyclic coalition $C$ in $M$, which can be found in $O(n_1)$ time by searching the envy graph restricted to reduced first-choice edges. After satisfying $C$, each applicant in $C$ is assigned to her reduced first choice. Therefore, no applicant is in more than one cyclic coalition, giving $O(n_1)$ cyclic coalitions overall. The runtime of this preliminary implementation then is $\Omega(m+n_1^2)$. However, with a careful choice of data structures we can achieve an $O(m)$ implementation.

To achieve this improvement, deletions of houses from applicants' preference lists are not explicitly carried out. Instead, a house that is no longer exchangeable is labelled (all houses are initially unlabelled). For each applicant $a_i \in A$ we maintain a pointer $p_i$ to the first unlabelled house on $a_i$'s preference list — this is equivalent to the first house on $a_i$'s reduced preference list. Initially $p_i$ points to the first house on $a_i$'s preference list, and subsequently $p_i$ traverses left to right. Also, in order to identify cyclic coalitions, we maintain a counter $z_i$ for each applicant $a_i$, which is initialised to 0. Then, we enter the main body of Algorithm Phase 3, as shown in Algorithm 6.2.

This algorithm repeatedly searches for cyclic coalitions, building a path $P$ of applicants (represented by a stack) in the envy graph restricted to reduced first-choice edges. At each iteration of the while loop, we pop an applicant $a_k$ from the stack and move $p_k$ down if necessary. If $P$ cycles (i.e., we find $z_k = 2$), there is a cyclic coalition $C$: the applicants involved in $C$ are identified and removed from consideration, and the houses assigned to these applicants are labelled, during stack-popping operations. $C$ is also satisfied (in practice $C$ can be satisfied during the stack popping operations). Alternatively, if $P$ reaches a dead-end (because $a_k$ is already assigned to her first choice), this applicant is removed from consideration and her assigned house is labelled. Otherwise, we keep extending the path by following the reduced first-choice edges.

At the termination of this phase we note that $M$ is cyclic coalition-free by the correctness of the TTC algorithm [527]. Also $M$ remains a maximum trade-in-free matching, since each applicant and house assigned at the end

---

**Algorithm 6.2** Algorithm Phase 3 for HA [18]

---

**Require:** HA instance $I$ and a maximal, trade-in-free matching $M$
**Ensure:** $M$ is a Pareto-optimal matching in $I$
 1: **for each** assigned applicant $a_i$ such that $p_i \neq M(a_i)$ **do**
 2:    $P := \{a_i\}$;   $\{P$ is a stack of applicants$\}$
 3:    $z_i := 1$;   $\{$number of times $a_i$ is in $P\}$
 4:    **while** $P \neq \emptyset$ **do**
 5:      $a_k := P.\text{pop}()$;
 6:      $p_k :=$ most-preferred unlabelled house on preference list of $a_k$;
 7:      **if** $z_k = 2$ **then** $\{$cyclic coalition identified$\}$
 8:        $C :=$ cyclic coalition in $P$ containing $a_k$;
 9:        satisfy $C$;
10:        **for each** $a_t \in C$ **do**
11:          label $M(a_t)$;
12:          $z_t := 0$;
13:          $P.\text{pop}()$;
14:        **end for**
15:      **else if** $p_k = M(a_k)$ **then** $\{$dead end reached$\}$
16:        label $M(a_k)$;
17:        $z_k := 0$;
18:      **else** $\{$extend the path$\}$
19:        $P.\text{push}(a_k)$;
20:        $a_t := M(p_k)$;
21:        $z_t\text{++}$;
22:        $P.\text{push}(a_t)$;
23:      **end if**
24:    **end while**
25: **end for**

---

of Phase 2 is also assigned at the end of Phase 3. Finally, it is clear this phase runs in $O(m)$ time given the data structures described above. We summarise the preceding discussion with the following theorem.

**Theorem 6.5 ([18]).** *Given an* HA *instance $I$, a maximum Pareto optimal matching can be found in $O(\sqrt{n_1}m)$ time, where $n_1$ is the number of applicants and $m$ is the number of acceptable applicant–house pairs. Such a matching is also a maximum matching of applicants to houses in the underlying graph of $I$.*

Note that any improvement to the complexity of the above algorithm would imply an improved algorithm for finding a maximum matching in a bipartite graph. For, without loss of generality, let $G = (A, H, E)$ be an arbitrary bipartite graph with no isolated vertices. Construct an instance $I$

of HA where $G$ is the underlying graph by letting each applicant's preference list in $I$ be an arbitrary permutation of her neighbours in $G$. By Theorem 6.5, any maximum Pareto optimal matching in $I$ is also a maximum matching in $G$. Since $I$ may be constructed from $G$ in $O(m)$ time, the complexity of finding a maximum matching in a bipartite graph is bounded above by the complexity of finding a maximum Pareto optimal matching.

**Initial endowment.** Suppose that a subset $A'$ of the applicants already own a house. Abraham *et al.* [18] described an *individually rational* modification of the algorithm, which ensures that every applicant in $A'$ ends up with the same house or better.

We begin with a matching $M$ that pre-assigns every applicant $a_i \in A'$ to her existing house. We then truncate the preference list of each such $a_i$ by removing all houses less preferable to her than $M(a_i)$. Now, we enter Phase 1, where we use the Hopcroft–Karp algorithm to exhaustively augment $M$ into some matching $M'$. Members of $A'$ must still be assigned in $M'$, and since their preference lists were truncated, their new assignments must be at least as good as those in $M$. Note that $M'$ may not be a maximum matching of $A$ to $H$, however $M'$ does have maximum cardinality among all matchings that respect the initial endowment. The remaining two phases do not move any applicant from being assigned to unassigned, and so the result follows immediately.

### 6.2.1.3 *Other results for Pareto optimal matchings*

In contrast to the existence of an efficient algorithm for finding a maximum Pareto optimal matching, it turns out that such an algorithm is unlikely to exist for the minimisation problem, as we now show.

For a given instance $I$ of HA, we denote by $p^-(I)$ and $p^+(I)$ the sizes of a minimum and maximum Pareto optimal matching in $I$ respectively. Similarly, we denote by $\beta^-(G)$ and $\beta^+(G)$ the sizes of a minimum maximal and a maximum matching in the underlying graph $G$ of $I$ respectively. Using Algorithm **Phase 2** and Algorithm **Phase 3**, we can transform any maximal matching $M$ in $G$ to a Pareto optimal matching $M'$ in $I$ where $|M'| = |M|$. Hence $p^-(I) = \beta^-(G)$[1] and $p^+(I) = \beta^+(G)$. Note that the problem of computing $\beta^-(G)$ is NP-hard even for subdivision graphs of cubic graphs, as indicated by Theorem 1.7. The following result (a weaker version of which was proved in Ref. [18]), is therefore immediate.

---

[1] In Ref. [18] only the inequality $p^-(I) \geq \beta^-(G)$ was observed.

**Theorem 6.6.** *Given an* HA *instance* $I$*, the problem of finding a minimum Pareto optimal matching is NP-hard. The result holds even if each applicant finds two houses acceptable and each house finds at most three applicants acceptable, or vice versa.*

It is also known that $\beta^-(G) \geq \beta^+(G)/2$ [399]. Hence as $p^-(I) = \beta^-(G)$ and $p^+(I) = \beta^+(G)$, the following result is immediate.

**Theorem 6.7** ([18]). *Given an* HA *instance, the problem of finding a minimum Pareto optimal matching is approximable within a factor of 2.*

It is also possible to prove an interpolation result for Pareto optimal matchings: namely, for a given HA instance $I$, there are Pareto optimal matchings of all sizes between $p^-(I)$ and $p^+(I)$. One way of proving this is to observe that the underlying graph $G$ admits a maximal matching $M_k$ for each $k$ $(\beta^-(G) \leq k \leq \beta^+(G))$ [276]. Again, using the fact that $M_k$ can be transformed to a Pareto optimal matching $M_k'$ such that $|M_k'| = |M_k|$, we obtain the following result.

**Theorem 6.8** ([18]). *For a given instance* $I$ *of* HA*, there exists a Pareto optimal matching of size* $k$*, for each* $k$ $(p^-(I) \leq k \leq p^+(I))$.

The final result that we present in this subsection gives a necessary and sufficient condition, checkable in linear time, for an HA instance to admit a unique Pareto optimal matching.

**Theorem 6.9** ([18]). *An instance* $I$ *of* HA *admits a unique Pareto optimal matching* $M$ *if and only if every applicant is assigned in* $M$ *to her first choice.*

Fleischer and Wang [219] considered the problem of finding Pareto optimal matchings in a dynamic setting where applicants and houses can both enter and leave the market. Let $I$ be an HA instance and let $M$ be a maximum Pareto optimal matching in $I$. The authors showed that, after a single applicant or house arrives or leaves, $M$ can be updated to yield a maximum Pareto optimal matching for the new HA problem instance in $O(m)$ time. Furthermore, given two Pareto optimal matchings $M_1$ and $M_2$ in $I$, the authors gave an $O(m)$ algorithm to transform $M_1$ into $M_2$ by alternating along a sequence of disjoint cycles in a graph that is similar to the envy graph for $M_1$. The authors remarked that this latter algorithm could lead to a method for listing all Pareto optimal matchings in $I$.

### 6.2.1.4 Matchings in the core

Recall the definition of HM from Sec. 1.5.2. The notion of Pareto optimality in HA is closely related to the concept of *core* matchings in the HM context [508]. In the literature, definitions of core outcomes are invariably expressed in terms of a cooperative game-theoretic setting, however we shall adapt the notation in order to be a closer fit with the context of matching problems with preferences. In what follows, we recall definitions of notation and terminology relating to HM from Sec. 1.5.2.

**Definition 6.10 ([508]).** *Let $I$ be an instance of* HM *where $M_0$ is the initial endowment, and let $M$ be an individually rational matching in $I$. Let $M'$ be a matching in $I$, and let $S$ be the set of applicants who are assigned in $M'$. Then $M'$ weakly blocks $M$ with respect to the coalition $S$ if:*

*(1) $\{M'(a_i) : a_i \in S\} = \{M_0(a_i) : a_i \in S\}$;*
*(2) some $a_i \in S$ prefers $M'(a_i)$ to $M(a_i)$;*
*(3) no $a_i \in S$ prefers $M(a_i)$ to $M'(a_i)$.*

*$M$ is a* strict core matching, *or $M$ is in the* strict core[2], *if there is no other matching in $I$ that weakly blocks $M$.*

In the above definition, Condition 1 states that the members of the coalition can only improve by exchanging the resources that they bring to the market (via their initial endowment $M_0$); Condition 2 states that some member of the coalition is better off $M'$ than in $M$; whilst Condition 3 states that no member of the coalition is worse off in $M'$ than in $M$. Note that $M$ is Pareto optimal if and only if $M$ is not weakly blocked by any matching $M'$ such that $|M'| = n_1$ (here the coalition comprises all applicants and is referred to as the *grand coalition*). Hence a strict core matching is Pareto optimal.

Fig. 6.2 gives an example HM instance (where each applicant's initial endowment in $M_0$ is the last house on her preference list) that has a Pareto optimal matching that is not in the strict core. To see this, observe that the matching

$$M = \{(a_1, h_4), (a_2, h_1), (a_3, h_2), (a_4, h_3)\}$$

is Pareto optimal, but is weakly blocked by $M' = \{(a_3, h_4), (a_4, h_3)\}$.

---

[2]The strict core is sometimes referred to as the *strong core* [129].

$$a_1 : h_4 \quad h_3 \quad h_1$$
$$a_2 : h_4 \quad h_1 \quad h_2$$
$$a_3 : h_4 \quad h_1 \quad h_2 \quad h_3$$
$$a_4 : h_3 \quad h_2 \quad h_4$$

Fig. 6.2   An HM instance having a Pareto optimal matching that is not in the strict core

Roth and Postlewaite [508] showed that every HM instance $I$ admits a unique strict core matching, which can be found using Gale's TTC algorithm [527]. This algorithm is essentially Algorithm **Phase 3** shown in Algorithm 6.2 as applied to the initial endowment $M_0$ (which must be maximal and trade-in-free in $I$ as $|M_0| = n_1$ and $n_1 = n_2$, where $n_2$ is the number of houses). We thus obtain the following result.

**Theorem 6.11 ([508]).** *Let $I$ be an instance of* HM. *A strict core matching can be found in $O(m)$ time, where $m$ is the number of acceptable applicant–house pairs in $I$.*

Roth [497] proved that the TTC algorithm is strategy-proof.

Notice that, given an HA instance $I$, the algorithm given in Sec. 6.2.1.2 produces a strict core matching relative to an HM instance $I'$ obtained as follows. Firstly we let $M$ be a matching produced at the end of Phase 2. Then, for every applicant $a_i$ who is unassigned in $M$, extend $a_i$'s preference list so that, in $I'$, we append a unique dummy house $h_i'$ to the end of $a_i$'s list in $I$. For every applicant $a_i$ who is assigned in $M$, $a_i$'s preference list in $I'$ is obtained by truncating her preference list in $I$ after $M(a_i)$. We then remove from $I'$ the houses that do not feature in any applicant's preference list; as $M$ is maximal and trade-in-free, it follows that the numbers of applicants and houses in $I'$ (including the dummy houses) are now equal. The initial endowment in $I'$ is then $M$ together with the pairs $(a_i, h_i')$ for every applicant $a_i$ who is unassigned in $M$. Note that the matching $M$ produced at the end of Phase 2 is not unique, and hence the same is true for $I'$.

### 6.2.2   *Preference lists with ties*

#### 6.2.2.1   *Characterisation of Pareto optimal matchings*

In this subsection we consider the House Allocation problem with Ties (HAT) as defined in Sec. 1.5.7. In this setting, the definition of a Pareto optimal matching is unchanged from Sec. 1.5.3. We now give a series of definitions that work towards a characterisation of Pareto optimal matchings in a given HAT instance $I$. Let $M$ be an arbitrary matching in $I$.

An *alternating path coalition* with respect to $M$ comprises a sequence $C = \langle a_{i_0}, a_{i_1}, \ldots, a_{i_{r-1}}, h_k \rangle$, for some $r \geq 1$, where $a_{i_j}$ is an assigned applicant $(0 \leq j \leq r-1)$ and $h_k$ is an unassigned house. If $r = 1$ then $a_{i_0}$ prefers $h_k$ to $M(a_{i_0})$. Otherwise if $r \geq 2$ then $a_{i_0}$ prefers $M(a_{i_1})$ to $M(a_{i_0})$, $a_{i_j}$ prefers $M(a_{i_{j+1}})$ to $M(a_{i_j})$ or is indifferent between them $(1 \leq j \leq r-2)$, and $a_{i_{r-1}}$ prefers $h_k$ to $M(a_{i_{r-1}})$ or is indifferent between them.

An *augmenting path coalition* with respect to $M$ comprises a sequence $C = \langle a_{i_0}, a_{i_1}, \ldots, a_{i_{r-1}}, h_k \rangle$, for some $r \geq 1$, where $a_{i_j}$ is an assigned applicant $(1 \leq j \leq r-1)$, $h_k$ is an unassigned house, $a_{i_0}$ is an unassigned applicant and $M(a_{i_1}) \in A(a_{i_0})$, $a_{i_j}$ prefers $M(a_{i_{j+1}})$ to $M(a_{i_j})$ or is indifferent between them $(1 \leq j \leq r-2)$, and $a_{i_{r-1}}$ prefers $h_k$ to $M(a_{i_{r-1}})$ or is indifferent between them. Note that $M$ is maximal if and only if $M$ admits no augmenting path coalition with $r = 1$.

A *cyclic coalition* with respect to $M$ is a sequence of applicants $C = \langle a_{i_0}, a_{i_1}, \ldots, a_{i_{r-1}} \rangle$, for some $r \geq 2$, all assigned in $M$, such that $a_{i_j}$ prefers $M(a_{i_{j+1}})$ to $M(a_{i_j})$ or is indifferent between them for each $j$ $(0 \leq j \leq r-1)$, and $a_{i_j}$ prefers $M(a_{i_{j+1}})$ to $M(a_{i_j})$ for some $j$ $(0 \leq j \leq r-1)$ (all subscripts are taken modulo $r$ when reasoning about cyclic coalitions).

We define an *improving coalition* to be an alternating path coalition, an augmenting path coalition or a cyclic coalition. A matching $M$ is *improving coalition-free* if it admits no improving coalition.

Given an improving coalition $C$, the matching

$$M' = (M \backslash \{(a_{i_j}, M(a_{i_j})) : \delta \leq j \leq r-1\}) \cup \{(a_{i_j}, M(a_{i_{j+1}})) : 0 \leq j \leq r-1\}$$

is defined to be the matching obtained from $M$ by *satisfying* $C$ ($\delta = 1$ in the case that $C$ is an augmenting path coalition, otherwise $\delta = 0$).

The following proposition gives a necessary and sufficient condition for a matching to be Pareto optimal. In the proof of the result, we recall the definition of $\lhd$ from Sec. 1.5.3.

**Proposition 6.12.** *Let $I$ be an instance of* HAT *and let $M$ be a matching in $I$. Then $M$ is Pareto optimal if and only if $M$ is improving coalition-free.*

**Proof.** Let $M$ be a matching in $I$ that is improving coalition-free, and suppose for a contradiction that $M$ is not Pareto optimal. Then there exists some matching $M' \neq M$ such that $M' \lhd M$. Let $a_t$ be an applicant who prefers $M'$ to $M$. Consider the graph $G = M \oplus M'$. Every connected component of $G$ is an alternating path or an alternating cycle. In particular there exists a connected component $C$ of $G$ containing $a_t$. We consider three cases.

- *Case (i):* $C$ is an alternating path with an even number of edges. If both endpoint vertices of $C$ are applicants, then there is an applicant who is assigned in $M$ and unassigned in $M'$, a contradiction since $M' \lhd M$. Hence both endpoint vertices of $C$ are houses, so that every applicant in $C$ is assigned in both $M$ and $M'$. We build a sequence $P = \langle a_{i_0}, a_{i_1}, \ldots, a_{i_{r-1}}, h_k \rangle$, for some $r \geq 1$, where $a_{i_0} = a_t$, $a_{i_{j+1}} = M(M'(a_{i_j}))$ $(0 \leq j \leq r - 2)$, $h_k = M'(a_{i_{r-1}})$ and $h_k$ is unassigned in $M$. That is, $P$ is a sub-path of $C$ starting at $a_t$. It follows that $P$ is an alternating path coalition with respect to $M$, a contradiction.

- *Case (ii):* $C$ is an alternating path with an odd number of edges. If both end edges of $C$ are in $M$, then there is an applicant who is assigned in $M$ and unassigned in $M'$, a contradiction since $M' \lhd M$. Hence both end edges of $C$ are in $M'$, so that $C$ gives rise to an augmenting path coalition with respect to $M$, a contradiction.

- *Case (iii):* $C$ is an alternating cycle. Then clearly the applicants in $C$ form a cyclic coalition with respect to $M$, a contradiction.

Hence $M$ is Pareto optimal.

Conversely let $M$ be a Pareto optimal matching in $I$. If $M$ admits an improving coalition $C$, let $M'$ be the matching obtained by satisfying $C$. Then $M' \lhd M$, a contradiction. Hence $M$ is improving coalition-free.   $\square$

The characterisation of Pareto optimal matchings in $I$ given by Proposition 6.12 leads to the following linear-time algorithm for checking a given matching $M$ for Pareto optimality. As a pre-processing step, we form a new instance $I'$ of HAT from $I$ by truncating the preference list of each applicant $a_i \in A$ who is assigned in $M$ in such a way that $a_i$ removes any house $h_k$ that she finds less preferable than $M(a_i)$. Let $G = (A, H, E)$ be the underlying graph of $I'$.

The search for an alternating path coalition is a modified breadth-first search in $G$ that fans out from each applicant $a_i$ who is assigned in $M$ and prefers some house to $M(a_i)$. In general, the search traverses from left to right along edges not in $M$ (i.e., for a given applicant $a_i$, across all edges $(a_i, h_j) \notin M$ such that either $a_i$ prefers $h_j$ to $M(a_i)$ or is indifferent between them), and from right to left along edges in $M$. If we reach an unassigned house, we have found an alternating path coalition. Clearly each edge in $G$ is traversed at most once, and hence this step takes $O(m)$ time.

The process is similar when searching for augmenting path coalitions — this time the breadth-first search in $G$ fans out from each applicant who is unassigned in $M$. As before, if we reach an unassigned house, we have found an augmenting path coalition. Again, the time taken is $O(m)$.

In order to detect cyclic coalitions, create a directed graph $D$ from $G$ by orienting edges in $M$ from left to right, and orienting edges in $E \backslash M$ from right to left. As $G$ is bipartite, clearly any directed cycle in $D$ gives an alternating cycle in $G$ and vice versa. If $(h_j, a_i)$ is an arc of $D$ where $a_i$ prefers $h_j$ to $M(a_i)$, colour this arc red. Let $S$ be the set of red arcs in $D$. Now create the strongly connected components of $D$ in $O(m)$ time [561]. For each red arc $(h_j, a_i)$ in $S$, simply test whether $a_i$ and $h_j$ are in the same strongly connected component. If so, $M$ admits a cyclic coalition. If this property is not satisfied for all red arcs, then $M$ admits no cyclic coalition. Clearly this step can be carried out in $O(m)$ overall time.

We thus obtain the following result.

**Proposition 6.13.** *Let $I$ be an instance of* HAT *and let $M$ be a matching in $I$. Then $M$ can be tested for Pareto optimality in $O(m)$ time, where $m$ is the number of acceptable applicant–house pairs in $I$.*

An arbitrary Pareto optimal matching in $I$ can be found in $O(m)$ time by breaking the ties arbitrarily and applying Algorithm SDM (see Proposition 6.4). We thus obtain the following.

**Proposition 6.14.** *Let $I$ be an instance of* HAT. *Then we may find a Pareto optimal matching in $I$ in $O(m)$ time, where $m$ is the number of acceptable applicant–house pairs in $I$.*

Clearly the Pareto optimal matchings in $I$ can have different sizes (as the same is true even for HA), and this approach cannot guarantee to maximise the size of a Pareto optimal matching.

A maximum Pareto optimal matching in $I$ can be found by constructing a minimum weight maximum cardinality matching in $I$ in $O(\sqrt{n} m \log n)$ time as described in Sec. 1.5.4 (the same definition of *rank* applies in the HAT case), where $n = n_1 + n_2$ is the total number of applicants and houses. However it remains open as to whether an $O(\sqrt{n} m)$ algorithm exists for this problem. Achieving such a bound may involve extending the TTC algorithm to handle ties, and to this end the papers of Yılmaz [591], Alcalde-Unzu and Molis [36], Jaramillo and Manjunath [348], and Aziz and de Keijzer [56] may be relevant.

$$a_1 : h_2 \quad h_3 \quad h_1$$
$$a_2 : (h_1 \quad h_3) \quad h_2$$
$$a_3 : h_2 \quad h_1 \quad h_3$$

Fig. 6.3   An instance of HMT with no strict core matching due to Shapley and Scarf [527]

### 6.2.2.2   *Matchings in the core*

Just as we relaxed the requirement for preference lists to be strict in an instance of HA in order to obtain HAT, we can do likewise for housing markets. The *Housing Market with Ties* (HMT) is the generalisation of HM in which preference lists can include ties. In the HMT setting, we can define weak blocking and the strict core in exactly the same way as in Definition 6.10.

In contrast to the case for HM, an HMT instance need not admit a strict core matching. To illustrate this, consider the HMT instance $I$, due to Shapley and Scarf [527], given in Fig. 6.3 (each applicant's initial endowment $M_0$ is the last house on her preference list). It may be verified that every individually rational matching in $I$ is weakly blocked by some matching. Quint and Wako [487] gave an $O(n^3)$ algorithm for finding a strict core matching or reporting that none exists, given an instance of HMT, where $n = n_1 (= n_2)$ is the number of applicants.

A weaker notion of the strict core exists in the HMT context (and indeed in the HM setting too), which we now define.

**Definition 6.15 ([508]).**   *Let $I$ be an instance of HMT where $M_0$ is the initial endowment, and let $M$ be an individually rational matching in $I$. Let $M'$ be a matching in $I$, and let $S$ be the set of applicants who are assigned in $M'$. Then $M'$ strongly blocks $M$ with respect to the coalition $S$ if:*

*(1) $\{M'(a_i) : a_i \in S\} = \{M_0(a_i) : a_i \in S\}$;*
*(2) each $a_i \in S$ prefers $M'(a_i)$ to $M(a_i)$.*

*$M$ is a core matching, or $M$ is in the core[3], if there is no other matching in $I$ that strongly blocks $M$.*

A core matching always exists in a given HMT instance $I$ (though need not be unique), and can be found in $O(m)$ time, where $m$ is the number of acceptable applicant–house pairs. To see this, simply break the ties

---

[3]The core is sometimes referred to as the *weak core* [105].

arbitrarily to obtain an HM instance $I'$. As in Sec. 6.2.1.4, find a strict core matching in $I'$ in $O(m)$ time. Such a matching is then in the core in $I$. We thus arrive at the following result.

**Theorem 6.16 ([508]).** *Let $I$ be an instance of* HMT. *A core matching can be found in $O(m)$ time, where $m$ is the number of acceptable applicant–house pairs in $I$.*

Clearly however, the different ways in which the ties in $I$ are broken will in general affect the core matching that is produced. Sotomayor [545] also gave a non-constructive proof of the existence of a core matching in $I$. We remark that a core matching need not be Pareto optimal.

## 6.3 Capacitated House Allocation problem

In this section we now consider the case where houses may be assigned more than one applicant, though all preferences are again strict — thus we are given an instance $I$ of CHA as defined in Sec. 1.5.7. Again, the problem we mainly focus on is that of finding a maximum Pareto optimal matching in $I$. Pareto optimal matchings in $I$ were defined in Sec. 1.5.7. A straightforward adaptation of Algorithm SDM will yield a Pareto optimal matching in $I$, but again different processing orders for the applicants will in general lead to Pareto optimal matchings of different sizes.

As in Sec. 1.5.4, a maximum Pareto optimal matching can be found by constructing a minimum weight maximum cardinality matching $M$ in the underlying graph $G$ of $I$, however note that (i) $G$ is a capacitated bipartite graph (where each vertex corresponding to an applicant has capacity 1, and each vertex corresponding to a house $h_j \in H$ has capacity $c_j$), and (ii) the time complexity for constructing $M$ is $O(C \min\{m \log n_1, n_1^2\})$ [226], where $n_1$ is the number of applicants, $m$ is the number of acceptable applicant–house pairs, and $C$ is the sum of the house capacities.

Sng [535] gave a faster algorithm for finding a maximum Pareto optimal matching in $I$, with running time $O(\sqrt{n_1}m)$. In order to describe this algorithm, we again require a characterisation of Pareto optimal matchings. The necessary and sufficient conditions for a matching to be Pareto optimal are similar to those in Proposition 6.2 for the HA case, however we need to interpret *maximal* and *trade-in-free* slightly differently in the CHA context. A matching $M$ in $I$ is *maximal* if there is no applicant–house pair $(a_i, h_j)$ such that $a_i$ is unassigned in $M$ and finds $h_j$ acceptable, and $h_j$

is undersubscribed in $M$. Also $M$ is *trade-in-free* if there is no applicant–house pair $(a_i, h_j)$ such that $a_i$ is unassigned in $M$ and finds $h_j$ acceptable, and $h_j$ is undersubscribed in $M$. With these definitions, Sng [535] proved the following results.

**Proposition 6.17 ([535]).** *Let $M$ be a matching in a given instance of* CHA. *Then $M$ is Pareto optimal if and only if $M$ is maximal, trade-in-free and cyclic coalition-free.*

**Proposition 6.18 ([535]).** *Let $M$ be a matching in a given instance of* CHA. *Then we may check whether $M$ is Pareto optimal in $O(m)$ time, where $m$ is the number of acceptable applicant–house pairs.*

Sng's algorithm for finding a maximum Pareto optimal matching in $I$ again has three phases, as in the HA case. Phase 1 involves finding a maximum cardinality matching $M$ in the underlying capacitated bipartite graph $G$. This can be accomplished in $O(\sqrt{n_1}m)$ time as indicated by Theorem 1.6. Clearly $M$ is maximal.

Phase 2 involves modifying $M$ to ensure that it is trade-in-free. The algorithm for carrying out this step is very similar to Algorithm Phase 2 for HA shown in Algorithm 6.1. We simply outline the modifications here for CHA. Firstly, between lines 8 and 9 of Algorithm Phase 2, we need to test whether $h_j$ needs to be pushed back onto the stack. This would be required if (i) $h_j$ is still undersubscribed even after $a_i$ is promoted to $h_j$, and (ii) $L_j$ remains non-empty. Also at line 11 of Algorithm Phase 2, we must determine whether $h_j$ is not already in the stack before the push operation (it is straightforward to maintain a boolean for each house which indicates whether it is in the stack). This situation could arise, for example, when a house $h_k$ with capacity 2 is preferred by some applicant $a_j$ to $M(a_j)$, and $M(h_k) = \{a_i\}$. It is possible that $a_i$ is promoted from $h_k$ to some other house before $a_j$ is promoted, meaning that $h_k$ loses $a_i$ as an assignee. However $h_k$ (now empty) was already on the stack, so should not be added again.

Phase 3 is a generalisation of Algorithm Phase 3 for HA shown in Algorithm 6.2. It is essentially an $O(m)$ implementation of the TTC algorithm extended to CHA[4]. In this phase, we transform the maximal and trade-in-

---

[4]Abdulkadiroğlu and Sönmez [7] extended the TTC algorithm to the school choice context, which is a many–one bipartite matching problem with preferences of students over schools. However their problem differs from CHA in that the students have *priorities* at schools. Also the authors did not consider the computational complexity of their algorithm.

---

**Algorithm 6.3** Algorithm Phase 3 for CHA [535]

---

**Require:** CHA instance $I$ and a maximal, trade-in-free matching $M$
**Ensure:** $M$ is a Pareto-optimal matching in $I$
 1: $M' := M$;
 2: $Q := \emptyset$;
 3: **for each** applicant $a_i \in A$ **do**
 4:    **if** $f(a_i) = M(a_i)$ **then**
 5:       $Q$.add($a_i$);
 6:    **end if**
 7: **end for**
 8: Process($Q$);
 9: **for each** unlabelled applicant $a_i \in A$ **do**
10:    $P := \{a_i\}$;   {$P$ is a stack of applicants}
11:    $z_i := 1$;   {counter records the number of times an applicant is in $P$}
12:    **while** $P \neq \emptyset$ **do**
13:       $a_j := P$.pop();
14:       **if** $z_j = 2$ **then**
15:          $a_k := a_j$;
16:          **repeat**
17:             $Q$.add($a_k$);
18:             $a_k := P$.pop();
19:          **until** $a_k = a_j$
20:          Process($Q$);
21:       **else**
22:          $P$.push($a_j$);
23:          choose any $a_k \in M'(f(a_j))$;
24:          $z_k$++;
25:          $P$.push($a_k$);
26:       **end if**
27:    **end while**
28: **end for**

---

free matching $M$ constructed after Phases 1 and 2 into a matching $M$ that admits no cyclic coalition and is therefore Pareto optimal. Algorithm Phase 3 for CHA is shown in Algorithm 6.3.

Throughout Phase 3, we maintain a stack of applicants $P$ (which implicitly represents vertices in the envy graph) in order to enable cyclic coalitions to be detected and satisfied. Also, for each applicant $a_i$, we maintain a pointer $f(a_i)$ to the first house on $a_i$'s preference list (this pointer is assumed to be implicitly updated as necessary after any deletions from $a_i$'s preference list). We also maintain a queue $Q$ of applicants $a_i$ who are waiting to be assigned to $f(a_i)$ in $M$. Initially each applicant $a_i$ such that

---

**Algorithm 6.4** Algorithm Process($Q$) [535]

---
**Require:** a queue of applicants $Q$
**Ensure:** applicants in $Q$ are processed (see accompanying description)
 1: **while** $Q \neq \emptyset$ **do**
 2:    $a_i := Q.\text{removeHead}()$;
 3:    $h_j := f(a_i)$;
 4:    $h_k := M(a_i)$;   {possibly $h_j = h_k$}
 5:    $M := (M \backslash \{(a_i, h_k)\}) \cup \{(a_i, h_j)\}$;
 6:    $M' := M' \backslash \{(a_i, h_k)\}$;
 7:    label $a_i$;
 8:    **if** $a_i \in P$ **then**
 9:       remove $a_i$ from $P$;
10:    **end if**
11:    **if** $M'(h_k) = \emptyset$ **then**
12:       **for each** unlabelled $a_t \in L_k$ **do**
13:          delete $h_k$ from the preference list of $a_t$;
14:          **if** $f(a_t) = M(a_t)$ **then**
15:             $Q.\text{add}(a_t)$;
16:          **end if**
17:       **end for**
18:    **end if**
19: **end while**

---

$f(a_i) = M(a_i)$ is added to $Q$. We take a copy $M'$ of matching $M$ at the outset of Phase 3; after we assign a given applicant $a_i$ to a house in $M$ during Phase 3, we remove the pair $(a_i, M(a_i))$ from $M'$. Finally, we also maintain a linked list $L_j$ for each house $h_j$ containing applicants $a_i$ who prefer $h_j$ to their assigned house at the end of Phase 2.

After $Q$ is initialised at the beginning of Algorithm Phase 3, we call an additional subroutine called Algorithm Process($Q$), shown in Algorithm 6.4. This subroutine makes use of the observation (as in Ref. [18]) that no applicant $a_i$ assigned to her first choice house $h_j$ in $M$ can be involved in a cyclic coalition. This subroutine (which may also be called subsequently during Phase 3) considers each applicant $a_i$ in $Q$ in turn, removing $a_i$ from $Q$ and promoting $a_i$ from $h_k = M(a_i)$ to $h_j = f(a_i)$ in $M$ (note that, possibly $h_j = h_k$ in general (certainly this is true when Algorithm Process($Q$) is first called), in which case $a_i$'s assignment in $M$ is unchanged). The pair $(a_i, h_k)$ is then removed from $M'$. Applicant $a_i$ is then labelled to ensure that she is not subsequently added to the stack $P$ (all applicants are initially unlabelled at the outset of Phase 3). Thus, an agent is added to $Q$ at most once during an execution of Algorithm Phase 3.

When Algorithm Process($Q$) is first called, $P$ must be empty. However, this need not be true during a subsequent execution of Process($Q$). We check whether $a_i$ lies in $P$, and if so, $a_i$ is removed from $P$ so that she is not considered further by Algorithm Phase 3.[5] If $M'(h_k)$ is empty after the assignment of $a_i$ to $h_j$, then we remove $h_k$ from the preference lists of the remaining unlabelled applicants in $L_k$, since $h_k$ can no longer be involved in a cyclic coalition. We refer to those preference lists from which houses have been removed as *reduced preference lists*. We then apply the earlier observation, namely that an applicant assigned to her first choice house cannot be involved in a cyclic coalition, recursively to the reduced lists by adding to $Q$ any applicant $a_t$ for whom $f(a_t) = M(a_t)$.

During the main loop of Algorithm Phase 3, for each applicant $a_i$ who is unlabelled (and therefore not assigned to $f(a_i)$ in $M$), we use the inner while loop to build a path of applicants (represented by $P$) starting from $a_i$ and check if $P$ cycles. To do so, we use a counter $z_i$, for each applicant $a_i$, which we assume is initialised to 0 at the outset of Phase 3.

If $z_j \neq 2$ for some applicant $a_j$ in $P$ during an iteration of the while loop, then we extend $P$ by implicitly adding an edge in the envy graph, from $a_j$ to some agent in $M'(f(a_j))$ who $a_j$ envies, in line 23. Note that $M'(f(a_j))$ is non-empty by the execution of Algorithm Process($Q$).

Otherwise if $z_j = 2$, it follows that we have a cyclic coalition in $P$ starting from $a_j$. We satisfy $C$ by popping each applicant $a_k$ in $C$ from $P$ until we remove $C$, and add each such $a_k$ to $Q$. We then call Process($Q$) to assign each $a_k$ to $f(a_k)$ in $M$, to label each $a_k$ in order to remove the applicant from further consideration by the algorithm, as well as to remove $M(a_k)(= M'(a_k))$ from the preference lists of the remaining unlabelled applicants if the house becomes empty in $M'$.

Sng [535] established the correctness of Algorithm Phase 3 for CHA, and also described suitable data structures for its efficient implementation. We summarise his arguments with the following result.

**Theorem 6.19 ([535]).** *Given a* CHA *instance $I$, a maximum Pareto optimal matching can be found in $O(\sqrt{n_1}m)$ time[6], where $n_1$ is the number of*

---

[5]If $Q = \langle a_{i_1}, \ldots, a_{i_t} \rangle$ at the beginning of a call to Algorithm Process($Q$), and $a_{i_j}$ is removed from $P$ by line 9 for some $j$ ($1 \leq j \leq t$) during this call, then $a_{i_k}$ will also be removed from $P$ by this line during the same execution of Algorithm Process($Q$), for each $k$ ($j + 1 \leq k \leq t$).

[6]In Ref. [535], the weaker upper bound of $O(\sqrt{C}m)$ was given as the complexity for this algorithm, where $C$ is the total capacity of the houses. The improved upper bound follows by the remark in Footnote 6 on Page 16.

*applicants and m is the number of acceptable applicant–house pairs. Such a matching is also a maximum matching of applicants to houses in the underlying graph of I.*

Again, the algorithm can be extended in a similar way to that described in Sec. 6.2.1.2 if a subset $A'$ of the applicants already own a house, and we must find a maximum Pareto optimal matching subject to the condition that each applicant in $A'$ obtains either her initial endowment or better.

## 6.4  Hospitals / Residents problem

It is also possible to study Pareto optimal matchings in the variant of CHA in which houses have preferences over applicants. We thus arrive at the HR setting. The concept of a Pareto optimal matching in an HR instance was defined on Page 147 in the subsection on Pareto stable matchings (recall that a hospital's preferences over sets of residents was defined to be *responsive* to its preferences over individual residents, where the term *responsive* was also defined in that subsection).

Recall the definitions of a *Pareto improvement cycle* and a *Pareto improvement chain* from Definition 3.12. These definitions were given for the HRT context; in the HR setting we can delete all occurrences of "is indifferent between" when reasoning about residents' preferences in each case (of course it is still possible, even in the HR setting, for a hospital to be indifferent between two sets of residents). Erdil and Ergin [192] showed that a matching $M$ in an HR instance is Pareto optimal if and only if it admits no Pareto improvement cycle or Pareto improvement chain. Moreover, Sng [535] proved that the existence of each of these structures can be checked in linear time. This leads to the following result.

**Proposition 6.20 ([192, 535]).** *Let I be an instance of HR and let M be a matching in I. Then M is Pareto optimal if and only if M admits no Pareto improvement cycle or Pareto improvement chain. On the basis of this characterisation, we may check whether M is Pareto optimal in $O(m)$ time, where m is the number of acceptable resident–hospital pairs in I.*

Clearly a stable matching in $I$ is Pareto optimal. However Sng [535] gave a family of arbitrarily large HR instances that admit a Pareto optimal matching that is twice the size of a stable matching. This motivates the problem of finding a maximum Pareto optimal matching. Sng [535] proved that this problem can be solved by essentially truncating the preference lists of the agents and using Phases 1 and 3 of the corresponding algorithm for CHA. This led him to the following result.

**Theorem 6.21 ([535]).** *Let $I$ be an instance of* HR. *A maximum Pareto optimal matching in $I$ can be found in $O(\sqrt{n_1}m)$ time, where $n_1$ is the number of residents and $m$ is the number of acceptable resident–hospital pairs in $I$.*

## 6.5 Stable Roommates problem

### 6.5.1 *Introduction*

Pareto optimality can also be defined in instances of SRI [28, 455]: indeed, the definition of a Pareto optimal matching in SRI is completely analogous to the definition given for HA in Sec. 1.5.3. In this section we study Pareto optimal matchings in instances of SRI and its variants. We begin with some preliminary observations in Sec. 6.5.2 and then give a characterisation of Pareto optimal matching in an SRI instance in Sec. 6.5.3. In Sec. 6.5.4 we deal with the problem of finding a mxaimum Pareto optimal matching in an SRI instance, whilst Sec. 6.5.5 considers Pareto optimal matchings in CFG instances.

### 6.5.2 *Preliminary observations*

A straightforward adaptation of Algorithm SDM for HA leads to an $O(m)$ algorithm for finding a Pareto optimal matching in a given SRI instance $I$, where $m$ is the number of acceptable pairs of agents. This algorithm, which we call the Serial Dictatorship Mechanism for SRI, denoted by Algorithm SDM-SRI for short, can be found in Algorithm 6.5. The correctness and complexity of this algorithm is established by the following proposition.

**Proposition 6.22 ([28]).** *Let $I$ be an instance of* SRI. *Then Algorithm* SDM-SRI *finds a Pareto optimal matching in $I$ in $O(m)$ time, where $m$ is the number of acceptable pairs of agents in $I$.*

It is not difficult to see that, in a given SRI instance, Pareto optimal matchings may be of different sizes. For, consider the SRI instance $I_1$ shown in Fig. 6.4. It may be verified that each of $M_1 = \{\{a_2, a_3\}\}$ and $M_2 = \{\{a_1, a_2\}, \{a_3, a_4\}\}$ is Pareto optimal in $I_1$. Note that Algorithm SDM-SRI

---

**Algorithm 6.5** Algorithm SDM-SRI [28]

---

**Require:** SRI instance $I$
**Ensure:** return $M$, a Pareto optimal matching in $I$
 1:   $M := \emptyset$;
 2:   **for each** agent $a_i \in A$ **do**
 3:     set $a_i$ to be unlabelled;
 4:   **end for**
 5:   **while** there exists an unlabelled agent $a_i \in A$ **do**
 6:     **if** $a_i$ finds an unlabelled agent acceptable **then**
 7:       $a_j := $ most-preferred unlabelled agent on $a_i$'s list;
 8:       $M := M \cup \{\{a_i, a_j\}\}$;
 9:       label $a_i$ and $a_j$;
10:     **end if**
11:   **end while**
12:   **return** $M$;

---

constructs $M_1$ if $a_1$ is processed first, and constructs $M_2$ if $a_2$ is processed first. Moreover, by replicating $I_1$ we can obtain an arbitrarily large family of SRI instances for which the size of a maximum Pareto optimal matching is twice the size of a stable matching.

Note that Algorithm SDM-SRI may not be capable of finding *all* Pareto optimal matchings in a given SRI instance. For example, consider the SRI instance $I_2$ shown in Fig. 6.4. It is straightforward to verify that $M = \{\{a_1, a_2\}, \{a_3, a_4\}\}$ is Pareto optimal in $I_2$, but no execution of Algorithm SDM-SRI will construct $M$. This observation contrasts with the situation for a given HA instance $I$, where, for any given Pareto optimal matching $M$ in $I$, there is some execution of Algorithm SDM that constructs $M$ [18, Lemma 1].

The concept of exchange-stability (see Sec. 5.7) superficially resembles Pareto optimality in the SRI context, however they are distinct properties. To see this, consider the SR instances $I_3$ and $I_4$ shown in Fig. 6.4 (in $I_4$, the symbol "..." denotes all remaining agents in arbitrary order). It is not difficult to see that $M = \{\{a_1, a_3\}, \{a_2, a_4\}\}$ is Pareto optimal but not exchange-stable in $I_3$, whilst $M = \{\{a_1, a_2\}, \{a_3, a_4\}, \{a_5, a_6\}\}$ is exchange-stable but not Pareto optimal in $I_4$.

### 6.5.3   *Characterising Pareto optimal matchings*

We next give a characterisation of Pareto optimal matchings in SRI that leads to a convenient necessary and sufficient condition for a matching to

| $a_1$ : $a_2$ | $a_1$ : $a_3$ $a_2$ $a_4$ | $a_1$ : $a_4$ $a_3$ $a_2$ | $a_1$ : $a_6$ $a_2$ ... |
|---|---|---|---|
| $a_2$ : $a_3$ $a_1$ | $a_2$ : $a_4$ $a_1$ $a_3$ | $a_2$ : $a_3$ $a_4$ $a_1$ | $a_2$ : $a_3$ $a_1$ ... |
| $a_3$ : $a_2$ $a_4$ | $a_3$ : $a_2$ $a_4$ $a_1$ | $a_3$ : $a_1$ $a_2$ $a_4$ | $a_3$ : $a_2$ $a_4$ ... |
| $a_4$ : $a_3$ | $a_4$ : $a_1$ $a_3$ $a_2$ | $a_4$ : $a_2$ $a_1$ $a_3$ | $a_4$ : $a_5$ $a_3$ ... |
| | | | $a_5$ : $a_4$ $a_6$ ... |
| Instance $I_1$ | Instance $I_2$ | Instance $I_3$ | $a_6$ : $a_1$ $a_5$ ... |

Instance $I_4$

Fig. 6.4  Four instances of SRI

be Pareto optimal. This condition involves a structure defined as follows (in the definition, recall that $bp(M)$ denotes the set of blocking pairs with respect to a given matching $M$).

**Definition 6.23 ([28]).** *Let $M$ be a matching in an instance $I$ of* SRI. *An* improving coalition *with respect to $M$ is a sequence of distinct agents $C = \langle a_{i_0}, a_{i_1}, \ldots, a_{i_{2r-1}} \rangle$, for some $r \geq 1$, such that:*

*(1) $\{a_{i_{2j-1}}, a_{i_{2j}}\} \in M$ $(1 \leq j \leq r-1)$;*
*(2) $\{a_{i_{2j}}, a_{i_{2j+1}}\} \in bp(M)$ $(0 \leq j \leq r-1)$;*
*(3) Either (a) $a_{i_0}$, $a_{i_{2r-1}}$ are unmatched in $M$, or (b) $r \geq 2$ and $\{a_{i_0}, a_{i_{2r-1}}\} \in M$.*

*If $C$ satisfies Condition 3(a), we also refer to $C$ as an* augmenting coalition, *otherwise we also refer to $C$ as a* cyclic coalition. *The matching*

$$M' = (M \backslash \{\{a_{i_{2j-1}}, a_{i_{2j}}\} : 1 \leq j \leq r\}) \cup \{\{a_{i_{2j}}, a_{i_{2j+1}}\} : 0 \leq j \leq r-1\}$$

*is defined to be the matching obtained from $M$ by* satisfying $C$, *where addition is taken modulo $2r$. (Note that if $C$ is an augmenting coalition then $\{a_{i_0}, a_{i_{2r-1}}\} \notin M$.)*

We remark that Definition 6.23 was also given independently by Morrill [455] for the special case that $I$ is an instance of SR. The following proposition indicates that Pareto optimality is equivalent to the absence of an improving coalition.

**Proposition 6.24 ([28]).** *Let $M$ be a matching in a given instance $I$ of* SRI. *Then $M$ is Pareto optimal in $I$ if and only if $M$ admits no improving coalition.*

Again, we note that Proposition 6.24 was proved independently by Morrill [455] for the special case that $I$ is an instance of SR.

We now show that Proposition 6.24 leads to an $O(m)$ algorithm for checking a matching for Pareto optimality in an instance $I$ of SRI, where $m$ is the number of acceptable pairs of agents in $I$. Let $G$ be the underlying graph of $I$ (i.e., $G$ has a vertex for each agent and an edge between each pair of acceptable agents). We form a subgraph $G_M$ of $G$ by letting $G_M$ contain only those edges that belong to $M \cup bp(M)$; any isolated vertices are removed from $G_M$. By Proposition 6.24, $M$ is Pareto optimal in $I$ if and only if $M$ admits no augmenting path or alternating cycle in $G_M$. We may test for the existence of the former structure in $O(m)$ time [225, 229]. For the latter structure, we remove any unmatched vertices from $G_M$ (and any edges incident to them) and apply the $O(m)$ alternating cycle detection algorithm of Gabow *et al.* [228]. This discussion leads to the following conclusion.

**Proposition 6.25 ([28]).** *Let $I$ be an instance of* SRI *and let $M$ be a matching in $I$. Then we may check whether $M$ is Pareto optimal in $O(m)$ time, where $m$ is the number of acceptable pairs of agents in $I$.*

Given an instance $I$ of SR with $n$ agents, and given a matching $M$ in $I$, Morrill [455] gave an explicit $O(n^2)$ algorithm for checking $M$ for Pareto optimality in $I$. Essentially his algorithm involves detecting alternating cycles, as above.

We next observe that a stable matching $M$ in an SRI instance $I$ must be Pareto optimal. For, suppose not. Then $M$ admits an improving coalition by Proposition 6.24, which implies that $bp(M) \neq \emptyset$ by Definition 6.23, a contradiction. We thus obtain the following result.

**Proposition 6.26 ([28]).** *Let $I$ be an instance of* SRI *and let $M$ be a stable matching in $I$. Then $M$ is Pareto optimal.*

### 6.5.4 *Maximum Pareto optimal matchings*

We now turn to the problem of finding a maximum Pareto optimal matching in an SRI instance $I$. This problem may be solved by imposing weights on the edges in the underlying graph $G = (A, E)$ of $I$, where for each edge $\{a_i, a_j\} \in E$, the weight of this edge is $rank(a_i, a_j) + rank(a_j, a_i)$ where $rank(a_i, a_j)$ is as defined in Sec. 1.4.2. We may construct a minimum weight maximum cardinality matching $M$ in $G$ in $O(\sqrt{n}\alpha(n, m) m \log^{3/2} n)$ time

[231], where $n = |A|$, $m = |E|$ and $\alpha$ is the inverse Ackermann function. The following result indicates that $M$ is a maximum Pareto optimal matching.

**Proposition 6.27 ([28]).** *Let $M$ be a minimum weight maximum cardinality matching in the weighted graph $G$ defined above. Then $M$ is a maximum Pareto optimal matching in $I$.*

Note that the above proposition also indicates that the size of a maximum Pareto optimal matching in $I$ is equal to the size of a maximum matching in $G$. An alternative way to find a maximum Pareto optimal matching in $I$ is to start with a maximum matching in $G$, which may be found in $O(\sqrt{n}m)$ time [451, 577]. Then, we could try to transform $M$ into a maximum Pareto optimal matching in $I$ by finding and satisfying a sequence of cyclic coalitions ($M$ cannot admit an augmenting coalition as it is of maximum cardinality). A naïve complexity bound for this step is $O(m^2)$, which follows by again observing that we can find a cyclic coalition relative to $M$, if one exists, in $O(m)$ time [228]. Satisfying a cyclic coalition involves at least two agents improving by at least one position on their preference lists, so the number of cyclic coalitions in the sequence must be bounded above by $m/2$. Since each cyclic coalition can be satisfied in $O(m)$ time, this leads to an $O(m^2)$ algorithm for transforming an arbitrary maximum matching into a maximum Pareto optimal matching.

Morrill [455] gave an $O(n^3)$ algorithm that transforms an arbitrary matching $M$ in an SR instance $I$ with $n$ agents to a Pareto optimal matching through a sequence of Pareto improvements. This is done using an $O(n^2)$ algorithm to find a Pareto improvement of $M$ (which essentially amounts to satisfying a cyclic coalition) if one exists. Morrill's algorithm may easily be extended to the case that $I$ is an SRI instance and $M$ is a maximum Pareto optimal matching in $I$ (simply discard the agents in $I$ who are unassigned after computing $M$). It then follows that Morrill's approach yields an $O(m)$ algorithm for finding a Pareto improvement of $M$ if one exists. However for every matching in Morrill's sequence of Pareto improvements, some agent obtains her highest achievable partner (which means that neither she nor her partner can subsequently be involved in a cyclic coalition). This implies a tighter bound of $n/2$ for the number of cyclic coalitions in the sequence, and thus an $O(nm)$ algorithm for transforming an arbitrary maximum matching into a maximum Pareto optimal matching.

Unfortunately though, neither of the approaches described in the two preceding paragraphs gives a better time complexity for the problem of finding a maximum Pareto optimal matching than is obtained by simply constructing a minimum weight maximum cardinality matching in the underlying weighted graph. Indeed, even Morrill's algorithm for transforming an arbitrary matching $M$ to a Pareto optimal matching in an SR instance $I$ can be improved upon by this technique (simply discard the unassigned agents in $I$, truncate the preference list of each assigned agent $a_i$ by deleting every agent inferior to $M(a_i)$, and construct a minimum weight maximum cardinality matching in the corresponding underlying weighted graph). It remains open as to whether there is an $O(\sqrt{n}m)$ algorithm for finding a maximum Pareto optimal matching in $I$.

The above discussions do however indicate that any matching $M$ in an SRI instance $I$ can be transformed into a Pareto optimal matching $M'$ such that $|M| = |M'|$. This implies that Theorems 6.6, 6.7 and 6.8 carry over to the SRI case. We remark that Theorem 6.6 holds even for an instance of SMI where each man finds two women acceptable, and each woman finds at most three men acceptable. Also Theorems 6.7 and 6.8 hold because the relevant results for maximal matchings (i.e., any two maximal matchings differ in size by at most a factor of 2, and there are maximal matchings of each size between the minimum and the maximum sizes) hold for general, and not just bipartite, graphs [399, 276].

### 6.5.5  *Coalition formation games*

In Sec. 4.8.8 we introduced CFG as an extension of SRI. Most of the main results in the papers referenced in that section relate to problems associated with computing core and strong core partitions of the agents. However it is equally possible to define the weaker concepts of Pareto optimal and strongly Pareto optimal partitions respectively. Whilst core and strong core partitions guarantee the absence of a blocking or weakly blocking coalition $A'$ respectively (see Sec. 4.8.8 for an informal description of these terms), Pareto and strongly Pareto optimal partitions ensure that this is true only for the special case that $A'$ is the grand coalition.

Aziz et al. [55] presented algorithmic results for deciding the existence of Pareto optimal and strongly Pareto optimal partitions, and testing whether a given partition is Pareto optimal or strongly Pareto optimal, in a given CFG instance. Cechlárová et al. [124] and Cechlárová and Borbel'ová [105] did likewise for the case that the agents within a coalition are ordered (this case is motivated by kidney exchange as described in the final paragraph of Sec. 4.8.8).

## 6.6 Conclusions and open problems

Whilst economists have studied Pareto optimal matchings for many years, and have long recognised their importance as a fundamental solution concept in a cooperative game, it is only relatively recently that computer scientists and others have begun to study Pareto optimal matchings from an algorithmic point of view. As discussed in Sec. 6.2.1 and 6.2.2, Pareto optimal matchings are closely associated with the concept of matchings in the core of an associated housing market. The classical TTC algorithm [527] is able to find the unique matching in the core in the HM case, and can also be used to find a Pareto improvement of a given matching, but only relatively recently was its efficient implementation considered [18].

Given the relative scarcity of algorithmic results concerning Pareto optimal matchings, some interesting open problems remain. Perhaps the most intriguing are the following two:

- Is there an $O(\sqrt{n_1}m)$ algorithm for finding a maximum Pareto optimal matching in a given HAT instance, where $n_1$ is the number of applicants and $m$ is the number of acceptable applicant–house pairs (see Sec. 6.2.2)? Such an algorithm may incorporate an $O(m)$ extension of the TTC algorithm to transform an arbitrary maximum matching into a maximum Pareto optimal matching in a given HAT instance.
- Similarly, is there an $O(\sqrt{n}m)$ algorithm for finding a maximum Pareto optimal matching in a given SRI instance, where $n$ is the number of agents and $m$ is the number of acceptable pairs of agents (see Sec. 6.5)? Again, such an algorithm may incorporate an $O(m)$ extension of the TTC algorithm to transform an arbitrary maximum matching into a maximum Pareto optimal matching in a given SRI instance.

# Chapter 7

# Popular matchings

## 7.1 Introduction

Popular matchings, defined in Sec. 1.5.5, have been an exciting area of research in the last few years. The notion of a popular matching was introduced by Gärdenfors [240] in 1975, who used the terms *majority assignment* and *strong majority assignment* rather than popular matching and strongly popular matching respectively. However, as we will see in this chapter, in fact the concept of popularity has its roots all the way back to 1785, when the notion of a *Condorcet winner* was proposed [154].

Between 1975 and 2005, there was, to the best of the author's knowledge, no published research on popular matchings, at least from an algorithms and complexity standpoint. The paper of Abraham *et al.* [20], appearing in *Proceedings of the 10th ACM–SIAM Symposium on Discrete Algorithms* in 2005, was the first to present algorithmic results on popular matchings.

Interestingly, the notion of a popular matching was presented by Rob Irving in December 2002 at an open problems session attended by members of the Formal Analysis, Theory and Algorithms research group of the School of Computing Science, University of Glasgow. There, he coined the term "popular matching" (unaware at that time of the previous work of Gärdenfors [240]), and posed the question as to whether there was a polynomial-time algorithm to find a popular matching or report that none exists, given an instance of HA.

Sitting in the audience was David Abraham, at that time a research student in the School. He did not work on the problem immediately, but kept it in mind and mentioned it when visiting the research group of Kurt Mehlhorn at Max-Planck Institut für Informatik in Saarbrücken in the spring of 2004. There, he collaborated with Kurt Mehlhorn and also Telikepalli Kavitha, at that time a postdoctoral researcher within the group. Following a collaboration involving the four authors, the end result was Ref. [20]. If ever there was a paper with a good example of the "book proof" [34] for the polynomial-time solvability of a particular problem, this is one.

The 2005 paper of Abraham et al. [20] led to a flurry of subsequent papers exploring further aspects and extensions of popular matchings in HA, HAT and in other matching problems. In this chapter we survey structural and algorithmic results for popular matchings in a range of types of matching problem instances, extending and updating the brief surveys on this topic that appeared in the *Encyclopedia of Algorithms* [361, 450].

The remainder of this chapter is structured as follows. We consider popular matchings in HA, CHA, SRI and SMI in Secs. 7.2, 7.3, 7.5 and 7.6 respectively. In Sec. 7.4 we study the case where the agents have weights indicating their priorities, and popularity is defined in terms of these weights.

Our main focus throughout these sections is on presenting structural and algorithmic results concerning characterisations of popular matchings, testing a matching for popularity and finding a popular matching or reporting that none exists.

Throughout this chapter we will say that an instance of a matching problem (such as HA or SRI) is *solvable* if it admits a popular matching, and *unsolvable* otherwise. In the case where an instance is unsolvable, we will consider matchings that have low "unpopularity". We also give similar results for strongly popular matchings throughout the chapter. Finally, in Sec. 7.7, we present some concluding remarks and open problems.

## 7.2   House Allocation problem

### 7.2.1   *Introduction*

We begin with popular matchings in HA: recall from instance $I_1$ shown in Fig. 1.3 that an HA instance need not admit a popular matching. In Sec. 7.2.2 we present a characterisation of popular matchings, leading on to an efficient algorithm for finding a popular matching or reporting that none exists, presented in Sec. 7.2.3. As we will see, popular matchings can have different sizes, so in Sec. 7.2.4 we see how to find a maximum cardinality popular matching in a solvable HA instance.

The set of popular matchings in an HA instance has some nice structural properties which are captured using the so-called *switching graph*. We

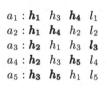

$a_1 : \boldsymbol{h_1} \quad h_3 \quad \boldsymbol{h_4} \quad l_1$
$a_2 : \boldsymbol{h_1} \quad \boldsymbol{h_4} \quad h_2 \quad l_2$
$a_3 : \boldsymbol{h_2} \quad h_1 \quad h_3 \quad \boldsymbol{l_3}$
$a_4 : \boldsymbol{h_2} \quad h_3 \quad \boldsymbol{h_5} \quad l_4$
$a_5 : \boldsymbol{h_3} \quad \boldsymbol{h_5} \quad h_1 \quad l_5$

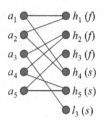

Fig. 7.1 An HA instance with $f(a_i)$ and $s(a_i)$ highlighted for each applicant $a_i$

Fig. 7.2 The reduced graph for the HA instance shown in Fig. 7.1

describe this graph and some algorithmic results that may be derived with the aid of it in Sec. 7.2.5. We then move on to the case where preference lists can include ties in Sec. 7.2.6. For an unsolvable instance of HA or HAT, we might try to find a matching that is "as popular as possible" in a precise sense. Suitable definitions of matchings with low unpopularity are given in Sec. 7.2.7, together with algorithmic results for computing such matchings. We consider strongly popular matchings in HA in Sec. 7.2.8. Finally, additional results concerning popular matchings in instances of HA and HAT are presented in Sec. 7.2.9.

### 7.2.2 *Characterising popular matchings*

Abraham *et al.* [21] arrived at a very neat characterisation of popular matchings in a given instance $I$ of HA. In particular they showed that, no matter the length of an applicant $a_i$'s preference list, she has at most two houses on her list to which she can be assigned in a given popular matching in $I$.

To describe this characterisation, we require to adjust the definition of instance $I$. That is, for each applicant $a_i \in A$ we create a new unique *last resort house* $l_i$ and append $l_i$ to $a_i$'s preference list. This ensures that henceforth we can assume that every applicant is assigned in a given matching in $I$. However in reality, $(a_i, l_i) \in M$ intuitively signifies that $a_i$ is unassigned. Henceforth throughout Secs. 7.2-7.4, we assume that this transformation has automatically been made for any HA or HAT instance, and that last resort houses are always included in the set of houses $H$.

For each applicant $a_i \in A$, define $f(a_i)$ to be the first house in $a_i$'s preference list. Given any house $h_j \in H$, define $f(h_j)$ to be the set of applicants $a_i$ such that $f(a_i) = h_j$. Then $h_j$ is called an *$f$-house* if $f(h_j) \neq \emptyset$. For a given applicant $a_i \in A$, define $s(a_i)$ to be the first non $f$-house in

$a_i$'s preference list (such a house is bound to exist because of $a_i$'s last resort $l_i$, which appears on no other preference list). A house $h_j \in H$ is called an *s-house* if $h_j = s(a_i)$ for some $a_i \in A$. By definition, the sets of $f$-houses and $s$-houses are disjoint. Fig. 7.1 shows an example HA instance where $f(a_i)$ and $s(a_i)$ are highlighted in the preference list of each applicant $a_i$. The $f$-houses are $\{h_1, h_2, h_3\}$ and the $s$-houses are $\{h_4, h_5, l_3\}$.

Abraham *et al.* [21] proved an important result concerning $f$ and $s$-houses, indicating that $f(a_i)$ and $s(a_i)$ are the only candidate houses to which an applicant $a_i$ can be assigned in a given popular matching $M$. They also showed that every $f$-house must be assigned in $M$, and furthermore that these two necessary conditions are also sufficient conditions for a matching to be popular.

**Lemma 7.1 ([21]).** *Let I be an instance of* HA *and let M be a matching in I. Then M is popular if and only if:*

*(i) every f-house is assigned in M;*
*(ii) for each applicant $a_i \in A$, $M(a_i) \in \{f(a_i), s(a_i)\}$.*

Lemma 7.1 leads to the definition of the *reduced graph $G'$*, which is a subgraph of the underlying graph $G$ of $I$ containing the vertices in $A \cup H$ together with the edges $\{a_i, f(a_i)\}$ and $\{a_i, s(a_i)\}$ for each $a_i \in A$. The reduced graph for the HA instance shown in Fig. 7.1 is illustrated in Fig. 7.2. In the latter figure, isolated house vertices are omitted, each $f$-house is labelled $(f)$, and each $s$-house is labelled $(s)$.

Given a set $S \subseteq A \cup H$, define $M$ to be *S-complete* if every agent in $S$ is assigned in $M$. An $A$-complete matching is also referred to as being *applicant-complete*. These definitions imply the following restatement of Lemma 7.1:

**Theorem 7.2 ([21]).** *Let I be an instance of* HA *and let M be a matching in I. Then M is popular if and only if:*

*(i) every f-house is assigned in M;*
*(ii) M is an applicant-complete matching in the reduced graph $G'$.*

**Corollary 7.3 ([21]).** *Let I be an instance of* HA *and let M be a matching in I. There is an $O(m)$ algorithm to test whether M is popular in I, where m is the number of acceptable applicant–house pairs in I.*

---

**Algorithm 7.1** Algorithm Popular-HA [21]

---

**Require:** HA instance $I$
**Ensure:** return a popular matching $M$ or "no popular matching exists"
1: $G' :=$ reduced graph of $I$;
2: **if** $G'$ admits an applicant-complete matching $M$ **then**
3:     **for each** $f$-house $h_j$ that is unassigned in $M$ **do**
4:        choose any $a_i \in f(h_j)$;
5:        $M := (M \backslash \{(a_i, M(a_i))\}) \cup \{(a_i, h_j)\}$;
6:     **end for**
7:     **return** $M$;
8: **else**
9:     **return** "no popular matching exists";
10: **end if**

---

*Proof.* Clearly the $f$-houses and $s$-houses in $I$ can be identified in $O(m)$ time. In order to obtain an overall time complexity of $O(m)$, it is sufficient to add only the $f$-houses and $s$-houses to $G'$. Clearly Parts (i) and (ii) of Theorem 7.2 can be verified in $O(m)$ time. $\square$

### 7.2.3 *Finding a popular matching*

Theorem 7.2 leads to an $O(n + m)$ algorithm, due to Abraham *et al.* [21], for finding a popular matching in $I$ or reporting that none exists, where $n = n_1 + n_2$ is the number of applicants and houses, and $m$ is the number of acceptable applicant–house pairs. This method is described by Algorithm Popular-HA in Algorithm 7.1.

This algorithm tries to construct an applicant-complete matching in $G'$, given the characterisation of Theorem 7.2. Note that even if $G'$ admits such a matching $M$, it is possible that $M$ might leave one or more $f$-houses unassigned, contrary to Part (i) of Theorem 7.2. Hence lines 3-6 of Algorithm Popular-HA perform the necessary promotions of applicants as many times as necessary in order to satisfy the required condition. Following this, the resultant matching $M$ is returned (note that, strictly speaking, pairs involving last resort houses should ultimately be deleted from $M$, but for brevity we will not carry out this step explicitly in this chapter).

Clearly $G'$ can be constructed in $O(n + m)$ time. Using the Hopcroft–Karp algorithm for finding a maximum matching in $G'$ would imply an $O(n_1^{1.5})$ time complexity for line 2 of Algorithm Popular-HA, since $m = 2n_1$. However this step can be carried out in $O(n + m)$ time using the algorithm shown in Algorithm 7.2. Note that in lines 5 and 8, $G' = G' \backslash \{v\}$ refers to

---

**Algorithm 7.2** Finding a maximum matching in $G'$ [21]

---

**Require:** reduced graph $G'$
**Ensure:** return an applicant-complete matching $M$ or "no applicant-complete matching exists"

1:   $M := \emptyset$;
2: **while** some house $h_j \in H$ has degree 1 in $G'$ **do**
3:     $a_i :=$ unique applicant adjacent to $h_j$ in $G'$;
4:     $M := M \cup \{(a_i, h_j)\}$;
5:     $G' := G' \backslash \{a_i, h_j\}$;
6: **end while**
7: **while** some house $h_j \in H$ has degree 0 in $G'$ **do**
8:     $G' := G' \backslash \{h_j\}$;
9: **end while**
10: **if** $|H| < |A|$ **then**
11:     **return** "no applicant-complete matching exists";
12: **else**
13:     $M' :=$ maximum matching in $G'$;
14:     **return** $M \cup M'$;
15: **end if**

---

the deletion of a vertex $v$, together with its incident edges, from $G'$. The corresponding vertex is also removed from $A$ or $H$ as appropriate.

This algorithm begins by adding to $M$ every pair $(a_i, h_j)$ such that $h_j$ has degree 1 in $G'$. Vertices $a_i$, $h_j$ and their incident edges can then be removed from $G'$ as they cannot subsequently belong to an augmenting path of $M$. Similarly we next delete every isolated house vertex from $G'$. Once we reach line 10, every house vertex has degree at least 2 and every applicant vertex has degree exactly 2 in $G'$. If $|H| < |A|$ then $G'$ cannot admit an applicant-complete matching by Theorem 1.3. Otherwise $|H| \geq |A|$, which implies that $|A| = |H|$ and every house vertex has degree exactly 2 in $G'$. That in turn implies that $G'$ is a disjoint collection of cycles, from which a perfect matching $M'$ can be easily found by selecting alternate edges during a traversal of each cycle. The edges in $M'$, together with the earlier edges added to $M$, can then be returned as an applicant-complete matching in $G'$. Clearly this step runs in $O(n + m)$ time overall. We thus obtain the following result.

**Theorem 7.4 ([21]).** *Let $I$ be an instance of* HA. *There is an $O(n + m)$ algorithm to find a popular matching $I$, or report that $I$ is unsolvable, where $n$ is the number of applicants and houses, and $m$ is the number of acceptable applicant–house pairs.*

To illustrate the algorithm, the matching

$$M_1 = \{(a_1, h_1), (a_2, h_4), (a_3, l_3), (a_4, h_2), (a_5, h_5)\}$$

could be constructed as an applicant-complete matching in the reduced graph for the HA instance $I'$ shown in Fig. 7.1. However $h_3$ is an unassigned $f$-house, so after promoting $a_5$ to $h_3$ we arrive at the following popular matching of size 4:

$$M_2 = \{(a_1, h_1), (a_2, h_4), (a_4, h_2), (a_5, h_3)\}.$$

To estimate the probability of a popular matching existing, Abraham et al. [21] generated random instances of HA and measured the proportion of these that were solvable.indexsolvability probability The experiments constructed instances where $n_1 = n_2$ and all preference lists were of the same length $k$. For given values of $n_1$ and $k$, 1000 random instances were created. For $n_1 = 10$, $k$ varied from $1, 2, \ldots, 10$ and the proportion of solvable instances fell from 1000 (for $k = 1$) to 556 (for $k = 10$). For $n_1 = 100$, $k$ varied over $1, 2, \ldots, 10, 20, 30, \ldots, 100$. The solvability proportion decreased from 1000 (when $k = 1$) to 2 (when $k = 10$), and none of the generated instances was solvable for $k \geq 20$.

### 7.2.4 *Maximum popular matchings*

It turns out that the matching $M_2$ illustrated at the end of the previous subsection is not a maximum cardinality popular matching (henceforth a maximum popular matching) in the HA instance shown in Fig. 7.1. In this subsection we show how to find a maximum popular matching, and indeed a popular matching of any size between the smallest and largest.

In a general solvable HA instance $I$, let $pop^-(I)$ and $pop^+(I)$ denote the sizes of a minimum and maximum popular matching in $I$ respectively. For any $k$ ($pop^-(I) \leq k \leq pop^+(I)$), we can efficiently construct a popular matching of size $k$. To see this, we begin by demonstrating how to find a minimum popular matching. Let $A_1$ be the set of applicants $a_i \in A$ satisfying $s(a_i) = l_i$, and let $A_2 = A \backslash A_1$. Now use Algorithm 7.2 on the subgraph $G''$ of $G'$ induced by the vertices in $A_2 \cup H$. If $G''$ admits no $A_2$-complete matching then $I$ has no popular matching by Theorem 7.2. Otherwise let $M$ be an $A_2$-complete matching in $G''$.

If any $f$-house $h_j$ where $A_2 \cap f(h_j) \neq \emptyset$ is unassigned in $M$, then promote some $a_i \in A_2 \cap f(h_j)$ to $h_j$ in $M$. Repeat this step until each $f$-house that is unassigned in $M$ satisfies $f(h_j) \subseteq A_1$. Now, for any $f$-house $h_j$ that is unassigned in $M$, add $(a_i, h_j)$ to $M$ for some $a_i \in f(h_j)$ (by necessity $a_i \in A_1$). Repeat this step until no unassigned $f$-houses remain. Note that the choice of applicant $a_i$ at each step can be completely arbitrary since $f(h_{j_1}) \cap f(h_{j_2}) = \emptyset$ for any two $f$-houses $h_{j_1}$ and $h_{j_2}$. At this point $M$ is a minimum popular matching in $I$ by Theorem 7.2 (where the remaining applicants in $A_1$ are implicitly assigned to their last resort houses).

To obtain larger popular matchings, proceed as follows. Let $G''$ now be the subgraph of $G'$ obtained by deleting all the last resort houses and their incident edges. Search for an augmenting path in $G''$ relative to $M$. If such a path $P$ exists then augment $M$ along $P$ to obtain a matching with one additional edge. Continuing in this way, we can arrive at a popular matching of size $k$ (where the remaining unassigned applicants in $A_1$ are simply allocated to their last resort houses). The overall time complexity is $O(n + m)$, since the applicants and houses involved in one alternating path search need never be considered in a subsequent alternating path search, as all applicants have degree at most 2. This leads to the following result.

**Theorem 7.5 ([21]).** *Let $I$ be a solvable instance of* HA. *Given any $k$ $(pop^-(I) \leq k \leq pop^+(I))$, there is an $O(n + m)$ algorithm to find a popular matching of size $k$ in $I$, where $n$ is the number of applicants and houses, and $m$ is the number of acceptable applicant–house pairs.*

$M = \{(a_1, h_1), (a_2, h_4), (a_3, h_2), (a_4, h_5), (a_5, h_3)\}$ is a maximum popular matching in the HA instance shown in Fig. 7.1. A corollary of Theorem 7.5 is that, in an instance of HA, popular matchings interpolate, i.e., there are popular matchings of all sizes between the minimum and maximum.

### 7.2.5   Structure of popular matchings

McDermid and Irving [442] characterised the structure of the set of popular matchings for an HA instance $I$, and gave efficient algorithms to count and enumerate the popular matchings, and to find several kinds of optimal popular matchings. Their characterisation was in terms of a structure called the *switching graph*, which is a directed graph that is defined relative to a popular matching $M$ in $I$.

To define the switching graph, we require some additional notation that is defined relative to $M$ and $I$. Recall from Theorem 7.2 that the only possible pairs in $M$ involving an applicant $a_i \in A$ are $(a_i, f(a_i))$ and $(a_i, s(a_i))$. If $(a_i, f(a_i)) \in M$, define $O_M(a_i) = s(a_i)$, whilst if $(a_i, s(a_i)) \in M$, define

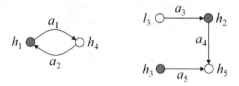

Fig. 7.3   The switching graph for a given popular matching

$O_M(a_i) = f(a_i)$. Intuitively, $O_M(a_i)$ is the "other" house that $a_i$ might be assigned to in some other popular matching. We are now in a position to define the switching graph.

**Definition 7.6 ([442]).** *Let $I$ be a solvable instance of* HA *and let $M$ be a popular matching in $I$. The switching graph $G_M$ of $M$ in $I$ is a directed graph whose vertices are precisely the houses $H$. There is an arc $(h_j, h_k)$ in $G_M$ whenever $h_j = M(a_i)$ and $h_k = O_M(a_i)$ for some applicant $a_i \in A$. In such a case $a_i$ is the label of arc $(h_j, h_k)$. A component of $G_M$ is any maximal weakly connected subgraph of $G_M$. An applicant $a_i \in A$ is said to be in a component $C$ of $G_M$, denoted $a_i \in C$, if the arc labelled by $a_i$ is in $C$.*

It is clear that the switching graph of a popular matching in a given HA instance $I$ can be constructed in $O(n + m)$ time, where $n$ is the number of applicants and houses, and $m$ is the number of acceptable applicant–house pairs in $I$. McDermid and Irving [442] remarked that Mahdian [412] defined a similar structure to the switching graph (though only as a means for examining the probability of a popular matching existing in a random HA instance).

Fig. 7.3 illustrates the switching graph relative to the HA instance shown in Fig. 7.1 and the matching

$$M = \{(a_1, h_1), (a_2, h_4), (a_3, l_3), (a_4, h_2), (a_5, h_3)\}.$$

Each arc is labelled with the applicant that represents it. Also shaded vertices correspond to $f$-houses, whilst clear vertices refer to $s$-houses. An example of the switching graph for a popular matching in a much larger HA instance is given in Ref. [442].

The following useful properties of switching graphs were observed by McDermid and Irving [442].

**Lemma 7.7 ([442]).** *Let $I$ be a solvable instance of* HA, *let $M$ be a popular matching in $I$ and let $G_M$ be the switching graph of $M$ in $I$. Then:*

(i)   *each vertex in $G_M$ has out-degree at most 1;*

(ii)  *the sink vertices in $G_M$ correspond to the houses that are unassigned in $M$; each sink vertex is an s-house;*

(iii) *each component of $G_M$ contains either a single sink vertex or a single cycle.*

Define a component in $G_M$ to be a *tree component* (respectively *cycle component*) if it contains a sink vertex (respectively cycle). By Lemma 7.7, this characterisation is well-defined. The unique cycle in a cycle component of $G_M$ is called a *switching cycle*. If $h_k$ is the unique sink vertex in a tree component $C$ of $G_M$, and $h_j$ is any other *s*-house in $C$, the (unique) path from $h_j$ to $h_k$ in $C$ is called a *switching path*. The switching graph in Fig. 7.3 clearly has one cycle component and one tree component, containing one switching path.

Given a switching cycle or a switching path $C$ in $G_M$, the operation of *applying* $C$ to $M$ results in the following matching:

$$M \cdot C = (M \setminus \{(a_i, M(a_i)) : a_i \in C\}) \cup \{(a_i, O_M(a_i)) : a_i \in C\}.$$

Intuitively, if an arc $(h_j, h_k)$ labelled by an applicant $a_i$ is in $C$, then $a_i$ moves from $h_j$ to $h_k$ when we apply $C$ to $M$. In the case that $C$ is a switching path from $h_p$ to $h_q$, $h_p$ is assigned in $M$ and unassigned in $M \cdot C$, whilst $h_q$ is unassigned in $M$ and assigned in $M \cdot C$. McDermid and Irving [442] proved that switching cycles and switching paths maintain the popularity of a given matching, as follows.

**Theorem 7.8 ([442]).** *Let $I$ be a solvable instance of* HA *and let $M$ be a popular matching in $I$. Let $C$ be a switching cycle or a switching path in the switching graph of $M$ in $I$. Then $M \cdot C$ is a popular matching in $I$.*

The authors also proved that, given any two popular matchings $M$ and $M'$ in an HA instance $I$, $M'$ can be obtained from $M$ by successively applying at most one switching path in each tree component of the switching graph $G_M$ of $M$ in $I$, and by either applying or not applying the switching cycle in each cycle component of $G_M$. This led to the following fundamental result concerning the set of popular matchings in $I$.

**Theorem 7.9 ([442]).** *Let $I$ be a solvable instance of* HA, *let $M$ be a popular matching in $I$ and let $G_M$ be the switching graph of $M$ in $I$. Suppose that the tree components of $G_M$ are $T_1, \ldots, T_r$ and the cycle components of $G_M$ are $C_1, \ldots, C_s$. Then the set of popular matchings in $I$ corresponds to* *precisely those matchings that are obtained by applying at most one switching path in $T_i$, for each $i$ ($1 \leq i \leq r$), and by either applying or not applying the switching cycle in $C_i$, for each $i$ ($1 \leq i \leq s$).*

Theorem 7.9 has a number of consequences for the problems of counting, sampling and enumerating all popular matchings, and finding various types of optimal popular matchings. Firstly, we obtain a nice closed-form formula for the number of popular matchings in a given HA instance. This follows from the observation that, in the context of the notation in Theorem 7.9, for each cycle component of $G_M$ we have two choices: either apply or do not apply the unique switching cycle. These choices are independent of one another, so there are $2^s$ possible combinations of decisions. Now suppose that $s(T_i)$ denotes the set of $s$-houses in each tree component $T_i$ ($1 \leq i \leq r$). One vertex in $s(T_i)$ is the sink vertex, so there are $|s(T_i)| - 1$ switching paths in $T_i$. If we are to apply at most one switching path in $T_i$ then there are $s(T_i)$ possible ways of doing this (including the choice not to apply any switching path). Again, all the choices in the various tree components are independent of one another (and independent of the earlier decisions in the cycle components) and thus we have the following result.

**Theorem 7.10 ([442]).** *Let $I$ be a solvable instance of HA, let $M$ be a popular matching in $I$ and let $G_M$ be the switching graph of $M$ in $I$. Suppose that the tree components of $G_M$ are $T_1, \ldots, T_r$ and the cycle components of $G_M$ are $C_1, \ldots, C_s$. Then the number of popular matchings in $I$ is $2^s \times \prod_{i=1}^{r} |s(T_i)|$.*

The switching graph illustrated in Fig. 7.3 for the HA instance $I'$ of Fig. 7.1 has one $s$-house other than the sink vertex in the tree component, and given that there is one cycle component, Theorem 7.10 implies that $I'$ admits 4 popular matchings.

The discussion prior to Theorem 7.10 suggests that, given a suitable representation of the various independent decisions that we require to make in each tree component and in each cycle component of the switching graph (see Ref. [442] for further details), we can obtain algorithms for sampling and enumerating the set of popular matchings.

**Theorem 7.11 ([442]).** *Let $I$ be an instance of HA, and let $\mathcal{M}_{pop}$ denote the set of popular matchings in $I$. There is an $O(n + m)$ algorithm for sampling $\mathcal{M}_{pop}$ (i.e., for generating a popular matching in $I$ uniformly at random), where $n$ is the number of applicants and houses, and $m$ is the* *number of acceptable applicant–house pairs in $I$. Also, there is an $O(n + m + |\mathcal{M}_{pop}|)$ algorithm for listing all the popular matchings in $I$.*

Define an acceptable applicant–house pair $(a_i, h_j)$ to be *popular* if it belongs to some popular matching. The switching graph $G_M$ for a given popular matching $M$ can be used to characterise the set of popular pairs in $I$. McDermid and Irving [442] proved that $(a_i, h_j)$ is popular if and only if either $(a_i, h_j) \in M$, or $(h_k, h_j)$ is an arc in $G_M$ labelled by $a_i$, where $(h_k, h_j)$ belongs to either a switching cycle or a switching path in $G_M$. From this, the following result can be deduced immediately.

**Theorem 7.12 ([442]).** *Let $I$ be an instance of* HA. *The set of popular pairs in $I$ can be found in $O(n+m)$ time, where $n$ is the number of applicants and houses, and $m$ is the number of acceptable applicant–house pairs in $I$.*

The switching graph can also be used to efficiently find various kinds of optimal popular matchings: for example a *minimum weight maximum cardinality popular matching*, a *rank-maximal popular matching* and a *generous popular matching*. The definitions of these matchings combine the concept of a popular matching with that of a minimum weight maximum cardinality matching (defined in Sec. 1.5.4), a rank-maximal matching and a generous maximum matching (both defined in Sec. 1.5.6) respectively.

Formally (and in terms of the notation defined in Secs. 1.5.4 and 1.5.6), let $I$ be an HA instance $I$, let $\mathcal{M}_{pop}$ denote the set of popular matchings in $I$, and let $\mathcal{M}_{pop}^+$ denote the set of maximum popular matchings in $I$. For each $a_i \in A$, assume that $rank(a_i, l_i) = m + 1$, where $m$ is the number of houses in $I$, regardless of the length $a_i$'s original preference list prior to last resorts being added.

Given an acceptable applicant–house pair $(a_i, h_j)$, define $wt(a_i, h_j) = rank(a_i, h_j)$. Then a matching $M$ is a *minimum weight maximum cardinality popular matching* if $M \in \mathcal{M}_{pop}^+$ and $wt(M)$ is minimum, taken over all matchings in $\mathcal{M}_{pop}^+$. $M$ is a *rank-maximal popular matching* if $M \in \mathcal{M}_{pop}$ and $p(M)$ is lexicographically maximum, taken over all matchings in $\mathcal{M}_{pop}$. Finally, $M$ is a *generous popular matching* if $M \in \mathcal{M}_{pop}$ and $p_R(M)$ is lexicographically minimum, taken over all matchings in $\mathcal{M}_{pop}$ (note that as $M$ minimises the number of $(m+1)$th choices, $M$ is automatically a maximum popular matching). McDermid and Irving proved the following results regarding the algorithmic complexity of computing these types of optimal popular matchings.

**Theorem 7.13 ([442]).** *Let $I$ be a solvable instance of* HA. *A minimum weight maximum cardinality popular matching can be found in $O(n + m)$ time, and both a rank-maximal popular matching and a generous popular matching can be found in $O(n \log n + m)$ time, where $n$ is the number of applicants and houses, and $m$ is the number of acceptable applicant–house pairs in $I$.*

Kavitha and Nasre [367] described an alternative method for computing minimum weight maximum cardinality popular matchings, rank-maximal popular matchings and generous popular matchings. However the complexity of their algorithm in each case is $O(n^2 + m)$, and thus slower than McDermid and Irving's approach.

### 7.2.6   *Popular matchings in* HAT

Recall from Sec. 1.5.7 that the definition of a popular matching carries over naturally to the extension of HA in which preference lists may include ties. Abraham *et al.* [21] extended their characterisation of popular matchings to the HAT case. This characterisation is more complex than for HA and utilises the Dulmage–Mendelsohn Decomposition of a graph (see Sec. 1.2). This is required in order to define $s(a_i)$ for each applicant $a_i \in A$: note that in general each of $f(a_i)$ and $s(a_i)$ is now a set of size greater than 1.

The definition of $f(a_i)$ for an applicant $a_i \in A$ is straightforward: it is the set of $a_i$'s most-preferred houses (that is, $f(a_i) = \{h_j \in H : rank(a_i, h_j) = 1\}$). Now define $G_1$ to be the subgraph of the underlying graph of $I$ with vertex set $A \cup H$ and edge set $\{\{a_i, h_j\} : a_i \in A \wedge h_j \in f(a_i)\}$. Relative to an EOU labelling of $G_1$ (see Definition 1.4), for each applicant $a_i \in A$, define $s(a_i)$ to be the most-preferred set of houses in $a_i$'s preference list that are even in $G_1$. Note that $s(a_i) \neq \emptyset$ since $l_i$ is an isolated vertex in $G_1$ and therefore even. Also, in contrast to the HA case, it is possible that $s(a_i) \subseteq f(a_i)$.

Define a house $h_j \in H$ to be an *$f$-house* (respectively *$s$-house*) if $h_j \in f(a_i)$ (respectively $h_j \in s(a_i)$) for some $a_i \in A$. Note that, again in contrast to the case for HA, the set of $f$-houses need not be disjoint from the set of $s$-houses. Let $H'$ denote the set of $f$-houses, and let $H''$ denote the union of the sets of $f$-houses and $s$-houses.

Again, as in the HA case, the members of $f(a_i) \cup s(a_i)$ are the only candidate houses to which an applicant $a_i \in A$ can be assigned in a given popular matching. Additionally, any popular matching must be a maximum matching in $G_1$. These two necessary conditions for a matching to be popular are also sufficient, as Abraham *et al.* [21] proved.

**Lemma 7.14 ([21]).** *Let $I$ be an instance of* HAT *and let $M$ be a matching in $I$. Then $M$ is popular if and only if:*

*(i) $M$ is a maximum matching in $G_1$;*
*(ii) for each applicant $a_i \in A$, $M(a_i) \in f(a_i) \cup s(a_i)$.*

As in the HA case, Lemma 7.14 leads to the definition of a *reduced graph* (as before, a subgraph of the underlying graph $G$ of $I$), which we again denote by $G'$. The vertices in $G'$ are $A \cup H$ and the edges in $G'$ are $\{\{a_i, h_j\} : a_i \in A \wedge h_j \in f(a_i) \cup s(a_i)\}$. We can then restate Lemma 7.14 in terms of $G'$, as follows:

**Theorem 7.15 ([21]).** *Let $I$ be an instance of* HAT *and let $M$ be a matching in $I$. Then $M$ is popular if and only if:*

*(i) $M$ is a maximum matching in $G_1$;*
*(ii) $M$ is an applicant-complete matching in the reduced graph $G'$.*

**Corollary 7.16.** *Let $I$ be an instance of* HAT *and let $M$ be a matching in $I$. There is an $O(m)$ algorithm to test whether $M$ is popular in $I$, where $l$ is the number of acceptable applicant–house pairs in $I$.*

**Proof.** Clearly $G_1$ can be built in $O(m)$ time (assuming that we add only the houses in $H'$ to $G_1$, rather than the whole set $H$). Clearly we can test whether $M$ is maximum in $G_1$ in $O(m)$ time by searching for an augmenting path relative to $M$ in $G_1$.

We then find an EOU labelling of $G_1$ relative to $M$, which can be accomplished in $O(m)$ time using a modified breadth-first search (this step can be combined with the verification that $M$ is maximum in $G_1$). This allows us to identify $s(a_i)$ for each applicant $a_i \in A$ in $O(m)$ overall time. Following this we construct the reduced graph $G'$, again in $O(m)$ time (assuming that we add only the houses in $H''$ to the vertex set, rather than all houses in $H$). Clearly we can verify whether $M$ is applicant-complete in $G'$ in $O(m)$ time.                                                                   $\square$

Theorem 7.15 leads to an $O(\sqrt{n_1}m)$ algorithm, due to Abraham *et al.* [21], for finding a popular matching in $I$ or reporting that none exists, where $n_1$ is the number of applicants and $m$ is the number of acceptable applicant–

---

**Algorithm 7.3** Algorithm Popular-HAT [21]

---

**Require:** HAT instance $I$
**Ensure:** return a popular matching $M$ or "no popular matching exists"
1: $G_1 := (A \cup H', E_1)$, where $E_1 = \{(a_i, h_j) : a_i \in A \land h_j \in f(a_i)\}$;
2: $M_1 :=$ maximum matching in $G_1$;
3: compute an EOU labelling of $G_1$;
4: $\{\mathcal{E}, \mathcal{O}, \mathcal{U}$ are the sets of even, odd and unreachable vertices in $G_1$ respectively$\}$

5: $G' := (A \cup H'', E')$, where $E' = \{(a_i, h_j) : a_i \in A \land h_j \in f(a_i) \cup s(a_i)\}$;
6: $\{G'$ is the reduced graph of $I\}$
7: $E' := E' \backslash \{(a_i, h_j) : (a_i \in \mathcal{O} \land h_j \in \mathcal{O} \cup \mathcal{U}) \lor (a_i \in \mathcal{O} \cup \mathcal{U} \land h_j \in \mathcal{O})\}$;
8: $\{M_1 \subseteq E'$ still holds at this point$\}$
9: augment $M_1$ to a maximum matching $M$ in $G'$;
10: **if** $M$ is applicant-complete in $G'$ **then**
11:     **return** $M$;
12: **else**
13:     **return** "no popular matching exists";
14: **end if**

---

house pairs. This method is described by Algorithm Popular-HAT, shown in Algorithm 7.3.

The algorithm begins by constructing $G_1$, and a maximum matching $M_1$ in $G_1$, which can be accomplished in $O(m)$ time and $O(\sqrt{n_1}m)$ time respectively (see the proof of Corollary 7.16). As in the proof of Corollary 7.16, we then find an EOU labelling of $G_1$ relative to $M_1$ in $O(m)$ time. Following this we construct the reduced graph $G'$, again in $O(m)$ time (see the proof of Corollary 7.16).

The algorithm then deletes any edge in $G'$ connecting an odd vertex with either an odd vertex or an unreachable vertex (which can be accomplished in $O(m)$ time). We then augment $M_1$ to a maximum matching $M$ in $G'$ in $O(\sqrt{n_1}m)$ time.

Note that the edges deleted in line 7 cannot belong to any maximum matching of $G_1$ by Part (iii) of Theorem 1.5, and therefore cannot belong to any popular matching in $I$ by Theorem 7.15. Thus if $M$ is not an applicant-complete matching in $G'$, then $I$ admits no popular matching by Theorem 7.15. On the other hand if $M$ is an applicant-complete matching in $G'$, then the edge deletions carried out in line 7 ensure that $M$ remains a maximum matching in $G_1$ (see Lemma 3.7 of Ref. [21] for further details) and hence $M$ is a popular matching in $I$ by Theorem 7.15.

We thus obtain the following result.

**Theorem 7.17 ([21]).** *Let $I$ be an instance of* HAT. *There is an* $O(\sqrt{n_1}m)$ *algorithm to find a popular matching $I$, or report that $I$ is unsolvable, where $n_1$ is the number of applicants and $m$ is the number of acceptable applicant–house pairs.*

An example execution of Algorithm Popular-HAT for a given HAT instance may be found in Ref. [21]. Abraham *et al.* also showed how to find a maximum popular matching or report that none exists, given an instance of HAT, in $O(\sqrt{n_1}m)$ time — see Ref. [21] for further details.

The problem of finding a popular matching (assuming one exists) in a given HAT instance is at least as hard as the problem of finding a maximum matching in a bipartite graph $G = (V, E)$. For, if $G$ has bipartition $A \cup H$ then $G$ gives rise to an HAT instance where each applicant $a_i \in A$ finds acceptable the houses in $H$ that she is adjacent to in $G$; moreover $a_i$'s preference list is a single tie involving these acceptable houses (followed by her last resort). Clearly a popular matching in $I$ is a maximum matching in $G$ and vice versa. Hence any improvement to the $O(\sqrt{n_1}m)$ algorithm for finding a popular matching, or reporting that none exists, given an HAT instance, would imply a faster algorithm for finding a maximum matching in a bipartite graph.

Given an instance of HAT, Kavitha and Shah [371] presented an $O(n^{\omega})$ randomised algorithm for the problem of finding a popular matching or reporting that none exists, where $n = n_1 + n_2$ is the number of applicants and houses, and $\omega \leq 2.3727$ is the best current exponent of an algorithm for matrix multiplication [586] (see also Ref. [550]). This algorithm is faster than that of Abraham *et al.* whenever $m > n^{1.8727}$.

As in the HA case, Abraham *et al.* [21] estimated the probability of a popular matching existing by generating random instances of HAT and measuring the proportion of solvable instances. As before, instances where $n_1 = n_2$ and all preference lists were of the same length $k$ were generated. Moreover a probability $t$ that a given preference list entry was tied with its successor was varied from 0 to 0.8 in steps of 0.2. For each value of $n_1$, $k$ and $t$, 1000 random instances were generated. For $n_1 = 10$, $k$ varied from $1, 2, \ldots, 10$, whilst when $n_1 = 100$, $k$ varied over $1, 2, \ldots, 10, 20, 30, \ldots, 100$. The experiments indicated that, for fixed values of $n_1$ and $k$, the solvability proportion tended to increase as $t$ increased. Similarly for fixed values of $n_1$ and $t$, the proportion of solvable instances tended to decrease as $k$ increased.

For example, when $n_1 = 100$, $k = 10$ and $t = 0.2$, only 28 instances were solvable, however this number increased to 531 when $t = 0.6$ (for the same values of $n_1$ and $k$). When $n_1 = 100$, $k = 20$ and $t = 0.2$, no instance was solvable, however when $t = 0.6$ (again for the same values of $n_1$ and $k$), the number of solvable instances was 346.

### 7.2.7 *Matchings with small unpopularity*

7.2.7.1 *Defining the unpopularity factor and unpopularity margin*

As a given HA or HAT instance need not be solvable, it is of interest to weaken the notion of popularity, and look for matchings that are "as popular as possible" in cases where a popular matching does not exist. To this end, McCutchen [436] defined two versions of "near-popular" matchings in HAT instances, namely a *least unpopularity factor matching* and a *least unpopularity margin matching*. We now define these two concepts in a given instance $I$ of HAT.

**Definition 7.18 ([436]).** *Let $I$ be an instance of* HAT *and let $\mathcal{M}$ denote the set of matchings in $I$. Given two matchings $M, M' \in \mathcal{M}$ we define $\Delta(M, M')$ to be a measure of the factor by which $M'$ is more popular than $M$. That is,*

$$\Delta(M, M') = \begin{cases} \frac{|P(M', M)|}{|P(M, M')|}, & \text{if } |P(M, M')| > 0 \\ 1, & \text{if } |P(M, M')| = |P(M', M)| = 0 \\ \infty, & \text{otherwise.} \end{cases}$$

*The* unpopularity factor *of a matching $M \in \mathcal{M}$, denoted by $u(M)$, is defined as follows:*

$$u(M) = \max_{M' \in \mathcal{M}} \Delta(M, M').$$

*A matching $M \in \mathcal{M}$ is a* least unpopularity factor matching *if $u(M)$ is minimum, taken over all matchings in $\mathcal{M}$.*

*Also, given two matchings $M, M' \in \mathcal{M}$ we define $\delta(M, M')$ to be a measure of the margin by which $M'$ is more popular than $M$. That is,*

$$\delta(M, M') = |P(M', M)| - |P(M, M')|.$$

*The* unpopularity margin *of a matching $M \in \mathcal{M}$, denoted by $g(M)$, is defined as follows:*

$$g(M) = \max_{M' \in \mathcal{M}} \delta(M, M').$$

*A matching $M \in \mathcal{M}$ is a* least unpopularity margin matching *if $g(M)$ is minimum, taken over all matchings in $\mathcal{M}$.*

Note that, for any $M \in \mathcal{M}$, $\Delta(M, M) = 1$ and $\delta(M, M) = 0$, and hence $u(M) \geq 1$ and $g(M) \geq 0$. Also $M$ is popular if and only if $u(M) = 1$, and similarly $M$ is popular if and only if $g(M) = 0$. Furthermore, given any $M' \in \mathcal{M}$, observe that $\Delta(M, M') = \infty$ if and only if $|P(M, M')| = 0$ and $|P(M', M)| > 0$. Thus $u(M)$ is finite if and only if $M$ is Pareto optimal.

McCutchen [436] proved that if $u(M)$ is finite, then $u(M)$ is equal to the maximum $k$ for which there exists an alternating path

$$P = \langle h_{i_1}, a_{i_1}, h_{i_2}, a_{i_2}, \ldots, h_{i_r}, a_{i_r}, h_{i_{r+1}}, a_{i_{r+1}} \rangle$$

relative to $M$ in the underlying graph of $I$, where:

(1) $r \geq k$;
(2) for each $j$ $(1 \leq j \leq r + 1)$, $(a_{i_j}, h_{i_j}) \in M$;
(3) for each $j$ $(1 \leq j \leq r)$, $a_{i_j}$ prefers $h_{i_{j+1}}$ to $h_{i_j}$ or is indifferent between them;
(4) there exists some $1 \leq s_1 < s_2 < \cdots < s_k \leq r$ such that, for each $j$ $(1 \leq j \leq k)$, $a_{i_{s_j}}$ prefers $h_{i_{s_j+1}}$ to $h_{i_{s_j}}$.

Intuitively, the matching $M' = M \oplus P$ will promote $k$ applicants (namely $a_{i_{s_1}}, \ldots, a_{i_{s_k}}$) and demote one applicant to her last resort (namely $a_{i_{r+1}}$), leaving the remaining applicants in $P$ indifferent between $M$ and $M'$. Thus $\Delta(M, M') = k$. McCutchen [436] defined $h_{i_{r+1}}$ to have *pressure* $k$, intuitively meaning that $r \geq k$ applicants are stacked up behind $a_{i_{r+1}}$, waiting for her to vacate $h_{i_{r+1}}$ so that $k$ of them can be promoted. Thus if $M$ is Pareto optimal, $u(M)$ is equal to the maximum pressure of an assigned house in $M$.

A consequence of the discussion in the preceding paragraph is that if $u(M)$ is finite then $u(M)$ is integral. Hence if $I$ is an HAT instance that admits a popular matching $M$ then $u(M) = 1$, otherwise any matching $M$ in $I$ satisfies $u(M) \geq 2$.

To give an example of the unpopularity factor and the unpopularity margin of a matching, in the HA instance $I'$ of Fig. 7.1, it may be verified that the matching

$$M = \{(a_1, l_1), (a_2, h_4), (a_3, h_1), (a_4, h_2), (a_5, h_3)\}$$

is Pareto optimal. House $h_2$ has pressure 3 due to the alternating path $\langle l_1, a_1, h_4, a_2, h_1, a_3, h_2, a_4 \rangle$. Clearly the maximum pressure of any assigned house must be 3 as two applicants have their first choice in $M$. Hence $u(M) = 3$.

Similarly $g(M) = 2$, which may be deduced by considering the matching

$$M' = \{(a_1, h_4), (a_2, h_1), (a_3, h_2), (a_4, l_4), (a_5, h_3)\}.$$

Clearly $|P(M', M)| = 3$ and $|P(M, M')| = 1$, so $g(M) \geq 2$. However since two applicants have their first choice in $M$, it is clear that $|P(M'', M| \leq 3$ for any matching $M''$ in $I'$. Moreover if $|P(M'', M)| = 3$ then $a_3$ moves into $h_2$, which is only possible if $a_4$ is demoted from $h_2$, hence $|P(M, M'')| \geq 1$. Thus $g(M) = 2$.

### 7.2.7.2   *Finding the unpopularity factor and unpopularity margin*

McCutchen [436] gave polynomial-time algorithms for computing the unpopularity factor and the unpopularity margin of a given matching $M$ in an instance $I$ of HAT. We describe slightly modified versions here with improved time complexity.

Firstly, we show how to find $u(M)$. The initial step is to test whether $M$ is Pareto optimal — this step can be accomplished in $O(m)$ time by Proposition 6.13, where $m$ is the number of acceptable applicant–house pairs in $I$. If $M$ is not Pareto optimal then $u(M) = \infty$. Now suppose that $M$ is Pareto optimal.

Build a weighted directed graph $D$ with a vertex for each house that is assigned in $M$. There is an arc $(h_i, h_j)$ of weight 1 if $M(h_i)$ prefers $h_j$ to $h_i$, and an arc $(h_i, h_j)$ of weight 0 if $M(h_i)$ is indifferent between $h_j$ to $h_i$. Every cycle in $D$ must have weight 0, otherwise $M$ is not Pareto optimal. Moreover, any arc $(h_i, h_j)$ where $h_i$ and $h_j$ are in the same strongly connected component of $D$ must have weight 0, as $h_i$ and $h_j$ are contained in some cycle in $D$.

These observations suggest the following approach. Construct the strongly connected components of $D$ in $O(m)$ time [561]. Let these be $C_1, \ldots, C_r$ for some $r \geq 1$. Now create (in $O(m)$ time) a new weighted directed graph $D'$ which contains a vertex $v_i$ for each strongly connected component $C_i$ in $D$, and an arc $(v_i, v_j)$ whenever $D$ contains an arc from a vertex in $C_i$ to a vertex in $C_j$. If there is such an arc in $D$ with weight 1, then the weight of $(v_i, v_j)$ in $D'$ is 1, otherwise the weight of $(v_i, v_j)$ in $D'$ is 0.

Clearly $D'$ is a directed acyclic graph. The problem of finding a longest path in $D'$ can be solved in $O(m)$ time using topological ordering (see e.g., Ref. [533, p.491]). It may be verified that the length of this longest path is then $u(M)$ (as it gives the maximum pressure of an assigned house).

We summarise this discussion as follows.

$$a_1 : h_1 \quad h_2 \quad \ldots \quad h_r \quad l_1$$
$$a_2 : h_1 \quad h_2 \quad \ldots \quad h_r \quad l_2$$
$$\ldots$$
$$a_r : h_1 \quad h_2 \quad \ldots \quad h_r \quad l_r$$

Fig. 7.4   An instance of HA

**Proposition 7.19.** *Given an instance $I$ of* HAT *and a matching $M$ in $I$, the unpopularity factor $u(M)$ of $M$ may be computed in $O(m)$ time, where $m$ is the number of acceptable applicant–house pairs in $I$.*

McCutchen's algorithm for computing $u(M)$ [436] has $O(\sqrt{n_1}m)$ complexity, where $n_1$ is the number of applicants in $I$.

We now turn to the problem of computing the unpopularity margin $g(M)$ for a given matching $M$. The technique is somewhat simpler than that used for the computation of $u(M)$, though the time complexity is superlinear. The method involves applying weights to the underlying graph $G$ of $I$. An edge $(a_i, h_j)$ of $G$ has weight -1 if $a_i$ prefers $h_j$ to $M(a_i)$; $(a_i, h_j)$ has weight 0 if $a_i$ is indifferent between $h_j$ and $M(a_i)$, or if $M(a_i) = h_j$; and $(a_i, h_j)$ has weight 1 if $a_i$ prefers $M(a_i)$ to $h_j$.

It may be verified that $g(M)$ is equal to the weight of a minimum weight applicant-complete matching $M'$ in this weighted graph. $M'$ may be computed in $O(\sqrt{n_1}m \log n_1)$ time (see Sec. 1.5.4). We thus have the following result.

**Proposition 7.20.** *Given an instance $I$ of* HAT *and a matching $M$ in $I$, the unpopularity margin $g(M)$ may be computed in $O(\sqrt{n_1}m \log n_1)$ time, where $n_1$ is the number of applicants and $m$ is the number of acceptable applicant–house pairs in $I$.*

McCutchen's approach for computing $g(M)$ [436] has $O(\sqrt{n}m(g+1))$ complexity, where $n$ is the number of applicants and houses in $I$, and $g = g(M)$.

### 7.2.7.3   *Least unpopularity factor and least unpopularity margin matchings*

We now turn to the problems of computing matchings with small unpopularity factor and unpopularity margin. Given a Pareto optimal matching $M$, the maximum pressure of any house is clearly $n_1 - 1$, where $n_1$ is the number of applicants. This implies that $u(M) \leq n_1 - 1$ (recall that $u(M) = \infty$ if $M$ is not Pareto optimal). However there is an HA instance $I$ where a least unpopularity factor matching $M'$ satisfies $u(M') = n_1 - 1$. To see this, consider the HA instance $I'$ illustrated in Fig. 7.4 (a generalisation of instance $I_1$ shown in Fig. 1.3), which is defined for any $r \geq 2$. Clearly the unique Pareto optimal matching up to symmetry is $M' = \{(a_i, h_i) : 1 \leq i \leq r\}$. However $h_1$ has pressure $r - 1 = n_1 - 1$, so $u(M') = n_1 - 1$.

Similarly, if a matching $M$ is Pareto optimal and $M''$ is any other matching where $|P(M'', M)| > 0$, the inequality $|P(M, M'')| > 0$ must also hold. Since $|P(M'', M)| \leq n_1$, it follows that $g(M) \leq n_1 - 1$. Again the Pareto optimal matching $M'$ described in the preceding paragraph satisfies $g(M') = n_1 - 1$ (and $M'$ is a least unpopularity margin matching). A trivial upper bound for $g(M)$ for a matching $M$ that is not Pareto optimal is $g(M) \leq n_1$. A matching that achieves this upper bound in $I'$ is $\{(a_i, l_i) : 1 \leq i \leq n_1\}$.

McCutchen [436] considered the algorithmic complexity of each of the problems of computing a least unpopularity factor matching and a least unpopularity margin matching and proved the following hardness results, which hold even in the absence of ties.

**Theorem 7.21 ([436]).** *Given an instance $I$ of* HA, *the problem of finding a least unpopularity factor matching is NP-hard. In particular, the problem of deciding whether $I$ admits a matching $M$ such that $u(M) = 2$ is NP-complete. Hence there is no approximation algorithm for the problem of finding a least unpopularity factor matching in $I$ with performance guarantee $3/2 - \varepsilon$, for any $\varepsilon > 0$, unless P=NP.*

**Theorem 7.22 ([436]).** *Given an instance $I$ of* HA, *the problem of finding a least unpopularity margin matching is NP-hard.*

Huang *et al.* [299] gave an algorithm that aims to find matchings with low unpopularity factor and unpopularity margin, given an instance of HAT. In particular, they proved the following.

**Theorem 7.23 ([299]).** *Let $I$ be an instance of* HAT *and let $k \geq 2$ be a given integer. Then a sequence of bipartite graphs $H_2, \ldots, H_k$ may be constructed in $O(k\sqrt{n_1}m)$ overall time, where $n_1$ is the number of applicants and $m$ is the number of acceptable applicant–house pairs, with the following properties: if $H_k$ admits an applicant-complete matching $M_k$, then $u(M_k) \leq k - 1$ and $g(M_k) \leq n_1 \left(1 - \frac{2}{k}\right)$.*

A trivial adaptation to the algorithm of Huang *et al.* [299] will guarantee that the series is bound to terminate in a polynomial number of steps with a matching that satisfies the two given inequalities. For, suppose that $k = 2n_1$ and $H_k$ does not admit an applicant-complete matching. In this case, simply output an arbitrary Pareto optimal matching $M$ in $I$ (which may be found in $O(m)$ time by Proposition 6.14). Then as observed above, the Pareto optimality of $M$ implies that $u(M) \leq n_1 - 1 \leq k - 1$ and $g(M) \leq n_1 - 1 = n_1 \left(1 - \frac{2}{k}\right)$.

Huang et al. [299] proved that, for a random HAT instance with $n_1$ applicants and $n_2 = n_1$ houses, where each preference list is a permutation of the $n_1$ houses generated uniformly at random, the expected number of iterations taken by their algorithm (i.e., the expected number of bipartite graphs generated until we arrive at an applicant-complete matching) is at most $\ln n_1 + 1$.

The authors also implemented their algorithm and carried out an empirical analysis based on randomly generated instances of HAT and measured the number of iterations of the algorithm that were required for each instance. Instances where $n_1 = n_2 = 100$ and $n_1 = n_2 = 500$ were considered. For each value of $n_1$, instances where all preference lists were of the same length $k$ were constructed. Also a probability $t$ of an entry being tied with its predecessor was used. For each value of $n_1$, $k$ and $t$, 1000 instances were created. For every instance generated, the algorithm terminated within 4 iterations, i.e., it constructed a matching $M$ with $u(M) \leq 5$.

Huang et al. [299] also generated *highly correlated instances* of HAT (where each applicant's preference list is derived from a master list of the houses). Fig. 7.4 gives an example of such an instance. In the authors' experimental trials, applicants chose uniformly at random a subset of size $n_2 p$ of the houses, where $p$ is a density parameter that was varied. For such instances the number of iterations taken was considerably larger in some cases, in comparison with the trials described in the previous paragraph.

An empirical analysis of the algorithm of Huang et al. [299] was also conducted by Michail [452]: he compared the unpopularity factors of its constructed matchings with those of algorithms to compute rank-maximal matchings. The (surprising) results are described in more detail in Sec. 8.2.2.4.

### 7.2.8 *Strongly popular matchings*

Recall the definition of a strongly popular matching from Sec. 1.5.5. It turns out that there is a straightforward necessary and sufficient condition for an HAT instance $I$ to admit a strongly popular matching. This is based on the graph $G_1$ that was defined in Sec. 7.2.6.

**Proposition 7.24.** *Let $I$ be an instance of* HAT. *A matching $M$ in $I$ is strongly popular if and only if $M$ is the unique perfect matching in $G_1$.*

**Proof.** Suppose that $I$ admits a strongly popular matching $M$. Firstly, we claim that $M$ is a perfect matching in $G_1$. For, suppose that $M(a_i) \notin f(a_i)$ for some $a_i \in A$. Choose an arbitrary $h_j \in f(a_i)$. If $h_j$ is unassigned in $M$, clearly $(M \backslash \{(a_i, M(a_i))\}) \cup \{(a_i, h_j)\}$ is more popular than $M$, a contradiction. Hence let $a_k = M(h_j)$. Now let

$$M' = (M \backslash \{(a_i, M(a_i)), (a_k, h_j)\}) \cup \{(a_i, h_j), (a_k, l_k)\}.$$

Then $|P(M', M)| = |P(M, M')| = 1$, a contradiction. Thus $M$ is a perfect matching in $G_1$. If $M''$ is some other perfect matching in $G_1$ then clearly $|P(M'', M)| = |P(M, M'')| = 0$, a contradiction. Hence $M$ is the unique perfect matching in $G_1$.

Conversely suppose that $M$ is the unique perfect matching in $G_1$. If $M$ is not strongly popular then there exists some matching $M' \neq M$ such that $|P(M', M)| \geq |P(M, M')|$. Now $|P(M', M)| = 0$, since every agent $a_i \in A$ satisfies $M(a_i) \in f(a_i)$. Hence $|P(M, M')| = 0$, which implies that $M'$ is also a perfect matching in $G_1$, a contradiction. Hence $M$ is strongly popular in $I$. □

It follows by Proposition 7.24 that it is easy to test a matching for strong popularity in $I$ (and indeed trivial if $I$ is an instance of HA).

**Proposition 7.25.** *Let $I$ be an instance of* HAT *and let $M$ be a matching in $I$. There is an $O(m)$ algorithm to test whether $M$ is strongly popular, where $m$ is the number of acceptable applicant–house pairs. This bound improves to $O(n_1)$ if $I$ is an instance of* HA, *where $n_1$ is the number of applicants in $I$.*

**Proof.** If $I$ is an instance of HA, the result follows easily. Hence suppose that $I$ is a general HAT instance. Clearly $G_1$ can be constructed in $O(m)$ time, and we can check that $M$ is a perfect matching in $G_1$ within the same time bound. A straightforward variant of breadth-first search, also taking $O(m)$ time, can be used to check for an alternating cycle in $G_1$ with respect to $M$. □

Finding a strongly popular matching or reporting that none exists is also straightforward.

**Proposition 7.26.** *Let $I$ be an instance of* HAT *and let $M$ be a matching in $I$. There is an $O(\sqrt{n_1}m)$ algorithm to find a strongly popular matching in $I$ or report that none exists, where $n_1$ is the number of applicants and $m$ is the number of acceptable applicant–house pairs. This bound improves to $O(n_1)$ if $I$ is an instance of* HA.

**Proof.** If $I$ is an instance of HA, the result follows easily. Hence suppose that $I$ is a general HAT instance. Clearly $G_1$ can be constructed in $O(m)$ time, and we can find a maximum matching $M$ in $G_1$ in $O(\sqrt{n_1}m)$ time. If $M$ is not a perfect matching in $G_1$ then we report that no strongly popular matching exists. Otherwise, as in the proof of Proposition 7.25, we determine whether $M$ admits an alternating cycle in $G_1$ in $O(m)$ time. If so, we report that no strongly popular matching exists, otherwise $M$ is strongly popular. $\qquad\qquad\square$

### 7.2.9  Further results for popular matchings in HA and HAT

In this subsection we review further results in the literature concerning popular matchings in instances of HA and HAT.

**Voting paths**  Abraham and Kavitha [25] studied a dynamic version of HAT, viewed as a matching market, where applicants and houses are allowed to enter and leave the market, and applicants can arbitrarily change their preference lists. In this case, a previously computed matching $M$ that was popular prior to changes of this nature occurring may no longer be popular. Abraham and Kavitha argued that one cannot simply recompute a popular matching $M'$ in the new instance and expect the applicants to agree to the switch from $M$ to $M'$, because unless $M' \blacktriangleright M$, there will not be a majority consensus to move directly from $M$ to $M'$ (recall that $\blacktriangleright$ is the "more popular than" relation from Sec. 1.5.5).

However, the authors argued that the applicants will agree to move from $M$ to $M'$ by consensus if there exists a sequence $\langle M = M_0, M_1, \ldots, M_k = M' \rangle$, for some $k \geq 0$, where $M_{i+1} \blacktriangleright M_i$ for each $i$ $(0 \leq i \leq k)$. Such a sequence is called a *length-$k$ voting path* from $M$ to a popular matching.

By way of illustration, Abraham and Kavitha [25] gave an example HA instance that admits two popular matchings $M'$ and $M''$, and they also gave an example matching $M$ that is not popular. They showed that neither $M' \blacktriangleright M$ nor $M'' \blacktriangleright M$ holds, yet there is a matching $M_1$ such that

$M' = M_2 \blacktriangleright M_1 \blacktriangleright M_0 = M$. Thus $M$ has a length-2 voting path to a popular matching, but not a length-0 or a length-1 voting path.

In general the problem then is to determine whether there is a length-$k$ voting path from some matching $M$ to a popular matching, for some $k \geq 1$, and if so, to determine the minimum value of $k$ for which this is the case. Of course, what makes this problem non-trivial is that the relation $\blacktriangleright$ is not transitive in general.

Surprisingly, Abraham and Kavitha [25] showed that every matching in an HAT instance $I$ admits a length-$k$ voting path to a popular matching for some $k$ where $0 \leq k \leq 2$. Moreover, they gave an $O(\sqrt{n}m)$ algorithm to compute such a voting path, where $n$ is the number of applicants and houses, and $m$ is the number of acceptable applicant–house pairs. The complexity of their algorithm reduces to $O(n + m)$ in the case that $I$ is an instance of HA.

Abraham and Kavitha's results can be regarded as an analogue of the "paths to stability" results for SMI presented in Sec. 2.6. This earlier section dealt with decentralised algorithms for arriving at a stable matching, starting from an arbitrary matching, via a process of successively satisfying blocking pairs. In this context, the aim is to arrive at a popular matching, starting from an arbitrary matching, by successively gaining agreement from a majority of the applicants to change the current matching.

**Existence of popular matchings** As mentioned in Sec. 7.2.3 and Sec. 7.2.6, Abraham *et al.* [21] carried out an empirical evaluation with the aim of estimating the likelihood of a popular matching existing in random instances of HA and HAT. Mahdian [412] conducted a theoretical investigation into the same problem. He showed that, for random instances of HA, popular matchings exist with high probability when the number of houses is at least a factor of $\alpha$ larger than the number of applicants, where $\alpha$ is the solution of $x^2 e^{-1/x} = 1$ ($\alpha \approx 1.42$).

**Popular mixed matchings** Kavitha *et al.* [366] studied the concept of a *popular mixed matching*, which is a probability distribution over matchings that is popular in a precise sense. Towards a definition of this concept, let $I$ be an instance of HAT and let $\mathcal{M}$ be the set of matchings in $I$. A *mixed matching* is a set of matching–probability pairs $\mathfrak{M} = \{(M_1, p_1), \ldots, (M_k, p_k)\}$ where $M_i \in \mathcal{M}$ ($1 \leq i \leq k$), $p_i \geq 0$ and $\sum_{i=1}^{k} p_i = 1$. Thus $\mathfrak{M}$ is a probability distribution over matchings in

$\mathcal{M}$. A "standard" matching $M \in \mathcal{M}$ corresponds to the mixed matching $\mathfrak{M} = \{(M, 1)\}$.

Recall that $|P(M, M')|$ is the number of applicants who prefer $M$ to $M'$, given two matchings $M, M' \in \mathcal{M}$. We can generalise this quantity to mixed matchings as follows: for two mixed matchings $\mathfrak{M}, \mathfrak{M}'$ in $I$, where $\mathfrak{M} = \{(M_1, p_1), \ldots, (M_k, p_k)\}$ and $\mathfrak{M}' = \{(M'_1, q_1), \ldots, (M'_k, q_l)\}$, the expected number of applicants who prefer $\mathfrak{M}$ to $\mathfrak{M}'$, denoted $\phi(\mathfrak{M}, \mathfrak{M}')$, is defined as follows:

$$\phi(\mathfrak{M}, \mathfrak{M}') = \sum_{i=1}^{k} \sum_{j=1}^{l} p_i q_j P(M_i, M'_j).$$

We can then extend the "more popular than" relation ▶ to mixed matchings as follows: $\mathfrak{M}'$ is *more popular than* $\mathfrak{M}$, denoted $\mathfrak{M}' \blacktriangleright \mathfrak{M}$, if $\phi(\mathfrak{M}', \mathfrak{M}) > \phi(\mathfrak{M}, \mathfrak{M}')$. $\mathfrak{M}$ is *popular* if it is ▶-maximal (i.e., there is no other mixed matching $\mathfrak{M}'$ such that $\mathfrak{M}' \blacktriangleright \mathfrak{M}$).

Kavitha *et al.* [366] showed that every instance $I$ of HAT admits a popular mixed matching, and they gave a polynomial-time algorithm based on linear programming for computing such a matching in $I$.

**Self-stabilising algorithm**  Shi [529] gave a self-stabilising algorithm[1] for the problem of finding a maximum popular matching or reporting that none exists, given an instance $I$ of HA. The algorithm stabilises in $O(n^5)$ moves given any scheduler, where $n$ is the number of applicants and houses in $I$.

**Condorcet's voting paradox**  The potential absence of a popular matching in a given HA instance can be related all the way back to the observation of Condorcet [154] that, given $k$ voters who each rank $n$ candidates in strict order of preference, there may not exist a "winner", namely a candidate who beats all others in a pairwise majority vote.

For example, suppose the "voters" are $a_1$, $a_2$ and $a_3$, the "candidates" are $h_1$, $h_2$ and $h_3$, and each voter has the preference list shown in Fig. 7.5. Then if, for example, $h_1$ is declared the winner, a majority of voters, namely $a_2$ and $a_3$, would vote for an alternative, namely $h_3$. A similar argument holds if either $h_2$ or $h_3$ is declared as winner, and hence there is no outright winner under a pairwise majority voting rule. This phenomenon is referred to as *Condorcet's voting paradox* [154] because the collective preferences

---

[1]See Ref. [169] for an introduction to self-stabilising algorithms.

$$a_1 : h_1 \quad h_2 \quad h_3$$
$$a_2 : h_2 \quad h_3 \quad h_1$$
$$a_3 : h_3 \quad h_1 \quad h_2$$

Fig. 7.5    An instance of HA

of the voters are cyclic (since $h_3$ is preferred to $h_1$ by a majority; $h_2$ is preferred to $h_3$ by a majority; and $h_1$ is preferred to $h_2$ by a majority) even though the individual preference list of each voter is strict and transitive.

In general, given an HAT instance $I$, we can extend the notation $P(M, M')$ to the case that $M$ and $M'$ are assignments in $I$ that assign each applicant to the same house. One such assignment $M$ is then called a *weak Condorcet winner* [154] if for every other assignment $M'$ that assigns every applicant to the same house, $|P(M, M')| \geq |P(M', M)|$. $M$ is a *strong Condorcet winner* [154] if for every other assignment $M'$ that assigns every applicant to the same house, $|P(M, M')| > |P(M', M)|$. Thus it follows that the HA instance shown in Fig. 7.5 has no weak Condorcet winner that assigns every applicant to the same house. It is then clear that the notions of popular and strongly popular matchings are the analogues of weak and strong Condorcet winners in the case that matchings in $I$ are considered. Various papers [136, 137, 135, 575] focus on the problems of finding weak and strong Condorcet winners for specific preference models.

## 7.3  Capacitated House Allocation problem

### 7.3.1  *Introduction*

In this section we generalise some of the results from the previous section to the capacitated case, both with and without ties. Thus our focus is on instances of CHA, dealt with in Sec. 7.3.2, and CHAT, considered in Sec. 7.3.3. We also survey additional results concerning variants of CHAT that typically involve choosing house capacities so as to arrive either at a solvable instance of CHAT, or to ensure that the weight of a popular matching (assuming that the houses have weights) is minimised. These problems are considered in Secs. 7.3.3 and 7.3.4.

### 7.3.2  *Strictly-ordered preference lists*

As indicated in Sec. 1.5.7, popular matchings can be defined in instances of CHA and CHAT, where each house $h_j$ has a capacity $c_j > 0$. Manlove and Sng [425] studied popular matchings in these more general problem contexts. In each of the CHA and CHAT cases, they gave necessary and sufficient conditions for a matching to be popular, leading to a polynomial-time algorithm for finding a popular matching or reporting that none exists. As in the 1–1 case, the equivalent conditions for popularity hinge on the definitions of $f(a_i)$ and $s(a_i)$ for each applicant $a_i \in A$. As before, we assume that a unique last resort house $l_i$ with capacity 1 is appended to the preference list of each applicant $a_i \in A$.

Starting with the CHA case, define $f(a_i)$ and $f(h_j)$ as in the HA case (see Sec. 7.2.2), for any $a_i \in A$ and $h_j \in H$. For any house $h_j \in H$, define $f_j = |f(h_j)|$; $h_j$ is called an $f$-*house* if $f_j > 0$. For any $a_i \in A$, we then define $s(a_i)$ to be the most-preferred house $h_j$ on $a_i$'s preference list such that *either* (i) $f_j = 0$, *or* (ii) $0 < f_j < c_j$ and $h_j \neq f(a_i)$ (such a house is guaranteed to exist due to $l_i$). A house $h_j \in H$ is an $s$-*house* if $h_j = s(a_i)$ for some $a_i \in A$. Clearly $s(a_i) \neq f(a_i)$ for each applicant $a_i \in A$, but in general the set of $s$-houses need not be disjoint from the set of $f$-houses. The reduced graph $G'$ may be defined in an analogous way to the HA case (see Sec. 7.2.2), though note that $G'$ is now a capacitated bipartite graph, with the capacity of each house $h_j$ being $c_j$.

Manlove and Sng [425] established the following characterisation of popular matchings in CHA.

**Theorem 7.27 ([425]).** *Let $I$ be an instance of* CHA *and let $M$ be a matching in $I$. Then $M$ is popular if and only if:*

(i) *for every $f$-house $h_j \in H$,*

    (a) *if $f_j \leq c_j$ then $f(h_j) \subseteq M(h_j)$;*
    (b) *if $f_j > c_j$ then $|M(h_j)| = c_j$ and $M(h_j) \subseteq f(h_j)$;*

(ii) *$M$ is an applicant-complete matching in the reduced graph $G'$.*

Theorem 7.27 is clearly equivalent to Theorem 7.2 in the case that $I$ is an instance of HA. Theorem 7.27 led the authors to an efficient algorithm for computing a maximum popular matching (or reporting that none exists).

**Theorem 7.28 ([425]).** *Let $I$ be an instance of* CHA. *There is an $O(n_1^{1.5} + m)$ algorithm[2] to find a maximum popular matching in $I$, or*

---

[2]In Ref. [425], the weaker upper bound of $O(\sqrt{C}n_1 + m)$ was given as the complexity for this algorithm, where $C$ is the total capacity of the houses. The improved upper bound follows by the remark in Footnote 6 on Page 16.

report that $I$ is unsolvable, where $n_1$ is the number of applicants and $m$ is the number of acceptable applicant–house pairs.

### 7.3.3   Preference lists with ties

We now turn to case that $I$ is an instance of CHAT. We define $f(a_i)$, for any $a_i \in A$, as in the HAT case (see Sec. 7.2.6). The definition of $G_1$ is analogous to the HAT case, though note that $G_1$ is now a capacitated bipartite graph, with the capacity of each house $h_j$ being $c_j$. Manlove and Sng [425] proved that a popular matching in $I$ must be a maximum matching in $G_1$.

Crucial to the definition of $s(a_i)$ is the extension of the Dulmage–Mendelsohn Decomposition (Theorem 1.4) to the capacitated bipartite graph case, and in particular to the capacitated graph $G_1$. Manlove and Sng [425] showed that one can indeed arrive at such an extension. One way to deduce this is to "clone" each house $h_j$ with capacity $c_j$ in $G_1$ into $c_j$ multiple copies $h_j^1, \ldots, h_j^{c_j}$ in the "cloned graph" $C(G_1)$, whose vertex set also includes the applicants in $A$, and whose edge set includes the edges $(a_i, h_j^k)$, for $1 \le k \le c_j$, whenever $(a_i, h_j)$ is an edge of $G_1$. The authors showed that in the EOU labelling $\mathcal{L}$ of the unit-capacity graph $C(G_1)$, any two "clones" of the same house $h_j$ have the same EOU label. Thus $\mathcal{L}$ gives rise to a well-defined EOU labelling of $G_1$.

On the basis of this observation, we can define $s(a_i)$, for any $a_i \in A$, as in Sec. 7.2.6: that is, $s(a_i)$ is the most-preferred set of houses in $a_i$'s preference list that are even in $G_1$. As in the HAT case, it need not follow that $f(a_i) \cap s(a_i) = \emptyset$. The reduced graph $G'$ may be defined in an analogous way to the HAT case (see Sec. 7.2.6), though note that $G'$ is now a capacitated bipartite graph, with the capacity of each house $h_j$ being $c_j$.

Manlove and Sng [425] established the following characterisation of popular matchings in CHAT.

**Theorem 7.29 ([425]).** *Let $I$ be an instance of CHAT and let $M$ be a matching in $I$. Then $M$ is popular if and only if:*

*(i) $M$ is a maximum matching in $G_1$;*

*(ii) $M$ is an applicant-complete matching in the reduced graph $G'$.*

Again, it is clear that Theorem 7.29 reduces to Theorem 7.15 in the case that $I$ is an instance of HA. Theorem 7.29 led to the following algorithmic result for computing maximum popular matchings (or reporting that none exists), given an instance of CHAT.

**Theorem 7.30 ([425]).** *Let $I$ be an instance of* CHAT. *There is an $O(n_1 m)$ algorithm[3] to find a maximum popular matching in $I$, or report that $I$ is unsolvable, where $n_1$ is the number of applicants and $m$ is the number of acceptable applicant–house pairs.*

Recently Paluch [473] considered the many–many extension of CHAT in which both applicants and houses can be multiply assigned (up to some given capacity, which is now specified for all agents). She characterised popular matchings in this setting and proved a range of algorithmic results for problems involving finding a popular matching or reporting that none exists.

### 7.3.4 *Variable house capacities*

Kavitha and Nasre [369] considered an interesting problem concerning popular matchings in CHAT. Suppose we are given an HAT instance $I$ that admits no popular matching. Intuitively, this is because a subset of the houses $H'$ are desirable for a set of applicants $A'$ whose cardinality exceeds that of $H'$. Kavitha and Nasre suggested that we regard $I$ as an instance of CHAT (initially with all houses having unit capacity) and then attempt to increase the capacities of certain houses in the hope that a popular matching might then exist.

For example, we have already seen that the HA instance $I_1$ shown in Fig. 1.3 admits no popular matching. However if we regard $I_1$ as an instance of CHA, setting $c_1 = 2$ and $c_2 = c_3 = 1$, then the matching $M = \{(a_1, h_1), (a_2, h_1), (a_3, h_2)\}$ is popular. In general, a trivial way to ensure that a popular matching always exists is to let $c_j = \max\{f_j, 1\}$, where $f_j$ is defined in Sec. 7.3.2. However this may be infeasible if each house has a limit on the extent to which its capacity may be increased.

These observations led Kavitha and Nasre [369] to define the following decision problem:

---

[3]In Ref. [425], the weaker upper bound of $O((\sqrt{C} + n_1)m)$ was given as the complexity for this algorithm, where $C$ is the total capacity of the houses. The improved upper bound follows by the remark in Footnote 6 on Page 16.

*Name:* POP CHAT VAR CAPS-1

*Instance:* an instance $I$ of HAT, with a bound $b_j \in \mathbb{Z}^+$ for each house $h_j \in H$ [4]

*Question:* is there a solvable instance $J$ of CHAT whose applicants, houses and preference lists are derived from $I$, where the capacity $c_j$ of each house $h_j \in H$ satisfies $1 \le c_j \le b_j$?

A special case of POP CHAT VAR CAPS-1 is the following:

*Name:* POP CHAT (1,2) CAPS

*Instance:* an instance $I$ of HAT, with a subset $H' \subseteq H$ [5]

*Question:* is there a solvable instance $J$ of CHAT whose applicants, houses and preference lists are derived from $I$, where $c_j = 1$ for each $h_j \in H \backslash H'$, and $c_j \in \{1,2\}$ for each $h_j \in H'$?

Thus in the POP CHAT (1,2) CAPS problem, we are given a subset $H'$ of houses, any of whose capacities could be raised to 2, whilst the remaining houses must have capacity 1. It is tempting to believe that, to maximise the chance of a popular matching existing, we should automatically increase the capacity of all houses in $H'$ to 2. However Kavitha and Nasre [369] gave an example to show that giving each house its maximum possible capacity does not always help to ensure that a popular matching exists. Specifically, they gave a solvable CHA instance where each house has capacity 1, but if we raise the capacity of every house to 2 then the instance is no longer solvable.

Kavitha and Nasre [369] proved the following algorithmic results for POP CHAT (1,2) CAPS.

**Theorem 7.31 ([369]).** POP CHAT (1,2) CAPS *is NP-complete. The result holds even in the following separate cases:*

(i) *each applicant $a_i \in A$ has a unique first choice, at most two (tied) second choices, and no house of rank $> 2$ (apart from $l_i$);*

(ii) *each applicant $a_i \in A$ has a strictly-ordered preference list $\mathcal{L}_i$ of length at most 3 (excluding $l_i$), where $\mathcal{L}_i$ is derived from a master list $\mathcal{L}$ of all houses in $H$.*

---

[4] Assume that $b_j = 1$ for each last resort house $h_j$ in $H$.

[5] Assume that no last resort house belongs to $H'$.

**Theorem 7.32 ([369]).** POP CHAT (1,2) CAPS *is solvable in polynomial time if each applicant* $a_i \in A$ *has a set of tied first choices (of any positive size), a unique second choice, and no house of rank* $> 2$ *(apart from* $l_i$*). If the answer is "yes", the algorithm constructs a* CHAT *instance* $J$ *with the desired properties, together with a popular matching in* $J$.

The authors also considered the following variant of POP CHAT VAR CAPS-1, in which the *total* capacity of the houses (rather than each individual house's capacity) is bounded:

*Name:*      POP CHAT VAR CAPS-2
*Instance:*  an instance $I$ of HAT, and an integer $K \in \mathbb{Z}^+$
*Question:*  is there a solvable instance $J$ of CHAT whose applicants, houses and preference lists are derived from $I$, where $\sum_{h_j \in H} c_j \leq K$? [6]

They showed that, in constrast to POP CHAT VAR CAPS-1, this problem is solvable in polynomial time.

**Theorem 7.33 ([369]).** POP CHAT VAR CAPS-2 *is solvable in polynomial time. If the answer is "yes", the algorithm constructs a* CHAT *instance* $J$ *with the desired properties, together with a popular matching in* $J$.

### 7.3.5   *Popularity at minimum cost*

Kavitha *et al.* [370] studied three problems that are somewhat related to those detailed in the previous subsection, but differ in that they involve weights on the houses.

The first problem is similar to POP CHAT VAR CAPS-1 in that we are given an HAT instance (which can be regarded as a CHAT instance with unit house capacities), and the task is to augment the house capacities in such a way that we obtain a solvable instance. However, rather than the house capacities being bounded, we now have a weight $wt(h_j) \geq 0$ for each $h_j \in H$. For each *additional* unit (above 1) by which we raise the capacity of $h_j$, we must "pay" a contribution of $wt(h_j)$. The task is to determine how to augment the house capacities to obtain a solvable instance so that the overall payment is minimum.

We now give a formal definition of this problem.

---
[6]Assume that $c_j = 1$ for each last resort house $h_j$ in $H$.

*Name:*    POP CHAT VAR CAPS-3

*Instance:*    an instance $I$ of HAT, and a weight $wt(h_j)$ for each house $h_j \in H$

*Solution:*    a solvable instance $J$ of CHAT whose applicants, houses and preference lists are derived from $I$, such that $\sum_{h_j \in H}(c_j - 1)wt(h_j)$ is minimum. [7]

Kavitha *et al.* [370] proved the following algorithmic results for POP CHAT VAR CAPS-3.

**Theorem 7.34 ([370]).** POP CHAT VAR CAPS-3 *is solvable in* $O(n_1^2)$ *time if each applicant's preference list is strictly ordered and of length at most 2 (excluding last resort houses), where* $n_1$ *is the number of applicants. By contrast,* POP CHAT VAR CAPS-3 *is NP-hard even if each applicant's preference list is strictly ordered and of length at most 3 (again excluding last resorts). The result holds even if each applicant's list is derived from a single master list of the houses. In the latter case it is also NP-hard to approximate* POP CHAT VAR CAPS-3 *within a factor of* $\sqrt{n_1}/2$.

The second problem is similar to POP CHAT VAR CAPS-3. Here we assume that we are given an HAT instance $I$, together with a weight $wt(h_j) \geq 0$ for each house. The problem is to choose a capacity $c_j \geq 0$ for each house $h_j \in H$ in order to obtain a solvable CHAT instance $J$. Note that the case $c_j = 0$ is permitted. The capacities must be chosen so that $\sum_{h_j \in H} c_j wt(h_j)$ is minimum. A degenerate solution to this problem is simply to set $c_j = 0$ for each $h_j \in H$, in which case the empty matching is trivially popular in $J$. To prevent this possibility, the authors insist that $J$ must admit an applicant-complete popular matching. This is always possible, for example by setting $c_j = \max\{f_j, 1\}$ for each $h_j \in H$, but of course such a solution may have a large total weight.

We now define the second problem formally.

*Name:*    POP CHAT VAR CAPS-4

*Instance:*    an instance $I$ of HAT, and a weight $wt(h_j)$ for each house $h_j \in H$

*Solution:*    an instance $J$ of CHAT whose applicants, houses and preference lists are derived from $I$, where each house $h_j$ has capacity $c_j \geq 0$ in $J$,[8] such that $J$ admits an applicant-complete popular matching and $\sum_{h_j \in H} c_j wt(h_j)$ is minimum.

---

[7] See Footnote 6.
[8] See Footnote 6.

Kavitha *et al.* [370] proved the following algorithmic results for POP CHAT VAR CAPS-4.

**Theorem 7.35 ([370]).** POP CHAT VAR CAPS-4 *is NP-hard even if each applicant's preference list is strictly ordered and of length at most 2 (excluding last resort houses). The result holds even if each applicant's list is derived from a single master list of the houses.*

We refer to the third problem that the authors consider as POP CHAT MIN WEIGHT, and define it as follows. We are given given a CHAT instance $I$ in which each house $h_j \in H$ has a weight $wt(h_j) \geq 0$, and the problem is to find a popular matching $M$ in $I$ such that $wt(M) = \sum_{(a_i, h_j) \in M} wt(h_j)$ is minimum, or report that no popular matching exists. Note that (i) we assume that $wt(l_i) = 0$ for all $a_i \in A$, and (ii) $wt(M)$ can be equivalently expressed as $\sum_{h_j \in H} |M(h_j)| wt(h_j)$. It turns out that this problem is solvable in polynomial time.

**Theorem 7.36 ([370]).** POP CHAT MIN WEIGHT *is solvable in $O(n_1 m)$ time, where $n_1$ is the number of applicants and $m$ is the number of acceptable applicant–house pairs.*

We remark that Theorem 7.36 is a generalisation of Theorem 7.30.

## 7.4   Weighted House Allocation problem

Mestre [449] introduced the *Weighted House Allocation problem* (WHA), which is a generalisation of HA in which each applicant $a_i \in A$ has a positive weight $wt(a_i)$ indicating her priority when it comes to majority voting. The "more popular than" relation introduced in Sec. 1.5.5 can be generalised to weighted majorities as follows. Let $I$ be an instance of WHA and let $\mathcal{M}$ be the set of matchings in $I$. For two matchings $M, M' \in \mathcal{M}$, we say that $M'$ is *more popular than* $M$, denoted $M' \blacktriangleright M$, if

$$\sum_{a_i \in P(M', M)} wt(a_i) > \sum_{a_i \in P(M, M')} wt(a_i).$$

$M$ is *popular* if there is no matching $M' \in \mathcal{M}$ such that $M' \blacktriangleright M$. Thus intuitively $M$ is popular if there is no other matching that is preferred by a weighted majority of the applicants.

Applicant weights can equally be introduced into instances of HAT, CHA and CHAT, giving rise to WHAT, WCHA and WCHAT respectively. In all cases the definition of popularity is unchanged from the WHA case.

Mestre [449] proved the following results concerning the complexity of finding a maximum popular matching or reporting that none exists, given instances of WHA and WHAT.

**Theorem 7.37 ([449]).** *Given an instance I of* WHA, *there is an* $O(n + m)$ *algorithm to find a maximum popular matching in I or report that I is unsolvable, where n is the number of applicants and houses, and m is the number of acceptable applicant–house pairs.*

**Theorem 7.38 ([449]).** *Given an instance I of* WHAT, *there is an* $O(\min(k\sqrt{n}, n)m)$ *algorithm to find a maximum popular matching in I or report that I is unsolvable, where k is the number of distinct applicant weights, n is the number of applicants and houses, and m is the number of acceptable applicant–house pairs.*

Sng and Manlove [537] extended Mestre's algorithm for WHA to the CHA case, proving the following result.

**Theorem 7.39 ([537]).** *Given an instance I of* WCHA, *there is an* $O(n_1^{1.5} + m)$ *algorithm[9] to find a maximum popular matching in I or report that I is unsolvable, where $n_1$ is the number of applicants and m is the number of acceptable applicant–house pairs.*

Itoh and Watanabe [331] investigated the solvability probability for a random instance of WHA in which all applicant preference lists are complete. In particular, they considered the *two-weighted* case, where $\{wt(a_i) : a_i \in A\} = \{w_1, w_2\}$. Without loss of generality assume that $w_1 > w_2$. We can think of the applicants of weight $w_1$ as the high priority applicants, whilst all other applicants have low priority. The authors in fact considered the stronger restriction that $w_1 \geq 2w_2$.

For these restrictions, Itoh and Watanabe [331] showed that a random WHA instance is solvable with probability $P_L(n_1, n_2) = O(n_2^3/n_1^4)$, where $n_1$ is the number of applicants and $n_2$ is the number of houses. Thus if $n_2/n_1^{4/3} = o(1)$ then $P_L(n_1, n_2) = o(1)$. Similarly they showed that a random WHA instance (again under the above restrictions) is solvable

---

[9]In Ref. [537], the weaker upper bound of $O(\sqrt{C}n_1 + m)$ was given as the complexity for this algorithm, where $C$ is the total capacity of the houses. The improved upper bound follows by the remark in Footnote 6 on Page 16.

with probability $P_U(n_1, n_2) = 1 - O(n_1^4/n_2^3)$. Thus if $n_1^{4/3}/n_2 = o(1)$ then $P_U(n_1, n_2) = 1 - o(1)$.

## 7.5 Stable Roommates problem

### 7.5.1 *Introduction*

Popular matchings may be defined in instances of SRI using a straightforward adjustment of the notation and terminology introduced in Sec. 1.5.5 for HA. Let $I$ be an instance of SRI and let $\mathcal{M}$ denote the set of matchings in $I$. Given two matchings $M, M' \in \mathcal{M}$, let $P(M, M')$ denote the set of agents who prefer $M$ to $M'$. The "more popular than" relation ▶, and the concept of a popular matching, are then defined as in Sec. 1.5.5. Thus a matching $M \in \mathcal{M}$ is popular if there is no other matching that is preferred by a majority of the agents. The notion of a popular matching can be defined in the SRTI context in the same way as for SRI.

We outline structural and algorithm results for popular matchings in instances of SRI and SRTI in Secs. 7.5.2 and 7.5.3 respectively. We then consider least unpopularity factor matchings in SRI in Sec. 7.5.4. Finally, we study strongly popular matchings in instances of SRI and SRTI in Sec. 7.5.5.

### 7.5.2 *Strictly-ordered preference lists*

Chung [153] was the first to study popular matchings in the context of an SRI instance. He noted the following:

**Proposition 7.40 ([240, 153]).** *Let $I$ be an instance of* SRI *and let $M$ be a stable matching in $I$. Then $M$ is popular in $I$.*

Proposition 7.40 was proved in the SMI case by Gärdenfors [240]. We note that the converse to Proposition 7.40 need not be true. Also an SRI instance need not admit a popular matching. To see these facts, consider the SR instance $I'$ shown in Fig. 7.6. Gale and Shapley [235] observed that $I'$ admits no stable matching, but Biró *et al.* [86] noted that $M_1 = \{\{a_1, a_4\}, \{a_2, a_3\}\}$ and $M_2 = \{\{a_2, a_4\}, \{a_1, a_3\}\}$ are popular matchings in $I'$. If we remove $a_4$ then the resulting instance is unsolvable.

Biró *et al.* [86] gave an algorithm for the problem of testing a given matching $M$ for popularity in a given SRI instance $I$. Let $G = (A, E)$ be the underlying graph of $I$, where $A = \{a_1, \ldots, a_n\}$. We form a weighted graph

$$a_1 : a_2 \; a_3 \; a_4$$
$$a_2 : a_3 \; a_1 \; a_4$$
$$a_3 : a_1 \; a_2 \; a_4$$
$$a_4 : a_1 \; a_2 \; a_3$$

Fig. 7.6   An instance of SRI with a popular matching but no stable matching [235, 86].

$H_M$ as follows. The vertices of $H_M$ are $A \cup A'$, where $A' = \{a_1', \ldots, a_n'\}$. The edges of $H_M$ are $E \cup E' \cup E''$, where $E' = \{\{a_i', a_j'\} : \{a_i, a_j\} \in E\}$ and $E'' = \{\{a_i, a_i'\} : 1 \leq i \leq n\}$. For each edge $\{a_i, a_j\} \in E$, we define $\delta_{i,j}$ as follows:

$$\delta_{i,j} = \begin{cases} 0, & \text{if } \{a_i, a_j\} \in M \\ \frac{1}{2}, & \text{if } a_i \text{ is unassigned in } M \text{ or prefers } a_j \text{ to } M(a_i) \\ -\frac{1}{2}, & \text{otherwise.} \end{cases}$$

For each edge $\{a_i, a_j\} \in E$, we define the weight of each of the edges $\{a_i, a_j\}$ and $\{a_i', a_j'\}$ in $H_M$ to be $\delta_{i,j} + \delta_{j,i}$. Also for each $a_i \in A$, we define the weight of the edge $\{a_i, a_i'\}$ in $H_M$ to be -1 if $a_i$ is assigned in $M$, and 0 otherwise. It is clear that the weight of each edge belongs to the set $\{-1, 0, 1\}$. Now let

$$M' = M \cup \{\{a_i', a_j'\} : \{a_i, a_j\} \in M\} \cup \{\{a_i, a_i'\} : a_i \text{ is unassigned in } M\}.$$

Clearly $M'$ is a perfect matching in $H_M$ with weight 0. It turns out that $M$ is popular if and only if $M'$ is a maximum weight perfect matching in $H_M$, as the following lemma indicates.

**Lemma 7.41 ([86]).** *Let $I$ be an instance of* SRI *and let $M$ be a matching in $I$. Let $H_M$ be the weighted graph defined above. Then $M$ is popular if and only if a maximum weight perfect matching in $H_M$ has weight 0.*

**Theorem 7.42 ([86]).** *Let $I$ be an instance of* SRI *and let $M$ be a matching in $I$. There is an $O(\sqrt{n\alpha(n,m)}m\log^{3/2}n)$ algorithm to test whether $M$ is popular, where $n$ is the number of agents, $m$ is the number of acceptable pairs of agents, and $\alpha$ is the inverse Ackermann function.*

**Proof.**   Clearly $H_M$ has $O(n)$ vertices and $O(m)$ edges. For a weighted graph with weights in the set $\{-1, 0, 1\}$, Gabow and Tarjan's algorithm [231] for finding a maximum weight perfect matching has complexity $O(\sqrt{n\alpha(n,m)}m\log^{3/2}n)$. $\qquad\square$

It is clear that a perfect matching $M^*$ of positive weight exists in $H_M$ if and only if $H_M$ admits an alternating cycle (relative to $M'$) of positive weight. At present it is unknown as to whether testing for such an alternating cycle can be achieved in a better running time than finding a maximum weight perfect matching in the general case.[10]

Given an SRI instance $I$, the algorithmic complexity of the problem of finding a popular matching in $I$ or reporting that none exists is, at the time of writing, unknown.

### 7.5.3   *Preference lists with ties*

The algorithm for testing a matching for popularity in an instance of SRI, as presented in the previous subsection, can be extended to the ties case in a natural way. We simply need to modify the definition of the $\delta_{i,j} = 0$ case to additionally include the possibility that $a_i$ is indifferent between $a_j$ and $M(a_i)$. As a result we will have weights $\{-1, -\frac{1}{2}, 0, \frac{1}{2}, 1\}$ in $H_M$ in general, but the remainder of the technique and the complexity of the popularity checking algorithm is as before. Hence we have the following result.

**Theorem 7.43 ([86]).**  *Given an instance $I$ of SRTI and a matching $M$ in $I$, there is an $O(\sqrt{n\alpha(n,m)}m\log^{3/2} n)$ algorithm to test whether $M$ is popular, where $n$ is the number of agents, $m$ is the number of acceptable pairs of agents, and $\alpha$ is the inverse Ackermann function.*

Recall from Proposition 7.40 that a stable matching in a given SRI instance is popular. However the analogue of this result does not hold for SRTI in the case of weak stability: Chung [153] gave an example SRTI instance with 7 weakly stable matchings but no popular matching. Biró *et al.* [86] proved that finding a popular matching, or reporting that none exists, is NP-hard in the case of SRTI.

**Theorem 7.44 ([86]).**  *Given an instance $I$ of SRTI, the problem of deciding whether a popular matching exists is NP-complete. The same is true even if the popular matching is required to be complete. Both results hold even if $I$ is an instance of SMTI.[11]*

---

[10]Note that if $G$ (and therefore $H_M$) is bipartite, the edges in $H_M$ can be directed to yield a digraph $D_M$ satisfying the property that $H_M$ has an alternating cycle of positive weight if and only if $D_M$ has a positive weight cycle [227]. By negating the arc weights in $D_M$, we can test for a positive weight cycle in $D_M$ in $O(\sqrt{n}m)$ time [255]. However this transformation breaks down in the case that $G$ is non-bipartite.

[11]See Theorem 7.62.

It is open as to whether Theorem 7.44 holds even if $I$ is an instance of SRT (i.e., all preference lists are complete).

### 7.5.4  *Least unpopularity factor matchings*

Huang and Kavitha [296] studied matchings with low unpopularity factor in the context of SRI. All notation and terminology defined in Definition 7.18 (see Sec. 7.2.7) relating to the unpopularity factor of matchings in HAT carry over to SRI without modification. The authors proved the following results:

**Theorem 7.45 ([292]).** *Let $I$ be an instance of SR. The problem of computing a least unpopularity factor matching in $I$ is NP-hard. Moreover there is no approximation algorithm for the problem of finding a least unpopularity factor matching in $I$ with performance guarantee $4/3 - \varepsilon$, for any $\varepsilon > 0$, unless P=NP.*

**Theorem 7.46 ([292]).** *Let $I$ be an instance of SRI. There is an $O(m)$ algorithm to find a matching $M$ in $I$ such that $u(M) = 4\log n + O(1)$, where $n$ is the number of agents and $m$ is the number of acceptable pairs of agents.*

Theorem 7.46, which guarantees the existence, in an arbitrary SRI instance, of a matching $M$ where $u(M) = O(\log n)$ is surprising. For, recall that Fig. 7.4 illustrated an HA instance $I'$ where $u(M') = \Omega(n_1)$ for every matching $M'$ in $I'$, where $n_1$ is the number of applicants in $I'$.

### 7.5.5  *Strongly popular matchings*

Given an instance of SRI or SRTI, a strongly popular matching can be defined in exactly the same way as for HAT (see Sec. 1.5.5). Two straightforward facts about strongly popular matchings in instances of SRTI are as follows:

**Proposition 7.47 ([240, 86]).** *Let $I$ be an instance of SRTI and let $M$ be a strongly popular matching in $I$. Then*

*(i) $M$ is the only popular matching in $I$;*
*(ii) $M$ is weakly stable in $I$.*

Biró *et al.* [86] proved Proposition 7.47 for SRI, and Gärdenfors [240] proved Part (ii) of Proposition 7.47 for SMTI. A consequence of Part (i) is that an SRTI instance admits at most one strongly popular matching. Together with Proposition 7.40, it follows that in an SRI instance, strong popularity implies stability, which in turn implies popularity.

Biró *et al.* [85, Example 2] gave an example instance of SRI (in fact an instance of SMI) that admits one popular matching (which is the unique stable matching) but no strongly popular matching.

We now turn to the problem of testing a matching for strong popularity. We firstly consider the case that we are given an instance $I$ of SRI. Suppose that $M$ is a stable matching in $I$. Define the graph $H'_M = (A, E_M)$, where

$$E_M = \left\{ \{a_i, a_j\} \in E : \begin{array}{l} a_i \text{ is unmatched in } M \text{ or prefers } a_j \text{ to } M(a_i) \vee \\ a_j \text{ is unmatched in } M \text{ or prefers } a_i \text{ to } M(a_j) \end{array} \right\}.$$

The following lemma, proved by Biró *et al.* [86], leads to an $O(m)$ algorithm for testing a matching for strong popularity, where $m$ is the number of acceptable pairs of agents.

**Lemma 7.48 ([86]).** *Let $I$ be an instance of* SRI *and let $M$ be a stable matching in $I$. Let $H'_M$ be the graph defined above. Then $M$ is strongly popular in $I$ if and only if $H'_M$ contains no alternating cycle or augmenting path relative to $M$.*

**Theorem 7.49 ([86]).** *Let $I$ be an instance of* SRI *and let $M$ be a matching in $I$. There is an $O(m)$ algorithm to test whether $M$ is strongly popular, where $m$ is the number of acceptable pairs of agents in $I$.*

**Proof.** We firstly check whether $M$ is stable, which may be verified in $O(m)$ time [261]. If not, then $M$ cannot be strongly popular by Proposition 7.47. Now build $H'_M$ in $O(m)$ time. By Lemma 7.48, $M$ is strongly popular if and only if $H'_M$ contains no augmenting path or alternating cycle relative to $M$. We may test for the existence of each of these structures in $O(m)$ time (see Refs. [225, 229] and [228] respectively).                                                                 □

Now suppose that $I$ is an instance of SRTI. To test whether a matching $M$ is strongly popular in $I$, we can use a slight variation on the technique described in the lead-up to Theorems 7.42 and 7.43. Recall the definitions of $H_M$ and $M'$ as given in the preamble to Theorem 7.42. It is not difficult to verify that $M$ is strongly popular if and only if $M'$ is the *unique* maximum weight perfect matching in $H_M$. Thus, we must firstly verify that $M'$ is a maximum weight perfect matching in $H_M$ by checking that a maximum weight perfect matching constructed by the $O(\sqrt{n}\alpha(n, m)m \log^{3/2} n)$ algorithm of Gabow and Tarjan [231] has weight 0. Next, we use the $O(m)$

algorithm of Gabow *et al.* [228] to determine whether a given maximum weight perfect matching in a weighted graph is unique. We thus have the following result.

**Theorem 7.50 ([86]).** *Let I be an instance of* SRTI *and let M be a matching in I. There is an* $O(\sqrt{n\alpha(n,m)}m\log^{3/2}n)$ *algorithm*[12] *to test whether M is strongly popular, where n is the number of agents, m is the number of acceptable pairs of agents, and* $\alpha$ *is the inverse Ackermann function.*

We now consider the problem of finding a strongly popular matching or reporting that none exists. Biró *et al.* [86] showed that this problem is polynomial-time solvable in the case of SRI.

**Theorem 7.51 ([86]).** *Let I be an instance of* SRI*. There is an* $O(m)$ *algorithm to find a strongly popular matching in I, or report that none exists, where m is the number of acceptable pairs of agents in I.*

**Proof.** We firstly use Irving's algorithm [261, Section 4.5.2] to find a stable matching $M$ in $I$ or report that no such matching exists in $O(m)$ time. In the latter case, $M$ does not admit a strongly popular matching by Part (ii) of Proposition 7.47. Otherwise, by Part (i) of Proposition 7.47, $M$ is popular. By Theorem 7.49, we can test whether $M$ is strongly popular in $O(m)$ time. If so, we output $M$. Otherwise, by Part (i) of Proposition 7.47, $I$ admits no strongly popular matching. □

The algorithmic complexity of the problem of finding a strongly popular matching, or reporting that none exists, given an instance of SRTI, is currently open.

## 7.6 Stable Marriage problem

### 7.6.1 *Introduction*

In the previous section we considered popular matchings in instances of SRI and SRTI. In this section we restrict attention to the bipartite case, where we are given an instance of SMI or SMTI. In the case of SMI, there is a nice structural characterisation of popular matchings (described in Sec. 7.6.2), leading to an efficient algorithm for finding a maximum popular matching (outlined in Sec. 7.6.3). This characterisation breaks down in the SMTI

---

[12] A slightly different technique for solving this problem is described in Ref. [86], with the same time complexity.

case: indeed, as we will see in Sec. 7.6.4 (among other results for popular matchings in SMI and SMTI), the NP-completeness results of Theorem 7.44 hold even in this restricted setting.

We begin with the following straightforward result for SMI, which is an immediate consequence of Proposition 7.40.

**Proposition 7.52.** *Let I be an instance of* SMI. *Then I admits a popular matching, and such a matching can be found in $O(m)$ time using the Gale–Shapley algorithm, where m is the number of acceptable man–woman pairs.*

### 7.6.2 Characterising popular matchings

Huang and Kavitha [297] arrived at a neat characterisation of popular matchings in SMI. To describe this characterisation, we require to define some additional notation, as follows.

**Definition 7.53 ([297]).** *Let I be an instance of* SMI *and let $G = (V, E)$ be the underlying graph of I. Let M be a matching in I. For any edge $(m_i, w_j) \in E$, define the* label *of this edge to be $(\alpha, \beta)$, where*

$$\alpha = \begin{cases} 1, & if \begin{cases} either\ m_i\ is\ unassigned\ in\ M, \\ or\ m_i\ is\ assigned\ in\ M\ and\ prefers\ w_j\ to\ M(m_i) \end{cases} \\ 0, & if\ (m_i, w_j) \in M \\ -1, & if\ m_i\ is\ assigned\ in\ M\ and\ prefers\ M(m_i)\ to\ w_j. \end{cases}$$

*The definition of $\beta$ is analogous. We define the* reduced labelled graph of *M, denoted $G_M^+$, to be the subgraph of the labelled graph G that is obtained by deleting all edges labelled $(-1, -1)$ from G.*

On the basis of this definition, Huang and Kavitha proved the following.

**Theorem 7.54 ([297]).** *Let I be an instance of* SMI *and let M be a matching in I. Let $G_M^+$ be as defined in Definition 7.53. Then M is popular in I if and only if the following three conditions are satisfied in $G_M^+$:*

*(i) there is no alternating cycle relative to M that contains an edge labelled (1,1);*

*(ii) there is no alternating path relative to M from an unassigned agent that contains an edge labelled (1,1);*

*(iii) there is no alternating path relative to M that contains two or more edges labelled (1,1).*

| Men's preferences | Women's preferences | |
|---|---|---|
| $m_1 : w_2\ w_1$ | $w_1 : m_1$ | Instance $I_1$ |
| $m_2 : w_2$ | $w_2 : m_1\ m_2$ | |
| | | |
| $m_1 : w_1$ | $w_1 : m_2\ m_1$ | |
| $m_2 : w_1\ w_2$ | $w_2 : m_3\ m_2$ | Instance $I_2$ |
| $m_3 : w_2\ w_3$ | $w_3 : m_3$ | |

Fig. 7.7 Two instances of SMI due to Biró *et al.* [86]

On the basis of this characterisation, Huang and Kavitha [297] gave a linear-time algorithm for testing a given matching for popularity.

**Theorem 7.55 ([297]).** *Let $I$ be an instance of* SMI *and let $M$ be a matching in $I$. There is an $O(m)$ algorithm to test whether $M$ is popular in $I$, where $m$ is the number of acceptable man–woman pairs in $I$.*

An $O(\sqrt{n}m)$ algorithm for this problem was given by Biró *et al.* [86], where $n$ is the size of $I$.

Huang and Kavitha [297] remarked that the characterisation also holds in the case that $I$ is an instance of SRI. However they were not able to use it in order to formulate an algorithm for testing a matching for popularity in the SRI context that improves on the $O(\sqrt{n\alpha(n,m)}m\log^{3/2} n)$ method given by Theorem 7.42. Hence we have presented the characterisation given by Theorem 7.54 in this section, whilst describing instead the characterisation of popular matchings via maximum weight matchings in Sec. 7.5.

### 7.6.3 *Maximum popular matchings*

We have already seen from Proposition 7.52 that every instance of SMI admits at least one popular matching. However, popular matchings may be of different sizes. This is illustrated by instance $I_1$, due to Biró *et al.* [86], shown in Fig. 7.7. The matching $M_1 = \{(m_1, w_2)\}$ is the unique stable matching and hence popular. However the matching $M_2 = \{(m_1, w_1), (m_2, w_2)\}$ is also popular. Clearly $I_1$ can be replicated as many times as necessary to produce an arbitrarily large SMI instance for which the size of a popular matching can be twice the size of a stable matching.

In many applications we seek to assign as many agents as possible, so the observations in the previous paragraph motivate the problem of finding a maximum popular matching in an SMI instance. One strategy for finding a maximum popular matching could involve starting with a stable matching, and attempting to find augmenting paths that preserve popularity until we reach a maximum popular matching. However there are three complications with this approach.

Firstly, it turns out that a maximum popular matching can be smaller than the size of a maximum cardinality matching in the underlying graph. To see this, consider instance $I_2$, due to Biró *et al.* [86], shown in Fig. 7.7. The unique perfect matching $M_1 = \{(m_1, w_1), (m_2, w_2), (m_3, w_3)\}$ is not popular (the stable matching $M_2 = \{(m_2, w_1), (m_3, w_2)\}$ is more popular than $M_1$).

Secondly, stable matchings do not provide a very good starting point, as the following result shows.

**Proposition 7.56 ([297]).** *Let $I$ be an instance of SMI and let $M$ be a stable matching in $I$. Then $M$ is a minimum popular matching.*

The third, and most serious, complication is that popularity is not an interpolating invariant in the context of an SMI instance $I$. That is, if $pop^-(I)$ and $pop^+(I)$ denote the minimum and maximum cardinalities of a popular matching in $I$, then there need not be a popular matching of size $k$, for each $k$ such that $pop^-(I) \leq k \leq pop^+(I)$. This was illustrated by Huang and Kavitha [297] who gave an example SMI instance with popular matchings of sizes 4 and 6, but no popular matching of size 5. Recall that popularity is an interpolating invariant in the context of HA (see Theorem 7.5).

There is, however, an alternative method for finding a maximum popular matching that is based on the characterisation given by Theorem 7.54, together with the following additional result.

**Theorem 7.57 ([297]).** *Let $I$ be an instance of SMI and let $M$ be a matching in $I$. Let $G_M^+$ be as defined by Definition 7.53. Suppose that $M$ satisfies Conditions (i)-(iii) of Theorem 7.54. (Then $M$ is popular.) Now suppose in addition that the following condition is satisfied in $G_M^+$:*

*(iv) there is no augmenting path relative to $M$.*

*Then $M$ is a maximum popular matching.*

Huang and Kavitha [297] showed that in general, Condition (iv) is not a necessary condition for a popular matching to be of maximum size: they gave an example SMI instance with a maximum popular matching $M$ that admits an augmenting path in $G_M^+$. However they showed that a different maximum popular matching that admits no augmenting path in $G_M^+$ exists. Moreover they showed that for a general SMI instance, it is always possible to find a matching satisfying Conditions (i)-(iv) in Theorems 7.54 and 7.57, as the following result indicates.

**Theorem 7.58 ([297]).** *Let $I$ be an instance of* SMI. *There is an $O(nm)$ algorithm to find a matching that satisfies Conditions (i)-(iv) of Theorems 7.54 and 7.57, where $n$ is the size of $I$ and $m$ is the number of acceptable man–woman pairs in $I$. By Theorems 7.54 and 7.57, $M$ is a maximum popular matching in $I$.*

Kavitha [362] gave a faster algorithm for finding a maximum popular matching in an instance of SMI.

**Theorem 7.59 ([362]).** *Let $I$ be an instance of* SMI. *There is an $O(m)$ algorithm to find a maximum popular matching in $I$, where $m$ is the number of acceptable man–woman pairs in $I$.*

Huang and Kavitha [297] also observed that a maximum popular matching has size at least two-thirds times that of a maximum cardinality matching in an instance of SMI. If we consider the SMI instance $I_2$ shown in Fig. 7.7, $M = \{(m_2, w_1), (m_3, w_2)\}$ is a popular matching of size 2 in $I$. We have already seen that a maximum cardinality matching in $I_2$ has size 3. Clearly $I_2$ can be replicated as many times as desired to produce an arbitrarily large SMI instance showing that this bound is tight.

In the case of SRI, the complexity of the problem of finding a maximum popular matching is open (note that, as already observed at the end of Sec. 7.5.2, the complexity of the problem of finding an arbitrary popular matching in an SRI instance is open). It looks as though additional ideas are required over and above the characterisation given by Theorem 7.54: Huang and Kavitha [297] gave an example of a solvable SRI instance where *no* popular matching satisfies Condition (iv) of Theorem 7.57.

### 7.6.4 *Further results for* SMI *and* SMTI

Kavitha [362] proved some further results regarding popular matchings in instances of SMI.

Let $I$ be an instance of SMI and let $\mathcal{M}^+$ be the set of maximum matchings in the underlying graph of $I$. We say that a matching $M \in \mathcal{M}^+$ is *popular among maximum matchings* if there is no matching $M' \in \mathcal{M}^+$ such that $|P(M', M)| > |P(M, M')|$. That is, there is no other maximum matching that is preferred to $M$ by a majority of the agents. Kavitha [362] proved that a matching that is popular among maximum matchings always exists in $I$ and can be found efficiently.

**Theorem 7.60 ([362]).** *Let $I$ be an instance of SMI. A matching that is popular among maximum matchings in $I$ exists and can be found in $O(nm)$ time, where $n$ is the size of $I$ and $m$ is the number of acceptable man–woman pairs in $I$.*

Theorem 7.59 indicates the existence of an $O(m)$ algorithm for finding a maximum popular matching (which, as previously mentioned, is guaranteed to be of size at least two-thirds times that of a maximum matching), whilst Theorem 7.60 shows that we can find in $O(nm)$ time a maximum matching $M$ that is popular among the set of maximum matchings. However $M$ may have a large unpopularity factor when considering *all* matchings in $I$ — see Ref. [362] for further details). It is thus of interest to determine whether there are matchings in between these two extremes.

Kavitha [362] demonstrated the existence of a spectrum of matchings $M_k$ $(2 \leq k \leq n)$ such that $|M_k| \geq \frac{k}{k+1}\beta^+(G)$ and $u(M_k) \leq k - 1$, where $\beta^+(G)$ is the size of a maximum matching in the underlying graph $G$ of $I$. Moreover she showed that each such matching can be computed efficiently.

**Theorem 7.61 ([362]).** *Let $I$ be an instance of SMI. For each $k$ $(2 \leq k \leq n)$, there is an $O(km)$ algorithm to construct a matching $M_k$ such that $|M_k| \geq \frac{k}{k+1}\beta^+(G)$ and $u(M_k) \leq k - 1$, where $n$ is the size of $I$, $\beta^+(G)$ is the size of a maximum matching in the underlying graph $G$ of $I$ and $m$ is the number of acceptable man–woman pairs in $I$.*

In the context of Theorem 7.61, if $k = 2$ then we obtain Theorem 7.59, whilst if $k = n$ then we obtain Theorem 7.60.

The notion of a popular matching can be defined in the SMTI context in the same way as for SMI. However, in contrast to the case for SMI, an instance of SMTI need not be solvable. To see this, consider the SMTI instance $I'$ given in Fig. 7.8. It is not difficult to verify that $I'$ is unsolvable, for the same reason that the HA instance $I_1$ of Fig. 1.3 is unsolvable.

Men's preferences  Women's preferences

$m_1 : w_1\ w_2\ w_3$  $w_1 : (m_1\ m_2\ m_3)$

$m_2 : w_1\ w_2\ w_3$  $w_2 : (m_1\ m_2\ m_3)$

$m_3 : w_1\ w_2\ w_3$  $w_3 : (m_1\ m_2\ m_3)$

Fig. 7.8  An instance of SMTI with no popular matching

Biró et al. [86] proved that finding a popular matching, or reporting that none exists, is NP-hard in the case of SMTI — this result can be regarded as strengthening Theorem 7.44.

**Theorem 7.62 ([86]).** *Given an instance $I$ of* SMTI, *the problem of deciding whether a popular matching exists is NP-complete. The same is true even if the popular matching is required to be complete.*

As in the case of Theorem 7.44, it is open as to whether Theorem 7.62 holds even if all preference lists are complete.

Sng [535] considered popular matchings in instances of SMTI with symmetric preferences. He gave an $O(\sqrt{n}m)$ algorithm for testing a given matching for popularity in such an instance $I$, where $n$ is the size of $I$ and $m$ is the number of acceptable man–woman pairs in $I$. He left open the complexity of the problem of finding a popular matching in $I$ or reporting that no such matching exists.

## 7.7  Conclusions and open problems

As discussed in Sec. 7.1, the study of the structure of popular matchings, and the algorithmic complexity of computing these types of matchings, is still a relatively young area of research. Despite this, some impressive progress has been made in a relatively short space of time, which accounts for the length of this chapter. For a range of matching problems, we know how to efficiently test a matching for popularity, find a popular matching or report that the instance is unsolvable, and even find a maximum popular matching in the case of a solvable instance. However some notable open problems remain. These include:

- It remains open to obtain a structural characterisation of the set of popular matchings in an instance of HAT. Clearly the switching graph definition (presented in Sec. 7.2.5) will need to be extended if there is any hope of progress in this direction, since in the HAT context we would presumably define $O_M(a_i) = (f(a_i) \cup s(a_i)) \backslash \{M(a_i)\}$ for a given applicant $a_i \in A$, with respect to a given matching $M$.

- In Sec. 7.2.7 we stated that each of the problems of computing a least unpopularity factor matching and a least unpopularity margin matching, given an instance of HA, is NP-hard. Theorem 7.21 indicates that the former problem is not approximable within $3/2 - \varepsilon$, for any $\varepsilon > 0$, unless P=NP. It remains open to find a tighter lower bound and/or to give an upper bound. No approximability results are known for the least unpopularity margin matching problem. The results of Theorem 7.23 may be useful in this context.

- One of the most intriguing open problems is the following: given an SRI instance $I$, find a popular matching $M$ in $I$ or report that $I$ is unsolvable. Currently it is not known whether this problem is solvable in polynomial time or is NP-hard. Similar remarks apply if $M$ is required to be a maximum popular matching. On the positive side, we have already remarked that the characterisation of popular matchings in SMI given by Theorem 7.54 holds in the SRI case, although the same is not true for the characterisation of maximum popular matchings in SMI given by Theorem 7.57.

- It is also open as to whether each of the problems of finding a popular matching and a maximum popular matching in the context of SRTI and SMTI is NP-hard even if the preference lists are complete. That is, do Theorems 7.44 and 7.62 hold even if we are given an instance of SRT or SMT respectively?

- Our last open problem concerns the complexity of the problem of finding a strongly popular matching, or reporting that none exists, given an instance of SRTI, which is unknown at the time of writing.

# Chapter 8

# Profile-based optimal matchings

## 8.1 Introduction

In this chapter we study matchings that are optimal with respect to criteria that involve the profile of a matching. In Sec. 1.5.6 we defined the concepts of a rank-maximal matching, a greedy maximum matching and a generous maximum matching. These are very natural notions of optimality that, it could be imagined, may be desirable as solution criteria for an administrator of a centralised matching scheme.

Our main focus here is on how matchings of these types can be computed efficiently. We will present polynomial-time algorithms for finding rank-maximal, greedy maximum and generous maximum matchings. Much less is known about the structure of the sets of matchings that are optimal with respect to these criteria: we shall return to this point in the open problems section at the end of this chapter.

These types of optimal matchings have applications in a range of bipartite matching settings where there are preferences on one side only. These include assigning students to projects or courses, customers to DVD rentals and reviewers to conference papers. The application involving assigning students to projects and courses was described in Sec. 1.5.8, whilst the latter two applications will be discussed in this chapter.

It should be noted that mechanisms based on computing profile-based optimal matchings are not strategy-proof in general. To see this, consider the HA instance $I$ involving two houses, $h_1$ and $h_2$, and two applicants, $a_1$ and $a_2$, each of whom prefers $h_1$ to $h_2$. The matching $M_1 = \{(a_1, h_2), (a_2, h_1)\}$ is a rank-maximal, greedy maximum and generous maximum matching in $I$. However suppose that $a_1$ now truncates her preference list by deleting $h_2$, to try to ensure that she is matched to her first-choice house. Then $M_2 = \{(a_1, h_1), (a_2, h_2)\}$ is the unique rank-maximal, greedy maximum and generous maximum matching in the HA instance so obtained. Thus it is in $a_1$'s interests to misrepresent her true preferences. The potential for manipulation can be seen as a weakness of the profile-based optimality criteria, in contrast to their qualities that are otherwise attractive to a mechanism designer.

The remainder of this chapter is organised as follows. In Sec. 8.2 we describe polynomial-time algorithms for finding a rank-maximal matching in various problem contexts including HAT, CHAT, HRT and SRTI. We consider algorithmic results for greedy maximum and generous maximum matchings for the same problems in Sec. 8.3. Sec. 8.4 concerns matchings that are *weight-maximal*, a profile-based optimality property that involves weight functions. This concept generalises rank-maximal, greedy maximum and generous maximum matchings. We then consider two further profile-based optimal matching problems in Sec. 8.5, namely the *Rental Market problem* and the *Reviewer Assignment problem*. These problems provide two nice practical applications of profile-based optimal matchings (further examples were given in Sec. 1.5.8). Finally Sec. 8.6 presents some concluding remarks and open problems.

## 8.2 Rank-maximal matchings

### 8.2.1 *Introduction*

Let $I$ be an instance of CHAT and let $\mathcal{M}$ denote the set of matchings in $I$. Recall from Sec. 1.5.6 and Sec. 1.5.7 that the *profile* of a matching $M \in \mathcal{M}$, denoted $p(M)$, is a tuple $\langle p_1, \ldots, p_{r^*} \rangle$, where $r^* = r(M)$ is the regret of $M$, and for each $k$ $(1 \leq k \leq r^*)$, $p_k$ is the number of applicants who have their $k$th-choice house in $M$.

Throughout this chapter it will be convenient to use the term *profile* to refer to a tuple of (non-negative) integers of arbitrary (non-zero) length. Let $r$ denote the maximum rank of a house in an applicant's preference list, taken over all applicants in $I$. For any $k$ $(1 \leq k \leq r)$, let $O_k$ denote the $k$-tuple $\langle 0, 0, \ldots, 0 \rangle$ of zeros. The empty matching has profile $O_1$. We will also assume throughout this chapter that, with the exception of $O_k$ (for any $k$), the rightmost element of a profile vector is non-zero, otherwise it can be deleted (and the rule applied recursively).

We define a relation $\succ_L$ on profiles as follows. Let $\rho_1 = \langle p_1, \ldots, p_{l_1} \rangle$ and $\rho_2 = \langle q_1, \ldots, q_{l_2} \rangle$ be two profiles. We say that $\rho_1$ *left-dominates* $\rho_2$, denoted $\rho_1 \succ_L \rho_2$, if either (i) $l_1 > l_2$ and $p_k = q_k$ for $1 \leq k \leq l_2$, or (ii) there exists some $s$ $(1 \leq s \leq l_1)$ such that $l_2 \geq s$, $p_k = q_k$ $(1 \leq k \leq s - 1)$ and $p_s > q_s$. Clearly $\succ_L$ is a strict linear order. This linear order may be extended to $\mathcal{M}$ as follows. If $M, M' \in \mathcal{M}$, we say that $M$ *left-dominates* $M'$, denoted $M \succ_L M'$, if $p(M) \succ_L p(M')$.

Clearly a matching $M \in \mathcal{M}$ is rank-maximal if and only if there is no matching that left-dominates $M$. It is also obvious that all rank-maximal matchings in $I$ have the same profile, and hence the same cardinality.

In this section we present efficient algorithms for constructing rank-maximal matchings in a range of types of problem instances. In particular we consider HAT (Sec. 8.2.2), CHAT (Sec. 8.2.3), HRT (Sec. 8.2.4) and SRTI (Sec. 8.2.5).

## 8.2.2 House allocation problem with Ties

There are several methods in the literature for finding a rank-maximal matching $M$ in an HAT instance. In Sec. 8.2.2.1 we describe an $O(\min(n_1 + r^*, r^*\sqrt{n_1})m)$ time algorithm, based on the Dulmage–Mendelsohn Decomposition of a bipartite graph (see Definition 1.4), where $n_1$ is the number of applicants, $m$ is the number of acceptable applicant–house pairs, and $r^* = r(M)$ is the regret of $M$. An alternative method is to reduce the problem to that of finding a maximum weight matching in a weighted bipartite graph — this technique is outlined in Sec. 8.2.2.2.

*Weighted rank-maximal matchings* generalise the concept of a rank-maximal matching to the case where applicants have weights that indicate their priority levels (an analogue for popular matchings was considered in Sec. 7.4). We study such matchings in Sec. 8.2.2.3. Finally, Sec. 8.2.2.4 gives an overview of an empirical comparison of algorithms for computing rank-maximal matchings by Michail [453].

### 8.2.2.1 *Algorithm using the Dulmage–Mendelsohn decomposition*

Irving *et al.* [318] described a polynomial-time algorithm for finding a rank-maximal matching in an instance of HAT that is based on the Dulmage–Mendelsohn Decomposition of a bipartite graph (see Definition 1.4).

Let $I$ be an instance of HAT where $A$ is the set of applicants and $H$ is the set of houses, and let $G = (V, E)$ be the underlying graph of $I$. Let $r = \max\{rank(a_i, h_j) : (a_i, h_j) \in E\}$ be the maximum rank of a house in an applicant $a_i$'s list, taken over all $a_i \in A$. For each $k$ $(1 \leq k \leq r)$, define

---

**Algorithm 8.1** Algorithm Rank-maximal-HAT [318]

---

**Require:** HAT instance $I$ with underlying graph $G = (V, E)$
**Ensure:** return a rank-maximal matching $M_r$ in $I$
  1: **for** $k := 1$ to $r$ **do**
  2:     $E'_k := E_k$;
  3: **end for**
  4: $G'_1 := (V, E'_1)$;
  5: $M_1 :=$ maximum matching in $G'_1$;
  6: **for** $k := 1$ to $r - 1$ **do**
  7:     $\{G'_k := (V, E'_k)$ and $M_k$ is a maximum matching in $G'_k\}$;
  8:     compute an EOU labelling of $G'_k$; {let $\mathcal{E}_k, \mathcal{O}_k, \mathcal{U}_k$ be the sets of even, odd
          and unreachable vertices in $G'_k$ respectively}
  9:     $E'_k := E'_k \backslash \{(a_i, h_j) : (a_i \in \mathcal{O}_k \wedge h_j \in \mathcal{O}_k \cup \mathcal{U}_k) \vee (a_i \in \mathcal{U}_k \wedge h_j \in \mathcal{O}_k)\}$;
 10:     **for** $l := k + 1$ to $r$ **do**
 11:         $E'_l := E'_l \backslash \{(a_i, h_j) : (a_i \in \mathcal{O}_k \cup \mathcal{U}_k) \vee (h_j \in \mathcal{O}_k \cup \mathcal{U}_k)\}$;
 12:     **end for**
 13:     $E'_{k+1} := E'_k \cup E'_{k+1}$;
 14:     $G'_{k+1} := (V, E'_{k+1})$;     $\{M_k \subseteq E'_{k+1}\}$
 15:     augment $M_k$ to a maximum matching $M_{k+1}$ in $G'_{k+1}$;
 16: **end for**
 17: **return** $M_r$;

---

$E_k$, the set of *rank-k edges* in $G$, as follows:

$$E_k = \{(a_i, h_j) \in E : rank(a_i, h_j) = k\}.$$

Define also $E_{\leq k} = \bigcup_{l=1}^{k} E_l$. Let $G_k = (V, E_{\leq k})$ be the subgraph of $G$ containing the edges in $E_{\leq k}$, and let $I_k$ be the corresponding sub-instance of $I$ with underlying graph $G_k$.

The algorithm works by constructing a maximum matching $M_k$ in a certain subgraph of $G_k$, denoted $G'_k = (V, E'_k)$, for each $k$ $(1 \leq k \leq r)$. Initially $E'_k = E_k$ $(1 \leq k \leq r)$, $G'_1 = G_1$, and $M_1$ is a maximum matching in $G'_1$. During a subsequent for loop, which iterates over values of $k$ between 1 and $r - 1$, the objective is to form a maximum matching $M_{k+1}$ in $G'_{k+1}$. At the beginning of each loop iteration, two invariants are that (i) $E'_k \subseteq E_{\leq k}$, and (ii) $M_k$ is a maximum matching in $G'_k$.

During each loop iteration, certain edges may be deleted from the sets $E'_l$ for $k \leq l \leq r$. These edges are selected by constructing an EOU labelling of $G'_k$ (see Definition 1.4). We denote the sets of even, odd and unreachable vertices in this labelling by $\mathcal{E}_k$, $\mathcal{O}_k$ and $\mathcal{U}_k$ respectively.

As $M_k$ is a maximum matching in $G'_k$, it follows by Theorem 1.5 that $M_k$ contains no edge joining a vertex in $\mathcal{O}_k$ to a vertex in $\mathcal{O}_k \cup \mathcal{U}_k$. Hence any such edge is removed from $E'_k$. Also by Theorem 1.5, every vertex in

$\mathcal{O}_k \cup \mathcal{U}_k$ is incident to an edge in $M_k$. Hence any edge incident to such a vertex is removed from $E'_l$ $(k + 1 \leq l \leq r)$. Having carried out these deletions, $G'_{k+1} = (V, E'_{k+1})$ is then obtained from $G'_k$ by adding the edges in $E'_k$ to $E'_{k+1}$. No edge of $M_k$ is removed by these deletions, and hence $M_k$ remains a subset of $E'_{k+1}$. We then augment $M_k$ to a maximum matching $M_{k+1}$ in $G'_{k+1}$.

The algorithm of Irving *et al.* [318] for finding a rank-maximal matching is shown as Algorithm Rank-maximal-HAT in Algorithm 8.1.

The edge deletions carried out at iteration $k$ ensure that $M_{k+1}$ still contains a maximum matching of $G'_k$. For, each vertex in $\mathcal{O}_k \cup \mathcal{U}_k$ remains matched by some edge of $M_{k+1}$, since $M_{k+1}$ is obtained by augmenting $M_k$. Moreover, any such edge must belong to $E'_k$, given the edges removed from $E'_l$ for $l > k$. Further, given the edges removed from $E'_k$, any $\mathcal{O}_k$-vertex is matched in $M_{k+1}$ to an $\mathcal{E}_k$-vertex, and any $\mathcal{U}_k$-vertex is matched in $M_{k+1}$ to another $\mathcal{U}_k$-vertex. Thus $M_{k+1}$ contains a matching of $G'_k$ of size $|\mathcal{O}_k| + |\mathcal{U}_k|/2$, which is a maximum matching of $G'_k$ by Theorem 1.5.

This argument led Irving *et al.* to prove inductively that, for each $k$ $(1 \leq k \leq r)$, $M_k$ is a rank-maximal matching in $I_k$. By considering the case $k = r$, we may deduce that $M_r$ is a rank-maximal matching in $I$. Indeed, Irving *et al.* established the following result.

**Theorem 8.1 ([318]).** *Let $I$ be an instance of* HAT. *Algorithm* Rank-maximal-HAT *finds a rank-maximal matching $M$ in $I$ in* $O(\min(n_1 + r^*, r^*\sqrt{n_1})m)$ *time, where $n_1$ is the number of applicants, $m$ is the number of acceptable applicant–house pairs, and $r^* = r(M)$ is the regret of $M$.*

Note that, to achieve the running time stated in Theorem 8.1, Algorithm Rank-maximal-HAT should be altered as follows. At the beginning of each iteration $k$ of the main loop, the graph $G'_r$ should be constructed (recall that $r$ is the maximum rank of a house in an applicant's list, taken over all applicants) — this graph has vertex set $V$ and edge set $\bigcup_{k=1}^{r} E'_k$. If $M_k$ is a maximum matching in $G'_r$ then the loop can be terminated. Otherwise $G'_r$ contains an edge $(a_i, h_j)$ such that $rank(a_i, h_j) > k$ and the addition of this edge leads to a larger matching. It follows that $r^* > k$. The remainder of iteration $k$ is as described in Algorithm Rank-maximal-HAT.

**Illustrative example.** We now give an example that illustrates an execution of Algorithm Rank-maximal-HAT as applied to the HA instance $I_4$ shown in Fig. 1.5. $G'_1$ contains the edges $\{(a_i, h_1) : 1 \leq i \leq 4\} \cup \{(a_5, h_2)\}$. Initially the algorithm may choose $M_1 = \{(a_1, h_1), (a_5, h_2)\}$.

At iteration 1 of the main loop, $\mathcal{O}_1 = \{h_1\}$, $\mathcal{U}_1 = \{a_5, h_2\}$ and $\mathcal{E}_1$ contains all other vertices not in $\mathcal{O}_1 \cup \mathcal{U}_1$. As $a_5 \in \mathcal{U}_1$, the following edge deletions occur: $(a_5, h_5)$ from $E_2'$, $(a_5, h_4)$ from $E_3'$, $(a_5, h_3)$ from $E_4'$, and $(a_5, h_1)$ from $E_5'$. As $h_2 \in \mathcal{U}_1$, $(a_1, h_2)$, $(a_2, h_2)$ and $(a_3, h_2)$ are deleted from $E_2'$, and $(a_4, h_2)$ is deleted from $E_5'$. The edge in $E_2' = \{(a_4, h_3)\}$ is added to $G_1'$ to give $G_2'$. The algorithm will then augment $M_1$ to $M_2 = \{(a_1, h_1), (a_4, h_3), (a_5, h_2)\}$ in $G_2'$.

At iteration 2, $\mathcal{O}_2 = \{h_1\}$, $\mathcal{U}_2 = \{a_4, a_5, h_2, h_3\}$ and $\mathcal{E}_2 = \{a_1, a_2, a_3, h_4, h_5\}$. As $a_4 \in \mathcal{U}_2$ and $h_1 \in \mathcal{O}_2$, $(a_4, h_1)$ is deleted from $E_2'$. As $a_4 \in \mathcal{U}_2$, $(a_4, h_5)$ is deleted from $E_3'$ and $(a_4, h_4)$ is deleted from $E_4'$. As $h_3 \in \mathcal{U}_2$, $(a_1, h_3)$, $(a_2, h_3)$ and $(a_3, h_3)$ are deleted from $E_3'$. Then $E_3' = \emptyset$, so $G_3' = G_2'$ and $M_3 = M_2$.

At iteration 3, $\mathcal{O}_3 = \mathcal{O}_2$, $\mathcal{U}_3 = \mathcal{U}_2$ and $\mathcal{E}_3 = \mathcal{E}_2$. Thus there are no edge deletions at this iteration. The edges in $E_4' = \{(a_1, h_4), (a_2, h_4), (a_3, h_4)\}$ are added to $G_3'$ to give $G_4'$. The algorithm may then augment $M_3$ to $M_4 = \{(a_1, h_1), (a_3, h_4), (a_4, h_3), (a_5, h_2)\}$ in $G_4'$.

At iteration 4, $\mathcal{E}_4 = \{a_1, a_2, a_3, h_5\}$, $\mathcal{U}_4 = \{a_4, a_5, h_2, h_3\}$ and $\mathcal{O}_4 = \{h_1, h_4\}$. There are no edge deletions at this iteration. The edges in $E_5' = \{(a_1, h_5), (a_2, h_5), (a_3, h_5)\}$ are added to $G_4'$ to give $G_5'$. The algorithm may then augment $M_4$ to $M_5 = \{(a_1, h_1), (a_2, h_4), (a_3, h_5), (a_4, h_3), (a_5, h_2)\}$ in $G_5'$. $M_5$ is then a rank-maximal matching in $I_4$.

### 8.2.2.2 *Reduction to the Assignment problem*

Given an instance $I$ of HAT, a rank-maximal matching may also be found using a transformation to the Assignment problem. To see this, let $A$ be the set of applicants and let $H$ be the set of houses in $I$, where $n_1 = |A|$. Also let $G = (V, E)$ be the underlying graph of $I$, and let $r$ be the maximum rank of a house in an applicant $a_i$'s preference list, taken over all $a_i \in A$. Define a weight function $wt : E \longrightarrow \mathbb{N}$ as follows. Given an edge $(a_i, h_j)$ in $G$, let $wt(a_i, h_j) = n_1^{r-k}$, where $rank(a_i, h_j) = k$.

This steeply-decreasing sequence of edge weights ensures that rank-1 edges have highest priority, followed by rank-2 edges, and so on. In particular, the weights ensure that only the addition of $n_1$ rank-$(k+1)$ edges can compensate for the loss of a rank-$k$ edge (for some $1 \leq k \leq r-1$), assuming that only lower-rank edges are available. This is sufficient to ensure that a maximum weight matching in $G$ is a rank-maximal matching. This transformation was observed by Irving *et al.* [318] without proof, and for completeness we now prove its correctness.

**Proposition 8.2 ([318]).** *Let $I$ be an instance of* HAT *and let wt be the weight function in the underlying graph $G = (V, E)$ as defined above. Then a maximum weight matching in $G$ is a rank-maximal matching in $I$.*

**Proof.** Let $M$ be a maximum weight matching in $G$ with respect to $wt$. Suppose that $M' \succ_L M$ for some matching $M'$ in $I$. We lose no generality by assuming that $M'$ is a rank-maximal matching in $I$. Let $p(M) = \langle p_1, \ldots, p_{l_1} \rangle$ and $p(M') = \langle q_1, \ldots, q_{l_2} \rangle$. Then either (i) $l_2 > l_1$ and $p_k = q_k$ for $1 \leq k \leq l_1$, or (ii) there exists some $s$ $(1 \leq s \leq l_2)$ such that $l_1 \geq s$, $p_k = q_k$ $(1 \leq k \leq s - 1)$ and $q_s > p_s$. In case (i), we let $s = \min\{k : k > l_1 \wedge q_k > 0\}$ and let $p_k = 0$ $(l_1 + 1 \leq k \leq s)$. Also let $l'_1 = s$. In case (ii), we let $l'_1 = l_1$.

Letting $wt(M)$ denote the total weight of the edges in $M$, we obtain the following:

$$wt(M) = \sum_{k=1}^{l'_1} p_k n_1^{r-k}$$

$$\leq \sum_{k=1}^{s} p_k n_1^{r-k} + \left( \sum_{k=s+1}^{l'_1} p_k \right) n_1^{r-s-1} \tag{8.1}$$

Also, considering $wt(M')$, we have:

$$wt(M') = \sum_{k=1}^{l_2} q_k n_1^{r-k}$$

$$= \sum_{k=1}^{s-1} p_k n_1^{r-k} + q_s n_1^{r-s} + \sum_{k=s+1}^{l_2} q_k n_1^{r-k}$$

$$\geq \sum_{k=1}^{s-1} p_k n_1^{r-k} + (p_s + 1) n_1^{r-s} + \sum_{k=s+1}^{l_2} q_k n_1^{r-k}$$

$$= \sum_{k=1}^{s} p_k n_1^{r-k} + n_1^{r-s} + \sum_{k=s+1}^{l_2} q_k n_1^{r-k} \tag{8.2}$$

Thus, from Inequalities 8.1 and 8.2, $wt(M') > wt(M)$ unless $\sum_{k=s+1}^{l'_1} p_k \geq n_1$. In this case, $p_k = q_k = 0$ for $1 \leq k \leq s-1$, $p_s = 0$ and $\sum_{k=s+1}^{l'_1} p_k = n_1$. In fact, the only way that $wt(M) \geq wt(M')$ can occur is if (i) $p_{s+1} = n_1$ and

$p_k = 0$ for $s + 2 \leq k \leq l'_1$, and (ii) $q_s = 1$ and $q_k = 0$ for $s + 1 \leq k \leq l_2$. Indeed, $wt(M) = wt(M')$ in this case.

In such a setting, $M' = \{(a_i, h_j)\}$ for some $(a_i, h_j) \in E$ such that $rank(a_i, h_j) = s$. Also $M$ contains $n_1$ rank-$(s + 1)$ edges. Pick any $a_{i'} \in A \backslash \{a_i\}$. Then $(a_{i'}, h_{j'}) \in M$ for some $h_{j'} \in H$ where $rank(a_{i'}, h_{j'}) = s + 1$. If $h_j \neq h_{j'}$ then $M'' = \{(a_i, h_j), (a_{i'}, h_{j'})\}$ satisfies $M'' \succ_L M'$, a contradiction to the rank-maximality of $M'$. Otherwise pick any $h_{j''}$ on the preference list of $a_{i'}$ such that $rank(a_{i'}, h_{j''}) \leq s$. Then $h_{j''} \neq h_j$ and $M'' = \{(a_i, h_j), (a_{i'}, h_{j''})\}$ satisfies $M'' \succ_L M'$, a contradiction to the rank-maximality of $M'$.

Thus the conclusion is that $wt(M') > wt(M)$, which contradicts the fact that $M$ is a maximum weight matching in $G$.                              □

As mentioned in Sec. 1.5.4, a maximum weight matching in $G$ can be found in time $O(\sqrt{n}m \log(nW))$ time, where $n$ is the number of applicants and houses, $m$ is the number of acceptable applicant–house pairs, and $W$ is the largest edge weight, using Gabow and Tarjan's algorithm [230]. There is also a strongly polynomial-time algorithm for the problem, due to Fredman and Tarjan [223], with complexity $O(n(m + n \log n))$.

Both of these algorithms assume that arithmetic operations involving edge weights can be carried out in constant time. However in the context of the above transformation, arithmetic is carried out on numbers of magnitude $O(nW)$, where $W = n_1^{r-1}$. Each such arithmetic operation takes time $O((\log W)/ \log n)$ if we make the standard assumption that arithmetic on numbers of magnitude $O(n)$ takes constant time.

This implies that, as observed by Mehlhorn and Michail [447], the true running times of the maximum weight matching algorithms in the context of this transformation are $O(\sqrt{n}m(\log(nW))^2/ \log n) = O(r^2 \sqrt{n}m \log n)$ for the Gabow–Tarjan algorithm, and $O(rn(m + n \log n))$ for the Fredman–Tarjan algorithm.

Irving *et al.* [317] presented a further algorithm for finding a rank-maximal matching $M$ in an instance of HAT, running in $O(r^* nm)$ time, where $r^* = r(M)$ is the regret of $M$. This complexity was improved to $O(r^* \sqrt{n}m \log n)$ by Mehlhorn and Michail [447]. A further improvement was made by Michail [452], yielding a running time of $O(\min(n + r^*, r^* \sqrt{n})m)$, where $n$ is the number of applicants, which is the same as given by Theorem 8.1. All of these are scaling algorithms based on the transformation to maximum weight matching described above.

### 8.2.2.3 *Weighted rank-maximal matchings*

In Sec. 7.4, we considered the case where, in a given instance of HAT, each applicant $a_i \in A$ has a positive weight $wt(a_i)$ indicating her priority when it comes to majority voting. Such a weight can equally represent her priority in the context of a rank-maximal matching. Based on this observation, Kavitha and Shah [371] defined a *weighted rank-maximal matching*, which is the focus of this section.

Formally, let $I$ be an instance of the *Weighted House Allocation problem with Ties* (WHAT), which is the generalisation of HAT in which each applicant $a_i \in A$ has a positive weight $wt(a_i)$. Let $\mathcal{M}$ be the set of matchings in $I$. The *weighted profile* of a matching $M \in \mathcal{M}$, denoted by $p^w(M)$, is a vector $\langle p_1^w, \ldots, p_{r*}^w \rangle$, where $r^* = r(M)$ such that for each $k$ ($1 \leq k \leq r^*$),

$$p_k^w = \sum \{wt(a_i) : (a_i, h_j) \in M : rank(a_i, h_j) = k\}.$$

Intuitively, $p_k^w$ is the sum of the weights of the applicants who have their $k$th-choice house in $M$. A matching $M \in \mathcal{M}$ is a *weighted rank-maximal matching* if there is no other matching $M' \in \mathcal{M}$ such that $p^w(M') \succ_L p^w(M)$.

Assume that $\{wt(a_i) : a_i \in A\} = \{w_i : 1 \leq i \leq k\}$ for some $k \in \mathbb{Z}^+$. Without loss of generality assume that $w_1 > w_2 > \cdots > w_k$. Kavitha and Shah [371] gave an algorithm for finding a weighted rank-maximal matching $M$ in a given instance $I$ of WHAT by reducing $I$ to an instance $J$ of HAT such that an acceptable pair $(a_i, h_j)$ in $I$ where $rank(a_i, h_j) = p$ corresponds to an acceptable pair $(a_i, h_j)$ in $J$ of rank $(p-1)k + q$, where $w_q = wt(a_i)$. The authors showed that $M$ can be constructed in time $O(\min(n_1 + r^*, r^* \sqrt{n_1})m)$ time, where $n_1$ is the number of applicants, $m$ is the number of acceptable applicant–house pairs in $I$, and $r^* = r(M)$ is the regret of $M$ when viewed as a rank-maximal matching in $J$.

### 8.2.2.4 *Empirical analysis of rank-maximal matching algorithms*

Michail [453] compared experimentally the performance of several algorithms for computing rank-maximal matchings (see Sec. 8.2.2.1 and Sec. 8.2.2.2), popular matchings (see Sec. 7.2.3), matchings with bounded unpopularity factor (see Sec. 7.2.7.3) and minimum weight maximum cardinality matchings (see Sec. 1.5.4) in an instance of HAT.

The algorithms were compared empirically on a number of randomly-generated problem instances, and in addition, so-called "real data" were constructed, generated artifically from the Zillow website[1] using the 3-attribute model (see Sec. 2.10.6). A similar construction was carried out with information from the National Basketball Association.

A number of attributes of the various algorithms were measured with respect to multiple randomly-generated / real problem instances, including running time, size and unpopularity of the computed matching.

Some of the empirical results are quite noteworthy. For example,

(1) concerning the running time of the rank-maximal matching algorithms, in several cases it was better to use standard breadth-first search to perform the augmentations rather than the Hopcroft–Karp algorithm;

(2) for sparse instances, rank-maximal matchings and matchings produced by the bounded unpopularity matching algorithm were noticeably smaller than maximum cardinality matchings;

(3) certain rank-maximal matching algorithms often computed matchings with lower unpopularity factors than the bounded unpopularity matching algorithm, even though the latter algorithm was specifically designed for this purpose.

### 8.2.3    Capacitated House Allocation problem with Ties

A rank-maximal matching can be defined in an instance $I$ of CHAT in exactly the same way as for HAT. Sng [535] gave an algorithm for computing a rank-maximal matching in $I$. Essentially his algorithm is the same as Algorithm Rank-maximal-HAT, except that an EOU labelling of $G'_k$ is now sought in a capacitated bipartite graph rather than in the 1–1 case as before. (See Sec. 7.3.3 for a discussion on obtaining an EOU labelling in the many–one case.) This led Sng to deduce the following result.

**Theorem 8.3 ([535]).** *Let $I$ be an instance of* CHAT. *There is an* $O(\min(n_1 + r^*, r^*\sqrt{n_1})m)$ *algorithm[2] to find a rank-maximal matching in* $I$, *where $n_1$ is the number of applicants, $m$ is the number of acceptable applicant–house pairs, and $r^* = r(M)$ is the regret of $M$.*

Mehlhorn and Michail [447] gave a scaling algorithm for computing a rank-maximal matching $M$ in a given instance of the many–many general-

---

[1]http://www.zillow.com. Accessed 25 May 2012.

[2]In Ref. [535], the weaker upper bound of $O(\min(C + r^*, r^*\sqrt{C})m)$ was given as the complexity for this algorithm, where $C$ is the total capacity of the houses. The improved upper bound follows by the remark in Footnote 6 on Page 16.

isation of CHAT in which each applicant can be assigned to multiple houses up to some positive integral capacity. Hence both applicants and houses can have non-unitary capacities, but there are preference lists on one side only (as before, applicants rank houses in order of preference). We refer to this problem as the *Many–Many Capacitated House Allocation problem with Ties* (MM-CHAT). Here, the definition of a rank-maximal matching is the same as in the CHAT case, even though an applicant may be multiply assigned (up to her fixed capacity) in the more general problem. The complexity of their algorithm is $O(r^* nm \log(n^2/m) \log n)$, where $n$ is the number of applicants and houses, $m$ is the number of acceptable applicant–house pairs and $r^* = r(M)$ is the regret of $M$.

### 8.2.4 Hospitals / Residents problem with Ties

The concept of a rank-maximal matching may also be defined in an instance $I$ of HRT. In order to extend the definition from CHAT to HRT, we first require to redefine the *regret* of a matching $M$ in $I$. This quantity, denoted $r(M)$, is the maximum rank of an assignee of an agent in $M$, taken over all assignees of all agents. Formally $r(M)$ is defined as follows:

$$r(M) = \max(\{rank(r_i, h_j) : (r_i, h_j) \in M\} \cup \{rank(h_j, r_i) : (r_i, h_j) \in M\}).$$

The *profile* of a matching $M$ in $I$, denoted[3] $p_I(M)$, is now a vector $\langle p_1, \ldots, p_{r^*} \rangle$, where $r^* = r(M)$ is the regret of $M$, and for each $k$ $(1 \leq k \leq r^*)$,

$$p_k = |\{(r_i, h_j) \in M : rank(r_i, h_j) = k\}| + |\{(r_i, h_j) \in M : rank(h_j, r_i) = k\}|.$$

Intuitively, $p_k$ is the sum, taken over each agent $a_i \in R \cup H$, of the number of $k$th-choice assignees (possibly 0) of $a_i$ in $M$. The linear order $\succ_L$ over profiles is as defined for the CHAT case in Sec. 8.2.1. With these definitions, we can now define a matching $M$ to be *rank-maximal* if $p_I(M)$ is lexicographically maximum, taken over all matchings in $I$.

Suppose firstly that $I$ is an instance of SMTI. Huang and Kavitha [295] gave an efficient algorithm for computing a rank-maximal matching in $I$ by using their more general algorithm for the *weight-maximal matching* problem (see Sec. 8.4). Specifically, they proved the following result.

---

[3]It will be useful in this section to indicate the instance $I$ in which the profile of $M$ is defined, hence the subscript on $p(M)$ here.

**Theorem 8.4 ([295]).** *Let $I$ be an instance of* SMTI. *There is an* $O(r^*\sqrt{nm}\log n)$ *algorithm to find a rank-maximal matching in $I$, where $n$ is the size of $I$, $m$ is the number of acceptable man–woman pairs in $I$, and $r^* = r(M)$ is the regret of $M$.*

We now turn to the case that $I$ is a general instance of HRT. In this case, a rank-maximal matching in $I$ can be found by "cloning" the hospitals in $I$ to create an SMTI instance $I'$, and then by invoking Theorem 8.4. This procedure is similar to that described in the proof of Theorem 3.11, but with one subtle difference involving the construction of the men's lists in $I'$, as compared to their construction in the case of Theorem 3.11.

**Proposition 8.5.** *Given an instance $I$ of* HRT, *we may construct in $O(n_1 + c_{\max}m)$ time an instance $I'$ of* SMTI *such that a matching $M$ in $I$ can be transformed in $O(c_{\max}m)$ time to a matching $M'$ in $I'$ such that $p_I(M) = p_{I'}(M')$, and conversely, where $n_1$ is the number of residents, $c_{\max}$ is the maximum hospital capacity and $m$ is the number of acceptable resident–hospital pairs in $I$.*

**Proof.** Let $I$ be an instance of HRT in which $R = \{r_1, r_2, \ldots, r_{n_1}\}$ is the set of residents and $H = \{h_1, h_2, \ldots, h_{n_2}\}$ is the set of hospitals. Let $c_j$ be the capacity of hospital $h_j \in H$.

We form an instance $I'$ of SMTI as follows. Each resident in $I$ corresponds to a man in $I'$. Each hospital $h_j \in H$ gives rise to $c_j$ women (hospital "clones") in $I'$, denoted by $h_j^1, h_j^2, \ldots, h_j^{c_j}$, each of whom has the same preference list as $h_j$ in $I'$. Each man $r_i \in R$ starts off with the same preference list in $I'$ as he has in $I$. We then replace each entry $h_j$ on his list by the $c_j$ women $h_j^1, h_j^2, \ldots, h_j^{c_j}$. If $h_j$ was involved in a tie in $r_i$'s list in $I$, then these $c_j$ women are simply added to the corresponding tie in $r_i$'s list in $I'$. Otherwise we form a new tie in $r_i$'s list in $I'$, comprising these $c_j$ women, at the corresponding point at which $h_j$ appeared on $r_i$'s list in $I$.

Now let $M$ be a matching in $I$. We form a matching $M'$ in $I'$ as follows. For each $h_j \in H$, let $r_{j,1}, r_{j,2}, \ldots, r_{j,x_j}$ be the set of residents assigned to $h_j$ in $M$, where $x_j \leq c_j$. Add $(r_{j,k}, h_j^k)$ to $M'$ $(1 \leq k \leq x_j)$. Clearly $p_I(M) = p_{I'}(M')$.

Conversely let $M'$ be a matching in $I'$. We form a matching $M$ in $I$ as follows. For each $(r_i, h_j^k) \in M'$, add $(r_i, h_j)$ to $M$. Clearly $p_{I'}(M') = p_I(M)$.

The stated time complexities follow from the fact that $I'$ has $O(n_1 + C)$ agents and $O(c_{\max}m)$ acceptable man–woman pairs, where $C$ is the total capacity of the hospitals in $I$. □

**Corollary 8.6.** *Given an instance $I$ of* HRT, *we may construct in $O(n_1 + c_{\max}m)$ time an instance $I'$ of* SMTI *such that a rank-maximal matching $M'$ in $I'$ can be transformed in $O(c_{\max}m)$ time to a rank-maximal matching $M$ in $I$, where $n_1$ is the number of residents, $c_{\max}$ is the maximum hospital capacity and $m$ is the number of acceptable resident–hospital pairs in $I$.*

The following result is an immediate consequence of Corollary 8.6 and Theorem 8.4.

**Theorem 8.7.** *Let $I$ be an instance of* HRT. *There is an $O(r^*\sqrt{n_1 + C}\,c_{\max}m\log(n_1 + C))$ algorithm to find a rank-maximal matching in $I$, where $n_1$ is the number of residents, $m$ is the number of acceptable resident–hospital pairs, $C$ is the total capacity of the hospitals, $c_{\max}$ is the maximum capacity of a hospital in $I$, and $r^* = r(M)$ is the regret of $M$.*

An alternative method for finding a rank-maximal matching in an HRT instance $I$ is to use a transformation to the *Weighted Upper Degree Constrained Subgraph problem* (WUDCS) (the weighted version of UDCS as defined in Sec. 1.2), as observed by Sng [535]. The definition of the weight function is similar to that described in Sec. 8.2.2.2 for HAT.

Let $R$ be the set of residents and let $H$ be the set of hospitals in $I$, and let $n = |R| + |H|$. Let $G = (V, E)$ be the underlying (capacitated) graph of $I$ (with capacity function $c' : R \cup H \longrightarrow \mathbb{Z}^+$, where $c'(r_i) = 1$ for all $r_i \in R$ and $c'(h_j) = c_j$ for all $h_j \in H$) and let $r$ be the maximum rank of an agent $a_j$ in a given agent $a_i$'s preference list, taken over all agents $a_i$ in $I$.

Define a weight function $wt : E \longrightarrow \mathbb{N}$ as follows. Given an edge $(r_i, h_j)$ in $G$, let $wt(r_i, h_j) = n^{r-p} + n^{r-q}$, where $rank(r_i, h_j) = p$ and $rank(h_j, r_i) = q$. It turns out that a maximum weight matching in $\langle G, c' \rangle$ is a rank-maximal matching in $I$.

**Proposition 8.8 ([535]).** *Let $I$ be an instance of* HRT *and let $wt$ be the weight function in the underlying capacitated graph $G = (V, E)$ with capacity function $c'$ as defined above. Then a maximum weight matching in $G$ is a rank-maximal matching in $I$.*

**Proof.** Let $M$ be a maximum weight matching in $\langle G, c' \rangle$ with respect to $wt$. Suppose that $M' \succ_L M$ for some matching $M'$ in $I$. Let $p(M) = \langle p_1, \ldots, p_{l_1} \rangle$ and $p(M') = \langle q_1, \ldots, q_{l_2} \rangle$. Then either (i) $l_2 > l_1$ and $p_k = q_k$ for $1 \leq k \leq l_1$, or (ii) there exists some $s$ ($1 \leq s \leq l_2$) such that $l_1 \geq s$, $p_k = q_k$ ($1 \leq k \leq s - 1$) and $q_s > p_s$. In case (i), we let $s = \min\{k : k >$

$l_1 \wedge q_k > 0\}$ and let $p_k = 0$ $(l_1 + 1 \leq k \leq s)$. Also let $l_1' = s$. In case (ii), we let $l_1' = l_1$.

Letting $wt(M)$ denote the total weight of the edges in $M$, we obtain the following inequalities using a similar argument to that in the proof of Proposition 8.2:

$$wt(M) \leq \sum_{k=1}^{s} p_k n^{r-k} + \left( \sum_{k=s+1}^{l_1'} p_k \right) n^{r-s-1} \qquad (8.3)$$

$$wt(M') = \sum_{k=1}^{s-1} p_k n^{r-k} + q_s n^{r-s} + \sum_{k=s+1}^{l_2} q_k n^{r-k}$$

$$\geq \sum_{k=1}^{s} p_k n^{r-k} + n^{r-s} + \sum_{k=s+1}^{l_2} q_k n^{r-k} \qquad (8.4)$$

Thus, from Inequalities 8.3 and 8.4, $wt(M') > wt(M)$ unless $\sum_{k=s+1}^{l_1'} p_k \geq n$. In this case, $p_k = q_k = 0$ for $1 \leq k \leq s-1$, $p_s = 0$ and $\sum_{k=s+1}^{l_1'} p_k = n$. In fact, the only way that $wt(M) \geq wt(M')$ can occur is if (i) $p_{s+1} = n$ and $p_k = 0$ for $s+2 \leq k \leq l_1'$, and (ii) $q_s = 1$ and $q_k = 0$ for $s+1 \leq k \leq l_2$. Indeed, $wt(M) = wt(M')$ in this case.

Hence $\sum_{k=1}^{l_2} q_k = 1$. But this is impossible, since $M'$ is a matching and thus $\sum_{k=1}^{l_2} q_k$ must be an even quantity. Hence $wt(M') > wt(M)$, which contradicts the fact that $M$ is a maximum weight matching in $G$. $\qquad \square$

A maximum weight matching can be found in time $O((n_1 + C) \min(m \log n, n^2)$ using Gabow's algorithm [226], where $n_1 = |R|$, $C$ is the total capacity of the hospitals, $n = |R| + |H|$ and $m = |E|$, assuming $O(n)$ edge weights. However due to the large edge weights involved, the true complexity of Gabow's algorithm in $\langle G, c' \rangle$ is $O(r(n_1 + C) \min(m \log n, n^2))$. Hence in general Theorem 8.7 gives a stronger complexity bound.

In this section, we have considered rank-maximal matchings in instances of SMI, HR and their variants involving ties. We remark that in SMI, normally the stability of a matching is the key priority. We can however combine stability with rank-maximality to obtain the notion of a *rank-maximal stable matching* in an SMI instance $I$. To define this, let $\mathcal{S}$ be the set of stable matchings in $I$. Then $M \in \mathcal{S}$ is a *rank-maximal stable matching*[4] if there is no $M' \in \mathcal{S}$ such that $M' \succ_L M$. Irving *et al.* [320] showed that a rank-maximal stable matching in $I$ can be computed in $O(nm^2 \log^2 n)$ time,

---

[4]Feder [202] referred to a rank-maximal stable matching as a *lexicographic stable matching*.

where $n$ is the size of $I$ and $m$ is the number of acceptable man–woman pairs in $I$. This bound follows by transforming to the problem of finding a minimum weight stable matching (see Sec. 1.3.4.2). Feder [202] improved the time bound to $O(n^{1/2}m^{3/2})$ using a transformation to weighted 2-SAT.

### 8.2.5 Stable Roommates with Ties and Incomplete lists

Recall from Sec. 1.4.3 that an instance of SR need not admit a stable matching [235]. To cope with the possible non-existence of a stable matching, various alternative optimality criteria have been suggested in the literature: for example maximum stable matchings (Sec. 4.3.4), "almost stable" matchings (Sec. 4.6), Pareto optimal matchings (Sec. 6.5) and popular matchings (Sec. 7.5).

Rank-maximality provides a further optimality criterion that can be defined in the context of SR, and indeed in the more general SRTI setting, as an alternative to stability. It is straightforward to extend the definition of a rank-maximal matching from the HAT case to SRTI. To do this, we redefine the *regret* and *profile* of a matching. Let $I$ be an instance of SRTI and let $M$ be a matching in $I$. The *regret* of $M$ in $I$, denoted $r(M)$, is the maximum rank of an agent's partner in $M$, taken over all agents in $I$. Formally $r(M)$ is defined as follows:

$$r(M) = \max(\{rank(a_i, a_j) : \{a_i, a_j\} \in M\}.$$

Let $A_M$ denote the set of agents who are assigned in $M$. The *profile* of $M$ in $I$, denoted $p(M)$, is now a vector $\langle p_1, \ldots, p_{r^*} \rangle$, where $r^* = r(M)$ is the regret of $M$, and for each $k$ ($1 \leq k \leq r^*$),

$$p_k = |\{a_i \in A_M : rank(a_i, M(a_i)) = k\}|.$$

Intuitively, $p_k$ is the number of agents who have their $k$th-choice partner in $M$. The linear order $\succ_L$ is as defined for the HAT case in Sec. 8.2.1. With these definitions, we can now define a matching $M$ to be *rank-maximal* if $p(M)$ is lexicographically maximum, taken over all matchings in $\mathcal{M}$.

Abraham *et al.* [27] observed that, given an instance $I$ of SRTI, a rank-maximal matching can be found using a transformation to the problem of computing a maximum weight matching in a general weighted graph. To see this, let $G = (V, E)$ be the underlying graph of $I$, and let $r$ be the maximum rank of an agent $a_j$ in a given agent $a_i$'s preference list, taken over all agents $a_i$ in $I$. Also, let $A$ be the set of agents in $I$, where $n = |A|$. Define a weight function $wt : E \longrightarrow \mathbb{N}$ as follows. Given an edge

$e = \{a_i, a_j\} \in E$, let $wt(e) = n^{r-p} + n^{r-q}$, where $rank(a_i, a_j) = p$ and $rank(a_j, a_i) = q$.

The following proposition, whose proof is similar to that of Proposition 8.8 and is omitted, indicates that a maximum weight matching in $G$ is a rank-maximal matching in $I$.

**Proposition 8.9 ([318]).** *Let $I$ be an instance of SRTI and let $wt$ be the weight function in the underlying graph $G = (V, E)$ as defined above. Then a maximum weight matching in $G$ is a rank-maximal matching in $I$.*

A maximum weight matching in a weighted graph can be found in time $O(\sqrt{n\alpha(n, m)}m \log^{3/2} n)$ time using Gabow and Tarjan's algorithm [231], where $n$ is the number of vertices and $m$ is the number of edges, assuming $O(n)$ edge weights. However due to the large edge weights involved, the true complexity of Gabow and Tarjan's algorithm in $G$ is $O(r\sqrt{n\alpha(n, m)}m \log^{3/2} n)$. We summarise this discussion as follows.

**Theorem 8.10 ([27]).** *Let $I$ be an instance of SRTI. There is an $O(r\sqrt{n\alpha(n, m)}m \log^{3/2} n)$ algorithm[5] to find a rank-maximal matching in $I$, where $n$ is the number of agents, $m$ is the number of acceptable pairs of agents and $r$ is the maximum rank of an agent $a_j$ in a given agent $a_i$'s preference list, taken over all agents $a_i$ in $I$.*

At present it is open as to whether a combinatorial algorithm for computing a rank-maximal matching in a given SRTI instance, generalising Algorithm Rank-maximal-HAT, can be found. However, such a generalisation does exist if $I$ is an instance of SRTI-GRP (see Sec. 4.7.3, and Theorem 4.37) in particular).

Finally, as in Sec. 8.2.4, we can combine the notions of a stable matching and a rank-maximal matching in instances of SR and its variants. As mentioned in Sec. 4.9, the complexity of the problem of finding a rank-maximal stable matching, given a solvable instance of SR, is currently open.

## 8.3 Greedy and generous maximum matchings

### 8.3.1 *Introduction*

Recall the definitions of a greedy maximum matching and a generous maximum matching from Sec. 1.5.6. In this section we describe polynomial-time

---

[5]The weaker bound of $O(r^2\sqrt{n\alpha(n, m)}m \log^{3/2} n)$ was given by Abraham *et al.* [27].

algorithms for computing a greedy maximum matching and a generous maximum matching in various problem instances. We begin with the CHAT case and firstly outline some notation and terminology.

Let $I$ be an instance of CHAT and let $\mathcal{M}$ denote the set of matchings in $I$. Denote by $A$ the set of applicants, by $H$ the set of houses in $I$ and by $r$ the maximum rank of a house in an applicant's list, taken over all applicants in $I$. Let $G = (V, E)$ be the underlying capacitated graph of $I$, with capacity function $c' : A \cup H \longrightarrow \mathbb{Z}^+$, where $c'(a_i) = 1$ for all $a_i \in A$ and $c'(h_j) = c_j$ for all $h_j \in H$. Denote by $\beta^+(G)$ the size of a maximum matching in $\langle G, c' \rangle$.

For each $k$ $(0 \le k \le \beta^+(G))$, let $\mathcal{M}_k$ denote the set of matchings in $G$ of size $k$. For the special case that $k = \beta^+(G)$, we also denote $\mathcal{M}_k$ by $\mathcal{M}^+$.

We define a relation $\prec_R$ on profiles as follows. Let $\rho_1 = \langle p_1, \ldots, p_{l_1} \rangle$ and $\rho_2 = \langle q_1, \ldots, q_{l_2} \rangle$ be two profiles. Then, we say that $\rho_1$ *right-dominates* $\rho_2$, denoted $\rho_1 \prec_R \rho_2$, if either (i) $l_1 < l_2$, or (ii) $l_1 = l_2$ and there exists some $s$ $(1 \le s \le l_1)$ such that $p_s < q_s$ and $p_k = q_k$ $(s + 1 \le k \le l_1)$. Clearly $\prec_R$ is a strict linear order. This linear order may be extended to $\mathcal{M}$ as follows. If $M, M' \in \mathcal{M}$, we say that $M$ *right-dominates* $M'$, denoted $M \prec_R M'$, if $p(M) \prec_R p(M')$.

Let $k$ $(0 \le k \le \beta^+(G))$ be given. A matching $M \in \mathcal{M}_k$ is said to be a *greedy $k$-matching* if there is no matching $M' \in \mathcal{M}_k$ such that $M' \succ_L M$. In the case that $k = \beta^+(G)$, $M_k$ is a greedy maximum matching. Similarly a matching $M \in \mathcal{M}_k$ is said to be a *generous $k$-matching* if there is no matching $M' \in \mathcal{M}_k$ such that $M' \prec_R M$. In the case that $k = \beta^+(G)$, $M_k$ is a generous maximum matching.

To illustrate some of these definitions, consider the HA instance $I_3$ shown in Fig. 1.3. Then $M = \{(a_1, h_1), (a_2, h_2)\}$ is a greedy 2-matching in $I_3$ (and in fact a rank-maximal matching) with profile $\langle 2, 0 \rangle$, whilst $M' = \{(a_1, h_3), (a_2, h_1), (a_3, h_2)\}$ is a greedy 3-matching in $I_3$ (and indeed a greedy maximum matching) with profile $\langle 1, 2 \rangle$.

This section is organised as follows. We begin in Sec. 8.3.2 by describing a polynomial-time algorithm for computing a greedy maximum matching in a given instance of CHAT. The amendments that should be made to this algorithm in order to find a generous maximum matching in CHAT are outlined in Sec. 8.3.3. Sec. 8.3.4 describes how to find greedy maximum and generous maximum matchings in other problem instances, including HRT, SRTI, and MM-CHAT.

$$a_1 : h_1 \quad h_2$$
$$a_2 : h_1 \quad h_2$$
$$a_3 : h_4 \quad h_3$$
$$a_4 : h_4$$

$$M_3 = \{(a_1, h_1), (a_2, h_2), (a_3, h_4)\}$$
$$M_4^1 = \{(a_1, h_2), (a_2, h_1), (a_3, h_3), (a_4, h_4)\}$$
$$M_4^2 = \{(a_1, h_1), (a_2, h_2), (a_3, h_3), (a_4, h_4)\}$$

Fig. 8.1   An instance $I'$ of HA and three particular matchings in $I'$ due to Sng [535]

### 8.3.2   *Finding a greedy maximum matching*

In this subsection we present an efficient algorithm for finding a greedy maximum matching in the CHAT instance $I$ defined in the previous subsection. This algorithm was described by Sng [535] and extends an earlier algorithm that was derived for the HAT case by Irving [313]. Irving [313] introduced all the relevant concepts in the HAT context that were built upon by Sng's algorithm for CHAT.

The algorithm is essentially based on the following inductive strategy: for each $k$ $(0 \le k \le \beta^+(G) - 1)$, assume that we have found a greedy $k$-matching $M_k$, and find a "suitable" augmenting path $P$ in order to augment $M_k$ to a greedy $(k + 1)$-matching $M_{k+1}$. The starting point is the empty matching. Of course it is not obvious that such a path $P$ exists, nor is it clear how to find $P$.

To illustrate these issues, consider the HA instance $I'$ and the three matchings $M_3$, $M_4^1$ and $M_4^2$ in $I'$, due to Sng [535], shown in Fig. 8.1. Clearly $M_3$ is a greedy 3-matching in $I'$. However $M_4^1$ is a greedy 4-matching that cannot be obtained from $M_3$ by augmenting along a single augmenting path. On the other hand $M_4^2$ is a different greedy 4-matching in $I'$ that is obtainable from $M_3$ via a single augmenting path.

Returning to a general CHAT instance $I$, it turns out that a single augmenting path can always be found that enables a greedy $k$-matching to be augmented to a greedy $(k + 1)$-matching $(0 \le k \le \beta^+(G) - 1)$, as the following lemma, due to Irving [313] and Sng [535], indicates.

**Lemma 8.11 ([313, 535]).** *Let $I$ be an instance of CHAT and let $\beta^+(G)$ denote the size of a maximum matching in the underlying capacitated bipartite graph $G$. Let $k$ $(0 \le k \le \beta^+(G) - 1)$ be given and let $M_k$ be a greedy $k$-matching in $I$. Then there is a greedy $(k + 1)$-matching $M_{k+1}$ that can be obtained from $M_k$ via an augmenting path.*

---

**Algorithm 8.2** Algorithm Greedy-Max [313, 535]

---

**Require:** CHAT instance $I$
**Ensure:** return a greedy maximum matching $M$ in $I$
 1: $M := \emptyset$;
 2: $k := 0$; {$k$ is the cardinality of $M$}
 3: **loop**
 4:    $P := $ Max-Aug($I,k,M$);
 5:    **if** $P \neq$ **null then**
 6:       $M := M \oplus P$;
 7:    **else**
 8:       **return** $M$; {a greedy maximum matching}
 9:    **end if**
10:    $k$++; {$M$ is a greedy $k$-matching}
11: **end loop**

---

We can now present a pseudocode description of Irving and Sng's greedy maximum matching algorithm. Algorithm Greedy-Max, shown in Algorithm 8.2, starts off with the empty matching and enters a loop that repeatedly tries to augment the current greedy $k$-matching $M$ into a greedy $(k+1)$-matching, using a single augmenting path $P$, for $k \geq 0$. The path $P$ will be found by an auxiliary procedure called Algorithm Max-Aug, which we will describe shortly. If this procedure returns a null path, then $M$ was already a greedy maximum matching, so Algorithm Greedy-Max returns $M$. Otherwise $M$ is augmented along $P$.

What now requires further explanation is the particular choice of augmenting path $P$ that will enable us to augment a greedy $k$-matching to obtain a greedy $(k+1)$-matching. It turns out that we seek a *maximum profile augmenting path*. In order to define this concept we require some additional notation.

Given a profile $\rho = \langle p_1, p_2, \ldots, p_t \rangle$, and given a positive integer $\alpha$ such that $1 \leq \alpha \leq t$, define $\rho + \alpha$ to be the profile with vector

$$\langle p_1, \ldots, p_{\alpha-1}, p_\alpha + 1, p_{\alpha+1}, \ldots, p_t \rangle$$

(that is, one more applicant has her $\alpha$th choice in $\rho + \alpha$ than in $\rho$). Similarly $\rho - \alpha$ is the profile with vector

$$\langle p_1, \ldots, p_{\alpha-1}, p_\alpha - 1, p_{\alpha+1}, \ldots, p_t \rangle$$

(so one fewer applicant has her $\alpha$th choice in $\rho - \alpha$ as compared to $\rho$). Note that if $\alpha = t$ then any zero entries from the rightmost end of $\rho - \alpha$ should be deleted until the rightmost element is non-zero.

Let $M$ be a matching in $I$. Define a house to be *exposed* relative to $M$ if $|M(h_j)| < c_j$. Now suppose that $P = \langle a_{i_0}, h_{i_0}, a_{i_1}, h_{i_1}, ..., a_{i_t}, h_{i_t} \rangle$ is an alternating path[6] in $G$ from an exposed applicant vertex $a_{i_0}$ to a (not necessarily exposed) house vertex $h_{i_t}$, for some $t \geq 0$, such that $(a_{i_s}, h_{i_{s-1}}) \in M$ for $1 \leq s \leq t$. We then define the *profile* of $P$ to be[7]

$$p(P) = O_r + rank(a_{i_0}, h_{i_0}) + rank(a_{i_1}, h_{i_1}) + ... + rank(a_{i_t}, h_{i_t})$$
$$-rank(a_{i_1}, h_{i_0}) - rank(a_{i_2}, h_{i_1}) - ... - rank(a_{i_t}, h_{i_{t-1}})$$

Note that if the rightmost entry in $p(P)$ is zero, it should be deleted (and this rule applied recursively until the rightmost entry is non-zero). It follows that if $P$ is an augmenting path, then $p(P)$ corresponds to the net change in the profile of $M$ if we augment $M$ along $P$.

For each house $h_j \in H$, we define the *L-value* of $h_j$ relative to $M$, denoted by $L(h_j)$, to be the maximum (with respect to $\succ_L$) profile taken over all alternating paths from an exposed applicant vertex ending at $h_j$. We say that an alternating path $P$ is a *maximum profile augmenting path* relative to $M$ if $P$ is an augmenting path, and $p(P) = \max\{L(h_j) : h_j \in H\}$ where max is with respect to the $\succ_L$ order on profiles.

Irving [313] and Sng [535] proved that a maximum profile augmenting path is key to transforming a greedy $k$-matching into a greedy $(k + 1)$-matching.

**Lemma 8.12 ([313, 535]).** *Let $I$ be an instance of* CHAT *and let $\beta^+(G)$ denote the size of a maximum matching in the underlying capacitated bipartite graph $G$. Let $k$ $(0 \leq k \leq \beta^+(G) - 1)$ be given and let $M_k$ be a greedy $k$-matching in $I$. Let $P$ be a maximum profile augmenting path relative to $M_k$. Then $M_{k+1} = M_k \oplus P$ is a greedy $(k + 1)$-matching.*

We now present Algorithm Max-Aug, due to Irving [313] and Sng [535], given in Algorithm 8.3, for finding a maximum profile augmenting path, or reporting that none exists, relative to a matching $M$ in $I$. (Here, and henceforth in this subsection, when reasoning about profiles, "maximum" is always with respect to $\succ_L$.) This algorithm is a variant of the Bellman–Ford algorithm for finding shortest paths in a graph (see, e.g., Ref. [156]). Algorithm Max-Aug will be passed three parameters, namely a CHAT instance $I$, an integer $k$ (where $0 \leq k \leq \beta^+(G)$ and $G$ is the underlying

---

[6]Although $I$ is an instance of CHAT, it turns out that we can restrict attention to alternating paths in which no house is repeated in $P$.

[7]In the definition of the profile of $P$, the operations of $+$ and $-$ are assumed to associate to the left.

**Algorithm 8.3** Algorithm Max-Aug [313, 535]

**Require:** CHAT instance $I$, an integer $k$ and a greedy $k$-matching $M$
**Ensure:** return $P$, a maximum profile augmenting path relative to $M$,
　　　　 or **null** if $M$ is a maximum matching
1: {Initialisation}
2: **for each** house $h_j \in H$ **do**
3:　　$l(h_j) := O_r$;
4:　　$pred(h_j) := $ **null**;
5:　　**for each** exposed applicant $a_i \in A$ such that $(a_i, h_j) \in E$ **do**
6:　　　　$t := rank(a_i, h_j)$;
7:　　　　$\rho := O_r + t$;
8:　　　　**if** $\rho \succ_L l(h_j)$ **then**
9:　　　　　$l(h_j) := \rho$;
10:　　　　　$pred(h_j) := a_i$;
11:　　　　**end if**
12:　　**end for**
13: **end for**
14: {Main loop}
15: **for** $p$ in $1..k$ **do**
16:　　**for each** $(a_i, h_j) \in E \backslash M$ such that $a_i$ is assigned in $M$ **do**
17:　　　　$\rho := l(M(a_i)) + rank(a_i, h_j) - rank(a_i, M(a_i))$;
18:　　　　**if** $\rho \succ_L l(h_j)$ **then**
19:　　　　　$l(h_j) := \rho$;
20:　　　　　$pred(h_j) := a_i$;
21:　　　　**end if**
22:　　**end for**
23: **end for**
24: {Final phase}
25: $\rho := \max_{\succ_L}(\{O_r\} \cup \{l(h_j) : h_j \in H \text{ is exposed}\})$;
26: **if** $\rho \succ_L O_r$ **then**
27:　　$h_q :\in \arg\max_{\succ_L}\{l(h_j) : h_j \in H \text{ is exposed}\}$;
28:　　$P := $ augmenting path obtained by following *pred* values and
　　　　 matching edges from $h_q$ to an exposed applicant;
29:　　**return** $P$;
30: **else**
31:　　**return null;** {no augmenting path exists}
32: **end if**

graph of $I$) and a greedy $k$-matching $M$ in $I$. If $M$ is already a maximum matching then the algorithm returns **null** to signify that no augmenting path exists. Otherwise a maximum profile augmenting path $P$ relative to $M$ is returned.

The algorithm consists of three phases: an initialisation phase (lines 2–13), the main loop (lines 15–23) and the final phase (lines 25–32). A fundamental loop invariant that holds is that, after the $p$th iteration of the main loop ($0 \leq p \leq k$), $l(h_j)$ holds the maximum profile of an alternating path starting from an exposed applicant and ending at $h_j$, and of length at most $2p + 1$, where the case that $p = 0$ refers to the point directly after the end of the initialisation phase.

For each house $h_j \in H$, the algorithm uses two variables, namely $l(h_j)$ and $pred(h_j)$. The first of these, $l(h_j)$, will hold the maximum profile of an alternating path *so far computed* relative to $M$, starting from an exposed applicant and ending at $h_j$. The second variable, $pred(h_j)$, will hold the predecessor of $h_j$ on this alternating path.

The initialisation of these variables is as follows. Variable $l(h_j)$ is initialised to be the profile $O_r + t$, where $t$ is the minimum value of $rank(a_i, h_j)$ taken over all unassigned applicants such that $(a_i, h_j) \in E$ (assuming that such an applicant exists). Variable $pred(h_j)$ is then assigned to equal an applicant $a_i$ where $rank(a_i, h_j) = t$. If no such applicant exists then $l(h_j)$ is initialised to $O_r$ and $pred(h_j)$ is initialised to **null**.

The main loop has $k$ iterations, where $k = |M|$. It uses an edge relaxation operation similar to that of the Bellman–Ford algorithm, but bases this operation in terms of the order $\succ_L$. The edge relaxation operation is defined in lines 17-21 of the algorithm. Let $a_i$ be any applicant vertex assigned in $M$ and let $(a_i, h_j) \in E \backslash M$. Also, let $P = \langle a_z, \dots, M(a_i) \rangle$ be an alternating path relative to $M$, starting from an exposed applicant vertex $a_z$ and ending at $M(a_i)$, whose profile is equal to $l(M(a_i))$. The essence of the edge relaxation operation is the following: if the profile of the alternating path $P' = \langle a_z, \dots, M(a_i), a_i, h_j \rangle$ left-dominates $l(h_j)$, then we update $l(h_j)$ to be the profile of $P'$ and similarly update the predecessor of $h_j$ to be $a_i$. As in the case of the Bellman–Ford algorithm, as the main loop iterates it allows for successively longer alternating paths to be checked to determine whether improvements in profile (relative to $\succ_L$) are possible.

The fundamental loop invariant mentioned above would imply that once the main loop terminates, for each house $h_j \in H$, $l(h_j)$ holds the maximum profile of an alternating path starting from an exposed applicant and ending at $h_j$, and of length at most $2k + 1$. Since $|M| = k$, the length restriction can be dropped, and hence the claim is that $l(h_j) = L(h_j)$ at this point. Irving [313] and Sng [535] proved that this is indeed the case.

**Lemma 8.13 ([313, 535]).** *At the end of the main loop of Algorithm* Max-Aug, *$l(h_j) = L(h_j)$ for each house vertex $h_j \in H$.*

We now enter the final phase of the algorithm. We let $\rho$ be the maximum profile $l(h_j)$, taken over all exposed $h_j \in H$, or $O_r$ if no exposed $h_j \in H$ exists. If $\rho = O_r$ then this implies that $M$ admits no augmenting path, so **null** is returned. Otherwise we let $h_q$ be an exposed house such that $\rho = l(h_q)$, and return the augmenting path $P$ that is obtained by tracing back the *pred* values and the edges in $M$ alternately, starting at $h_q$ and terminating at an exposed applicant. Sng [535] proved that this traceback cannot cycle. The proof also indicates that, as claimed in Footnote 6, we lose no generality in considering alternating paths that have no repeated house. These observations, together with Lemmas 8.12 and 8.13, imply the correctness of Algorithm Greedy-Max.

We now consider the time complexity of the algorithm. Algorithm Greedy-Max performs $\beta^+(G) = O(n_1)$ calls to Algorithm Max-Aug, where $n_1$ is the number of applicants. In the latter algorithm, each single operation involving a profile (assignment, addition, subtraction or comparison) takes $O(r)$ time. The initialisation phase thus has $O(rn_1n_2)$ complexity, where $n_2$ is the number of houses. The main loop iterates $k$ times, where $k = |M|$. Within an iteration of the main loop, the inner loop iterates $O(m)$ times, where $m = |E|$. Thus the main loop has overall complexity $O(rn_1m)$. The final phase involves computing $\rho$, which has $O(rn_2)$ complexity, and $P$, which takes $O(m)$ time, so overall the complexity for this phase is $O(rn_2 + m)$. Hence overall Algorithm Max-Aug has complexity $O(rn_1m)$, which gives an overall complexity of $O(rn_1^2m)$ for Algorithm Greedy-Max.

The following theorem summarises the above observations regarding the correctness and complexity of Algorithm Greedy-Max.

**Theorem 8.14 ([313, 535]).** *Let $I$ be an instance of* CHAT. *Algorithm* Greedy-Max *finds a greedy maximum matching in $I$ in $O(rn_1^2m)$ time, where $n_1$ is the number of applicants, $m$ is the number of acceptable applicant–house pairs, and $r$ is the maximum rank of a house in an applicant's list, taken over all applicants.*

A straightforward refinement to the algorithm is to observe that if no house $h_j \in H$ has an improvement in $l(h_j)$ during an iteration of the main loop, then no further improvements to the $l(h_j)$ values will be possible and the main loop can terminate at that point.

Irving [313] and Sng [535] also observed that Algorithm Greedy-Max gives an alternative method for computing a rank-maximal matching (albeit with a poorer time complexity than given by Theorem 8.3). This approach hinges on the following result.

**Lemma 8.15 ([313, 535]).** *Let $I$ be an instance of* CHAT, *let $G$ be the underlying capacitated bipartite graph, and let $\beta^+(G)$ denote the size of a maximum matching in $G$. For each $k$ $(1 \leq k \leq \beta^+(G))$, let $M_k$ be a greedy $k$-matching in $I$, and let $\rho_k$ be the profile of $M_k$. Define $\rho_{\beta+1} = O_r$, where $\beta = \beta^+(G)$ and $r$ is the maximum rank of a house in an applicant's list, taken over all applicants in $I$. Let $s$ be the minimum integer $(1 \leq s < \beta)$ such that (i) for each $k$ $(1 \leq k \leq s - 1)$, $\rho_{k+1} \succ_L \rho_k$, and (ii) $\rho_s \succ_L \rho_{s+1}$ (such an $s$ is bound to exist because of $\rho_{\beta+1}$). Then $M_s$ is a rank-maximal matching in $I$.*

### 8.3.3 *Finding a generous maximum matching*

Irving [313] and Sng [535] showed how to adapt Algorithm Greedy-Max in order to find a generous maximum matching in $I$. In this subsection we outline the necessary modifications. The pseudocode of Algorithm Greedy-Max itself is unchanged. In the context of Algorithm Max-Aug (which should strictly speaking be renamed Algorithm Min-Aug), we will be now seeking a minimum (with respect to $\prec_R$) profile augmenting path relative to $M$. We define this as follows.

For each house $h_j \in H$, we define the *R-value* of $h_j$ relative to $M$, denoted by $R(h_j)$, to be the minimum (with respect to $\prec_R$) profile taken over all alternating paths from an exposed applicant vertex ending at $h_j$. We say that an alternating path $P$ is a *minimum profile augmenting path* relative to $M$ if $P$ is an augmenting path, and $p(P) = \min \{R(h_j) : h_j \in H\}$ where min is with respect to the $\prec_R$ order on profiles.

Variable $l(h_j)$ should now be denoted $r(h_j)$ and will hold the minimum profile of an alternating path so far computed relative to $M$, starting from an exposed applicant and ending at $h_j$. Let $O'_r$ be the vector $\langle p_1, \ldots, p_r \rangle$, where $p_k = 0$ $(1 \leq k \leq r - 1)$ and $p_r = n + 1$. Then $p(M') \prec_R O'_r$, for any matching $M'$ in $I$.

Throughout the pseudocode of Algorithm Max-Aug, the following changes should be made:

- $l(h_j)$ should be replaced by $r(h_j)$;
- $O_r$ should be replaced by $O'_r$;
- max should be replaced by min;
- $\succ_L$ should be replaced by $\prec_R$.

Irving [313] and Sng [535] modified the proof of correctness for Algorithm Greedy-Max for the case of generous maximum matchings and arrived at the following result.

**Theorem 8.16 ([313, 535]).** *Let $I$ be an instance of* CHAT. *A generous maximum matching in $I$ can be found in $O(rn_1^2 m)$ time, where $n_1$ is the number of applicants, $m$ is the number of acceptable applicant–house pairs, and $r$ is the maximum rank of a house in an applicant's list, taken over all applicants in $I$.*

### 8.3.4 *Greedy and generous maximum matchings in other problem contexts*

A greedy maximum matching in an instance $I$ of HAT may be computed using a transformation to the Assignment problem, similar to that described in Sec. 8.2.2.2. Now, the weight of an edge $(a_i, h_j)$ such that $rank(a_i, h_j) = k$ should be $n_1^r + n_1^{r-k}$. It follows by a similar argument to that in the proof of Proposition 8.2 that a maximum weight matching corresponds to a greedy maximum matching. Analogous remarks apply if we wish to compute a generous maximum matching — in this case the required weight is $n_1^r - n_1^{k-1}$. As observed by Mehlhorn and Michail [447], the true running times of the maximum weight matching algorithms in the context of this transformation (either in the greedy or generous cases) are $O(r^2 \sqrt{n}m \log n)$ for the Gabow–Tarjan algorithm [231], and $O(rn(m + n \log n))$ for the Fredman–Tarjan algorithm [223].

Mehlhorn and Michail [447] gave a scaling algorithm for computing a greedy / generous maximum matching in $I$ in time $O(r\sqrt{n}m \log n)$. A further improvement was given by Huang and Kavitha, even for the SMTI case, using their more general algorithm for the *weight-maximal matching* problem (see Sec. 8.4), leading to the following result.

**Theorem 8.17 ([295]).** *Let $I$ be an instance of* SMTI. *There are $O(r^* \sqrt{n}m \log n)$ algorithms to find a greedy / generous maximum matching in $I$, where $n$ is the size of $I$, $m$ is the number of acceptable man–woman pairs, and $r^* = r(M)$ is the regret of $M$.*

Note that a greedy / generous maximum matching in an instance $I$ of HAT may be found be transforming $I$ into an instance $I'$ of SMTI in which each woman is indifferent between all of the men who find her acceptable, and by then applying Huang and Kavitha's algorithms [295] for finding a greedy / generous maximum matching in $I'$. Comparing Theorem 8.17 with Theorem 8.14, we see that this method for finding a greedy / generous maximum matching in an instance of HAT is faster than Irving and Sng's algorithms for the same problem class. However we chose to present Algorithm Greedy-Max in full for computing a greedy maximum matching (and indicate the changes required if we wish to compute a generous maximum matching) because it is conceptually straightforward and illustrates a fundamental technique for arriving at an optimal matching by augmenting along a series of optimal augmenting paths.

Sng [535] generalised Algorithm Greedy-Max to the case of HRT (indicating the modifications required if we wish to compute a generous maximum matching), with no change to the time complexity given by Theorem 8.14.

In the case of an SRTI instance $I$, we assume the notation given in Sec. 8.2.5. A greedy maximum or generous maximum matching can be found using a transformation to the problem of computing a maximum weight matching in a general weighted graph. Given an edge $\{a_i, a_j\}$ where $rank(a_i, a_j) = p$ and $rank(a_j, a_i) = q$, let $wt(\{a_i, a_j\}) = n^r + n^{r-p} + n^{r-q}$ in the case of a greedy maximum matching, and let $wt(\{a_i, a_j\}) = n^r - n^{p-1} - n^{q-1}$ in the case of a generous maximum matching. We then obtain a similar bound for the time taken to compute a greedy maximum or generous maximum matching as given in Theorem 8.10.

Mehlhorn and Michail [447] also generalised their scaling algorithm to MM-CHAT. The definitions of a greedy / generous maximum matching are the same as in the CHAT case, but clearly an applicant may be multiply assigned (up to her fixed capacity) in the more general problem. The complexity of their algorithm is $O(rnm \log(n^2/m) \log n)$, where $n$ is the number of applicants and houses, $m$ is the number of acceptable applicant–house pairs and $r$ is the maximum rank of a house in an applicant's list, taken over all applicants.

## 8.4   Weight-maximal matchings

Huang and Kavitha [295] studied *weight-maximal matchings* in instances of SMTI — this concept generalises the notions of a rank-maximal, greedy maximum and generous maximum matching. Let $I$ be an instance of SMTI and let $G = (V, E)$ be the underlying bipartite graph of $I$. Let $U$ be the set of men and let $W$ be the set of women in $I$.

We are given a tuple of weight functions $\langle f_1, \ldots, f_s \rangle$, for some $s \geq 1$, where $f_k : E \longrightarrow \{0, 1, \ldots, F\}$ for each $k$ $(1 \leq k \leq s)$. It is assumed that $F$ (the largest edge weight) is a constant. Intuitively, weight function $f_1$ has highest priority, followed by $f_2$, and then $f_3$, etc. Informally, the problem is to compute a matching $M$ in $G$ such that (i) $M$ is a maximum weight matching relative to $f_1$, (ii) subject to (i), $M$ is a maximum weight matching relative to $f_2$, (iii) subject to (i) and (ii), $M$ is a maximum weight matching relative to $f_3$, etc.

Formally, define the *weight-function profile* of a matching $M$ in $I$, denoted $p^f(M)$, to be the vector $\langle p_1, \ldots, p_s \rangle$, where $p_i = f_k(M)$ and $f_k(M) = \sum_{e \in M} f_k(e)$ $(1 \leq k \leq s)$. We define a matching $M$ to be a *weight-maximal matching* if there is no other matching $M'$ such that $p^f(M') \succ_L p^f(M)$. Equivalently, $M$ is weight-maximal if $p^f(M)$ is lexicographically maximum, taken over all matchings in $I$. Define the *weighted regret* of $M$, denoted $r^w(M)$, to be the maximum $k$ $(1 \leq k \leq s)$ such that $f_k(M) > 0$.

Let $r$ be the maximum rank of an agent $a_j$ in a given agent $a_i$'s preference list, taken over all agents $a_i$. Huang and Kavitha [295] showed that weight-maximal matchings can be used to compute profile-based optimal matchings in $I$, which can be seen as follows.

- *Rank-maximal matchings*: let $s = r$, and for each $k$ $(1 \leq k \leq r)$, let $f_k$ be defined as follows. For each edge $e = (m_i, w_j)$, where $m_i \in U$ and $w_j \in W$, define $f_k(e) = \delta_{rank(m_i, w_j), k} + \delta_{rank(w_j, m_i), k}$, where $\delta_{p,q}$ is the Kronecker delta function.
- *Greedy maximum matchings*: let $s = r + 1$. Define $f_1(e) = 1$ for all $e \in E$. For each $k$ $(2 \leq k \leq r + 1)$, let $f_k$ be defined as follows. For each edge $e = (m_i, w_j)$, where $m_i \in U$ and $w_j \in W$, define $f_k(e) = \delta_{rank(m_i, w_j), k-1} + \delta_{rank(w_j, m_i), k-1}$.
- *Generous maximum matchings*: let $s = r$. Define $f_1(e) = 1$ for all $e \in E$. For each $k$ $(2 \leq k \leq r)$, let $f_k$ be defined as follows. For each edge $e = (m_i, w_j)$, where $m_i \in U$ and $w_j \in W$, define $f_k(e) = \delta'_{rank(m_i, w_j), r-k+1} + \delta'_{rank(w_j, m_i), r-k+1}$, where $\delta'_{p,q} = 1$ if $p \leq q$, and $\delta'_{p,q} = 0$ otherwise.

Huang and Kavitha [295] remarked that a weight-maximal matching can be computed using a similar transformation to the Assignment problem to the one described in Sec. 8.2.2.2. Essentially, the $s$ separate weight functions are combined into a single steeply decreasing weight function, which ensures that the priorities of the individual weight functions are respected. However computing a maximum weight matching in the presence of such large weights takes $O(r^2 \sqrt{n} m \log n)$ time using the Gabow–Tarjan algorithm [231], and $O(rn(m + n \log n))$ time using the Fredman–Tarjan algorithm [223], where $n$ is the size of $I$ and $m$ is the number of acceptable man–woman pairs in $I$. Huang and Kavitha [295] described a faster algorithm for computing a weight-maximal matching, which leads to the following generalisation of Theorems 8.4 and 8.17.

**Theorem 8.18 ([295]).** *Let $I$ be an instance of* SMTI, *and let* $\langle f_1, \ldots, f_s \rangle$ *be a given tuple of weight functions, for some $s \geq 1$, where $f_k : E \longrightarrow \{0, 1, \ldots, F\}$ for each $k$ $(1 \leq k \leq s)$. There is an $O(r^* \sqrt{n} m \log n)$ algorithm to find a weight-maximal matching in $I$, where $n$ is the size of $I$, $m$ is the number of acceptable man–woman pairs in $I$, and $r^* = r^w(M)$ is the weighted regret of $M$.*

## 8.5 Other profile-based optimal matching problems

### 8.5.1 *Rental Market problem*

Abraham et al. [19] considered profile-based optimal matchings in the context of the *Rental Market* problem. We are given an instance of HAT in which the houses intuitively correspond to DVDs that can be rented by the applicants (customers). There are $t$ time steps $1, 2, \ldots, t$, and the aim is to compute $t$ matchings $M_1, M_2, \ldots, M_t$ in $I$ such that, for each applicant $a_i$ and house $h_j$,

$$|\{M_k : 1 \leq k \leq t \wedge (a_i, h_j) \in M_k\}| \leq 1.$$

That is, a given house (DVD) can be assigned to a given applicant (customer) at most once over the $t$ time steps. It is assumed that if a house is assigned to an applicant at time step $k$, it is then available for use by another applicant at time step $k + 1$.

The authors considered a range of different criteria for an optimal matching in a given instance of the Rental Market problem. Among these are rank maximal matchings and generous maximum matchings. They also proposed both offline and online variants of the problem — in the latter case, applicants can add or remove elements from their preference lists between successive time steps. The authors proved theoretical bounds concerning the performance of various optimal matching algorithms for both the offline and online versions. They also compared the performance of the algorithms on real data, noting among other things the total number of *skips* that each algorithm encounters (this is the number of higher-ranked houses that an applicant misses out on in a given matching, taken over all applicants and over the matchings in all time steps). Interestingly, the authors found that generous maximum matchings performed best with respect to this particular measure.

### 8.5.2 *Assigning papers to reviewers*

Refereed conferences often involve a Programme Committee (PC) which is collectively in charge of reviewing (or arranging sub-reviewers for) the submitted papers. Many conference management software systems, such as EasyChair[8], automate the task of assigning papers to reviewers, taking into account information such as the following:

(1) *Reviewers' interest*: each reviewer can specify a level of interest in reviewing a particular paper;
(2) *Reviewers' expertise*: each reviewer can specify their level of expertise in the topic of a given paper;
(3) *Conflicts of interest*: each reviewer may have a number of papers for which she has a conflict of interest;
(4) *Coverage*: there will be a value of $t \geq 1$ such each paper must be reviewed $t$ times;
(5) *Load balancing*: each reviewer should be given roughly the same number of papers. If there are $n_1$ reviewers, $n_2$ papers and each should be reviewed $t$ times, then each reviewer will be assigned either $b$ or $b - 1$ papers, where $b = \left\lceil \frac{tn_2}{n_1} \right\rceil$.

In some cases, data mining techniques are used to predict (1) and (2), using comparisons between keywords from the submitted papers and keywords from the reviewers' prior publications. We do not pursue this direction any further here. We focus on the case where (1) and (2) are given, and are combined into a single score, or *valuation* function, which assumes that a reviewer's interest level for a given paper will be in close correlation with their expertise for reviewing that paper.

Formally, the valuation function is defined as follows. Let $R$ be the set of reviewers, let $P$ be the set of papers, and let $\Delta \geq 1$ be some positive constant. Define $v : R \times P \longrightarrow \{0, 1, \ldots, \Delta\}$ to be the *valuation function*

---

[8]http://www.easychair.org. Accessed 25 May 2012.

such that $v(r_i, p_j)$ denotes the valuation of a given paper $p_j \in P$ by a given reviewer $r_i \in R$. (The valuation function can also be regarded as a utility function, and $v(r_i, p_j) > v(r_i, p_k)$ implies that $r_i$ prefers $p_j$ to $p_k$.) We assume that $v(r_i, p_j) = 0$ indicates that either $r_i$ has a conflict of interest for $p_j$, or $r_i$ has no interest in, or insufficient expertise for, reviewing $p_j$.

We are thus given a many–many matching problem, where each reviewer must be assigned $b$ or $b-1$ papers, and each paper must be reviewed exactly $t$ times. We refer to the problem of finding an optimal matching in this context (subject to a suitable definition of optimality) as the *Reviewer Assignment problem* (RA).

To give an example from Ref. [242] of the typical dimensions of the problem, EasyChair was used to assign papers to reviewers for the Design and Analysis Track of the ESA 2008 conference (the 16th Annual European Symposium on Algorithms). In this case, $n_1 = |R| = 14$, $n_2 = |P| = 202$, $t = 4$, $\Delta = 3$ and $b = 58$. Thus in a load-balanced matching of reviewers to papers, we require to find the minimum integer $k$ such that $k$ reviewers are assigned $b$ papers and $n - k$ reviewers are assigned $b - 1$ papers, and $kb + (n_1 - k)(b - 1) \geq tn_2$. That is, $k = tn_2 - n_1(b - 1)$. In this case $k = 10$.

There has been a great deal of previous work on modelling and solving RA, with references appearing in the Artificial Intelligence, Operations Research, Algorithms and Complexity, and Decision Theory literature. The solution techniques used include heuristics, approximation algorithms, integer programming, polynomial-time algorithms (based on graph matching, network flow and other techniques, both exact and approximate) and strategy-proof mechanisms. Wang *et al.* [583] gave a relatively recent survey of the literature.

We may as well regard the valuation function as giving rise to a preference list (with ties in general) for each reviewer, as follows. Given a reviewer $r_i \in R$, for each $k$ $(1 \leq k \leq \Delta)$, let the $k$th tied batch of papers in $r_i$'s list comprise those papers $p_j$ for which $v(r_i, p_j) = \Delta - k + 1$. For example, suppose that $\Delta = 3$, $n_2 = 10$ and the following list indicates the $v(r_1, p_j)$ values for a particular reviewer $r_1$:

$$p_1 : 0 \quad p_2 : 2 \quad p_3 : 3 \quad p_4 : 1 \quad p_5 : 3 \quad p_6 : 0 \quad p_7 : 0 \quad p_8 : 3 \quad p_9 : 1 \quad p_{10} : 0.$$

Then $r_1$'s preference list is as follows:

$$r_1 : \quad (p_3 \quad p_5 \quad p_8) \quad p_2 \quad (p_4 \quad p_9).$$

It is useful in this subsection to redefine the notion of *rank*, such that, for any reviewer $r_i \in R$ and for any paper $p_j \in P$ such that $v(r_i, p_j) > 0$,

Reviewers' preferences      Paper coverage $t = 1$
$r_1 : (p_1 \quad p_2) \quad (p_3 \quad p_4)$
$r_2 : (p_1 \quad p_2) \quad (p_3 \quad p_4)$

$M_1 = \{(r_1, p_1), (r_1, p_2), (r_2, p_3), (r_2, p_4)\}$
$M_2 = \{(r_1, p_1), (r_1, p_3), (r_2, p_2), (r_2, p_4)\}$

Fig. 8.2    An instance of MM-CHAT together with two matchings due to Garg *et al.* [242].

$rank(r_i, p_j) = \Delta - v(r_i, p_j) + 1$. Thus in the above example, $rank(r_1, p_2) = 2$ and $rank(r_1, p_9) = 3$. The profile of a matching is now defined relative to this revised definition of *rank*.

In general, for a given reviewer $r_i \in R$, if there are any "missing" values in the set $\{v(r_i, p_j) : p_j \in P\}$ then we can always introduce a sufficient number of "dummy" papers that need to be reviewed 0 times in order to "pad out" $r_i$'s preference list so that the $k$th tied batch corresponds to the papers $p_j$ for which $rank(r_i, p_j) = k$. For simplicity, in what follows we assume that such dummy papers are not required.

Given the remarks above, it is clear that an instance of RA can be modelled as an MM-CHAT instance (as defined on Page 391), in which the applicants correspond to reviewers and the houses correspond to papers. Each reviewer $r_i \in R$ has capacity $c(r_i) = b$, and each paper $p_j \in P$ has capacity $c(p_j) = t$. In such a setting it is natural to consider profile-based optimal matchings. In general, we discount rank-maximality as a potential solution criterion because we assume that the overriding requirement is to ensure that every paper is reviewed the required number of times. This implies that maximising the cardinality of the matching should be the top priority. In particular, we will restrict attention to the set $\mathcal{M}^+$ of maximum matchings in $I$ (each of which ideally has size $tn_2$, meaning that all papers are reviewed sufficiently many times). However there is a sense in which a greedy / generous maximum matching, or a maximum utility maximum cardinality matching need not be fair to all reviewers, as observed by Garg *et al.* [242].

To illustrate what we mean, consider the MM-CHAT instance $I'$ and the two matchings $M_1$ and $M_2$ shown in Fig. 8.2, due to Garg *et al.* [242]. Then $p(M_1) = p(M_2) = \langle 2, 2 \rangle$, and each of $M_1$ and $M_2$ has total utility 6. Hence each of $M_1$ and $M_2$ is a greedy maximum, generous maximum and maximum utility maximum cardinality matching in $I'$. However $M_2$ is fairer overall, giving each reviewer one first-choice paper and one second-choice paper, whilst $M_1$ favours $r_1$ at the expense of $r_2$. Hence a tighter definition of optimality is required if we are to truly optimise the overall social welfare of the reviewers.

Garg *et al.* [242] gave an alternative definition of an optimal matching, called a *leximin optimal matching*, that better models the idea of fairness. In order to define this concept, we require some preliminary definitions. Let $I$ be an instance of MM-CHAT and let $M$ be a maximum matching in $I$. Given a resident $r_i \in M$, define the *profile of $M$ for $r_i$*, denoted $p_{r_i}(M)$, to be a $\Delta$-tuple $\langle q_1, q_2, \ldots, q_\Delta \rangle$, where, for each $k$ $(1 \le k \le \Delta)$,

$$q_k = |\{(r_i, p_j) \in M : rank(r_i, p_j) = k\}.$$

Hence, relative to the MM-CHAT instance $I'$ illustrated in Fig. 8.2, $p_{r_1}(M_1) = \langle 2, 0 \rangle$ and $p_{r_2}(M_1) = \langle 0, 2 \rangle$, whilst $p_{r_1}(M_2) = p_{r_2}(M_2) = \langle 1, 1 \rangle$.

The concept of a leximin optimal matching was defined by Garg *et al.* [242] with respect to two models of preference over profiles, namely *lexicographic preferences* and *weighted preferences*. We consider each of these models separately. In what follows, we use the term *profile* to mean any $\Delta$-tuple of integers $\rho = \langle q_1, q_2, \ldots, q_\Delta \rangle$, and we use the term *meta-profile* to mean any $n_1$-tuple of profiles $\langle \rho_1, \rho_2, \ldots, \rho_{n_1} \rangle$. In practice, a meta-profile will contain the profile of a given matching for each of the $n_1$ reviewers.

- *Lexicographic preferences.* Given a meta-profile $\mathcal{P} = \langle \rho_1, \rho_2, \ldots, \rho_{n_1} \rangle$, define $sort_{\succ_L}(\mathcal{P})$ to be the *sorted* meta-profile $\langle \rho_{\pi_1}, \rho_{\pi_2}, \ldots, \rho_{\pi_{n_1}} \rangle$, where $\pi$ is a permutation of $1, 2, \ldots, n_1$, such that, for each $k$ $(1 \le k \le n_1 - 1)$, either $\rho_{\pi_{k+1}} \succ_L \rho_{\pi_k}$ or $\rho_{\pi_{k+1}} = \rho_{\pi_k}$.
  The definition of $sort_{\succ_L}(\mathcal{P})$ ensures that the profiles are ordered from left to right as "worst" to "best" under $\succ_L$. For example, with respect to the MM-CHAT instance $I'$ illustrated in Fig. 8.2, if $\rho_1 = p_{r_1}(M_1) = \langle 2, 0 \rangle$ and $\rho_2 = p_{r_2}(M_1) = \langle 0, 2 \rangle$, then $sort_{\succ_L}(\langle \rho_1, \rho_2 \rangle) = \langle \rho_2, \rho_1 \rangle$.
  Given two (not necessarily sorted) meta-profiles $\mathcal{P}_1 = \langle \sigma_1, \ldots, \sigma_{n_1} \rangle$ and $\mathcal{P}_2 = \langle \sigma'_1, \ldots, \sigma'_{n_1} \rangle$, we say that $\mathcal{P}_1 \succ_L \mathcal{P}_2$ if there exists some $s$ $(1 \le s \le n_1)$ such that $\sigma_k = \sigma'_k$ $(1 \le k \le s - 1)$ and $\sigma_s \succ_L \sigma'_s$.
  For example, if $\mathcal{P}_1 = \langle \rho_2, \rho_1 \rangle$, where $\rho_1, \rho_2$ are as defined above, and $\mathcal{P}_2 = \langle \rho_3, \rho_4 \rangle$, where $\rho_3 = p_{r_1}(M_2) = \langle 1, 1 \rangle$ and $\rho_4 = p_{r_2}(M_2) = \langle 1, 1 \rangle$ then $\mathcal{P}_2 \succ_L \mathcal{P}_1$.
  We define a maximum matching $M$ in $I$ to be a *leximin optimal matching under lexicographic preferences* if the meta-profile $sort(\langle p_{r_1}(M), \ldots, p_{r_{n_1}}(M) \rangle)$ of $M$ is maximum under $\succ_L$, taken over all

maximum matchings in $I$. Intuitively, a leximin optimal matching under lexicographic preferences maximises (with respect to $\succ_L$) the profile of the worst-off reviewer. For example, $M_2$ is the unique leximin optimal matching under lexicographic preferences in the MM-CHAT instance illustrated in Fig. 8.2.

• *Weighted preferences.* We assume the existence of a decreasing weight function $w$ that maps ranks to real numbers, i.e., $w : \{1, 2, \ldots, \Delta\} \longrightarrow \mathbb{R}$, such that $w(k) > w(k + 1)$ for each $k$ ($1 \leq k \leq \Delta - 1$). For example, one possibily is $w'(k) = \Delta - k + 1$ ($1 \leq k \leq \Delta - 1$). This ensures that first-choice papers have largest weight, followed by second-choice papers, and so on (as in the case of the valuation function). Given a profile $\rho = \langle q_1, q_2, \ldots, q_\Delta \rangle$, define the *weight* of $\rho$, denoted $w(\rho)$, to be $\sum_{k=1}^{\Delta} w(q_k)$.

Given a meta-profile $\mathcal{P} = \langle \rho_1, \rho_2, \ldots, \rho_{n_1} \rangle$, define $sort_w(\mathcal{P})$ to be the *sorted* meta-profile $\langle \rho_{\pi_1}, \rho_{\pi_2}, \ldots, \rho_{\pi_{n_1}} \rangle$, where $\pi$ is a permutation of $1, 2, \ldots, n_1$, such that, for each $k$ ($1 \leq k \leq n_1 - 1$), $w(\rho_{\pi_{k+1}}) \geq w(\rho_{\pi_k})$. The definition of $sort_w(\mathcal{P})$ ensures that the profiles are ordered from left to right as "worst" to "best" under weighted preferences. For example, with respect to the MM-CHAT instance $I'$ illustrated in Fig. 8.2, $\Delta = 2$, and if $\rho_1 = p_{r_1}(M_1) = \langle 2, 0 \rangle$ and $\rho_2 = p_{r_2}(M_1) = \langle 0, 2 \rangle$, and $w'$ is the weight function where $w'(k) = \Delta - k + 1$ ($1 \leq k \leq \Delta - 1$), then $w'(\rho_1) = 4$, $w'(\rho_2) = 2$, and $sort_{w'}(\langle \rho_1, \rho_2 \rangle) = \langle \rho_2, \rho_1 \rangle$.

Given two (not necessarily sorted) meta-profiles $\mathcal{P}_1 = \langle \sigma_1, \ldots, \sigma_{n_1} \rangle$ and $\mathcal{P}_2 = \langle \sigma'_1, \ldots, \sigma'_{n_1} \rangle$, we say that $\mathcal{P}_1 >_w \mathcal{P}_2$ if there exists some $s$ ($1 \leq s \leq n_1$) such that $w(\sigma_k) = w(\sigma'_k)$ ($1 \leq k \leq s - 1$) and $w(\sigma_s) > w(\sigma'_s)$. For example, if $\mathcal{P}_1 = \langle \rho_2, \rho_1 \rangle$, where $\rho_1, \rho_2$ are as defined above, and $\mathcal{P}_2 = \langle \rho_3, \rho_4 \rangle$, where $\rho_3, \rho_4$ are also as defined above, then $\mathcal{P}_2 >_{w'} \mathcal{P}_1$.

We define a maximum matching $M$ in $I$ to be a *leximin optimal matching under weighted preferences* if the meta-profile $sort(\langle p_{r_1}(M), \ldots, p_{r_{n_1}}(M) \rangle)$ of $M$ is maximum under $>_w$, taken over all maximum matchings in $I$. A leximin optimal matching under weighted preferences maximises the weight of the profile of the worst-off reviewer. As above, $M_2$ is the unique leximin optimal matching under weighted preferences (assuming weight function $w'$) in the MM-CHAT instance illustrated in Fig. 8.2.

Garg *et al.* [242] proved the following results concerning the computational complexity of computing leximin optimal matchings relative to both lexicographic and weighted preferences.

**Theorem 8.19 ([242]).** *Let $I$ be an instance of* MM-CHAT *and let $\Delta$ be the maximum rank of a paper in a reviewer's preference list, taken over all reviewers. The following results hold:*

*(i)* *if $\Delta = 2$, each of the problems of finding a leximin optimal matching in $I$ under lexicographic and weighted preferences is solvable in polynomial time;*

*(ii)* *if $\Delta = 3$, each of the problems of finding a leximin optimal matching in $I$ under lexicographic and weighted preferences is NP-hard;*

*(iii)* *if $\Delta = 3$, the problem of finding a leximin optimal fractional matching in $I$ under weighted preferences is solvable in polynomial time. A fractional assignment can be rounded to give an approximation to a leximin optimal matching under weighted preferences in polynomial time.*

With respect to the dataset described on Page 410 for the ESA 2008 conference, Garg *et al.* [242] showed that their approximation algorithm, referred to in Part (iii) of Theorem 8.19, performed very favourably compared to the software used within the EasyChair conference management system at that time.

## 8.6 Conclusions and open problems

Despite their very intuitive definitions, profile-based optimal matchings have not been extensively studied in the literature. However within the last few years, a small number of papers have formulated efficient algorithms for generating such matchings in various problem instances.

One striking omission is SRTI. We have described polynomial-time algorithms for computing rank maximal, greedy maximum and generous maximum matchings in a given SRTI instance by transforming to the problem of finding a maximum weight matching in a general weighted graph. However it remains open to formulate combinatorial algorithms for these problems that avoid such a reduction. Also, an intriguing open problem is obtained by combining stability with rank-maximality in an SR instance $I$, as mentioned in Sec. 4.9.

Very little is known about the structure of profile-based optimal matchings. For example, it is an open problem to characterise the set of rank-maximal matchings in a given instance $I$ of HAT. Such a characterisation could lead to efficient algorithms for problems such as counting rank-maximal matchings in $I$, generating all rank-maximal matchings in $I$, and sampling a rank-maximal matching in $I$ uniformly at random. Similar remarks apply to greedy maximum and generous maximum matchings, and to other problem instances.

Finally, we recall that, as discussed in Sec. 8.1, mechanisms for computing profile-based optimal matchings are not strategy-proof in general. To mitigate this, an alternative is to design a strategy-proof mechanism that provides an approximate solution, following Procaccia and Tennenholtz [484]. For example, the Algorithm SDM (see Sec. 6.1) is a strategy-proof mechanism for HAT. One challenge is to formulate an appropriate notion of approximation when considering the profile of a matching.

# Bibliography

[1] Abdulkadiroğlu, A., Pathak, P.A. and Roth, A.E. (2005a). The Boston public school match, *American Economic Review* **95**, 2, pp. 368–371. *Cited on page(s):* 4, 12, 31

[2] Abdulkadiroğlu, A., Pathak, P.A. and Roth, A.E. (2005b). The New York City high school match, *American Economic Review* **95**, 2, pp. 364–367. *Cited on page(s):* 4, 12, 31

[3] Abdulkadiroğlu, A., Pathak, P.A. and Roth, A.E. (2009). Strategy-proofness versus efficiency in matching with indifferences: Redesigning the NYC high school match, *American Economic Review* **99**, 5, pp. 1954–1978. *Cited on page(s):* 149

[4] Abdulkadiroğlu, A., Pathak, P.A., Roth, A.E. and Sönmez, T. (2006). Changing the Boston school-choice mechanism, NBER working paper 11965. *Cited on page(s):* 4, 12, 31

[5] Abdulkadiroğlu, A. and Sönmez, T. (1998). Random serial dictatorship and the core from random endowments in house allocation problems, *Econometrica* **66**, 3, pp. 689–701. *Cited on page(s):* 5, 38, 304

[6] Abdulkadiroğlu, A. and Sönmez, T. (1999). House allocation with existing tenants, *Journal of Economic Theory* **88**, pp. 233–260. *Cited on page(s):* 38, 308

[7] Abdulkadiroğlu, A. and Sönmez, T. (2003). School choice: A mechanism design approach, *American Economic Review* **93**, 3, pp. 729–747. *Cited on page(s):* 100, 320

[8] Abdulkadiroğlu, A. and Sönmez, T. (2012). Matching markets: Theory and practice, in D. Acemoglu, M. Arello and E. Dekel (eds.), *Advances in Economics and Econometrics, Econometric Society Monographs*, Vol. 49 (Cambridge University Press). *Cited on page(s):* 8

[9] Abeledo, H. and Blum, Y. (1996). Stable matchings and linear programming, *Linear Algebra and its Applications* **245**, pp. 321–333. *Cited on page(s):* 180

[10] Abeledo, H.G., Blum, Y. and Rothblum, U.G. (1996). Canonical monotone decomposition of fractional stable matchings, *International Journal of Game Theory* **25**, pp. 161–176. *Cited on page(s):* 73

417

[11] Abeledo, H.G. and Rothblum, U.G. (1994). Stable matchings and linear inequalities, *Discrete Applied Mathematics* **54**, pp. 1–27. *Cited on page(s):* 32,
180

[12] Abeledo, H.G. and Rothblum, U.G. (1995a). Courtship and linear programming, *Linear Algebra and its Applications* **216**, 111-124. *Cited on
page(s):* 73

[13] Abeledo, H. and Rothblum, U.G. (1995b). Paths to marriage stability, *Discrete Applied Mathematics* **63**, pp. 1–12. *Cited on page(s):* 80, 88

[14] Abraham, D.J. (2003). *Algorithmics of two-sided matching problems*, Master's thesis, University of Glasgow, Department of Computing Science.
*Cited on page(s):* 8, 176

[15] Abraham, D.J. (2009). *Matching Markets: Design and Analysis*, Ph.D.
thesis, Carnegie-Mellon University, School of Computer Science. *Cited on
page(s):* 8

[16] Abraham, D.J., Biró, P. and Manlove, D.F. (2006a). "Almost stable"
matchings in the Roommates problem, in *Proceedings of WAOA '05: the
3rd Workshop on Approximation and Online Algorithms, Lecture Notes in
Computer Science*, Vol. 3879 (Springer), pp. 1–14. *Cited on page(s):* 204,
205, 206, 207, 208

[17] Abraham, D.J., Blum, A. and Sandholm, T. (2007a). Clearing algorithms
for barter exchange markets: enabling nationwide kidney exchanges, in
*Proceedings of EC '07: the 8th ACM Conference on Electronic Commerce*
(ACM), pp. 295–304. *Cited on page(s):* 4, 12, 37

[18] Abraham, D.J., Cechlárová, K., Manlove, D.F. and Mehlhorn, K. (2004).
Pareto optimality in house allocation problems, in *Proceedings of ISAAC
'04: the 15th Annual International Symposium on Algorithms and Computation, Lecture Notes in Computer Science*, Vol. 3341 (Springer), pp. 3–15.
*Cited on page(s):* 305, 306, 308, 310, 311, 312, 322, 326, 331

[19] Abraham, D.J., Chen, N., Kumar, V. and Mirrokni, V.S. (2006b). Assignment problems in rental markets, in *Proceedings of WINE '06: the
2nd International Workshop on Internet and Network Economics, Lecture
Notes in Computer Science*, Vol. 4286 (Springer), pp. 198–213. *Cited on
page(s):* 4, 44, 47, 408

[20] Abraham, D.J., Irving, R.W., Kavitha, T. and Mehlhorn, K. (2005).
Popular matchings, in *Proceedings of SODA '05: the 16th ACM-SIAM
Symposium on Discrete Algorithms* (ACM-SIAM), pp. 424–432. *Cited on
page(s):* 333, 334, 418

[21] Abraham, D.J., Irving, R.W., Kavitha, T. and Mehlhorn, K. (2007b). Popular matchings, *SIAM Journal on Computing* **37**, pp. 1030–1045, preliminary version appeared as [20]. *Cited on page(s):* 41, 335, 336, 337, 338, 339,
340, 345, 346, 347, 348, 357

[22] Abraham, D.J., Irving, R.W. and Manlove, D.F. (2003). The Student-
Project Allocation Problem, in *Proceedings of ISAAC '03: the 14th Annual
International Symposium on Algorithms and Computation, Lecture Notes in
Computer Science*, Vol. 2906 (Springer), pp. 474–484. *Cited on page(s):* 419

[23] Abraham, D.J., Irving, R.W. and Manlove, D.F. (2007c). Two algorithms

for the Student-Project allocation problem, *Journal of Discrete Algorithms* **5**, 1, pp. 79–91, preliminary version appeared as [22]. *Cited on page(s):* 100, 260, 263, 265, 266, 267, 271

[24] Abraham, D.J. and Kavitha, T. (2006). Dynamic matching markets and voting paths, in *Proceedings of SWAT '06: the 10th Scandinavian Workshop on Algorithm Theory, Lecture Notes in Computer Science,* Vol. 4059 (Springer), pp. 65–76. *Cited on page(s):* 419

[25] Abraham, D.J. and Kavitha, T. (2010). Voting paths, *SIAM Journal on Discrete Mathematics* **24**, 2, pp. 520–537, preliminary version appeared as [24]. *Cited on page(s):* 356, 357

[26] Abraham, D.J., Levavi, A., Manlove, D.F. and O'Malley, G. (2007d). The Stable Roommates Problem with Globally-Ranked Pairs, in *Proceedings of WINE '07: the 3rd International Workshop on Internet and Network Economics, Lecture Notes in Computer Science,* Vol. 4858 (Springer), pp. 431–444. *Cited on page(s):* 419

[27] Abraham, D.J., Levavi, A., Manlove, D.F. and O'Malley, G. (2008). The Stable Roommates Problem with Globally-Ranked Pairs, *Internet Mathematics* **5**, 4, pp. 493–515, preliminary version appeared as [26]. *Cited on page(s):* 43, 135, 146, 158, 210, 211, 212, 395, 396

[28] Abraham, D.J. and Manlove, D.F. (2004). Pareto optimality in the Roommates problem, Tech. Rep. TR-2004-182, University of Glasgow, Department of Computing Science. *Cited on page(s):* 325, 326, 327, 328, 329

[29] Abu El-Atta, A.H. and Moussa, M.I. (2009). Student project allocation with preference lists over (student,project) pairs, in *Proceedings of ICCEE 09: the 2nd International Conference on Computer and Electrical Engineering* (IEEE), pp. 375–379. *Cited on page(s):* 271, 273

[30] Ackermann, H., Goldberg, P. W., Mirrokni, V. S., Röglin, H. and Vöcking, B. (2011). Uncoordinated two-sided matching markets, *SIAM Journal on Computing* **40**, 1, pp. 92–106. *Cited on page(s):* 88, 89

[31] Adachi, H. (2000). On a characterization of stable matchings, *Economics Letters* **68**, 1, pp. 43–49. *Cited on page(s):* 71

[32] Aggarwal, G., S, Muthukrishnan, Pál, D. and Pál, M. (2009). General auction mechanism for search advertising, in *Proceedings of WWW '09: the 18th International World Wide Web conference* (ACM), pp. 241–250. *Cited on page(s):* 31

[33] Aharoni, R. and Fleiner, T. (2003). On a lemma of Scarf, *Journal of Combinatorial Theory, Series B* **87**, pp. 72–80. *Cited on page(s):* 190, 191

[34] Aigner, M. and Ziegler, G.M. (2010). *Proofs from THE BOOK*, 4th edn. (Springer). *Cited on page(s):* 113, 334

[35] Alcalde, J. (1995). Exchance-proofness or divorce-proofness? Stability in one-sided matching markets, *Economic Design* **1**, pp. 275–287. *Cited on page(s):* 287, 288

[36] Alcalde-Unzu, J. and E.Molis (2011). Exchange of indivisible goods and indifferences: The Top Trading Absorbing Sets mechanisms, *Games and Economic Behavior* **73**, pp. 1–16. *Cited on page(s):* 317

[37] Aldershof, B. and Carducci, O.M. (1996). Stable matching with couples, *Discrete Applied Mathematics* **68**, pp. 203–207. *Cited on page(s):* 247

[38]  Aldershof, B. and Carducci, O.M. (1999). Stable marriage and genetic al-
      gorithms: a fertile union, *Journal of Heuristics* **5**, pp. 29–46. *Cited on
      page(s):* 124, 251

[39]  Aldershof, B., Carducci, O.M. and Lorenc, D.C. (1999). Refined inequalities
      for stable marriage, *Constraints* **4**, pp. 281–292. *Cited on page(s):* 73

[40]  Alkan, A. (1988). Non-existence of stable threesome matchings, *Mathemat-
      ical Social Sciences* **16**, pp. 207–209. *Cited on page(s):* 274

[41]  Alkan, A. (1999). On the properties of stable many-to-many matchings un-
      der responsive preferences, in A. Alkan, C.D. Aliprantis and N.C.Yannelis
      (eds.), *Current Trends in Economics: Theory and Applications* (Springer),
      pp. 29–40. *Cited on page(s):* 260

[42]  Alkan, A. (2001). On preferences over subsets and the lattice structure of
      stable matchings, *Review of Economic Design* **6**, 1, pp. 99–111. *Cited on
      page(s):* 260

[43]  Alkan, A. (2002). A class of multipartner matching markets with a strong
      lattice structure, *Economic Theory* **19**, pp. 737–746. *Cited on page(s):* 260

[44]  Amira, N., Giladi, R. and Lotker, Z. (2010). Distributed weighted sta-
      ble marriage problem, in *Proceedings of SIROCCO '10: the 17th Interna-
      tional Colloquium on Structural Information and Communication Complex-
      ity, Lecture Notes in Computer Science*, Vol. 6058 (Springer), pp. 29–40.
      *Cited on page(s):* 59

[45]  Ando, K. and Kanemaru, S. (2006). A linear time algorithm for the sta-
      ble $b$-matching problem, *The Institute of Statistical Mathematics Coop-
      erative Research Report* **19**, pp. 77–84, available from http://www.ism.
      ac.jp/~tsuchiya/sympo/shukai04/shukai04-papers/077andoh.pdf (ac-
      cessed 25 May 2012). *Cited on page(s):* 219

[46]  Anshelevich, E., Das, S. and Naamad, Y. (2009). Anarchy, stability and
      utopia: creating better matchings, in *Proceedings of SAGT '09: the 2nd
      International Symposium on Algorithmic Game Theory, Lecture Notes in
      Computer Science*, Vol. 5814 (Springer), pp. 159–170. *Cited on page(s):* 125

[47]  Anwar, A.A. and Bahaj, A.S. (2003). Student project allocation using in-
      teger programming, *IEEE Transactions on Education* **46**, 3, pp. 359–367.
      *Cited on page(s):* 260

[48]  Apt, K. (2003). *Principles of Constraint Programming* (Cambridge Univer-
      sity Press). *Cited on page(s):* 73, 74

[49]  Arcaute, E. and Vassilvitskii, S. (2009). Social networks and stable match-
      ings in the job market, in *Proceedings of WINE '09: the 5th International
      Workshop on Internet and Network Economics, Lecture Notes in Computer
      Science*, Vol. 5929 (Springer), pp. 220–231. *Cited on page(s):* 122

[50]  Arkin, E.M., Bae, S.W., Efrat, A., Mitchell, J.S.B., Okamoto, K. and Pol-
      ishchuk, V. (2009). Geometric Stable Roommates, *Information Processing
      Letters* **109**, pp. 219–224. *Cited on page(s):* 37, 124, 125, 212, 286, 287

[51]  Avann, S.P. (1961). Metric ternary distributive semi-lattices, *Proceed-
      ings of the American Mathematical Society* **12**, 3, pp. 407–414. *Cited on
      page(s):* 195

[52]  Avery, C., Jolls, C., Posner, R.A. and Roth, A.E. (2001). The market for

federal judicial law clerks, *University of Chicago Law Review* **68**, 793-902. *Cited on page(s):* 31

[53] Avery, C., Jolls, C., Posner, R.A. and Roth, A.E. (2007). The new market for federal judicial law clerks, *University of Chicago Law Review* **74**, 448-486. *Cited on page(s):* 31

[54] Aziz, H. (2012). Stable marriage and roommate problems with individual-based stability, Tech. Rep. 1204.1628, Computing Research Repository, Cornell University Library, available from http://arxiv.org/abs/1204. 1628 (accessed 14 August 2012). *Cited on page(s):* 223

[55] Aziz, H., Brandt, F. and Harrenstein, P. (2011). Pareto optimality in coalition formation, in *Proceedings of SAGT '11: the 4th Symposium on Algorithmic Game Theory, Lecture Notes in Computer Science*, Vol. 6982 (Springer), pp. 93–104. *Cited on page(s):* 330

[56] Aziz, H. and de Keijzer, B. (2012). Housing markets with indifferences: a tale of two mechanisms, in *Proceedings of COMSOC '12: the 4th International Workshop on Computational Social Choice*. *Cited on page(s):* 317

[57] Baïou, M. and Balinski, M. (2000a). Many-to-many matching: stable polyandrous polygamy (or polygamous polyandry), *Discrete Applied Mathematics* **101**, pp. 1–12. *Cited on page(s):* 117, 255, 256

[58] Baïou, M. and Balinski, M. (2000b). The stable admissions polytope, *Mathematical Programming, Series A* **87**, pp. 427–439. *Cited on page(s):* 73, 117

[59] Baïou, M. and Balinski, M. (2002). Erratum: the stable allocation (or ordinal transportation) problem, *Mathematics of Operations Research* **27**, 4, pp. 662–680. *Cited on page(s):* 219, 220

[60] Baïou, M. and Balinski, M. (2003). Admissions and recruitment, *American Mathematical Monthly* **110**, 5, pp. 386–399. *Cited on page(s):* 31, 117

[61] Baïou, M. and Balinski, M. (2004). Student admissions and faculty recruitment, *Theoretical Computer Science* **322**, 2, pp. 245–265. *Cited on page(s):* 31

[62] Balinski, M. and Ratier, G. (1997). Of stable marriages and graphs, and strategy and polytopes, *SIAM Review* **39**, 4, pp. 575–604. *Cited on page(s):* 117

[63] Balinski, M. and Ratier, G. (1998). Graphs and marriages, *American Mathematical Monthly* **105**, 5, pp. 430–445. *Cited on page(s):* 117

[64] Balinski, M. and Sönmez, T. (1999). A tale of two mechanisms: student placement, *Journal of Economic Theory* **84**, 1, pp. 73–94. *Cited on page(s):* 4, 31

[65] Ballester, C. (2004). NP-completeness in hedonic games, *Games and Economic Behaviour* **49**, 1, pp. 1–30. *Cited on page(s):* 222

[66] Banerjee, S., Konishi, H. and Sönmez, T. (2001). Core in a simple coalition formation game, *Social Choice and Welfare* **18**, pp. 135–153. *Cited on page(s):* 222

[67] Bansal, V., Agrawal, A. and Malhotra, V.S. (2007). Polynomial time algorithm for an optimal stable assignment with multiple partners, *Theoretical Computer Science* **379**, 3, pp. 317–328. *Cited on page(s):* 121, 255, 257

[68] Bartholdi, III, J. and Trick, M.A. (1986). Stable matchings with prefer-

ences derived from a psychological model, *Operations Research Letters* **5**, pp. 165–169. *Cited on page(s):* 212

[69] Beimel, A., Malkin, T., Nissim, K. and Weinreb, E. (2008). How should we solve search problems privately? *Journal of Cryptology* **23**, 2, pp. 344–371. *Cited on page(s):* 124

[70] Benjamin, A.T., Converse, C. and Krieger, H.A. (1995). How do I marry thee? Let me count the ways, *Discrete Applied Mathematics* **59**, pp. 285–292. *Cited on page(s):* 55

[71] Berge, C. (1957). Two theorems in graph theory, *Proceedings of the National Academy of Sciences of the USA* **43**, pp. 842–844. *Cited on page(s):* 14

[72] Bessière, C. and Régin, J.-C. (1997). Arc consistency for general constraint networks: Preliminary results, in *Proceedings of IJCAI '97: the 15th International Joint Conference on Artificial Intelligence*, Vol. 1 (Morgan Kaufmann), pp. 398–404. *Cited on page(s):* 74

[73] Bhatnagar, N., Greenberg, S. and Randall, D. (2008). Sampling stable marriages: why spouse-swapping won't work, in *Proceedings of SODA '08: the 19th ACM/SIAM Symposium on Discrete Algorithms* (ACM-SIAM), pp. 1223–1232. *Cited on page(s):* 117, 118

[74] Bianco, D., Hartke, S. and Larimer, A. (2001). Stable matchings in the couples problem, *Morehead Electronic Journal of Applicable Mathematics* **2**, MATH-2001-06. *Cited on page(s):* 251

[75] Biermann, F.M. (2011). A measure to compare matchings in marriage markets, *Games and Economic Behavior*, submitted for publication . *Cited on page(s):* 102

[76] Birkhoff, G. (1937). Rings of sets, *Duke Mathematical Journal* **3**, 443-454. *Cited on page(s):* 198

[77] Biró, P. (2008). Student admissions in Hungary as Gale and Shapley envisaged, Tech. Rep. TR-2008-291, University of Glasgow, Department of Computing Science. *Cited on page(s):* 4, 13, 31

[78] Biró, P. (ed.) (2012). *Proceedings of Match-UP '12: the 2nd International Workshop on Matching Under Preferences. Cited on page(s):* x, 8

[79] Biró, P., Bomhoff, M., Golovach, P.A., Kern, W. and Paulusma, D. (2012a). Solutions for the stable roommates problem with payments, in *Proceedings of WG '12: the 38th International Workshop on Graph Theoretic Concepts in Computer Science*, Vol. 7551, pp. 69–80. *Cited on page(s):* 12

[80] Biró, P. and Cechlárová, K. (2007). Inapproximability of the kidney exchange problem. *Information Processing Letters* **101**, 5, pp. 199–202. *Cited on page(s):* 223

[81] Biró, P., Cechlárová, K. and Fleiner, T. (2008). The dynamics of stable matchings and half-matchings, *International Journal of Game Theory* **36**, pp. 333–352. *Cited on page(s):* 80, 81, 87, 89, 181, 184, 187, 190

[82] Biró, P. and Fleiner, T. (2010). The integral stable allocation problem on graphs, *Discrete Optimization* **7**, 1-2, pp. 64–73. *Cited on page(s):* 219, 220

[83] Biró, P., Fleiner, T., Irving, R.W. and Manlove, D.F. (2009a). The College Admissions problem with lower and common quotas, Tech. Rep. TR-2009-303, University of Glasgow, School of Computing Science. *Cited on page(s):* 231, 232, 423

[84] Biró, P., Fleiner, T., Irving, R.W. and Manlove, D.F. (2010a). The College Admissions problem with lower and common quotas, *Theoretical Computer Science* **411**, pp. 3136–3153, full version available as [83]. *Cited on page(s):* 227, 228, 229, 230, 231, 234, 235, 236, 237, 238, 239

[85] Biró, P., Irving, R.W. and Manlove, D.F. (2009b). Popular matchings in the Marriage and Roommates problems, Tech. Rep. TR-2009-306, University of Glasgow, Department of Computing Science. *Cited on page(s):* 372, 423

[86] Biró, P., Irving, R.W. and Manlove, D.F. (2010b). Popular matchings in the Marriage and Roommates problems, in *Proceedings of CIAC '10: the 7th International Conference on Algorithms and Complexity, Lecture Notes in Computer Science*, Vol. 6078 (Springer), pp. 97–108, full version available as [85]. *Cited on page(s):* 368, 369, 370, 371, 372, 373, 375, 376, 379

[87] Biró, P., Irving, R.W. and Schlotter, I. (2011). Stable matching with couples: an empirical study, *ACM Journal of Experimental Algorithmics* **16**, section 1, article 2, 27 pages. *Cited on page(s):* 245, 246, 248, 249, 252

[88] Biró, P., Kern, W. and Paulusma, D. (2010c). On solution concepts for matching games, in *Proceedings of TAMC '10: the 7th Annual Conference on Theory and Applications of Models of Computation, Lecture Notes in Computer Science*, Vol. 6108 (Springer), pp. 117–127. *Cited on page(s):* 423

[89] Biró, P., Kern, W. and Paulusma, D. (2012b). Computing solutions for matching games, *International Journal of Game Theory* **41**, 1, pp. 75–90, preliminary version appeared as [88]. *Cited on page(s):* 12

[90] Biró, P. and Kiselgof, S. (2012). College admissions with stable score-limits, in *Proceedings of COMSOC '12: the 4th International Workshop on Computational Social Choice*. *Cited on page(s):* 2, 13, 31, 246

[91] Biró, P. and Klijn, F. (2013). Matching with couples: a multidisciplinary survey, *International Game Theory Review*, to appear. *Cited on page(s):* 243, 244, 245, 246, 247, 251

[92] Biró, P., Manlove, D.F. and McDermid, E.J. (2012c). "Almost stable" matchings in the Roommates problem with bounded preference lists, *Theoretical Computer Science* **432**, pp. 10–20. *Cited on page(s):* 207, 208, 209

[93] Biró, P., Manlove, D.F. and Mittal, S. (2010d). Size versus stability in the marriage problem, *Theoretical Computer Science* **411**, pp. 1828–1841. *Cited on page(s):* 101, 102

[94] Biró, P. and McDermid, E. (2010). Three-sided stable matchings with cyclic preferences, *Algorithmica* **58**, 1, pp. 5–18. *Cited on page(s):* 279, 280, 285, 286

[95] Biró, P. and Norman, G. (2013). Analysis of stochastic matching markets, International Journal of Game Theory, *to appear*. *Cited on page(s):* 80, 193

[96] Bistarelli, S., Foley, S.N., O'Sullivan, B. and Santini, F. (2008). From Marriages to Coalitions: A soft CSP approach, in *Proceedings of CSCLP '08: the 13th Annual ERCIM International Workshop on Constraint Solving and Constraint Logic Programming, Lecture Notes in Computer Science*, Vol. 5655 (Springer), pp. 1–15. *Cited on page(s):* 74

[97] Blair, C. (1988). The lattice structure of the set of stable matchings with multiple partners, *Mathematics of Operations Research* **13**, 4, pp. 619–628. *Cited on page(s):* 259, 260

[98] Blum, Y., Roth, A.E. and Rothblum, U.G. (1997). Vacancy chains and equilibration in senior-level labor markets, *Journal of Economic Theory* **76**, pp. 362–411. *Cited on page(s):* 87, 89

[99] Blum, Y. and Rothblum, U.G. (2002). "Timing is everything" and marital bliss, *Journal of Economic Theory* **103**, pp. 429–443. *Cited on page(s):* 87, 89

[100] Bodin, L. and Panken, A. (2003). High tech for a higher authority: the placement of graduating rabbis from Hebrew Union College–Jewish Institute of Religion, *Interfaces* **33**, 3, pp. 1–11. *Cited on page(s):* 31

[101] Bodine-Baron, E., Lee, C., Chong, A., Hassibi, B. and Wierman, A. (2011). Peer effects and stability in matching markets, in *Proceedings of SAGT '11: the 4th International Symposium on Algorithmic Game Theory, Lecture Notes in Computer Science*, Vol. 6982, pp. 117–129. *Cited on page(s):* 289

[102] Bogomolnaia, A. and Jackson, M.O. (2002). The stability of hedonic coalition structures, *Games and Economic Behavior* **38**, pp. 201–230. *Cited on page(s):* 222

[103] Bogomolnaia, A. and Laslier, J.-F. (2007). Euclidean preferences, *Journal of Mathematical Economics* **43**, p. 8798. *Cited on page(s):* 118

[104] Borbel'ová, V. and Cechlárová, K. (2008a). On the stable *b*-matching problem in multigraphs, *Discrete Applied Mathematics* **156**, pp. 673–684. *Cited on page(s):* 218, 219

[105] Borbel'ová, V. and Cechlárová, K. (2008b). Pareto optimality in the kidney exchange problem, *Kybernetika* **44**, 3, pp. 373–384. *Cited on page(s):* 38, 318, 330

[106] Borbel'ová, V. and Cechlárová, K. (2010). Rotations in the stable *b*-matching problem, *Theoretical Computer Science* **411**, pp. 1750–1762. *Cited on page(s):* 219

[107] Boros, E., Fedzhora, L., Gurvich, V. and Jaslar, S. (2012a). On rank-profiles of stable matchings, in *Proceedings of MATCH-UP '12: the 2nd International Workshop on Matching Under Preferences*, pp. 27–38, full version available as [108]. *Cited on page(s):* 123

[108] Boros, E., Fedzhora, L., Gurvich, V. and Jaslar, S. (2012b). On rank-profiles of stable matchings, Tech. Rep. RRR 16-2012, RUTCOR: Rutgers Center for Operations Research, Rutgers University. *Cited on page(s):* 424

[109] Boros, E., Gurvich, V., Jaslar, S. and Krasner, D. (2004). Stable matchings in three-sided systems with cyclic preferences, *Discrete Mathematics* **289**, pp. 1–10. *Cited on page(s):* 279

[110] Boyle, E. and Echenique, F. (2009). Sequential entry in many-to-one matching markets, *Social Choice and Welfare* **33**, pp. 87–99. *Cited on page(s):* 89

[111] Braun, S., Dwenger, N. and Kübler, D. (2010). Telling the truth may not pay off: An empirical study of centralized university admissions in Germany, *The B.E. Journal of Economic Analysis and Policy* **10**, 1, article 22. *Cited on page(s):* 4, 31

[112] Brito, I. and Meseguer, P. (2005). Distributed stable matching problems, in *Proceedings of CP '05: the 11th International Conference on Principles and Practice of Constraint Programming, Lecture Notes in Computer Science*, Vol. 3705 (Springer), pp. 152–166. *Cited on page(s):* 59, 74

[113] Brito, I. and Meseguer, P. (2006). Distributed stable matching problems with ties and incomplete lists, in *Proceedings of CP '06: the 12th International Conference on Principles and Practice of Constraint Programming, Lecture Notes in Computer Science*, Vol. 4204 (Springer), pp. 675–679. *Cited on page(s):* 59, 74

[114] Buhrman, H. and de Wolf, R. (2002). Complexity measures and decision tree complexity: A survey, *Theoretical Computer Science* **288**, pp. 21–43. *Cited on page(s):* 64

[115] Cantala, D. (2004a). Matching markets: the particular case of couples, *Economics Bulletin* **3**, 45, pp. 1–11. *Cited on page(s):* 245, 250

[116] Cantala, D. (2004b). Restabilizing matching markets at senior level, *Games and Economic Behavior* **48**, pp. 1–17. *Cited on page(s):* 89

[117] Cantala, D. (2011). Agreement toward stability in matching markets, *Review of Economic Design* **15**, 4, pp. 293–316. *Cited on page(s):* 260

[118] Cechlárová, K. (2002a). On the complexity of exchange-stable roommates, *Discrete Applied Mathematics* **116**, 3, pp. 279–287. *Cited on page(s):* 87, 288

[119] Cechlárová, K. (2002b). Randomized matching mechanism revisited, Preprint, PJ Šafárik University, Faculty of Science, Institute of Mathematics. *Cited on page(s):* 80

[120] Cechlárová, K. (2008). Stable Partition Problem, in M.-Y. Kao (ed.), *Encyclopedia of Algorithms* (Springer), pp. 885–888. *Cited on page(s):* 7, 222, 223

[121] Cechlárová, K. and Ferková, S. (2004). The stable crews problem, *Discrete Applied Mathematics* **140**, 1-3, pp. 1–17. *Cited on page(s):* 216

[122] Cechlárová, K. and Fleiner, T. (2005). On a generalization of the stable roommates problem, *ACM Transactions on Algorithms* **1**, 1, pp. 143–156. *Cited on page(s):* 214, 218, 219

[123] Cechlárová, K. and Fleiner, T. (2009). Stable roommates with free edges, Tech. Rep. 2009-01, Egerváry Research Group on Combinatorial Optimization, Operations Research Department, Eötvös Loránd University. *Cited on page(s):* 215, 216

[124] Cechlárová, K., Fleiner, T. and Manlove, D. (2005). The kidney exchange game, in *Proceedings of SOR '05: the 8th International Symposium on Operations Research in Slovenia*, pp. 77–83, IM Preprint series A, no. 5/2005, PJ Šafárik University, Faculty of Science, Institute of Mathematics. *Cited on page(s):* 223, 330

[125] Cechlárová, K. and Hajduková, J. (1999). Stability testing in coalition formation games, in V. Rupnik, Z. Stirn and S. Drobne (eds.), *Proceedings of SOR '99: the 5th International Symposium on Operations Research in Slovenia*, pp. 111–116. *Cited on page(s):* 38, 222

[126] Cechlárová, K. and Hajduková, J. (2002). Computational complexity of stable partitions with *B*-preferences, *International Journal of Game Theory* **31**, pp. 353–364. *Cited on page(s):* 222

[127] Cechlárová, K. and Hajduková, J. (2004a). Stability of partitions under *WB*-preferences and *BW*-preferences, *International Journal of Information Technology and Decision Making* **3**, 4, pp. 605–618. *Cited on page(s):* 222

[128] Cechlárová, K. and Hajduková, J. (2004b). Stable partitions with $\mathcal{W}$-preferences, *Discrete Applied Mathematics* **138**, pp. 333–347. *Cited on page(s):* 222

[129] Cechlárová, K. and Lacko, V. (2012). The kidney exchange problem: How hard is to find a donor? *Annals of Operations Research* **193**, pp. 255–271. *Cited on page(s):* 223, 313

[130] Cechlárová, K. and Manlove, D.F. (2005). The Exchange-Stable Marriage Problem, *Discrete Applied Mathematics* **152**, 1-3, pp. 109–122. *Cited on page(s):* 288, 289, 290, 291, 292

[131] Cechlárová, K. and Romero-Medina, A. (2001). Stability in coalition formation games, *International Journal of Game Theory* **29**, 4, pp. 487–494. *Cited on page(s):* 38, 222

[132] Chebolu, P., Goldberg, L.A. and Martin, R. (2010). The complexity of approximately counting stable matchings, in *Proceedings of APPROX '10: the 13th International Workshop on Approximation Algorithms for Combinatorial Optimization Problems, Lecture Notes in Computer Science*, Vol. 6302 (Springer), pp. 81–94, full version available at http://arxiv.org/abs/1004.1836. *Cited on page(s):* 426

[133] Chebolu, P., Goldberg, L.A. and Martin, R. (2012a). The complexity of approximately counting stable matchings, *Theoretical Computer Science* **437**, pp. 35–68, preliminary version appeared as [132]. *Cited on page(s):* 117, 119

[134] Chebolu, P., Goldberg, L.A. and Martin, R. (2012b). The complexity of approximately counting stable roommate assignments, *Journal of Computer and System Sciences* **78**, 5, pp. 1579–1605. *Cited on page(s):* 117, 119

[135] Chen, L. (2007). An optimal utility value method with majority equilibrium for public facility allocation, *Pacific Journal of Optimization* **3**, 2, pp. 227–234. *Cited on page(s):* 359

[136] Chen, L., Deng, X., Fang, Q. and Tian, F. (2003). Majority equilibrium for public facility allocation (preliminary version), in *Proceedings of COCOON '03: the 9th Annual International Computing and Combinatorics Conference, Lecture Notes in Computer Science*, Vol. 2697 (Springer), pp. 435–444. *Cited on page(s):* 359

[137] Chen, L. and Wang, Q. (2005). Majority equilibrium of distribution centers allocation in supply chain management, in *Proceedings of WINE '05: the First International Workshop on Internet and Network Economics, Lecture Notes in Computer Science*, Vol. 3828 (Springer), pp. 793–800. *Cited on page(s):* 359

[138] Chen, N. (2012). On computing Pareto stable assignments, in *Proceedings of STACS '12: the 29th Annual Symposium on Theoretical Aspects of Computer Science, Leibniz International Proceedings in Informatics*, Vol. 14 (Schloss Dagstuhl - Leibniz-Zentrum für Informatik), pp. 384–395. *Cited on page(s):* 148

[139] Chen, N. and A.Ghosh (2010). Strongly stable assignment, in *Proceedings of ESA '10: the 18th Annual European Symposium on Algorithms, Lecture Notes in Computer Science*, Vol. 6347 (Springer), pp. 147–158. *Cited on page(s):* 168

[140] Chen, N. and Ghosh, A. (2010). Algorithms for Pareto stable assignment, in V. Conitzer and J. Rothe (eds.), *Proceedings of COMSOC '10: the 3rd International Workshop on Computational Social Choice* (Düsseldorf University Press), pp. 343–354. *Cited on page(s):* 148

[141] Chen, X., Ding, G., Hu, X. and Zang, W. (2012). The maximum-weight stable matching problem: duality and efficiency, *SIAM Journal on Discrete Mathematics* **26**, 3, pp. 1346–1360. *Cited on page(s):* 73

[142] Chen, Y. and Sönmez, T. (2002). Improving efficiency of on-campus housing: An experimental study, *American Economic Review* **92**, 5, pp. 1669–1686. *Cited on page(s):* 4, 5, 46

[143] Cheng, C.T. (2008). The generalized median stable matchings: Finding them is not that easy, in *Proceedings of LATIN '08: the 8th Latin-American Theoretical INformatics symposium, Lecture Notes in Computer Science,* Vol. 4957 (Springer), pp. 568–579. *Cited on page(s):* 93, 94, 97, 99

[144] Cheng, C.T. (2010). Understanding the generalized median stable matchings, *Algorithmica* **58**, 1, pp. 34–51. *Cited on page(s):* 91, 93, 94, 96, 97, 98, 99

[145] Cheng, C.T. (2011). Personal communication. *Cited on page(s):* 98

[146] Cheng, C.T. and Lin, A. (2011). Stable roommates matchings, mirror posets, median graphs, and the local/global median phenomenon in stable matchings, *SIAM Journal on Discrete Mathematics* **25**, 1, pp. 72–94. *Cited on page(s):* 34, 98, 173, 194, 195, 196, 197, 198

[147] Cheng, C.T. and McDermid, E. (2012). Maximum locally stable matchings, in *Proceedings of MATCH-UP '12: the 2nd International Workshop on Matching Under Preferences*, pp. 51–62. *Cited on page(s):* 123

[148] Cheng, C., McDermid, E. and Suzuki, I. (2008). A unified approach to finding good stable matchings in the hospitals/residents setting, *Theoretical Computer Science* **400**, 1-3, pp. 84–99. *Cited on page(s):* 120, 121, 251

[149] Cheng, C., McDermid, E. and Suzuki, I. (2011). Center stable matchings and centers of cover graphs of distributive lattices, in *Proceedings of ICALP '11: the 38th International Colloquium on Automata, Languages and Programming, Lecture Notes in Computer Science*, Vol. 6755 (Springer), pp. 678–689. *Cited on page(s):* 98

[150] Chevaleyre, Y., Endriss, U., Lang, J. and Maudet, N. (2007). A short introduction to computational social choice, in *Proceedings of SOFSEM '07: the 33rd International Conference on Current Trends in Theory and Practice of Computer Science, Lecture Notes in Computer Science*, Vol. 4362 (Springer), pp. 51–69. *Cited on page(s):* 9

[151] Chou, J.-H. and Lu, C.-J. (2010). Communication requirements for stable marriages, in *Proceedings of CIAC '10: the 7th International Conference on Algorithms and Complexity, Lecture Notes in Computer Science*, Vol. 6078 (Springer), pp. 371–382. *Cited on page(s):* 59

[152] Chuang, S.-T., Goel, A., McKeown, N. and Prabhakar, B. (1998). Matching output queueing with a combined input output queued switch, *IEEE Journal on Selected Areas in Communication* **17**, 6, pp. 1030–1039. *Cited on page(s):* 59

[153] Chung, K.S. (2000). On the existence of stable roommate matchings, *Games and Economic Behavior* **33**, 2, pp. 206–230. *Cited on page(s):* 192, 200, 368, 370

[154] Condorcet, M.-J.-A.-N. de C. (Marquis de) (1785). *Essai sur l'application de l'analyse à la probabilité des décisions rendues à la pluralité des voix* (L'Imprimerie Royale). *Cited on page(s):* 333, 358, 359

[155] Cook, S.A., Filmus, Y. and Lê, D.T.M. (2012). The complexity of the comparator circuit value problem, Tech. Rep. 1208.2721, Computing Research Repository, Cornell University Library, available from http://arxiv.org/abs/1208.2721 (accessed 3 December 2012). *Cited on page(s):* 68

[156] Cormen, T.H., Leiserson, C.E. and Rivest, R.L. (2009). *Introduction to Algorithms*, 3rd edn. (McGraw-Hill/MIT). *Cited on page(s):* 400

[157] Cui, L. and Jia, W. (2013). Cyclic stable matching for three-sided networking services, *Computer Networks* **57**, 1, pp. 351–363. *Cited on page(s):* 281

[158] Dabney, J. and Dean, B.C. (2010). An efficient algorithm for batch stability testing, *Algorithmica* **58**, 1, pp. 52–58. *Cited on page(s):* 60

[159] Damianidis, I. (2011). *The Stable Marriage Problem — Optimizing Different Criteria Using Genetic Algorithms*, Master's thesis, University of Borås, School of Business and Informatics, available from http://bada.hb.se/handle/2320/7987 (accessed 3 December 2012). *Cited on page(s):* 8, 124

[160] Danilov, V.I. (2003). Existence of stable matchings in some three-sided systems, *Mathematical Social Sciences* **46**, pp. 145–148. *Cited on page(s):* 278

[161] Dean, B.C., Goemans, M.X. and Immorlica, N. (2006). The unsplittable stable marriage problem, in *Proceedings of IFIP TCS '06: the 4th IFIP International Conference on Theoretical Computer Science, IFIP — International Federation for Information Processing*, Vol. 209 (Springer), pp. 65–75. *Cited on page(s):* 254

[162] Dean, B.C. and Munshi, S. (2010). Faster algorithms for stable allocation problems, *Algorithmica* **58**, 1, pp. 59–81. *Cited on page(s):* 176, 179, 220, 224

[163] Dean, B.C. and Swar, N. (2009). The Generalized Stable Allocation Problem, in *Proceedings of WALCOM '09: the 3rd International Workshop on Algorithms and Computation, Lecture Notes in Computer Science*, Vol. 5431 (Springer), pp. 238–249. *Cited on page(s):* 220

[164] Demange, G., Gale, D. and Sotomayor, M. (1987). A further remark on the stable matching problem, *Discrete Applied Mathematics* **16**, pp. 217–222. *Cited on page(s):* 103, 104

[165] Demange, M. and Ekim, T. (2008). Minimum maximal matching is NP-hard in regular bipartite graphs, in *Proceedings of TAMC '08: the 5th Annual Conference on Theory and Applications of Models of Computation, Lecture Notes in Computer Science*, Vol. 4978 (Springer), pp. 364–374. *Cited on page(s):* 17

[166] Deng, X., Papadimitriou, C. and Safra, S. (2003). On the complexity of equilibria, *Journal of Computer and System Sciences* **67**, 2, pp. 311–324. *Cited on page(s):* 38

[167] Diamantoudi, E., Miyagawa, E. and Xue, L. (2004). Random paths to

stability in the roommate problem, *Games and Economic Behavior* **48**, pp. 18–28. *Cited on page(s):* 192, 193

[168] Dias, V.M.F., da Fonseca, G.D., de Figueiredo, C.M.H. and Szwarcfiter, J.L. (2003). The stable marriage problem with restricted pairs, *Theoretical Computer Science* **306**, 1-3, pp. 391–405. *Cited on page(s):* 107, 108, 123

[169] Dolev, S. (2000). *Self-Stabilization* (MIT Press). *Cited on page(s):* 358

[170] Dougherty, D.J. and Selkow, S.M. (2004). The complexity of the certification of properties of Stable Marriage, *Information Processing Letters* **92**, pp. 275–277. *Cited on page(s):* 64, 123

[171] Drèze, J.H. and Greenberg, J. (1980). Hedonic coalitions: optimality and stability, *Econometrica* **48**, 4, pp. 987–1003. *Cited on page(s):* 222

[172] Duan, R. and Su, H.-H. (2012). A scaling algorithm for maximum weight matching in bipartite graphs, in *Proceedings of SODA '12: the 23rd ACM-SIAM Symposium on Discrete Algorithms* (ACM-SIAM), pp. 1413–1424. *Cited on page(s):* 41

[173] Dubins, L.E. and Freedman, D.A. (1981). Machiavelli and the Gale-Shapley algorithm, *American Mathematical Monthly* **88**, 7, pp. 485–494. *Cited on page(s):* 103, 104

[174] Dulmage, A.L. and Mendelsohn, N.S. (1958). Coverings of bipartite graphs, *Canadian Journal of Mathematics* **10**, pp. 517–534. *Cited on page(s):* 15

[175] Dulmage, A.L. and Mendelsohn, N.S. (1959). A structure theory of bipartite graphs of finite exterior dimension, *Transactions of the Royal Society of Canada, Series III* **53**, pp. 1–13. *Cited on page(s):* 15

[176] Dulmage, A.L. and Mendelsohn, N.S. (1967). Graphs and matrices, in F. Harary (ed.), *Graph Theory and Theoretical Physics* (Academic Press), pp. 167–177. *Cited on page(s):* 15

[177] Dur, U.M. and Ünver, M.U. (2012). Tuition exchange, in *Proceedings of MATCH-UP '12: the 2nd International Workshop on Matching Under Preferences*, p. 127. *Cited on page(s):* 31

[178] Dutta, B. and Massó, J. (1997). Stability of matchings when individuals have preferences over colleagues, *Journal of Economic Theory* **75**, pp. 464–475. *Cited on page(s):* 245

[179] Dye, J. (2001). A constraint logic programming approach to the stable marriage problem and its application to student-project allocation, BSc Honours project report, University of York, Department of Computer Science. *Cited on page(s):* 74, 260

[180] Dyer, M.E., Goldberg, L.A., Greenhill, C. and Jerrum, M. (2004). The relative complexity of approximate counting problems, *Algorithmica* **38**, pp. 471–500. *Cited on page(s):* 118, 119

[181] Echenique, F. (2008). What matchings can be stable? The testable implications of matching theory, *Mathematics of Operations Research* **33**, 3, pp. 757–768. *Cited on page(s):* 111

[182] Echenique, F. and Oviedo, J. (2006). A theory of stability in many-to-many matching markets, *Theoretical Economics* **1**, 2, pp. 233–273. *Cited on page(s):* 255, 259

[183] Edmonds, J. (1965). Paths, trees and flowers, *Canadian Journal of Mathematics* **17**, pp. 449–467. *Cited on page(s):* 14, 15, 16

[184] Edwards, K. and Farr, G. (2008). Planarization and fragmentability of some classes of graphs, *Discrete Mathematics* **308**, 12, pp. 2396–2406. *Cited on page(s):* 63

[185] Eguchi, A., Fujishige, S. and Tamura, A. (2003). A generalized Gale-Shapley algorithm for a discrete-concave stable marriage model, in *Proceedings of ISAAC '03: the 14th Annual International Symposium on Algorithms and Computation, Lecture Notes in Computer Science*, Vol. 2906 (Springer), pp. 495–504. *Cited on page(s):* 72, 264

[186] Ehlers, L. (2008). Truncation strategies in matching markets, *Mathematics of Operations Research* **33**, 2, pp. 327–335. *Cited on page(s):* 103

[187] Eilers, D. (1999). Tech. Rep. ICC TR1999-2, Irvine Compiler Corporation. *Cited on page(s):* 54

[188] Eirinakis, P., Magos, D., Mourtos, I. and Miliotis, P. (2007). Hyperarc consistency for the stable admissions problem, in *Proceedings of ICTAI '07: the 19th IEEE International Conference on Tools with Artificial Intelligence*, Vol. 1 (IEEE Computer Society), pp. 239–242. *Cited on page(s):* 73

[189] Eirinakis, P., Magos, D., Mourtos, I. and Miliotis, P. (2012). Finding all stable pairs and solutions to the many-to-many stable matching problem, *INFORMS Journal on Computing* **24**, 2, pp. 245–259. *Cited on page(s):* 74, 255, 257, 260

[190] Eirinakis, P., Magos, D., Mourtos, I. and Miliotis, P. (2013). Finding a minimum-regret many-to-many stable matching, *Optimization,* to appear. *Cited on page(s):* 257

[191] Emek, Y., Langner, T. and Wattenhofer, R. (2012). Stability vs. cost of matchings, Tech. Rep. 1112.4632, Computing Research Repository, Cornell University Library, available from http://arxiv.org/abs/1112.4632 (accessed 25 May 2012). *Cited on page(s):* 125

[192] Erdil, A. and Ergin, H. (2006). Two-sided matching with indifferences, Unpublished manuscript. *Cited on page(s):* 147, 148, 149, 324

[193] Erdil, A. and Erkin, H. (2008). What's the matter with tie-breaking? Improving efficiency in school choice, *American Economic Review* **98**, pp. 669–689. *Cited on page(s):* 149

[194] Erdös, P., Rubin, A.L. and Taylor, R. (1980). Choosability in graphs, *Congressus Numerantium* **26**, 122-157. *Cited on page(s):* 113

[195] Eriksson, K. and Häggström, O. (2008). Instability of matchings in decentralized markets with various preference structures, *International Journal of Game Theory* **36**, 3-4, pp. 409–420. *Cited on page(s):* 100, 203

[196] Eriksson, K. and Karlander, J. (2000). Stable matching in a common generalization of the marriage and assignment models, *Discrete Mathematics* **217**, pp. 135–156. *Cited on page(s):* 12

[197] Eriksson, K. and Karlander, J. (2001). Stable outcomes of the roommate game with transferable utility, *International Journal of Game Theory* **29**, 4, pp. 555–569. *Cited on page(s):* 12

[198] Eriksson, K., Sjostrand, J. and Strimling, P. (2006). Three dimensional stable matching with cyclic preferences, *Mathematical Social Sciences* **52**, pp. 77–87. *Cited on page(s):* 278, 279

[199] Eriksson, K. and Strimling, P. (2005). How unstable are matchings from decentralized mate search? Unpublished manuscript. *Cited on page(s):* 100, 203

[200] Feder, T. (1989). A new fixed point approach for stable networks and stable marriages, in *Proceedings of STOC '89: the 21st ACM Symposium on Theory of Computing* (ACM), pp. 513–522. *Cited on page(s):* 65, 431

[201] Feder, T. (1990). *Stable Networks and Product Graphs*, Ph.D. thesis, Stanford University, published in *Memoirs of the American Mathematical Society*, vol. 116, no. 555, 1995. *Cited on page(s):* 8, 65, 109, 110, 171

[202] Feder, T. (1992). A new fixed point approach for stable networks and stable marriages, *Journal of Computer and System Sciences* **45**, pp. 233–284, preliminary version appeared as [200]. *Cited on page(s):* 10, 51, 64, 65, 69, 70, 71, 177, 179, 394, 395

[203] Feder, T. (1994). Network flow and 2-satisfiability, *Algorithmica* **11**, 3, pp. 291–319. *Cited on page(s):* 64, 66, 70, 71, 177, 178, 179

[204] Feder, T., Megiddo, N. and Plotkin, S. (1994). A sublinear parallel algorithm for stable matching, in *Proceedings of SODA '94: the 5th ACM-SIAM Symposium on Discrete Algorithms* (ACM-SIAM), pp. 632–637. *Cited on page(s):* 431

[205] Feder, T., Megiddo, N. and Plotkin, S.A. (2000). A sublinear parallel algorithm for stable matching, *Theoretical Computer Science* **233**, 1–2, pp. 297–308, preliminary version appeared as [204]. *Cited on page(s):* 32, 58

[206] Fekete, S.P., Skutella, M. and Woeginger, G.J. (2003). The complexity of economic equilibria for house allocation markets, *Information Processing Letters* **88**, pp. 219–223. *Cited on page(s):* 38

[207] Fishburn, P. (1999). Preference structures and their numerical representations, *Theoretical Computer Science* **217**, pp. 359–383. *Cited on page(s):* 169

[208] Fleiner, T. (2000). *Stable and Crossing Structures*, Ph.D. thesis, Centrum voor Wiskunde en Informatica (CWI), Amsterdam. *Cited on page(s):* 8, 71, 72

[209] Fleiner, T. (2001). A matroid generalization of the stable matching polytope, in *Proceedings of IPCO '01: the 8th Conference on Integer Programming and Combinatorial Optimization, Lecture Notes in Computer Science*, Vol. 2081 (Springer), pp. 105–114. *Cited on page(s):* 71, 72, 263, 264

[210] Fleiner, T. (2002). Some results on stable matchings and fixed points, Tech. Rep. 2002-08, Egerváry Research Group on Combinatorial Optimization, Operations Research Department, Eötvös Loránd University. *Cited on page(s):* 71, 72, 99

[211] Fleiner, T. (2003a). A fixed-point approach to stable matchings and some applications, *Mathematics of Operations Research* **28**, 1, pp. 103–126. *Cited on page(s):* 32, 71, 263, 264

[212] Fleiner, T. (2003b). On the stable $b$-matching polytope, *Mathematical Social Sciences* **46**, pp. 149–158. *Cited on page(s):* 73, 218, 255, 257

[213] Fleiner, T. (2004). Personal communication. *Cited on page(s):* 264

[214] Fleiner, T. (2008). The Stable Roommates Problem with Choice Functions,

in *Proceedings of IPCO '08: the 13th Conference on Integer Programming and Combinatorial Optimization, Lecture Notes in Computer Science*, Vol. 5035 (Springer), pp. 385–400. *Cited on page(s):* 432

[215] Fleiner, T. (2010). The Stable Roommates Problem with Choice Functions, *Algorithmica* **58**, 1, pp. 82–101, preliminary version appeared as [214]. *Cited on page(s):* 220, 221

[216] Fleiner, T., Irving, R.W. and Manlove, D.F. (2007). Efficient algorithms for generalised stable marriage and roommates problems, *Theoretical Computer Science* **381**, 1-3, pp. 162–176. *Cited on page(s):* 71, 166, 213, 214, 215

[217] Fleiner, T., Irving, R.W. and Manlove, D.F. (2011). An algorithm for a super-stable roommates problem, *Theoretical Computer Science* **412**, 50, pp. 7059–7065. *Cited on page(s):* 215

[218] Fleiner, T. and Kamiyama, N. (2012). A matroid approach to stable matchings with lower quotas, in *Proceedings of SODA '12: the 23rd ACM/SIAM Symposium on Discrete Algorithms* (ACM-SIAM), pp. 135–142. *Cited on page(s):* 227, 237, 243

[219] Fleischer, R. and Wang, Y. (2008). Dynamic Pareto optimal matching, in *Proceedings of ISISE '08: International Symposium on Information Science and Engineering*, Vol. 2, pp. 797–802. *Cited on page(s):* 312

[220] Floréen, P., Kaski, P., Polishchuk, V. and Suomela, J. (2010). Almost stable matchings by truncating the Gale-Shapley algorithm, *Algorithmica* **58**, 1, pp. 102–118. *Cited on page(s):* 59, 102

[221] Fragiadakis, D., Iwasaki, A., Troyan, P., Ueda, S. and Yokoo, M. (2012). Strategyproof assignment with minimum and maximum quotas, in *Proceedings of AAMAS '12: the 11th International Conference on Autonomous Agents and Multiagent Systems*, Vol. 3 (International Foundation for Autonomous Agents and Multiagent Systems), pp. 1327–1328. *Cited on page(s):* 234

[222] Franklin, M., Gondree, M. and Mohassel, P. (2007). Improved efficiency for private stable matching, in *Topics in Cryptology – CT-RSA '07; Proceedings of the Cryptographers' Track at the RSA Conference '07, Lecture Notes in Computer Science*, Vol. 4377 (Springer), pp. 163–177. *Cited on page(s):* 124

[223] Fredman, M.L. and Tarjan, R.E. (1987). Fibonacci heaps and their uses in improved network optimization algorithms, *Journal of the ACM* **34**, 3, pp. 596–615. *Cited on page(s):* 12, 388, 405, 408

[224] Fujishige, S. and Tamura, A. (2006). A general two-sided matching market with discrete concave utility functions, *Discrete Applied Mathematics* **154**, 6, pp. 950–970. *Cited on page(s):* 12, 72

[225] Gabow, H.N. (1976). An efficient implementations of Edmonds' algorithm for maximum matching on graphs, *Journal of the ACM* **23**, 2, pp. 221–234. *Cited on page(s):* 328, 372

[226] Gabow, H.N. (1983). An efficient reduction technique for degree-constrained subgraph and bidirected network flow problems, in *Proceedings of STOC '83: the 15th Annual ACM Symposium on Theory of Computing* (ACM), pp. 448–456. *Cited on page(s):* 16, 65, 234, 319, 394

[227] Gabow, H.N. (2012). Personal communication. *Cited on page(s):* 370

[228] Gabow, H.N., Kaplan, H. and Tarjan, R.E. (2001). Unique maximum matching algorithms, *Journal of Algorithms* **40**, pp. 159–183. *Cited on page(s):* 328, 329, 372, 373

[229] Gabow, H.N. and Tarjan, R.E. (1985). A linear-time algorithm for a special case of disjoint set union, *Journal of Computer and System Sciences* **30**, pp. 209–221. *Cited on page(s):* 328, 372

[230] Gabow, H.N. and Tarjan, R.E. (1989). Faster scaling algorithms for network problems, *SIAM Journal on Computing* **18**, pp. 1013–1036. *Cited on page(s):* 12, 41, 388

[231] Gabow, H.N. and Tarjan, R.E. (1991). Faster scaling algorithms for general graph-matching problems, *Journal of the ACM* **38**, 4, pp. 815–853. *Cited on page(s):* 12, 329, 369, 372, 396, 405, 408

[232] Gai, A.-T., Lebedev, D., Mathieu, F., de Montgolfier, F., Reynier, J. and Viennot, L. (825-834). Acyclic preference systems in P2P networks, in *Proceedings of Euro-Par '07 (European Conference on Parallel and Distributed Computing): The 13th International Euro-Par Conference, Lecture Notes in Computer Science*, Vol. 4641 (Springer), p. 2007. *Cited on page(s):* 4, 13, 37, 210, 212

[233] Gai, A.-T., Mathieu, F., de Montgolfier, F. and Reynier, J. (2007). Stratification in P2P networks: Application to BitTorrent, in *Proceedings of ICDCS '07: the 27th IEEE International Conference on Distributed Computing Systems* (IEEE Computer Society). *Cited on page(s):* 4, 13, 37, 212

[234] Gale, D. (2001). The two-sided matching problem. Origin, development and current issues, *International Game Theory Review* **3**, 2-3, pp. 237–252. *Cited on page(s):* 8

[235] Gale, D. and Shapley, L.S. (1962). College admissions and the stability of marriage, *American Mathematical Monthly* **69**, pp. 9–15. *Cited on page(s):* viii, 4, 6, 7, 18, 19, 20, 22, 23, 32, 33, 34, 51, 67, 86, 114, 223, 227, 368, 369, 395

[236] Gale, D. and Sotomayor, M. (1985a). Ms. Machiavelli and the stable matching problem, *American Mathematical Monthly* **92**, 4, pp. 261–268. *Cited on page(s):* 103

[237] Gale, D. and Sotomayor, M. (1985b). Some remarks on the stable matching problem, *Discrete Applied Mathematics* **11**, pp. 223–232. *Cited on page(s):* 21

[238] Gallai, T. (1965). Maximale systeme unabhänginger kanten, *Magyar Tudományos Akadémia Matematikai Kutató Intézetének Közleményei* **9**, pp. 401–413. *Cited on page(s):* 15, 16

[239] Galvin, F. (1995). The list chromatic index of a bipartite multigraph, *Journal of Combinatorial Theory, Series B* **63**, pp. 153–158. *Cited on page(s):* 10, 113

[240] Gärdenfors, P. (1975). Match making: assignments based on bilateral preferences, *Behavioural Science* **20**, pp. 166–173. *Cited on page(s):* 333, 368, 371

[241] Garey, M.R. and Johnson, D.S. (1979). *Computers and Intractability* (Freeman, San Francisco, CA.). *Cited on page(s):* 62, 180

[242] Garg, N., Kavitha, T., Kumar, A., Mehlhorn, K. and Mestre, J. (2010). Assigning papers to referees, *Algorithmica* **58**, 1, pp. 119–136. *Cited on page(s):* 4, 43, 47, 410, 411, 412, 413, 414

[243] Gavril, F. (1974). Personal communication with M.R. Garey and D.S. Johnson. *Cited on page(s):* 180

[244] Gelain, M. (2010). *Reasoning with incomplete and imprecise preferences,* Ph.D. thesis, Universitá di Padova e Universitá di Bologna. *Cited on page(s):* 142

[245] Gelain, M., Pini, M. S., Rossi, F., Venable, K. B. and Walsh, T. (2009). Male optimal and unique stable marriages with partially ordered preferences, in C. Guttmann, F. Dignum and M. Georgeff (eds.), *Proceedings of CARE '09: International Workshop on Collaborative Agents – Research and Development. Cited on page(s):* 171

[246] Gelain, M., Pini, M. S., Rossi, F., Venable, K. B. and Walsh, T. (2010a). Local search for stable marriage problems, in V. Conitzer and J. Rothe (eds.), *Proceedings of COMSOC '10: the 3rd International Workshop on Computational Social Choice* (Düsseldorf University Press), pp. 367–378. *Cited on page(s):* 118, 142

[247] Gelain, M., Pini, M. S., Rossi, F., Venable, K. B. and Walsh, T. (2010b). Local search for stable marriage problems with ties and incomplete lists, in *Proceedings of PRICAI '10: the 11th Pacific Rim Conference on Artificial Intelligence, Lecture Notes in Artificial Intelligence,* Vol. 6230 (Springer), pp. 64–75. *Cited on page(s):* 142

[248] Gelain, M., Pini, M. S., Rossi, F., Venable, K. B. and Walsh, T. (2011). Procedural fairness in stable marriage problems, in K. Tumer, P. Yolum, L. Sonenberg and P. Stone (eds.), *Proceedings of AAMAS '11: the 10th International Conference on Autonomous Agents and Multiagent Systems,* Vol. 3 (International Foundation for Autonomous Agents and Multiagent Systems), pp. 1209–1210. *Cited on page(s):* 142

[249] Gelain, M., Rossi, F., Pini, M. S., Venable, K. B. and Walsh, T. (2010c). Male optimal and unique stable marriages with partially ordered preferences, in *Proceedings of AAMAS '10: the 9th International Conference on Autonomous Agents and Multiagent Systems* (International Foundation for Autonomous Agents and Multiagent Systems), pp. 1387–1388. *Cited on page(s):* 171

[250] Gent, I.P., Irving, R.W., Manlove, D.F., Prosser, P. and Smith, B.M. (2001). A constraint programming approach to the stable marriage problem, in *Proceedings of CP '01: the 7th International Conference on Principles and Practice of Constraint Programming, Lecture Notes in Computer Science,* Vol. 2239 (Springer), pp. 225–239. *Cited on page(s):* 73, 74, 78

[251] Gent, I.P. and Prosser, P. (2002a). An empirical study of the stable marriage problem with ties and incomplete lists, in *Proceedings of ECAI '02: the 15th European Conference on Artificial Intelligence* (IOS Press), pp. 141–145. *Cited on page(s):* 61, 73, 142

[252] Gent, I.P. and Prosser, P. (2002b). SAT encodings of the stable marriage problem with ties and incomplete lists, in *Proceedings of SAT '02: The*

*5th International Symposium on the Theory and Applications of Satisfiability Testing*, pp. 133–140, http://gauss.ececs.uc.edu/Conferences/SAT2002/Abstracts/gent.ps (accessed 25 May 2012). *Cited on page(s):* 61, 73, 142

[253] Georgiadis, G. and Papatriantafilou, M. (2010). Overlays with preferences: approximation algorithms for matching with preference lists, in *Proceedings of IPDPS '10: the 24th IEEE International Symposium on Parallel and Distributed Processing* (IEEE), pp. 1–10. *Cited on page(s):* 217, 218

[254] Georgiadis, G. and Papatriantafilou, M. (2012). Adaptive distributed b-matching in overlays with preferences, in *Proceedings of SEA '12: the 11th International Symposium on Experimental Algorithms, Lecture Notes in Computer Science*, Vol. 7276 (Springer), pp. 208–223. *Cited on page(s):* 217, 218

[255] Goldberg, A.V. (1995). Scaling algorithms for the shortest paths problem, *SIAM Journal on Computing* **24**, 3, pp. 494–504. *Cited on page(s):* 370

[256] Golle, P. (2006). A private stable matching algorithm, in *In Proceedings of FC '06: the 10th International Conference on Financial Cryptography and Data Security, Lecture Notes in Computer Science*, Vol. 4107 (Springer), pp. 65–80. *Cited on page(s):* 124

[257] Green, M.J. and Cohen, D.A. (2003). Tractability by approximating constraint languages, in *Proceedings of CP '03: the 9th International Conference on Principles and Practice of Constraint Programming, Lecture Notes in Computer Science*, Vol. 2833 (Springer-Verlag), pp. 392–406. *Cited on page(s):* 73

[258] Gusfield, D. (1985). Roommate stability leads to marriage: The structure of the stable roommate problem, Tech. Rep. YALEU/DCS/TR-435, Yale University, Department of Computer Science. *Cited on page(s):* 176

[259] Gusfield, D. (1987). Three fast algorithms for four problems in stable marriage, *SIAM Journal on Computing* **16**, 1, pp. 111–128. *Cited on page(s):* 24, 26, 51

[260] Gusfield, D. (1988). The structure of the stable roommate problem – efficient representation and enumeration of all stable assignments, *SIAM Journal on Computing* **17**, 4, pp. 742–769. *Cited on page(s):* 36, 173

[261] Gusfield, D. and Irving, R.W. (1989). *The Stable Marriage Problem: Structure and Algorithms* (MIT Press). *Cited on page(s):* viii, x, 2, 4, 6, 7, 8, 9, 10, 14, 18, 20, 21, 22, 23, 24, 25, 26, 27, 29, 32, 33, 34, 35, 36, 51, 54, 55, 58, 59, 60, 61, 64, 65, 66, 67, 71, 72, 73, 74, 75, 89, 94, 102, 103, 104, 108, 109, 116, 123, 138, 143, 153, 161, 162, 173, 174, 175, 177, 178, 179, 188, 194, 195, 196, 199, 201, 205, 206, 207, 213, 219, 227, 243, 245, 264, 292, 294, 296, 372, 373

[262] Gusfield, D., Irving, R.W., Leather, P. and Saks, M. (1987). Every finite distributive lattice is a set of stable matchings for a small stable marriage instance, *Journal of Combinatorial Theory, Series A* **44**, pp. 304–309. *Cited on page(s):* 51, 198

[263] Gusfield, D. and Pitt, L. (1992). A bounded approximation for the minimum cost 2-SAT problem, *Algorithmica* **8**, pp. 103–117. *Cited on page(s):* 179

[264]  Hajduková, J. (2006). Coalition formation games: a survey, *Interational Game Theory Review* **8**, 4, pp. 613–641. *Cited on page(s):* 222, 223

[265]  Hall, P. (1935). On representatives of subsets, *Journal of the London Mathematical Society* **10**, pp. 26–30. *Cited on page(s):* 15

[266]  Halldórsson, M., Irving, R.W., Iwama, K. and Manlove, D.F. (eds.) (2008). *Proceedings of Match-UP '08: Matching Under Preferences – Algorithms and Complexity, held at ICALP '08: the 38th International Colloquium on Automata, Languages and Programming. Cited on page(s):* ix, 8

[267]  Halldórsson, M.M., Irving, R.W., Iwama, K., Manlove, D.F., Miyazaki, S., Morita, Y. and Scott, S. (2003). Approximability results for stable marriage problems with ties, *Theoretical Computer Science* **306**, 1-3, pp. 431–447. *Cited on page(s):* 127, 137, 144, 145, 146

[268]  Halldórsson, M., Iwama, K., Miyazaki, S. and Morita, Y. (2002). Inapproximability results on stable marriage problems, in *Proceedings of LATIN '02: the Latin-American Theoretical INformatics symposium, Lecture Notes in Computer Science*, Vol. 2286 (Springer), pp. 554–568. *Cited on page(s):* 137

[269]  Halldórsson, M.M., Iwama, K., Miyazaki, S. and Yanagisawa, H. (2003a). Improved approximation of the stable marriage problem, in *Proceedings of ESA '03: the 11th Annual European Symposium on Algorithms, Lecture Notes in Computer Science*, Vol. 2832 (Springer), pp. 266–277. *Cited on page(s):* 436

[270]  Halldórsson, M.M., Iwama, K., Miyazaki, S. and Yanagisawa, H. (2003b). Randomized approximation of the stable marriage problem, in *Proceedings of COCOON '03: the 9th Annual International Computing and Combinatorics Conference, Lecture Notes in Computer Science*, Vol. 2697 (Springer), pp. 339–350. *Cited on page(s):* 436

[271]  Halldórsson, M.M., Iwama, K., Miyazaki, S. and Yanagisawa, H. (2004). Randomized approximation of the stable marriage problem, *Theoretical Computer Science* **325**, 3, pp. 439–465, preliminary version appeared as [270]. *Cited on page(s):* 127, 136, 141, 144

[272]  Halldórsson, M., Iwama, K., Miyazaki, S. and Yanagisawa, H. (2007). Improved approximation of the stable marriage problem, *ACM Transactions on Algorithms* **3**, 3, article number 30. Preliminary version appeared as [269]. *Cited on page(s):* 127, 136, 137, 141, 144

[273]  Hamada, K., Iwama, K. and Miyazaki, S. (2008). The hospitals/residents problem with quota lower bounds, in *Proceedings of Match-UP '08: Matching Under Preferences – Algorithms and Complexity, held at ICALP '08: the 38th International Colloquium on Automata, Languages and Programming*, pp. 55–66. *Cited on page(s):* 436

[274]  Hamada, K., Iwama, K. and Miyazaki, S. (2009). An improved approximation lower bound for finding almost stable stable maximum matchings, *Information Processing Letters* **109**, 18, pp. 1036–1040. *Cited on page(s):* 101

[275]  Hamada, K., Iwama, K. and Miyazaki, S. (2011). The hospitals/residents problem with quota lower bounds, in *Proceedings of ESA '11: the 19th European Symposium on Algorithms, Lecture Notes in Computer Science*, Vol. 6942 (Springer), pp. 180–191, preliminary version appeared as [273]. *Cited on page(s):* 227, 228, 232, 233, 234

[276] Harary, F. (1983). Maximum versus minimum invariants for graphs, *Journal of Graph Theory* **7**, pp. 275–284. *Cited on page(s):* 133, 312, 330

[277] Hitsch, G., Hortaçsu, A. and Ariely, D. (2010). Matching and sorting in online dating, *American Economic Review* **100**, 1, pp. 130–163. *Cited on page(s):* 31

[278] Hoefer, M. (2011). Local matching dynamics in social networks, in *Proceedings of ICALP '11: the 38th International Colloquium on Automata, Languages and Programming, Lecture Notes in Computer Science*, Vol. 6756 (Springer), pp. 113–124. *Cited on page(s):* 437

[279] Hoefer, M. (2013). Local matching dynamics in social networks, *Information and Computation* **222**, pp. 20–35, preliminary version appeared as [278]. *Cited on page(s):* 122

[280] Hoefer, M. and Wagner, L. (2012). Locally stable matching with general preferences, Tech. Rep. 1207.1265, Computing Research Repository, Cornell University Library, available from http://arxiv.org/abs/1207.1265 (accessed 23 August 2012). *Cited on page(s):* 123

[281] Hopcroft, J.E. and Karp, R.M. (1973). A $n^{5/2}$ algorithm for maximum matchings in bipartite graphs, *SIAM Journal on Computing* **2**, pp. 225–231. *Cited on page(s):* 15, 307

[282] Horton, J.D. and Kilakos, K. (1993). Minimum edge dominating sets, *SIAM Journal on Discrete Mathematics* **6**, pp. 375–387. *Cited on page(s):* 17

[283] Hsueh, Y.-C. (1991). A unifying approach to the structures of the stable matching problems, *Computers and Mathematics with Applications* **22**, 6, pp. 13–27. *Cited on page(s):* 176

[284] Huang, C.-C. (2005). How hard is it to cheat in the Gale-Shapley stable matching algorithm? Tech. Rep. TR2005-565, Department of Computer Science, Dartmouth College. *Cited on page(s):* 105

[285] Huang, C.-C. (2006). Cheating by men in the Gale-Shapley stable matching algorithm, in *Proceedings of ESA '06: the 14th Annual European Symposium on Algorithms, Lecture Notes in Computer Science*, Vol. 4168 (Springer), pp. 418–431. *Cited on page(s):* 104

[286] Huang, C.-C. (2007a). Cheating to get better roommates in a random stable matching, in *Proceedings of STACS '07: the 24th Annual Symposium on Theoretical Aspects of Computer Science, Lecture Notes in Computer Science*, Vol. 4393 (Springer), pp. 453–464. *Cited on page(s):* 107

[287] Huang, C.-C. (2007b). Two's company, three's a crowd: stable family and threesome roommate problems, in *Proceedings of ESA '07: the 15th Annual European Symposium on Algorithms, Lecture Notes in Computer Science*, Vol. 4698 (Springer), pp. 558–569, full version available as [288]. *Cited on page(s):* 276, 277, 282, 283

[288] Huang, C.-C. (2007c). Two's company, three's a crowd: stable family and threesome roommate problems, Tech. Rep. TR2007-598, Department of Computer Science, Dartmouth College. *Cited on page(s):* 276, 277, 282, 283, 437

[289] Huang, C.-C. (2009). Classified stable matching, Tech. Rep. 0907.1779, Computing Research Repository, Cornell University Library, available

from http://arxiv.org/abs/0907.1779 (accessed 25 May 2012). *Cited on page(s):* 241, 438

[290]  Huang, C.-C. (2010a). Circular stable matching and 3-way kidney transplant, *Algorithmica* **58**, 1, pp. 137–150. *Cited on page(s):* 237, 280, 281, 284, 285

[291]  Huang, C.-C. (2010b). Classified stable matching, in *Proceedings of SODA '10: the 21st ACM/SIAM Symposium on Discrete Algorithms* (ACM-SIAM), pp. 1235–1253, full version available as [289]. *Cited on page(s):* 227, 229, 239, 240, 241, 242

[292]  Huang, C.-C. and Kavitha, T. (2011a). Near-popular matchings in the roommates problem, in *Proceedings of ESA '11: the 19th European Symposium on Algorithms, Lecture Notes in Computer Science*, Vol. 6942 (Springer), pp. 167–179. *Cited on page(s):* 371, 438

[293]  Huang, C.-C. and Kavitha, T. (2011b). Popular matchings in the stable marriage problem, in *Proceedings of ICALP '11: the 38th International Colloquium on Automata, Languages and Programming, Lecture Notes in Computer Science*, Vol. 6755 (Springer), pp. 666–677. *Cited on page(s):* 438

[294]  Huang, C.-C. and Kavitha, T. (2012a). Efficient algorithms for maximum weight matchings in general graphs with small edge weights, in *Proceedings of SODA '12: the 23rd ACM-SIAM Symposium on Discrete Algorithms* (ACM-SIAM), pp. 1400–1412. *Cited on page(s):* 41

[295]  Huang, C.-C. and Kavitha, T. (2012b). Weight-maximal matchings, in *Proceedings of MATCH-UP '12: the 2nd International Workshop on Matching Under Preferences*, pp. 87–98. *Cited on page(s):* 43, 44, 391, 392, 405, 406, 407, 408

[296]  Huang, C.-C. and Kavitha, T. (2013a). Near-popular matchings in the roommates problem, *SIAM Journal on Discrete Mathematics* **27**, 1, pp. 43–62, preliminary version appeared as [292]. *Cited on page(s):* 371

[297]  Huang, C.-C. and Kavitha, T. (2013b). Popular matchings in the stable marriage problem, *Information and Computation* **222**, pp. 180–194, preliminary version appeared as [293]. *Cited on page(s):* 374, 375, 376, 377

[298]  Huang, C.-C., Kavitha, T., Michail, D. and Nasre, M. (2008). Bounded unpopularity matchings, in *Proceedings of SWAT '08: the 12th Scandinavian Workshop on Algorithm Theory, Lecture Notes in Computer Science*, Vol. 5124 (Springer), pp. 127–137. *Cited on page(s):* 438

[299]  Huang, C.-C., Kavitha, T., Michail, D. and Nasre, M. (2011). Bounded unpopularity matchings, *Algorithmica* **61**, 3, pp. 738–757, preliminary version appeared as [298]. *Cited on page(s):* 353, 354

[300]  Hwang, J.S. (1986). The algebra of stable marriages, *International Journal of Computer Mathematics* **20**, pp. 227–243. *Cited on page(s):* 55

[301]  Hylland, A. and Zeckhauser, R. (1979). The efficient allocation of individuals to positions, *Journal of Political Economy* **87**, 2, pp. 293–314. *Cited on page(s):* 5, 38

[302]  Immorlica, N. and Mahdian, M. (2005). Marriage, honesty and stability, in *Proceedings of SODA '05: the 16th Annual ACM-SIAM Symposium on Discrete Algorithms* (ACM-SIAM), pp. 53–62. *Cited on page(s):* 104

[303] Inarra, E., Larrea, C. and Molis, E. (2008a). Random paths to P-stability in the roommate problem, *International Journal of Game Theory* **36**, pp. 461–471. *Cited on page(s):* 192, 193

[304] Inarra, E., Larrea, C. and Molis, E. (2008b). The stability of the roommate problem revisited, in *Proceedings of Match-UP '08: Matching Under Preferences – Algorithms and Complexity, held at ICALP '08: the 38th International Colloquium on Automata, Languages and Programming*, pp. 114–125. *Cited on page(s):* 193, 194

[305] Inoshita, T., Irving, R. W., Iwama, K., Miyazaki, S. and Nagase, T. (2011). Improving man-optimal stable matchings by minimum change of preference lists, in *Proceedings of HJ '11: the 7th Hungarian-Japanese Symposium on Discrete Mathematics and Its Applications. Cited on page(s):* 106

[306] Irving, R.W. (1985). An efficient algorithm for the "stable roommates" problem, *Journal of Algorithms* **6**, pp. 577–595. *Cited on page(s):* 6, 32, 33, 34, 66, 173, 183, 184, 189, 201, 202, 205, 216, 217, 219, 223, 291

[307] Irving, R.W. (1986). On the stable room-mates problem, Tech. Rep. CSC/86/R5, University of Glasgow, Department of Computing Science. *Cited on page(s):* 173, 183

[308] Irving, R.W. (1994). Stable marriage and indifference, *Discrete Applied Mathematics* **48**, pp. 261–272. *Cited on page(s):* 27, 60, 127, 154, 163, 201, 202, 276

[309] Irving, R.W. (1998). Matching medical students to pairs of hospitals: a new variation on a well-known theme, in *Proceedings of ESA '98: the 6th European Symposium on Algorithms, Lecture Notes in Computer Science*, Vol. 1461 (Springer), pp. 381–392. *Cited on page(s):* 12

[310] Irving, R.W. (2002). Personal communication. *Cited on page(s):* 292

[311] Irving, R.W. (2007a). Personal communication. *Cited on page(s):* 280

[312] Irving, R.W. (2007b). The cycle roommates problem: a hard case of kidney exchange, *Information Processing Letters* **103**, 1, pp. 1–4. *Cited on page(s):* 296, 297

[313] Irving, R.W. (2007c). Greedy and generous matchings via a variant of the Bellman-Ford algorithm, Unpublished manuscript. *Cited on page(s):* 43, 398, 399, 400, 401, 402, 403, 404, 405

[314] Irving, R.W. (2008a). Optimal Stable Marriage, in M.-Y. Kao (ed.), *Encyclopedia of Algorithms* (Springer), pp. 606–609. *Cited on page(s):* 7, 64

[315] Irving, R.W. (2008b). Stable Marriage, in M.-Y. Kao (ed.), *Encyclopedia of Algorithms* (Springer), pp. 877–879. *Cited on page(s):* 7, 22, 23

[316] Irving, R.W. (2008c). Stable matching problems with exchange restrictions, *Journal of Combinatorial Optimization* **16**, pp. 344–360. *Cited on page(s):* 290, 292, 293

[317] Irving, R.W., Kavitha, T., Mehlhorn, K., Michail, D. and Paluch, K. (2004). Rank-maximal matchings, in *Proceedings of SODA '04: the 15th ACM-SIAM Symposium on Discrete Algorithms* (ACM-SIAM), pp. 68–75. *Cited on page(s):* 388, 440

[318] Irving, R.W., Kavitha, T., Mehlhorn, K., Michail, D. and Paluch, K. (2006a). Rank-maximal matchings, *ACM Transactions on Algorithms* **2**, 4,

pp. 602–610, preliminary version appeared as [317]. *Cited on page(s):* 43, 211, 212, 383, 384, 385, 386, 387, 396

[319]  Irving, R.W. and Leather, P. (1986). The complexity of counting stable marriages, *SIAM Journal on Computing* **15**, 3, pp. 655–667. *Cited on page(s):* 21, 25, 51, 54, 55, 67, 94, 118, 119, 198

[320]  Irving, R.W., Leather, P. and Gusfield, D. (1987). An efficient algorithm for the "optimal" stable marriage, *Journal of the ACM* **34**, 3, pp. 532–543. *Cited on page(s):* 13, 24, 51, 64, 90, 91, 93, 394

[321]  Irving, R.W. and Manlove, D.F. (2002). The Stable Roommates Problem with Ties, *Journal of Algorithms* **43**, pp. 85–105. *Cited on page(s):* 32, 36, 199, 201, 202, 205

[322]  Irving, R.W. and Manlove, D.F. (2008). Approximation algorithms for hard variants of the stable marriage and hospitals/residents problems, *Journal of Combinatorial Optimization* **16**, 3, pp. 279–292. *Cited on page(s):* 127, 137, 141, 142

[323]  Irving, R.W. and Manlove, D.F. (2009). Finding large stable matchings, *ACM Journal of Experimental Algorithmics* **14**, section 1, article 2, 30 pages. *Cited on page(s):* 127, 142

[324]  Irving, R.W., Manlove, D.F. and O'Malley, G. (2006b). Stable marriage with ties and bounded length preference lists, in *Proceedings of ACiD '06: the 2nd Algorithms and Complexity in Durham workshop, Texts in Algorithmics*, Vol. 7 (College Publications), pp. 95–106. *Cited on page(s):* 440

[325]  Irving, R.W., Manlove, D.F. and O'Malley, G. (2009). Stable marriage with ties and bounded length preference lists, *Journal of Discrete Algorithms* **7**, 2, pp. 213–219, preliminary version appeared as [324]. *Cited on page(s):* 135, 137

[326]  Irving, R.W., Manlove, D.F. and Scott, S. (2000). The Hospitals/Residents problem with Ties, in *Proceedings of SWAT '00: the 7th Scandinavian Workshop on Algorithm Theory, Lecture Notes in Computer Science*, Vol. 1851 (Springer), pp. 259–271. *Cited on page(s):* 27, 29, 160, 161, 164, 165

[327]  Irving, R.W., Manlove, D.F. and Scott, S. (2002). Strong stability in the Hospitals/Residents problem, Tech. Rep. TR-2002-123, University of Glasgow, Department of Computing Science, revised May 2005. *Cited on page(s):* 151, 155, 157, 158, 440

[328]  Irving, R.W., Manlove, D.F. and Scott, S. (2003). Strong stability in the Hospitals/Residents problem, in *Proceedings of STACS '03: the 20th Annual Symposium on Theoretical Aspects of Computer Science, Lecture Notes in Computer Science*, Vol. 2607 (Springer), pp. 439–450, full version available as [327]. *Cited on page(s):* 27, 29, 155, 157, 158, 168, 169, 201

[329]  Irving, R.W., Manlove, D.F. and Scott, S. (2008). The stable marriage problem with master preference lists, *Discrete Applied Mathematics* **156**, 15, pp. 2959–2977. *Cited on page(s):* 135, 137, 145, 146, 147

[330]  Irving, R.W. and Scott, S. (2007). The stable fixtures problem, *Discrete Applied Mathematics* **155**, pp. 2118–2129. *Cited on page(s):* 216, 217

[331]  Itoh, T. and Watanabe, O. (2010). Weighted random popular matchings, *Random Structures and Algorithms* **37**, 4, pp. 477–494. *Cited on page(s):* 367

[332]  Iwama, K., Manlove, D., Miyazaki, S. and Morita, Y. (1999). Stable marriage with incomplete lists and ties, in *Proceedings of ICALP '99: the 26th International Colloquium on Automata, Languages, and Programming, Lecture Notes in Computer Science*, Vol. 1644 (Springer), pp. 443–452. *Cited on page(s):* 27, 127, 132, 133

[333]  Iwama, K. and Miyazaki, S. (2008a). Stable Marriage with Ties and Incomplete lists, in M.-Y. Kao (ed.), *Encyclopedia of Algorithms* (Springer), pp. 883–885. *Cited on page(s):* 7, 27, 127

[334]  Iwama, K. and Miyazaki, S. (2008b). A survey of the stable marriage problem and its variants, in *Proceedings of ICKS '08: the International Conference on Informatics Education and Research for Knowledge-Circulating Society* (IEEE Computer Society), pp. 131–136. *Cited on page(s):* 8

[335]  Iwama, K., Miyazaki, S. and Okamoto, K. (2004). A $\left(2 - c\frac{\log n}{n}\right)$-approximation algorithm for the stable marriage problem, in *Proceedings of SWAT '04: the 9th Scandinavian Workshop on Algorithm Theory, Lecture Notes in Computer Science*, Vol. 3111 (Springer), pp. 349–361. *Cited on page(s):* 441

[336]  Iwama, K., Miyazaki, S. and Okamoto, K. (2006). A $\left(2 - c\frac{\log n}{n}\right)$-approximation algorithm for the stable marriage problem, *IEICE Transactions on Information and Systems* **E89-D**, 8, pp. 2380–2387, preliminary version appeared as [335]. *Cited on page(s):* 127, 136

[337]  Iwama, K., Miyazaki, S. and Okamoto, K. (2007a). Stable roommates problem with triple rooms, in *Proceedings of WAAC '07: the 10th Korea-Japan Workshop on Algorithms and Computation*, pp. 105–112. *Cited on page(s):* 283, 284

[338]  Iwama, K., Miyazaki, S. and Okamoto, K. (2008a). Inapproximability of stable roommates problem with triple rooms, in *Proceedings of AAAC '08: the 1st Annual Meeting of the Asian Association for Algorithms and Computation*, p. 31. *Cited on page(s):* 284

[339]  Iwama, K., Miyazaki, S. and Yamauchi, N. (2005). A $\left(2 - c\frac{1}{\sqrt{n}}\right)$-approximation algorithm for the stable marriage problem, in *Proceedings of ISAAC '05: the 16th Annual International Symposium on Algorithms and Computation, Lecture Notes in Computer Science*, Vol. 3827 (Springer), pp. 902–914. *Cited on page(s):* 441

[340]  Iwama, K., Miyazaki, S. and Yamauchi, N. (2007b). A 1.875–approximation algorithm for the stable marriage problem, in *Proceedings of SODA '07: the 18th ACM/SIAM Symposium on Discrete Algorithms* (ACM-SIAM), pp. 288–297. *Cited on page(s):* 127, 136, 141

[341]  Iwama, K., Miyazaki, S. and Yamauchi, N. (2008b). A $\left(2 - c\frac{1}{\sqrt{n}}\right)$-approximation algorithm for the stable marriage problem, *Algorithmica* **51**, 3, pp. 342–356, preliminary version appeared as [339]. *Cited on page(s):* 127, 136

[342]  Iwama, K., Miyazaki, S. and Yanagisawa, H. (2007c). Approximation algorithms for the sex-equal stable marriage problem, in *Proceedings of WADS '07: the 10th International Workshop on Algorithms and Data Structures,*

*Lecture Notes in Computer Science*, Vol. 4619 (Springer), pp. 201–213. *Cited on page(s):* 442

[343] Iwama, K., Miyazaki, S. and Yanagisawa, H. (2010a). A 25/17-approximation algorithm for the stable marriage with one-sided ties, in *Proceedings of ESA '10: the 18th Annual European Symposium on Algorithms, Lecture Notes in Computer Science*, Vol. 6347 (Springer), pp. 135–146. *Cited on page(s):* 141, 442

[344] Iwama, K., Miyazaki, S. and Yanagisawa, H. (2010b). Approximation algorithms for the sex-equal stable marriage problem, *ACM Transactions on Algorithms* **7**, 1, article number 2. Preliminary version appeared as [342]. *Cited on page(s):* 63, 127

[345] Iwama, K., Miyazaki, S. and Yanagisawa, H. (2011). Improved approximation bounds for the student-project allocation problem with preferences over projects, in *Proceedings of TAMC '11: the 8th Annual Conference on Theory and Applications of Models of Computation, Lecture Notes in Computer Science*, Vol. 6648 (Springer), pp. 440–451. *Cited on page(s):* 442

[346] Iwama, K., Miyazaki, S. and Yanagisawa, H. (2012). Improved approximation bounds for the student-project allocation problem with preferences over projects, *Journal of Discrete Algorithms* **13**, pp. 59–66, preliminary version appeared as [345]. *Cited on page(s):* 260, 270

[347] Iwama, K., Miyazaki, S. and Yanagisawa, H. (2013). A 25/17-approximation algorithm for the stable marriage with one-sided ties, *Algorithmica,* to appear Preliminary version appeared as [343]. *Cited on page(s):* 137

[348] Jaramillo, P. and Manjunath, V. (2012). The difference indifference makes in strategy-proof allocation of objects, *Journal of Economic Theory* **147**, 5, pp. 1913–1946. *Cited on page(s):* 317

[349] Jiang, I.H.-R. and Chang, H.-Y. (2012). ECOS: stable matching based metal-only ECO synthesis, *IEEE Transactions on Very Large Scale Integration (VLSI) Systems* **20**, 3, pp. 485–497. *Cited on page(s):* 31

[350] Johnson, D.S. and Niemi, K.A. (1983). On knapsacks, partitions, and a new dynamic programming technique for trees, *Mathematics of Operations Research* **8**, 1, pp. 1–14. *Cited on page(s):* 62

[351] Jones, N.D. (1975). Space-bounded reducibility among combinatorial problems, *Journal of Computer and System Sciences* **11**, pp. 68–85. *Cited on page(s):* 67

[352] Joshi, K. and Kumar, S. (2012). Matchmaking using fuzzy analytical hierarcy process, compatibility measure and stable matching for online matrimony in India, *Journal of Multi-Criteria Decision Analysis* **19**, 1-2, pp. 57–66. *Cited on page(s):* 31

[353] Kalyanaraman, S. and Umans, C. (2008). The complexity of rationalizing matchings, in *Proceedings of ISAAC '08: the 19th International Symposium on Algorithms and Computation, Lecture Notes in Computer Science*, Vol. 5369 (Springer), pp. 171–182. *Cited on page(s):* 112

[354] Kamada, Y. and Kojima, F. (2010). Improving efficiency in matching markets with regional caps: The case of the Japan Residency Matching Pro-

gram, SIEPR Discussion Paper no. 10-011, Stanford Institute For Economic Policy Research, Stanford University. *Cited on page(s):* 30, 228, 238, 239

[355] Kamada, Y. and Kojima, F. (2012). Stability and strategy-proofness for matching with constraints: A problem in the Japanese medical match and its solution, *American Economic Review: Papers and Proceedings* **102**, 3, pp. 366–370. *Cited on page(s):* 30, 228, 238, 239

[356] Kamiyama, N. (2012). A new approach to the Pareto-stable matching problem, MI Preprint Series, Faculty of Mathematics, Kyushu University, number MI 2012-8. *Cited on page(s):* 148

[357] Kao, M.Y. (ed.) (2008). *Encyclopedia of Algorithms* (Springer). *Cited on page(s):* 7

[358] Karp, R. and Luby, M. (1983). Monte-Carlo algorithms for enumeration and reliability problems, in *Proceedings of FOCS '83: the 24th Annual IEEE Symposium on Foundations of Computer Science* (IEEE Computer Society), pp. 56–64. *Cited on page(s):* 118

[359] Karp, R., Luby, M. and Madras, N. (1989). Monte-Carlo approximation algorithms for enumeration problems, *Journal of Algorithms* **10**, 3, pp. 429–448. *Cited on page(s):* 118

[360] Kato, A. (1993). Complexity of the sex-equal stable marriage problem, *Japan Journal of Industrial and Applied Mathematics* **10**, pp. 1–19. *Cited on page(s):* 62

[361] Kavitha, T. (2008). Ranked Matching, in M.-Y. Kao (ed.), *Encyclopedia of Algorithms* (Springer), pp. 744–748. *Cited on page(s):* 7, 334

[362] Kavitha, T. (2012). Popularity vs maximum cardinality in the stable marriage setting, in *Proceedings of SODA '12: the 23rd ACM-SIAM Symposium on Discrete Algorithms* (ACM-SIAM), pp. 123–134. *Cited on page(s):* 377, 378

[363] Kavitha, T., Mehlhorn, K., Michail, D. and Paluch, K. (2004). Strongly stable matchings in time $O(nm)$ and extension to the Hospitals-Residents problem, in *Proceedings of STACS '04: the 21st International Symposium on Theoretical Aspects of Computer Science, Lecture Notes in Computer Science*, Vol. 2996 (Springer), pp. 222–233. *Cited on page(s):* 443

[364] Kavitha, T., Mehlhorn, K., Michail, D. and Paluch, K.E. (2007). Strongly stable matchings in time $O(nm)$ and extension to the Hospitals-Residents problem, *ACM Transactions on Algorithms* **3**, 2, article number 15. Preliminary version appeared as [363]. *Cited on page(s):* 29, 150, 155, 156, 157, 158, 168, 201

[365] Kavitha, T., Mestre, J. and Nasre, M. (2009). Popular mixed matchings, in *Proceedings of ICALP '09: the 36th International Colloquium on Automata, Languages and Programming, Lecture Notes in Computer Science*, Vol. 5555 (Springer), pp. 574–584. *Cited on page(s):* 443

[366] Kavitha, T., Mestre, J. and Nasre, M. (2011). Popular mixed matchings, *Theoretical Computer Science* **412**, pp. 2679–2690, preliminary version appeared as [365]. *Cited on page(s):* 357, 358

[367] Kavitha, T. and Nasre, M. (2009a). Optimal popular matchings, *Discrete Applied Mathematics* **157**, pp. 3181–3186. *Cited on page(s):* 44, 345

[368] Kavitha, T. and Nasre, M. (2009b). Popular matchings with variable job capacities, in *Proceedings of ISAAC '09: the 20th International Symposium on Algorithms and Computation, Lecture Notes in Computer Science*, Vol. 5878 (Springer), pp. 423–433. *Cited on page(s): 444*

[369] Kavitha, T. and Nasre, M. (2011). Popular matchings with variable item copies, *Theoretical Computer Science* **412**, pp. 1263–1274, preliminary version appeared as [368]. *Cited on page(s): 362, 363, 364*

[370] Kavitha, T., Nasre, M. and Nimbhorkar, P. (2010). Popularity at minimum cost, in *Proceedings of ISAAC '10: the 21st International Symposium on Algorithms and Computation, Lecture Notes in Computer Science*, Vol. 6506 (Springer), pp. 145–156. *Cited on page(s): 364, 365, 366*

[371] Kavitha, T. and Shah, C.D. (2006). Efficient algorithms for weighted rank-maximal matchings and related problems, in *Proceedings of ISAAC '06: the 17th International Symposium on Algorithms and Computation, Lecture Notes in Computer Science*, Vol. 4288 (Springer), pp. 153–162. *Cited on page(s): 43, 211, 348, 389*

[372] Kazakov, D. (2002). Co-ordination of student-project allocation, Manuscript, University of York, Department of Computer Science. *Cited on page(s): 260*

[373] Keizer, K.M., de Klerk, M., Haase-Kromwijk, B.J.J.M. and Weimar, W. (2005). The Dutch algorithm for allocation in living donor kidney exchange, *Transplantation Proceedings* **37**, pp. 589–591. *Cited on page(s): 37*

[374] Kelso, A.S., Jr. and Crawford, V.P. (1982). Job matching, coalition formation and gross substitutes, *Econometrica* **50**, pp. 1483–1504. *Cited on page(s): 221, 258*

[375] Kennes, J., Monte, D. and Tumennasan, N. (2011). The day-care assignment problem, Economics working paper 2011-05, Department of Economics and Business, Aarhus University. Available from http://econ.au.dk/fileadmin/site_files/filer_oekonomi/Working_Papers/Economics/2011/wp11_05.pdf (accessed 25 May 2012). *Cited on page(s): 31*

[376] Khot, S. (2002). On the power of unique 2-prover 1-round games, in *Proceedings of STOC '02: the 34th Annual ACM Symposium on Theory of Computing* (ACM), pp. 767–755. *Cited on page(s): 138*

[377] Khot, S. and Regev, O. (2008). Vertex cover might be hard to approximate to within $2 - \varepsilon$, *Journal of Computer and System Sciences* **74**, 3, pp. 335–349. *Cited on page(s): 138, 180*

[378] Khuller, S., Mitchell, S.G. and Vazirani, V.V. (1994). On-line algorithms for weighted bipartite matching and stable marriages, *Theoretical Computer Science* **127**, pp. 255–267. *Cited on page(s): 100, 119*

[379] Kijima, S. and Nemoto, T. (2009). Finding a level ideal of a poset, in *Proceedings of COCOON '09: the 15th Annual International Computing and Combinatorics Conference, Lecture Notes in Computer Science*, Vol. 5609 (Springer), pp. 317–327. *Cited on page(s): 445*

[380] Kijima, S. and Nemoto, T. (2012). On randomized approximation for finding a level ideal of a poset and the generalized median stable matchings,

*Mathematics of Operations Research* **37**, 2, pp. 356–371, preliminary version appeared as [379]. *Cited on page(s):* 96, 97, 99

[381] Kimbrough, S.O. and Kuo, A. (2010). On heuristics for two-sided matching: revisiting the stable marriage problem as a multiobjective problem, in *GECCO '10: the Genetic and Evolutionary Computation Conference* (ACM), pp. 1283–1290. *Cited on page(s):* 124

[382] Kinjo, H., Nakamura, M. and Onaga, K. (1997). Autonomous mechanism for partner exchanging in distributed stable marriage problems, *IEICE Transactions on Fundamentals of Electronics, Communications and Computer Sciences* **E80-A**, 6, pp. 1040–1048. *Cited on page(s):* 59

[383] Kipnis, A. and Patt-Shamir, B. (2010). On the complexity of distributed stable matching with small messages, *Distributed Computing* **23**, 3, pp. 151–161. *Cited on page(s):* 59

[384] Király, Z. (2008). Better and simpler approximation algorithms for the stable marriage problem, in *Proceedings of ESA '08: the 16th Annual European Symposium on Algorithms, Lecture Notes in Computer Science*, Vol. 5193 (Springer), pp. 623–634. *Cited on page(s):* 445

[385] Király, Z. (2011). Better and simpler approximation algorithms for the stable marriage problem, *Algorithmica* **60**, pp. 3–20, preliminary version appeared as [384]. *Cited on page(s):* 127, 128, 136, 137, 138, 141, 142

[386] Király, Z. (2012). Linear time local approximation algorithm for maximum stable marriage, in *Proceedings of MATCH-UP '12: the 2nd International Workshop on Matching Under Preferences*, pp. 99–110. *Cited on page(s):* 127, 136, 138, 139, 140, 142

[387] Klaus, B. and Klijn, F. (2005). Stable matchings and preferences of couples, *Journal of Economic Theory* **121**, pp. 75–106. *Cited on page(s):* 245, 249, 250, 251

[388] Klaus, B. and Klijn, F. (2006a). Median stable matching for college admissions, *International Journal of Game Theory* **34**, pp. 1–11. *Cited on page(s):* 99

[389] Klaus, B. and Klijn, F. (2006b). Procedurally fair and stable matching, *Economic Theory* **27**, pp. 431–447. *Cited on page(s):* 89

[390] Klaus, B. and Klijn, F. (2007). Paths to stability for matching markets with couples, *Games and Economic Behavior* **58**, pp. 154–171. *Cited on page(s):* 89, 245, 247, 250

[391] Klaus, B. and Klijn, F. (2010). Smith and Rawls share a room: stability and medians, *Social Choice and Welfare* **35**, pp. 647–667. *Cited on page(s):* 99, 195

[392] Klaus, B., Klijn, F. and Nakamura, T. (2009). Corrigendum: Stable matchings and preferences of couples, *Journal of Economic Theory* **144**, 5, pp. 2227–2233. *Cited on page(s):* 245, 250

[393] Klerk, M. de, Keizer, K.M., Claas, F.H.J., Witvliet, M., Haase-Kromwijk, B.J.J.M. and Weimar, W. (2005). The Dutch national living donor kidney exchange program, *American Journal of Transplantation* **5**, pp. 2302–2305. *Cited on page(s):* 37

[394] Knuth, D.E. (1976). *Mariages Stables* (Les Presses de L'Université de

Montréal), English translation in *Stable Marriage and its Relation to Other Combinatorial Problems*, volume 10 of CRM Proceedings and Lecture Notes, American Mathematical Society, 1997. *Cited on page(s):* viii, 7, 10, 21, 22, 23, 32, 54, 55, 56, 61, 79, 108, 223, 227, 273

[395] Kobayashi, H. and Matsui, T. (2010). Cheating strategies for the Gale-Shapley algorithm with complete preference lists, *Algorithmica* **58**, 1, pp. 151–169. *Cited on page(s):* 105

[396] Kojima, F., Pathak, P.A. and Roth, A.E. (2013). Matching with couples: Stability and incentives in large matching markets, *Quarterly Journal of Economics,* to appear. *Cited on page(s):* 245

[397] Kojima, F. and Troyan, P. (2011). Matching and market design: an introduction to selected topics, *Japanese Economic Review* **62**, 1, pp. 82–98. *Cited on page(s):* 8

[398] Kojima, F. and Ünver, M.U. (2008). Random paths to pairwise stability in many-to-many matching problems: a study on market equilibration, *International Journal of Game Theory* **36**, 3-4, pp. 473–488. *Cited on page(s):* 89, 255, 259, 260

[399] Korte, B. and Hausmann, D. (1978). An analysis of the greedy heuristic for independence systems, in *Annals of Discrete Mathematics*, Vol. 2 (North-Holland), pp. 65–74. *Cited on page(s):* 17, 136, 312, 330

[400] Korte, B. and Vygen, J. (2012). *Combinatorial Optimization, Algorithms and Combinatorics*, Vol. 21, 5th edn. (Springer). *Cited on page(s):* 14

[401] Kujansuu, E., Lindberg, T. and Mäkinen, E. (1999). The stable roommates problem and chess tournament pairings, *Divulgaciones Matemáticas* **7**, 1, pp. 19–28. *Cited on page(s):* 4, 38, 100

[402] Lê, D.T.M., Cook, S.A. and Ye, Y. (2011). A formal theory for the complexity class associated with the stable marriage problem, in *Proceedings of CSL '11: the 25th International Workshop on Computer Science Logic, Leibniz International Proceedings in Informatics*, Vol. 12 (Schloss Dagstuhl - Leibniz-Zentrum für Informatik), pp. 381–395. *Cited on page(s):* 68

[403] Lebedev, D., Mathieu, F., Viennot, L., Gai, A.-T., Reynier, J. and de Montgolfier, F. (2007). On using matching theory to understand P2P network design, in *Proceedings of INOC '07: International Network Optimization Conference. Cited on page(s):* 4, 13, 37, 193, 212

[404] Lee, H. (1999). Online stable matching as a means of allocating distributed resources, *Journal of Systems Architecture* **45**, pp. 1345–1355. *Cited on page(s):* 120

[405] Lennon, C.T. (2012). Stable matchings with additional objectives, Unpublished manuscript. Available from http://www.math.usma.edu/people/Lennon/Lennondrft.pdf (accessed 25 May 2012). *Cited on page(s):* 124

[406] Liebowitz, J. and Simien, J. (2005). Computational efficiencies for multi-agents: a look at a multi-agent system for sailor assignment, *Electronic Government* **2**, 4, pp. 384–402. *Cited on page(s):* 31

[407] Lovász, L. and Plummer, M.D. (1986). *Matching Theory*, Annals of Discrete Mathematics (North-Holland). *Cited on page(s):* 15

[408] Lu, E. and Zheng, S.Q. (2003). A parallel iterative improvement stable

matching algorithm, in *Proceedings of HiPC '03: the 10th International Conference on High Performance Computing, Lecture Notes in Computer Science*, Vol. 2913 (Springer), pp. 55–65. *Cited on page(s):* 58

[409] Lustig, I.J. and Puget, J. (2001). Program does not equal program: constraint programming and its relationship to mathematical programming, *Interfaces* **31**, pp. 29–53. *Cited on page(s):* 73, 74

[410] Ma, J. (1996). On randomized matching mechanisms, *Economic Theory* **8**, pp. 377–381. *Cited on page(s):* 85, 86

[411] Maffray, F. (1992). Kernels in perfect line graphs, *Journal of Combinatorial Theory, Series B* **55**, pp. 1–8. *Cited on page(s):* 114, 115, 116

[412] Mahdian, M. (2006). Random popular matchings, in *Proceedings of EC '06: the 7th ACM Conference on Electronic Commerce* (ACM), pp. 238–242. *Cited on page(s):* 341, 357

[413] Malhotra, V.S. (2004). On the stability of multiple partner stable marriages with ties, in *Proceedings of ESA '04: the 12th Annual European Symposium on Algorithms, Lecture Notes in Computer Science*, Vol. 3221 (Springer), pp. 508–519. *Cited on page(s):* 168

[414] Manlove, D.F. (1999). Stable marriage with ties and unacceptable partners, Tech. Rep. TR-1999-29, University of Glasgow, Department of Computing Science. *Cited on page(s):* 27, 60, 127, 154, 163

[415] Manlove, D.F. (2002). The structure of stable marriage with indifference, *Discrete Applied Mathematics* **122**, 1-3, pp. 167–181. *Cited on page(s):* 61, 153, 154, 162, 163, 169, 171

[416] Manlove, D.F. (2008). Hospitals / Residents problem, in M.-Y. Kao (ed.), *Encyclopedia of Algorithms* (Springer), pp. 390–394. *Cited on page(s):* 7, 18

[417] Manlove, D.F., Irving, R.W. and Iwama, K. (eds.) (2010). *Special Issue on Matching Under Preferences*, Algorithmica (Springer). *Cited on page(s):* ix, 8

[418] Manlove, D.F., Irving, R.W., Iwama, K., Miyazaki, S. and Morita, Y. (1999). Hard variants of stable marriage, Tech. Rep. TR-1999-43, University of Glasgow, School of Computing Science. *Cited on page(s):* 129, 131, 447

[419] Manlove, D.F., Irving, R.W., Iwama, K., Miyazaki, S. and Morita, Y. (2002). Hard variants of stable marriage, *Theoretical Computer Science* **276**, 1-2, pp. 261–279, full version available as [418]. *Cited on page(s):* 27, 29, 127, 129, 133, 134, 136, 144, 145, 146, 167, 168, 200

[420] Manlove, D.F. and O'Malley, G. (2005a). Modelling and solving the stable marriage problem using constraint programming, in *Proceedings of the 5th Workshop on Modelling and Solving Problems with Constraints, held at IJ-CAI '05: the 19th International Joint Conference on Artificial Intelligence*, pp. 10–17. *Cited on page(s):* 73, 74, 75, 76, 77, 78

[421] Manlove, D.F. and O'Malley, G. (2005b). Student project allocation with preferences over projects, in *Proceedings of ACiD '05: the 1st Algorithms and Complexity in Durham workshop, Texts in Algorithmics*, Vol. 4 (KCL Publications), pp. 69–80. *Cited on page(s):* 448

[422]  Manlove, D.F. and O'Malley, G. (2008). Student project allocation with preferences over projects, *Journal of Discrete Algorithms* **6**, pp. 553–560, preliminary version appeared as [421]. *Cited on page(s): 260, 269, 270*

[423]  Manlove, D.F. and O'Malley, G. (2012). Paired and altruistic kidney donation in the UK: Algorithms and experimentation, in *Proceedings of SEA '12: the 11th International Symposium on Experimental Algorithms, Lecture Notes in Computer Science*, Vol. 7276 (Springer), pp. 271–282. *Cited on page(s): 4, 12, 37*

[424]  Manlove, D.F., O'Malley, G., Prosser, P. and Unsworth, C. (2007). A Constraint Programming Approach to the Hospitals / Residents Problem, in *Proceedings of CP-AI-OR '07: the 4th International Conference on Integration of AI and OR Techniques in Constraint Programming for Combinatorial Optimization, Lecture Notes in Computer Science*, Vol. 4510 (Springer), pp. 155–170. *Cited on page(s): 73*

[425]  Manlove, D.F. and Sng, C.T.S. (2006). Popular matchings in the Capacitated House Allocation problem, in *Proceedings of ESA '06: the 14th Annual European Symposium on Algorithms, Lecture Notes in Computer Science*, Vol. 4168 (Springer), pp. 492–503. *Cited on page(s): 16, 360, 361, 362*

[426]  Martínez, R., Massó, J., Neme, A. and Oviedo, J. (2004). An algorithm to compute the full set of many-to-many stable matchings, *Mathematical Social Sciences* **47**, 2, pp. 187–210. *Cited on page(s): 255, 259*

[427]  Marx, D. and Schlotter, I. (2010). Parameterized complexity and local search approaches for the stable marriage problem with ties, *Algorithmica* **58**, 1, pp. 170–187. *Cited on page(s): 135, 146*

[428]  Marx, D. and Schlotter, I. (2011). Stable assignment with couples: parameterized complexity and local search, *Discrete Optimization* **8**, pp. 25–40. *Cited on page(s): 245, 249*

[429]  Mathieu, F. (2007). Upper bounds for stabilization in acyclic preference-based systems, in *Proceedings of SSS '07: the 9th International Symposium on Stabilization, Safety, and Security of Distributed Systems, Lecture Notes in Computer Science*, Vol. 4838 (Springer), pp. 372–382. *Cited on page(s): 4, 13, 37, 193, 210, 212*

[430]  Mathieu, F. (2008). Self-stabilization in preference-based systems, *Peer-to-Peer Networking and Applications* **1**, 2, pp. 104–121. *Cited on page(s): 4, 13, 37, 193, 212, 217*

[431]  Mathieu, F. (2010). Acyclic preference-based systems: A self-stabilizing approach for P2P systems, in X. Shen, H. Yu, J. Buford and M. Akon (eds.), *Handbook of Peer-to-Peer Networking* (Springer), pp. 1165–1203. *Cited on page(s): 4, 13, 37, 193, 210*

[432]  Matsui, T. (2010). Algorithmic aspects of equilibria of stable marriage model with complete preference lists, in B. Hu, K. Morasch, S. Pickl and M. Siegle (eds.), *Operations Research Proceedings '10: Selected Papers of the Annual International Conference of the German Operations Research Society (GOR) at Universität der Bundeswehr München*, Vol. 2 (Springer), pp. 47–52. *Cited on page(s): 105*

[433] Mayr, E.W. and Subramanian, A. (1989). The complexity of circuit value and network stability, in *Proceedings of CoCo '89: the 4th Annual Structure in Complexity Theory Conference* (IEEE Computer Society), pp. 114–123. *Cited on page(s):* 65, 449

[434] Mayr, E.W. and Subramanian, A. (1992). The complexity of circuit value and network stability, *Journal of Computer and System Sciences* **44**, pp. 302–323, preliminary version appeared as [433]. *Cited on page(s):* 65, 66, 67, 68, 69, 71

[435] McBride, I. (2011). Personal communication. *Cited on page(s):* 87, 88

[436] McCutchen, R.M. (2008). The least-unpopularity-factor and least-unpopularity-margin criteria for matching problems with one-sided preferences, in *Proceedings of LATIN '08: the 8th Latin-American Theoretical INformatics symposium, Lecture Notes in Computer Science*, Vol. 4957 (Springer), pp. 593–604. *Cited on page(s):* 349, 350, 351, 352, 353

[437] McDermid, E. (2009). A 3/2 approximation algorithm for general stable marriage, in *Proceedings of ICALP '09: the 36th International Colloquium on Automata, Languages and Programming, Lecture Notes in Computer Science*, Vol. 5555 (Springer), pp. 689–700. *Cited on page(s):* 127, 136, 141

[438] McDermid, E.J. (2010a). Personal communication. *Cited on page(s):* 109, 110

[439] McDermid, E. (2010b). *A Structural Approach to Matching Problems with Preferences*, Ph.D. thesis, Department of Computing Science, University of Glasgow. *Cited on page(s):* 8

[440] McDermid, E., Cheng, C. and Suzuki, I. (2007). Hardness results on the man-exchange stable marriage problem with short preference lists, *Information Processing Letters* **101**, pp. 13–19. *Cited on page(s):* 290, 293

[441] McDermid, E. and Irving, R.W. (2009). Popular matchings: Structure and algorithms, in *Proceedings of COCOON '09: the 15th Annual International Computing and Combinatorics Conference, Lecture Notes in Computer Science*, Vol. 5609 (Springer), pp. 506–515. *Cited on page(s):* 449

[442] McDermid, E. and Irving, R.W. (2011). Popular matchings: Structure and algorithms, *Journal of Combinatorial Optimization* **22**, 3, pp. 339–358, preliminary version appeared as [441]. *Cited on page(s):* 44, 340, 341, 342, 343, 344, 345

[443] McDermid, E. and Irving, R.W. (2013). Sex-equal stable matchings: Complexity and exact algorithms, *Algorithmica,* to appear. *Cited on page(s):* 62

[444] McDermid, E.J. and Manlove, D.F. (2010). Keeping partners together: Algorithmic results for the hospitals / residents problem with couples, *Journal of Combinatorial Optimization* **19**, 3, pp. 279–303. *Cited on page(s):* 135, 245, 246, 251, 252, 253, 254

[445] McVitie, D.G. and Wilson, L.B. (1971a). The stable marriage problem, *Communications of the ACM* **14**, 7, pp. 486–490. *Cited on page(s):* 19, 20, 58

[446] McVitie, D.G. and Wilson, L.B. (1971b). Three procedures for the stable marriage problem, *Communications of the ACM* **14**, 7, pp. 491–492. *Cited on page(s):* 19, 20, 58

[447] Mehlhorn, K. and Michail, D. (2006). Network problems with non-polynomial weights and applications, Unpublished manuscript. Available from http://www.mpi-sb.mpg.de/~mehlhorn/ftp/HugeWeights.ps (accessed 25 May 2012). *Cited on page(s):* 44, 388, 390, 405, 406

[448] Mertens, S. (2005). Random stable matchings, *Journal of Statistical Mechanics: Theory and Experiment* **2005**, 10, p. P10008. *Cited on page(s):* 175, 202

[449] Mestre, J. (2006). Weighted popular matchings, in *Proceedings of ICALP '06: the 33rd International Colloquium on Automata, Languages and Programming, Lecture Notes in Computer Science*, Vol. 4051 (Springer), pp. 715–726, full version available at http://arxiv.org/abs/0707.0546 (accessed 25 May 2012). *Cited on page(s):* 366, 367

[450] Mestre, J. (2008). Weighted Popular Matchings, in M.-Y. Kao (ed.), *Encyclopedia of Algorithms* (Springer), pp. 1023–1024. *Cited on page(s):* 7, 334

[451] Micali, S. and Vazirani, V.V. (1980). An $O(\sqrt{|V|} \cdot |E|)$ algorithm for finding maximum matching in general graphs, in *Proceedings of FOCS '80: the 21st Annual IEEE Symposium on Foundations of Computer Science* (IEEE Computer Society), pp. 17–27. *Cited on page(s):* 14, 15, 211, 329

[452] Michail, D. (2007). Reducing rank-maximal to maximum weight matching, *Theoretical Computer Science* **389**, 1-2, pp. 125–132. *Cited on page(s):* 44, 354, 388

[453] Michail, D. (2011). An experimental comparison of single-sided preference matching algorithms, *ACM Journal of Experimental Algorithmics* **16**, 3, article 1.4, 16 pages. *Cited on page(s):* 43, 383, 389

[454] Mongell, S. and Roth, A.E. (1991). Sorority rush as a two-sided matching mechanism, *American Economic Review* **81**, pp. 441–464. *Cited on page(s):* 31

[455] Morrill, T. (2010). The roommates problem revisited, *Journal of Economic Theory* **145**, pp. 1739–1756. *Cited on page(s):* 325, 327, 328, 329

[456] Moussa, M.I. and El-Atta, A.H. Abu (2011). A visual implementation of student-project allocation, *International Journal of Computer Theory and Engineering* **3**, 2, pp. 178–184. *Cited on page(s):* 273

[457] Mulder, H.M. and Schrijver, A. (1979). Median graphs and Helly hypergraphs, *Discrete Mathematics* **25**, pp. 41–50. *Cited on page(s):* 195

[458] Mulder, M. (1987). The structure of median graphs, *Discrete Mathematics* **24**, pp. 197–204. *Cited on page(s):* 195

[459] Nakamura, M., Onaga, K., Kyan, S. and Silva, M. (1995). Genetic algorithm for sex-fair stable marriage problem, in *Proceedings of ISCAS '95: 1995 IEEE International Symposium on Circuits and Systems*, Vol. 1 (IEEE Computer Society), pp. 509–512. *Cited on page(s):* 63, 124

[460] Nebeský, L. (1971). Median graphs, *Commentationes Mathematicae Universitatis Carolinae* **12**, pp. 317–325. *Cited on page(s):* 195

[461] Nemoto, T. (2000). Some remarks on the median stable marriage problem, in *Proceedings of ISMP '00: the 17th International Symposium on Mathematical Programming*. *Cited on page(s):* 93

[462] Ng, C. (1990). An $o(n^3\sqrt{\log n})$ algorithm for the optimal stable marriage problem, Tech. Rep. 90-22, University of California, Irvine. *Cited on page(s):* 64

[463] Ng, C. and Hirschberg, D.S. (1988). Complexity of the stable marriage and stable roommate problems in three dimensions, Tech. Rep. UCI-ICS 88-28, Department of Information and Computer Science, University of California, Irvine. *Cited on page(s):* 248

[464] Ng, C. and Hirschberg, D.S. (1990). Lower bounds for the stable marriage problem and its variants, *SIAM Journal on Computing* **19**, pp. 71–77. *Cited on page(s):* 123, 177

[465] Ng, C. and Hirschberg, D.S. (1991). Three-dimensional stable matching problems, *SIAM Journal on Discrete Mathematics* **4**, pp. 245–252. *Cited on page(s):* xxix, 68, 274, 275, 276, 277, 279, 281, 282, 283

[466] Niederle, M. and Roth, A.E. (2004). Market culture: How norms governing exploding offers affect market performance, National Bureau of Economic Research (Cambridge, MA) working paper 10256. Available from `http://www.nber.org/papers/w10256` (accessed 25 May 2012). *Cited on page(s):* 100, 203

[467] Niederle, M., Roth, A.E. and Sönmez, T. (2008). Matching and market design, in S. Derlauf and L. Blume (eds.), *The New Palgrave Dictionary of Economics*, 2nd edn. (Palgrave-Macmillan). *Cited on page(s):* 2, 8

[468] Nisan, N. and Ronen, A. (2001). Algorithmic mechanism design, *Games and Economic Behavior* **35**, pp. 166–196. *Cited on page(s):* 9

[469] Nisan, N., Roughgarden, T., Tardos, E. and Vazirani, V. (eds.) (2007). *Algorithmic Game Theory* (Cambridge University Press). *Cited on page(s):* 9

[470] O'Malley, G. (2007). *Algorithmic Aspects of Stable Matching Problems*, Ph.D. thesis, University of Glasgow, Department of Computing Science. *Cited on page(s):* 8, 16, 61, 73, 135, 145, 146, 158, 165, 201, 202, 271

[471] Ostrovsky, M. (2008). Stability in supply chain networks, *American Economic Review* **98**, 3, pp. 897–923. *Cited on page(s):* 31

[472] Paluch, K. (2012a). Faster and simpler approximation of stable matchings, in *Proceedings of WAOA '11: 9th Workshop on Approximation and Online Algorithms*, Lecture Notes in Computer Science, Vol. 7164 (Springer), pp. 176–187. *Cited on page(s):* 127, 136, 141

[473] Paluch, K. (2012b). Popular and clan-popular b-matchings, in *Proceedings of ISAAC '12: the 23rd International Symposium on Algorithms and Computation*, Lecture Notes in Computer Science, Vol. 7676 (Springer), pp. 116–125. *Cited on page(s):* 362

[474] Perach, N., Polak, J. and Rothblum, U.G. (2008). A stable matching model with an entrance criterion applied to the assignment of students to dormitories at the Technion, *International Journal of Game Theory* **36**, 3-4, pp. 519–535. *Cited on page(s):* 4, 5, 37, 46

[475] Perach, N. and Rothblum, U.G. (2010). Incentive compatibility for the stable matching model with an entrance criterion, *International Journal of Game Theory* **39**, 4, pp. 657–667. *Cited on page(s):* 37

[476] Petersen, J. (1891). Die theorie der regulären graphs, *Acta Mathematica* **15**, p. 193220. *Cited on page(s):* 14

[477] Pini, M.S., Rossi, F., Venable, K.B. and Walsh, T. (2010). Stable marriage problems with quantitative preferences, in V. Conitzer and J. Rothe (eds.), *Proceedings of COMSOC '10: the 3rd International Workshop on Computational Social Choice* (Düsseldorf University Press), pp. 355–366. *Cited on page(s):* 125

[478] Pini, M.S., Rossi, F., Venable, K.B. and Walsh, T. (2011a). Manipulation complexity and gender neutrality in stable marriage procedures, *Autonomous Agents and Multi-Agent Systems* **22**, 1, pp. 183–199. *Cited on page(s):* 106

[479] Pini, M.S., Rossi, F., Venable, K.B. and Walsh, T. (2011b). Stability in matching problems with weighted preferences, in *Proceedings of ICAART '11: the 3rd International Conference on Agents and Artificial Intelligence. Cited on page(s):* 125

[480] Pini, M.S., Rossi, F., Venable, K.B. and Walsh, T. (2011c). Weights in stable marriage problems increase manipulation opportunities, in *Proceedings of the IJCAI workshop on Social Choice and Artificial Intelligence, held at IJCAI '11: the 22nd International Joint Conference on Artificial Intelligence. Cited on page(s):* 125

[481] Pittel, B. (1993). The "Stable Roommates" problem with random preferences, *Annals of Probability* **21**, 3, pp. 1441–1477. *Cited on page(s):* 175, 176

[482] Pittel, B.G. and Irving, R.W. (1994). An upper bound for the solvability probability of a random stable roommates instance, *Random Structures and Algorithms* **5**, pp. 465–486. *Cited on page(s):* 175, 181, 182, 183, 184, 185, 188, 223

[483] Podhradský, A. (2010). *Stable marriage problem algorithms*, Master's thesis, Masaryk University, Faculty of Informatics, available from http://is.muni.cz/th/172646/fi_m (accessed 25 May 2012). *Cited on page(s):* 8, 140

[484] Procaccia, A.D. and Tennenholtz, M. (2009). Approximate mechanism design without money, in *Proceedings of EC '09: the 10th ACM Conference on Electronic Commerce* (ACM), pp. 177–186. *Cited on page(s):* 415

[485] Proll, L.G. (1972). A simple method of assigning projects to students, *Operational Research Quarterly* **23**, 2, pp. 195–201. *Cited on page(s):* 260

[486] Quinn, M.J. (1985). A note on two parallel algorithms to solve the stable marriage problem, *BIT* **25**, pp. 473–476. *Cited on page(s):* 58

[487] Quint, T. and Wako, J. (2004). On houseswapping, the strict core, segmentation, and linear programming, *Mathematics of Operations Research* **29**, 4, pp. 861–877. *Cited on page(s):* 38, 318

[488] Rastegari, B., Condon, A., Immorlica, N. and Leyton-Brown, K. (2012). Two-sided matching with partial information, Unpublished manuscript. Available from http://www.cs.ubc.ca/~kevinlb/pub.php?u=2012-matching-partial-info.pdf (accessed 22 July 2012). *Cited on page(s):* 169, 170

[489] Ratier, G. (1996). On the stable marriage polytope, *Discrete Mathematics* **148**, pp. 141–159. *Cited on page(s):* 115, 116

[490] Robards, P.A. (2001). *Applying two-sided matching processes to the United States Navy enlisted assignment process*, Master's thesis, Naval Postgraduate School, Monterey, California. *Cited on page(s):* 31, 100

[491] Romero-Medina, A. (1998). Implementation of stable solutions in a restricted matching market, *Review of Economic Design* **3**, 2, pp. 137–147. *Cited on page(s):* 4, 31, 61, 62

[492] Ronn, E. (1986). *On the complexity of stable matchings with and without ties*, Ph.D. thesis, Yale University. *Cited on page(s):* 247, 248

[493] Ronn, E. (1990). NP-complete stable matching problems, *Journal of Algorithms* **11**, pp. 285–304. *Cited on page(s):* 29, 199, 205, 243, 245, 247, 248

[494] Rossi, F., van Beek, P. and Walsh, T. (eds.) (2006). *Handbook of constraint programming* (Elsevier). *Cited on page(s):* 73

[495] Rossi, F., Venable, B. and Walsh, T. (2011). *A Short Introduction to Preferences: Between Artificial Intellgience and Social Choice*, Synthesis Lectures on Artificial Intelligence and Machine Learning (Morgan and Claypool). *Cited on page(s):* 9

[496] Roth, A.E. (1982a). The economics of matching: Stability and incentives, *Mathematics of Operations Research* **7**, 4, pp. 617–628. *Cited on page(s):* 103, 106

[497] Roth, A.E. (1982b). Incentive compatibility in a market with indivisible goods, *Economics Letters* **9**, pp. 127–132. *Cited on page(s):* 38, 308, 314

[498] Roth, A.E. (1984a). The evolution of the labor market for medical interns and residents: a case study in game theory, *Journal of Political Economy* **92**, 6, pp. 991–1016. *Cited on page(s):* 3, 5, 21, 29, 130, 243, 245, 247, 259

[499] Roth, A.E. (1984b). Stability and polarization of interests in job matching, *Econometrica* **52**, 1, pp. 47–57. *Cited on page(s):* 258, 260

[500] Roth, A.E. (1985a). The college admissions problem is not equivalent to the marriage problem, *Journal of Economic Theory* **36**, pp. 277–288. *Cited on page(s):* 147, 258

[501] Roth, A.E. (1985b). Conflict and coincidence of interest in job matching: some new results and open questions, *Mathematics of Operations Research* **10**, 3, pp. 379–389. *Cited on page(s):* 260

[502] Roth, A.E. (1986). On the allocation of residents to rural hospitals: a general property of two-sided matching markets, *Econometrica* **54**, pp. 425–427. *Cited on page(s):* 21

[503] Roth, A.E. (1990). New physicians: A natural experiment in market organization, *Science* **250**, pp. 1524–1528. *Cited on page(s):* 5

[504] Roth, A.E. (1991). A natural experiment in the organization of entry level labor markets: Regional markets for new physicians and surgeons in the U.K, *American Economic Review* **81**, pp. 415–440. *Cited on page(s):* 5

[505] Roth, A.E. (2008). Deferred acceptance algorithms: history, theory, practice, and open questions, *International Journal of Game Theory* **36**, 3-4, pp. 537–569. *Cited on page(s):* 8, 20, 31

[506] Roth, A.E. and Peranson, E. (1997). The effects of the change in the NRMP matching algorithm, *Journal of the American Medical Association* **278**, 9, pp. 729–732. *Cited on page(s):* 30

[507] Roth, A.E. and Peranson, E. (1999). The redesign of the matching market for American physicians: Some engineering aspects of economic design, *American Economic Review* **89**, 4, pp. 748–780. *Cited on page(s):* 243

[508] Roth, A.E. and Postlewaite, A. (1977). Weak versus strong domination in a market with indivisible goods, *Journal of Mathematical Economics* **4**, pp. 131–137. *Cited on page(s):* 38, 286, 308, 313, 314, 318, 319

[509] Roth, A.E. and Rothblum, U.G. (1999). Truncation strategies for matching markets – in search of advice for participants, *Econometrica* **67**, 1, pp. 21–43. *Cited on page(s):* 103

[510] Roth, A.E., Rothblum, U.G. and Vate, J.H. Vande (1993). Stable matchings, optimal assignments, and linear programming, *Mathematics of Operations Research* **18**, 4, pp. 803–828. *Cited on page(s):* 72, 73, 180, 181

[511] Roth, A.E., Sönmez, T. and Ünver, M.U. (2004). Kidney exchange, *Quarterly Journal of Economics* **119**, 2, pp. 457–488. *Cited on page(s):* 4, 12, 37, 100

[512] Roth, A.E., Sönmez, T. and Ünver, M.U. (2005). Pairwise kidney exchange, *Journal of Economic Theory* **125**, pp. 151–188. *Cited on page(s):* 4, 12, 37, 100

[513] Roth, A.E., Sönmez, T. and Ünver, M.U. (2007). Efficient kidney exchange: Coincidence of wants in a market with compatibility-based preferences, *American Economic Review* **97**, 3, pp. 828–851. *Cited on page(s):* 4, 12, 37

[514] Roth, A.E. and Sotomayor, M.A.O. (1990). *Two-sided matching: a study in game-theoretic modeling and analysis*, Econometric Society Monographs, Vol. 18 (Cambridge University Press). *Cited on page(s):* 2, 8, 12, 18, 22, 23, 29, 32, 53, 89, 102, 103, 143, 221, 227, 243, 258, 304

[515] Roth, A.E. and Sotomayor, M. (1992). Two-sided matching, in R.J. Aumann and S. Hart (eds.), *Handbook of Game Theory*, Vol. 1, chap. 16 (Elsevier). *Cited on page(s):* 8

[516] Roth, A.E. and Vate, J.H. Vande (1990). Random paths to stability in two-sided matching, *Econometrica* **58**, 6, pp. 1475–1480. *Cited on page(s):* xxxi, 79, 81, 82, 84, 86, 89, 104, 190, 192, 200

[517] Roth, A.E. and Vate, J.H. Vande (1991). Incentives in two-sided matching with random stable mechanisms, *Economic Theory* **1**, pp. 31–44. *Cited on page(s):* 104

[518] Roth, A.E. and Xing, X. (1994). Jumping the gun: imperfections and institutions related to the timing of market transactions, *American Economic Review* **84**, 4, pp. 992–1044. *Cited on page(s):* 2, 5

[519] Roth, A.E. and Xing, X. (1997). Turnaround time and bottlenecks in market clearing: Decentralized matching in the market for clinical psychologists, *Journal of Political Economy* **105**, 2, pp. 284–329. *Cited on page(s):* 100

[520] Rothblum, U.G. (1992). Characterization of stable matchings as extreme points of a polytope, *Mathematical Programming* **54**, pp. 57–67. *Cited on page(s):* 72, 117, 124, 180, 181

[521] Scarf, H.E. (1967). The core of an N-person game, *Econometrica* **35**, 1, pp. 50–69. *Cited on page(s):* 191

[522] Schwarz, M. and Yenmez, M.B. (2011). Median stable matching for markets with wages, *Journal of Economic Theory* **146**, pp. 619–637. *Cited on page(s):* 99

[523] Scott, S. (2005). *A study of stable marriage problems with ties*, Ph.D. thesis, University of Glasgow, Department of Computing Science. *Cited on page(s):* 8, 61, 147, 152, 158, 201, 217, 218

[524] Sethuraman, J. (2011). Personal communication. *Cited on page(s):* 296

[525] Sethuraman, J. and Teo, C.-P. (2001). A polynomial-time algorithm for the bistable roommates problem, *Journal of Computer and System Sciences* **63**, 3, pp. 486–497. *Cited on page(s):* 295, 296

[526] Sethuraman, J., Teo, C.-P. and Qian, L.W. (2006). Many-to-one stable matching: Geometry and fairness, *Mathematics of Operations Research* **31**, 3, pp. 581–596. *Cited on page(s):* 73, 94, 99, 250

[527] Shapley, L. and Scarf, H. (1974). On cores and indivisibility, *Journal of Mathematical Economics* **1**, pp. 23–37. *Cited on page(s):* 38, 286, 308, 309, 314, 318, 331

[528] Shapley, L.S. and Shubik, M. (1971). The assignment game I: The core, *International Journal of Game Theory* **1**, 1, pp. 111–130. *Cited on page(s):* 12

[529] Shi, Z. (2009). A self-stabilizing algorithm for maximum popular matching of strictly ordered preference lists, *International Journal of Computational and Mathematical Sciences* **3**, pp. 331–340. *Cited on page(s):* 358

[530] Silaghi, M.-C., Abhyankar, A., Zanker, M. and Barták, R. (2005). Desk-mates (stable matching) with privacy of preferences, and a new distributed CSP framework, in *Proceedings of FLAIRS '05: the 18th International Conference of the Florida Artificial Intelligence Research Society* (AIII Press), pp. 671–677. *Cited on page(s):* 74, 124

[531] Silaghi, M.-C., Doshi, P., Matsui, T., Yokoo, M. and Zanker, M. (2008). Optimization in private stable matching with cost of privacy loss, in *Proceedings of OPTMAS '08: the 1st International Workshop on Optimisation in Multi-Agent Systems, held at AAMAS '08: the 7th Joint Conference on Autonomous and Multi-Agent Systems*. *Cited on page(s):* 124

[532] Silaghi, M.-C., Zanker, M. and Barták, R. (2004). Desk-mates (stable matching) with privacy of preferences, and a new distributed CSP framework, in *Proceedings of the Workshop on CSP Techniques with Immediate Application (CSPIA), held at CP '04: the 10th International Conference on Principles and Practice of Constraint Programming*, pp. 83–96. *Cited on page(s):* 59, 74, 124

[533] Skiena, S.S. (2008). *The Algorithm Design Manual*, 2nd edn. (Springer). *Cited on page(s):* 351

[534] Sng, C.T.S. (2005). Stable Roommates problem: experiments and results, Unpublished manuscript. *Cited on page(s):* 175

[535] Sng, C.T.S. (2008). *Efficient Algorithms for Bipartite Matching Problems with Preferences*, Ph.D. thesis, University of Glasgow, Department of Computing Science. *Cited on page(s):* 8, 16, 319, 320, 321, 322, 323, 324, 325, 379, 390, 393, 398, 399, 400, 401, 402, 403, 404, 405, 406

[536] Sng, C.T.S. and Manlove, D.F. (2007). Popular matchings in the weighted

capacitated house allocation problem, in *Proceedings of ACiD '07: the 3rd Algorithms and Complexity in Durham workshop, Texts in Algorithmics*, Vol. 9 (College Publications), pp. 129–140. *Cited on page(s):* 456

[537] Sng, C.T.S. and Manlove, D.F. (2010). Popular matchings in the weighted capacitated house allocation problem, *Journal of Discrete Algorithms* **8**, pp. 102–116, preliminary version appeared as [536]. *Cited on page(s):* 367

[538] Sönmez, T. (1997). Manipulation via capacities in two-sided matching markets, *Journal of Economic Theory* **77**, pp. 197–204. *Cited on page(s):* 107

[539] Sönmez, T. (1999). Can pre-arranged matches be avoided in two-sided matching markets? *Journal of Economic Theory* **86**, pp. 148–156. *Cited on page(s):* 107

[540] Sönmez, T. and Switzer, T.B. (2013). Matching with (branch-of-choice) contracts at United States Military Academy, *Econometrica,* to appear Boston College Working Papers in Economics, number 782. Available from http://fmwww.bc.edu/EC-P/wp782.pdf (accessed 21 July 2012). *Cited on page(s):* 31

[541] Sönmez, T. and Ünver, M.U. (2011). Matching, allocation and exchange of discrete resources, in J. Benhabib, A. Bisin and M. Jackson (eds.), *Handbook of Social Economics*, Vol. 1A (North-Holland), pp. 781–852. *Cited on page(s):* 8, 9

[542] Sotomayor, M. (1996). A non-constructive elementary proof of the existence of stable marriages, *Games and Economic Behavior* **13**, pp. 135–137. *Cited on page(s):* 67

[543] Sotomayor, M. (1999a). The lattice structure of the set of stable outcomes of the multiple partners assignment game, *International Journal of Game Theory* **28**, pp. 567–583. *Cited on page(s):* 255

[544] Sotomayor, M. (1999b). Three remarks on the many-to-many stable matching problem, *Mathematical Social Sciences* **38**, 1, pp. 55–70. *Cited on page(s):* 255, 259

[545] Sotomayor, M. (2005). An elementary non-constructive proof of the non-emptiness of the Housing Market of Shapley and Scarf, *Mathematical Social Sciences* **50**, pp. 298–303. *Cited on page(s):* 319

[546] Sotomayor, M. (2011). The Pareto-stability concept is a natural solution concept for discrete matching markets with indifferences, *International Journal of Game Theory* **40**, 3, pp. 631–644. *Cited on page(s):* 147

[547] Sotomayor, M. and Özak, Ö. (2009). Two-sided matching models, in R.A. Meyers (ed.), *Encyclopedia of Complexity and Systems Science* (Springer), pp. 9654–9678. *Cited on page(s):* 8

[548] Spieker, B. (1995). The set of super-stable marriages forms a distributive lattice, *Discrete Applied Mathematics* **58**, pp. 79–84. *Cited on page(s):* 61, 161

[549] Stathopoulos, G.K. (2011). *Variants of stable marriage algorithms, complexity and structural properties*, Master's thesis, University of Athens, Department of Mathematics–MPLA. *Cited on page(s):* 8, 55

[550] Stothers, A.J. (2010). *On the Complexity of Matrix Multiplication*, Ph.D. thesis, University of Edinburgh, School of Mathematics. *Cited on page(s):* 348

[551] Subramanian, A. (1994). A new approach to stable matching problems, *SIAM Journal on Computing* **23**, 4, pp. 671–700. *Cited on page(s):* 10, 32, 51, 58, 65, 66, 67, 68, 71, 126, 275, 276

[552] Sukegawa, N. and Yamamoto, Y. (2012). Preference profiles determining the proposals in the Gale–Shapley algorithm for stable matching problems, *Japan Journal of Industrial and Applied Mathematics* **29**, pp. 547–560. *Cited on page(s):* 105

[553] Svensson, L.-G. (1999). Strategy-proof allocation of indivisible goods, *Social Choice and Welfare* **16**, pp. 557–567. *Cited on page(s):* 304

[554] Tamura, A. (1993). Transformation from arbitrary matchings to stable matchings, *Journal of Combinatorial Theory, Series A* **62**, pp. 310–323. *Cited on page(s):* 56, 79

[555] Tamura, A. (2008). Stable Marriage and Discrete Convex Analysis, in M.-Y. Kao (ed.), *Encyclopedia of Algorithms* (Springer), pp. 880–883. *Cited on page(s):* 7, 264

[556] Tan, J.J.M. (1990). A maximum stable matching for the roommates problem, *BIT* **29**, pp. 631–640. *Cited on page(s):* 10, 173, 188, 189

[557] Tan, J.J.M. (1991a). A necessary and sufficient condition for the existence of a complete stable matching, *Journal of Algorithms* **12**, pp. 154–178. *Cited on page(s):* 10, 173, 177, 181, 182, 183, 184

[558] Tan, J.J.M. (1991b). Stable matchings and stable partitions, *International Journal of Computer Mathematics* **39**, pp. 11–20. *Cited on page(s):* 10, 173, 183, 188, 189

[559] Tan, J.J.M. and Hsueh, Y.-C. (1995). A generalization of the stable matching problem, *Discrete Applied Mathematics* **59**, pp. 87–102. *Cited on page(s):* 10, 89, 173, 184, 185, 186, 187

[560] Tan, J.J.M. and Su, W.C. (1995). On the divorce digraph of the stable marriage problem, *Proceedings of the National Science Council, Republic of China - Part A* **19**, 2, pp. 342–354. *Cited on page(s):* 57, 79

[561] Tarjan, R.E. (1972). Depth-first search and linear graph algorithms, *SIAM Journal on Computing* **1**, 2, pp. 146–160. *Cited on page(s):* 317, 351

[562] Tarski, A. (1955). A lattice-theoretic fixpoint theorem and its applications, *Pacific Journal of Mathematics* **5**, pp. 285–309. *Cited on page(s):* 67, 71, 72

[563] Teo, C.Y. and Ho, D.J. (1998). A systematic approach to the implementation of final year projectin an electrical engineering undergraduate course, *IEEE Transactions on Education* **41**, 1, pp. 25–30. *Cited on page(s):* 260

[564] Teo, C.P. and Sethuraman, J. (1997). LP based approach to optimal stable matchings, in *Proceedings of SODA '97: the 8th ACM-SIAM Symposium on Discrete Algorithms* (ACM-SIAM), pp. 710–719. *Cited on page(s):* 180

[565] Teo, C.-P. and Sethuraman, J. (1998). The geometry of fractional stable matchings and its applications, *Mathematics of Operations Research* **23**, 4, pp. 874–891. *Cited on page(s):* 73, 90, 94, 179, 180, 194, 296

[566] Teo, C.-P. and Sethuraman, J. (2000). On a cutting plane heuristic for the stable roommates problem and its applications, *European Journal of Operational Research* **123**, pp. 195–205. *Cited on page(s):* 181

[567] Teo, C.-P., Sethuraman, J. and Tan, W.-P. (1999). Gale-Shapley stable marriage problem revisited: strategic issues and applications, *Management Science* **47**, 9, pp. 1252–1267. *Cited on page(s):* 31, 104, 106

[568] Thorn, M. (2003). A constraint programming approach to the student-project allocation problem, BSc Honours project report, University of York, Department of Computer Science. *Cited on page(s):* 74, 260

[569] Thurber, E.G. (2002). Concerning the maximum number of stable matchings in the stable marriage problem, *Discrete Applied Mathematics* **248**, pp. 195–219. *Cited on page(s):* 55

[570] Tijs, S.H., Parthasarathy, T., Potters, J.A.M. and Prasad, V. Rajendra (1984). Permutation games: Another class of totally balanced games, *OR Spectrum* **6**, 2, pp. 119–123. *Cited on page(s):* 12

[571] Unsworth, C. (2008). *A Specialised Constraint Approach for Stable Matching Problems*, Ph.D. thesis, University of Glasgow, Department of Computing Science. *Cited on page(s):* 8, 73, 78

[572] Unsworth, C. and Prosser, P. (2005a). An *n*-ary constraint for the stable marriage problem, in *Proceedings of the 5th Workshop on Modelling and Solving Problems with Constraints, held at IJCAI '05: the 19th International Joint Conference on Artificial Intelligence*, pp. 32–38. *Cited on page(s):* 73

[573] Unsworth, C. and Prosser, P. (2005b). A specialised binary constraint for the stable marriage problem, in *Proceedings of SARA '05: Symposium on Abstraction, Reformulation and Approximation, Lecture Notes in Artificial Intelligence*, Vol. 3607 (Springer), pp. 218–233. *Cited on page(s):* 73

[574] van Hentenryck, P., Deville, Y. and Teng, C-M. (1992). A generic arc-consistency algorithm and its specializations, *Artificial Intelligence* **57**, pp. 291–321. *Cited on page(s):* 77

[575] van Zuylen, A., Schalekamp, F. and Williamson, D.P. (2013). Popular ranking, *Discrete Applied Mathematics*, to appear. *Cited on page(s):* 359

[576] Vate, J.E. Vande (1989). Linear programming brings marital bliss, *Operations Research Letters* **8**, 3, pp. 147–153. *Cited on page(s):* 72, 73, 117, 124, 180, 181

[577] Vazirani, V.V. (1994). A theory of alternating paths and blossoms for proving correctness of the $O(\sqrt{V}E)$ general graph maximum matching algorithm, *Combinatorica* **14**, 1, pp. 71–109. *Cited on page(s):* 14, 15, 211, 329

[578] Veskioja, T. (2005). *Stable Marriage Problem and College Admission*, Ph.D. thesis, Tallin University of Technology, Department of Informatics. *Cited on page(s):* 8

[579] Veskioja, T. and Võhandu, L. (2004a). A framework for solving hard variants of stable matching within a limited time, in *IADIS International Conference Applied Computing*, Vol. II, pp. 177–182. *Cited on page(s):* 251

[580] Veskioja, T. and Võhandu, L. (2004b). Majority voting in stable marriage problem with couples, in *Proceedings of ICEIS '04: the 6th International Conference on Enterprise Information Systems*, Vol. 2, pp. 442–447. *Cited on page(s):* 251

[581] Vien, N.A. and Chung, T.C. (2006). Multiobjective fitness functions for stable marriage problem using genetic algorithm, in *Proceedings of SICE-ICCAS '06: SICE-ICASE International Joint Conference* (IEEE), pp. 5500–5503. *Cited on page(s):* 124

[582] Vien, N.A., Viet, N.H., Kim, H., Lee, S.G. and Chung, T.C. (2007). Ant colony based algorithm for stable marriage problem, in K. Elleithy (ed.), *Advances and Innovations in Systems, Computing Sciences and Software Engineering* (Springer), pp. 457–461. *Cited on page(s):* 63, 124

[583] Wang, F., Chen, B. and Miao, Z. (2008). A survey on reviewer assignment problem, in *Proceedings of IEA/AIE '08: the 21st International Conference on Industrial, Engineering and Other Applications of Applied Intelligent Systems, Lecture Notes in Computer Science*, Vol. 5027 (Springer), pp. 718–727. *Cited on page(s):* 410

[584] Weems, B.P. (1998). Bistable Versions of the Marriages and Roommates Problems: Details of Bistable Roommates, Tech. rep., Department of Computer Science and Engineering, University of Texas at Arlington. *Cited on page(s):* 296

[585] Weems, B.P. (1999). Bistable Versions of the Marriages and Roommates Problems, *Journal of Computer and System Sciences* **59**, 3, pp. 504–520. *Cited on page(s):* 293, 294, 295

[586] Williams, V.V. (2011). Breaking the Coppersmith-Winograd barrier, Unpublished manuscript. Available from http://www.cs.berkeley.edu/~virgi/matrixmult.pdf (accessed 25 May 2012). *Cited on page(s):* 348

[587] Yanagisawa, H. (2003). *Approximation Algorithms for Stable Marriage Problems*, Master's thesis, Kyoto University, School of Informatics. *Cited on page(s):* 8, 144

[588] Yanagisawa, H. (2007). *Approximation Algorithms for Stable Marriage Problems*, Ph.D. thesis, Kyoto University, School of Informatics. *Cited on page(s):* 8, 137, 138

[589] Yang, W., Giampapa, J.A. and Sycara, K. (2003). Two-sided matching for the U.S. Navy Detailing Process with market complication, Tech. Rep. CMU-RI-TR-03-49, Robotics Institute, Carnegie-Mellon University. *Cited on page(s):* 31, 100

[590] Yannakakis, M. and Gavril, F. (1980). Edge dominating sets in graphs, *SIAM Journal on Applied Mathematics* **18**, 1, pp. 364–372. *Cited on page(s):* 17

[591] Yılmaz, Ö. (2009). Random assignment under weak preferences, *Games and Economic Behavior* **66**, pp. 546–558. *Cited on page(s):* 317

[592] Yuan, Y. (1996). Residence exchange wanted: a stable residence exchange problem, *European Journal of Operational Research* **90**, pp. 536–546. *Cited on page(s):* 46, 308

[593] Zhang, H. (2009). *An analysis of the Chinese college admission system*, Ph.D. thesis, University of Edinburgh, School of Economics. *Cited on page(s):* 30, 31

[594] Zhang, Y. (2011). *The determinants of national college entrance exam performance in China with an analysis of private tutoring*, Ph.D. thesis, Columbia University. *Cited on page(s):* 31

[595] Zhou, L. (1990). On a conjecture by Gale about one-sided matching problems, *Journal of Economic Theory* **52**, 1, pp. 123–135. *Cited on page(s):* 5,
38

[596] http://www.jrmp.jp (Japan Residency Matching Program website).
Accessed 25 May 2012. *Cited on page(s):* 4, 30, 228

[597] http://www.organdonation.nhs.uk/ukt/about_transplants/
organ_allocation/kidney_(renal)/living_donation/
paired_donation_matching_scheme.jsp (UK Paired Donation matching
scheme website). Accessed 25 May 2012. *Cited on page(s):* 37

[598] http://marketdesigner.blogspot.com (Al Roth's Market Design blog).
Accessed 23 May 2012. *Cited on page(s):* 31

[599] http://econ.core.hu/english/res/game_app.html (Matching Schemes
Worldwide, maintained by Péter Biró). Accessed 23 May 2012. *Cited on
page(s):* 31

[600] http://www.matching-in-practice.eu (Matching in Practice website).
Accessed 23 May 2012. *Cited on page(s):* 31

[601] http://www.tuitionexchange.org/ (The Tuition Exchange website).
Accessed 21 July 2012. *Cited on page(s):* 31

[602] http://www.nrmp.org (National Resident Matching Program website).
Accessed 25 May 2012. *Cited on page(s):* 3, 4, 132

[603] http://www.carms.ca (Canadian Resident Matching Service website).
Accessed 25 May 2012. *Cited on page(s):* 4, 26, 30

[604] http://www.nes.scot.nhs.uk/sfas (Scottish Foundation Allocation Scheme website). Accessed 25 May 2012. *Cited on page(s):* 4, 12,
26, 30, 99, 140

[605] http://www.paireddonation.org (Alliance for Paired Donation website).
Accessed 3 December 2012. *Cited on page(s):* 37

# Glossary of symbols

This glossary, organised alphabetically, indicates the first usage of major notation, together with the context in which the relevant symbol is defined and a brief description of the symbol's meaning. In the context column, the symbol "$G$" corresponds to an undirected graph, and the symbol "$\langle G, c \rangle$" corresponds to a capacitated graph.

| Symbol | Page | Context | Meaning |
|--------|------|---------|---------|
| $A$ | 281 | 3PSA | set of agents |
| $A$ | 38 | HA | set of applicants |
| $A$ | 32 | SRI | set of agents |
| $A_M$ | 32 | SRI | agents who are assigned in $M$ |
| $a_i$ | 281 | 3PSA | individual agent |
| $a_i$ | 38 | HA | individual applicant |
| $a_i$ | 32 | SRI | individual agent |
| $A(H_k)$ | 235 | HR-CQ | acceptable residents of $H_k \in \mathcal{H}$ |
| $A(a_i)$ | 39 | HA | acceptable houses of $a_i \in A$ |
| $A(a_i)$ | 32 | SRI | acceptable agents of $a_i \in A$ |
| $A(f_j)$ | 256 | WF | acceptable workers of $f_j \in F$ |
| $A(h_j)$ | 39 | HA | acceptable applicants of $h_j \in H$ |
| $A(h_j)$ | 18 | HR | acceptable residents of $h_j \in H$ |
| $A(r_i)$ | 18 | HR | acceptable hospitals of $r_i \in R$ |
| $A(s_i)$ | 261 | SPA | acceptable projects of $s_i \in S$ |
| $A(w_i)$ | 256 | WF | acceptable firms of $w_i \in W$ |
| $\alpha_j$ | 93 | SM | matching $\{(m_i, p_{j,\mathcal{S}}(m_i)) : m_i \in U\}$ |
| $\alpha_{j,\mathcal{T}}$ | 90 | SM | matching $\{(m_i, p_{j,\mathcal{T}}(m_i)) : m_i \in U\}$ |
| $\alpha(m, n)$ | 329 | all | inverse Ackermann function |
| $ba(M, I)$ | 100 | SMI | set of blocking agents of $M$ in $I$ |

461

| Symbol | Page | Context | Meaning |
|--------|------|---------|---------|
| $ba(M,I)$ | 203 | SRI | set of blocking agents of $M$ in $I$ |
| $ba(I)$ | 203 | SRI | $\min\{|ba(I,M)| : M \in \mathcal{M}\}$ |
| $ba^+(I)$ | 100 | SMI | $\min\{|ba(I,M)| : M \in \mathcal{M}^+\}$ |
| $bp(M,I)$ | 100 | SMI | set of blocking pairs of $M$ in $I$ |
| $bp(M,I)$ | 203 | SRI | set of blocking pairs of $M$ in $I$ |
| $bp(I)$ | 203 | SRI | $\min\{|bp(I,M)| : M \in \mathcal{M}\}$ |
| $bp^+(I)$ | 100 | SMI | $\min\{|bp(I,M)| : M \in \mathcal{M}^+\}$ |
| $\beta^+(G)$ | 133 | $G$ | max. size of a matching in $G$ |
| $\beta^-(G)$ | 311 | $G$ | min. size of a maximal matching in $G$ |
| $\beta_{j,\mathcal{T}}$ | 91 | SM | matching $\{(p_j,\mathcal{T}(w_i), w_i) : w_i \in W\}$ |
| | | | |
| $C$ | 16 | $\langle G, c \rangle$ | sum of vertex capacities |
| $C$ | 18 | HR | sum of hospital capacities |
| $C(G_1)$ | 361 | CHAT | counterpart of $G_1$ with cloned houses |
| $c$ | 16 | $\langle G, c \rangle$ | vertex capacity function |
| $c$ | 19 | HR | agent capacity function |
| $c_j$ | 18 | HR | capacity of $h_j \in H$ |
| $c_j$ | 262 | SPA | capacity of $p_j \in P$ |
| $c_{\max}$ | 143 | HR | maximum hospital capacity |
| $c(M)$ | 23 | SMI | cost of $M$ |
| $c^U(M)$ | 23 | SMI | cost of $M$ for the men |
| $c^W(M)$ | 23 | SMI | cost of $M$ for the women |
| $c(a_k)$ | 256 | WF | capacity of $a_k \in W \cup F$ |
| $Ch_{a_i}$ | 220 | SRCF | choice function of agent $a_i$ |
| $Ch(a_k, S)$ | 258 | WF-2 | most-preferred subset of $S$ in $\mathcal{S}(a_k)$ |
| | | | |
| $\|x\|_d$ | 286 | 3DSR | Euclidean norm of $x$ in $\mathbb{R}^d$ |
| $\Delta(M, M')$ | 349 | HAT | factor by which $M'$ is more popular than $M$ |
| $\delta(M, M')$ | 349 | HAT | margin by which $M'$ is more popular than $M$ |
| $D$ | 274 | 3GSM | set of dogs |
| $D_M$ | 290 | SMI | envy graph of $M$ |
| $D'_M$ | 306 | HA | envy graph of $M$ |
| $D(I)$ | 25 | SMI | rotation digraph of $I$ |
| $d(M)$ | 23 | SMI | sex-equality of $M$ |
| $d_k$ | 262 | SPA | capacity of $l_k \in L$ |
| $d_k$ | 274 | 3GSM | individual dog |
| $deg_M(t)$ | 276 | 3GSM | degree of triple $t$ in $M$ |

| Symbol | Page | Context | Meaning |
|---|---|---|---|
| $\mathcal{E}$ | 15 | $G$ | set of even vertices |
| $E$ | 14 | $G$ | edges of $G$ |
| $E$ | 38 | HA | set of acceptable pairs |
| $E$ | 18 | HR | set of acceptable pairs |
| $E$ | 32 | SRI | set of acceptable pairs |
| $E$ | 256 | WF | set of acceptable pairs |
| $E_k$ | 384 | HAT | set of rank-$k$ edges |
| $E_{\leq k}$ | 384 | HAT | $E_1 \cup E_2 \cup \cdots \cup E_k$ |
| | | | |
| $F$ | 255 | WF | set of firms |
| $f_j$ | 360 | CHA | $|f(h_j)|$ for $h_j \in H$ |
| $f_j$ | 255 | WF | individual firm |
| $f_i(S)$ | 283 | SR-TR | $a_i$'s most-preferred member of $S$ |
| $f(a_i)$ | 360 | CHA | first house on $a_i$'s list |
| $f(a_i)$ | 361 | CHAT | set of first-choice houses on $a_i$'s list |
| $f(a_i)$ | 335 | HA | first house on $a_i$'s list |
| $f(a_i)$ | 345 | HAT | set of first-choice houses on $a_i$'s list |
| $f(h_j)$ | 360 | CHA | $\{a_i \in A : f(a_i) = h_j\}$ for $h_j \in H$ |
| $f(h_j)$ | 335 | HA | $\{a_i \in A : f(a_i) = h_j\}$ for $h_j \in H$ |
| | | | |
| $G=(V,E)$ | 14 | $G$ | undirected graph |
| $G=(U,W,E)$ | 32 | $G$ | bipartite graph $G$ with bipartition $V = U \cup W$ |
| $G=(V,E)$ | 39 | HA | underlying graph |
| $G=(V,E)$ | 19 | HR | underlying graph |
| $G=(A,E)$ | 32 | SRI | underlying graph |
| $G'$ | 360 | CHA | (capacitated) reduced graph of $G$ |
| $G'$ | 361 | CHAT | (capacitated) reduced graph of $G$ |
| $G'$ | 336 | HA | reduced graph of $G$ |
| $G'$ | 346 | HAT | reduced graph of $G$ |
| $G_1$ | 361 | CHAT | subgraph of $G$ with first-choice edges |
| $G_1$ | 345 | HAT | subgraph of $G$ with first-choice edges |
| $G_k$ | 384 | HAT | subgraph $G_k = (V, E_{\leq k})$ of $G$ |
| $G_M$ | 341 | HA | switching graph of $M$ |
| $G_M^+$ | 374 | SMI | reduced labelled graph of $G$ |
| $g(M)$ | 349 | HAT | unpopularity margin of $M$ |
| | | | |
| $\mathcal{H}$ | 234 | HR-CQ | set system of hospitals $\mathcal{H} \subseteq \mathbb{P}(H)$ |

| Symbol | Page | Context | Meaning |
|---|---|---|---|
| $\mathcal{H}_I$ | 94 | SM | Hasse diagram of stable matchings in $I$ |
| $\mathcal{H}_M(I)$ | 129 | HRT | tie-breaking instances of $I$ relative to $M$ |
| $H$ | 38 | HA | set of houses |
| $H$ | 18 | HR | set of hospitals |
| $H'$ | 345 | HAT | set of $f$-houses |
| $H''$ | 345 | HAT | set of $f$-houses and $s$-houses |
| $H_k$ | 234 | HR-CQ | bounded set of hospitals in $\mathcal{H}$ |
| $h_j$ | 38 | HA | individual house |
| $h_j$ | 18 | HR | individual hospital |
| | | | |
| $\sim$ | 153 | HRT | equivalence relation for strong stability |
| $\sim_{a_i}$ | 26 | HRP | indifference relation for $a_i \in R \cup H$ |
| $\sim_{a_i}$ | 36 | SRPI | indifference relation for $a_i \in A$ |
| $I$ | 18 | all | problem instance |
| $\hat{I}$ | 293 | SR | preference lists in $I$ reversed |
| $I_k$ | 384 | HAT | sub-instance of $I$ with underlying graph $G_k$ |
| $I \backslash S$ | 188 | SRI | deletion of agents in $S$ from $I$ |
| | | | |
| $\mathcal{L}_k$ | 262 | SPA-S | preference list of $l_k \in L$ |
| $\mathcal{L}_k^j$ | 262 | SPA-S | projected preference list of $l_k$ for $p_j$ |
| $L$ | 261 | SPA | set of lecturers |
| $L(h_j)$ | 400 | CHAT | $L$-value of $h_j \in H$ |
| $l_i$ | 335 | HA | last resort house of $a_i \in A$ |
| $l_j$ | 230 | HR-LQ | lower quota of $h_j \in H$ |
| $l_k$ | 261 | SPA | individual lecturer |
| $l_j^k$ | 239 | HR-CR | lower quota of $R_j^k \in \mathcal{R}_j$ |
| | | | |
| $\mathcal{M}$ | 14 | $G$ | set of matchings |
| $\mathcal{M}$ | 40 | HA | set of matchings |
| $\mathcal{M}$ | 100 | SMI | set of matchings |
| $\mathcal{M}^+$ | 40 | HA | set of maximum matchings |
| $\mathcal{M}^+$ | 100 | SMI | set of maximum matchings |
| $\mathcal{M}_k$ | 397 | CHAT | set of matching of size $k$ |
| $\mathcal{M}_{pop}$ | 343 | HA | set of popular matchings |
| $\mathcal{M}_{pop}^+$ | 344 | HA | set of maximum popular matchings |
| $M$ | 14 | $G$ | matching |
| $M$ | 39 | HA | matching |

| Symbol | Page | Context | Meaning |
|--------|------|---------|---------|
| $M$ | 18 | HR | matching |
| $M$ | 32 | SRI | matching |
| $M \cdot C$ | 342 | HA | switching path/cycle $C$ applied to $M$ |
| $M/\rho$ | 25 | SMI | elimination of rotation $\rho$ from $M$ |
| $M_a$ | 20 | HR | resident-optimal stable matching |
| $M_z$ | 20 | HR | hospital-optimal stable matching |
| $M(H_k)$ | 235 | HR-CQ | set of assignees of $H_k \in \mathcal{H}$ in $M$ |
| $M(a)$ | 279 | 3DSM-CYC | next agent in $t \in M$ where $a \in t$ |
| $M(a)$ | 274 | 3GSM | pair formed by removing $a$ from $M[a]$ |
| $M[a]$ | 274 | 3GSM | triple of $M$ containing $a$ |
| $M(a_i)$ | 281 | 3PSA | $M[a_i] \backslash \{a_i\}$ where $a_i \in A$ |
| $M(a_i)$ | 284 | 3WKT | next element of triple of $M$ containing $a_i$ |
| $M(a_i)$ | 39 | HA | assignee of $a_i \in A$ if $a_i$ assigned in $M$ |
| $M(a_i)$ | 32 | SRI | assignee of $a_i \in A_M$ in $M$ |
| $M(a_k)$ | 256 | WF | set of assignees of $a_k \in W \cup F$ in $M$ |
| $M[a_i]$ | 281 | 3PSA | unordered triple of $M$ containing $a_i$ |
| $M(h_j)$ | 39 | HA | assignee of $h_j \in H$ if $h_j$ assigned in $M$ |
| $M(h_j)$ | 18 | HR | set of assignees of $h_j \in H$ in $M$ |
| $M(l_k)$ | 262 | SPA | set of assignees of $l_k \in L$ in $M$ |
| $M(p_j)$ | 262 | SPA | set of assignees of $p_j \in P$ in $M$ |
| $M(r_i)$ | 18 | HR | assignee of $r_i \in R$ if $r_i$ assigned in $M$ |
| $M(s_i)$ | 262 | SPA | assignee of $s_i \in S$ if $s_i$ assigned in $M$ |
| $M(v)$ | 14 | $G$ | assignee of $v \in V$ in $M$ |
| $M(v)$ | 16 | $\langle G, c \rangle$ | set of assignees of $v \in V$ in $M$ |
| $m$ | 14 | $G$ | number of edges |
| $m$ | 38 | HA | number of acceptable pairs |
| $m$ | 18 | HR | number of acceptable pairs |
| $m$ | 32 | SRI | number of acceptable pairs |
| $m_i$ | 274 | 3GSM | individual man |
| $m_i$ | 22 | SMI | individual man |
| $N(v)$ | 15 | $G$ | open neighbourhood of $v \in V$ |
| $n$ | 14 | $G$ | number of vertices of $G$ |
| $n$ | 41 | HA | total number of applicants and houses |
| $n$ | 393 | HRT | total number of residents and hospitals |
| $n$ | 22 | SMI | number of men (=number of women) |
| $n_1$ | 38 | HA | number of applicants |

| Symbol | Page | Context | Meaning |
|--------|------|---------|---------|
| $n_1$ | 18 | HR | number of residents |
| $n_2$ | 38 | HA | number of houses |
| $n_2$ | 18 | HR | number of hospitals |
| | | | |
| $\oplus$ | 14 | $G$ | symmetric difference operator |
| $\mathcal{O}$ | 15 | $G$ | set of odd vertices |
| $\mathcal{O}(I)$ | 188 | SRI | number of odd parties in a stable partition |
| $O_k$ | 382 | CHAT | $k$-tuple $\langle 0, 0, \ldots, 0 \rangle$ of zeros |
| $O_j^M$ | 253 | HRS | occupancy of $h_j$ in $M$ |
| $O_M(a_i)$ | 340 | HA | single member of $\{f(a_i), s(a_i)\} \backslash \{M(a_i)\}$ |
| | | | |
| $\lhd$ | 40 | HA | Pareto improvement relation on matchings |
| $\lhd$ | 25 | SMI | partial order on rotations |
| $\lhd$ | 35 | SRI | partial order on rotations |
| $\preceq$ | 21 | HR | dominance relation |
| $\preceq$ | 153 | HRT | dominance relation |
| $\prec_{a_i}$ | 282 | 3PSA | preference list of $a_i$ over agents in $A \backslash \{a_i\}$ |
| $\prec_{a_i}$ | 26 | HRP | preference poset of $a_i \in R \cup H$ |
| $\preceq_{a_i}$ | 191 | SRI | preference relation for $a_i$ |
| $\prec_{m_i}^D$ | 276 | 3GSM | linear order of $m_i \in U$ over dogs in $D$ |
| $\prec_{m_i}^W$ | 276 | 3GSM | linear order of $m_i \in U$ over women in $W$ |
| $\prec_R$ | 397 | CHAT | right-domination relation on profiles |
| $\Pi$ | 182 | SRI | stable partition |
| $\mathbb{P}(X)$ | 220 | all | power set of $X$ |
| $\mathbb{P}_2(X)$ | 281 | all | $\{X \in \mathbb{P}(X) : |X| = 2\}$ |
| $P$ | 409 | RA | set of papers |
| $P$ | 261 | SPA | set of projects |
| $P_{\mathcal{T}}(a_i)$ | 90 | SM | sorted multiset of $a_i$'s partners in $\mathcal{T} \subseteq \mathcal{S}$ |
| $P(M, M')$ | 41 | HA | set of applicants who prefer $M$ to $M'$ |
| $P(M, M')$ | 368 | SRI | set of applicants who prefer $M$ to $M'$ |
| $P_k$ | 261 | SPA | projects offered by $l_k \in L$ |
| $p(M)$ | 43 | HA | profile of matching $M$ |
| $p(M)$ | 395 | SRTI | profile of matching $M$ |
| $p(P)$ | 400 | CHAT | profile of alternating path $P$ |
| $p^+(I)$ | 311 | HA | max. size of a Pareto optimal matching in $I$ |
| $p^-(I)$ | 311 | HA | min. size of a Pareto optimal matching in $I$ |
| $p_I(M)$ | 391 | HRT | profile of matching $M$ in $I$ |

| Symbol | Page | Context | Meaning |
|--------|------|---------|---------|
| $p^R(M)$ | 44 | HA | reverse profile of matching $M$ |
| $p^f(M)$ | 407 | SMTI | weight-function profile of matching $M$ |
| $p^w(M)$ | 389 | WHAT | weighted profile of matching $M$ |
| $p_j$ | 410 | RA | individual paper |
| $p_j$ | 261 | SPA | individual project |
| $p_n$ | 174 | SR | solvability probability |
| $p_{j,\mathcal{T}}(a_i)$ | 90 | SM | $j$th element in $P_{\mathcal{T}}(a_i)$ |
| $pop^+(I)$ | 339 | HA | max. size of a popular matching in $I$ |
| $pop^+(I)$ | 376 | SMI | max. size of a popular matching in $I$ |
| $pop^-(I)$ | 339 | HA | min. size of a popular matching in $I$ |
| $pop^-(I)$ | 376 | SMI | min. size of a popular matching in $I$ |
| | | | |
| $\rho$ | 25 | SMI | rotation |
| $\rho$ | 35 | SRI | rotation |
| $\mathcal{R}_j$ | 239 | HR-CR | classification of $A(h_j)$ by $h_j \in H$ |
| $\mathcal{R}_M(I)$ | 129 | HRT | tie-breaking instances of $I$ relative to $M$ |
| $R$ | 18 | HR | set of residents |
| $R$ | 409 | RA | set of reviewers |
| $R(h_j)$ | 404 | CHAT | $R$-value of $h_j \in H$ |
| $R(I)$ | 25 | SMI | set of rotations in $I$ |
| $R_C$ | 244 | HRC | set of resident couples |
| $R_S$ | 244 | HRC | set of single residents |
| $R_j^k$ | 239 | HR-CR | class in classification of $A(h_j)$ by $h_j \in H$ |
| $r$ | 44 | HA | max. rank of a house in an applicant's list |
| $r(M)$ | 42 | HA | regret of matching $M$ |
| $r(M)$ | 391 | HRT | regret of matching $M$ |
| $r(M)$ | 23 | SMI | regret of matching $M$ |
| $r(M)$ | 395 | SRTI | regret of matching $M$ |
| $r_i$ | 18 | HR | individual resident |
| $r_i$ | 410 | RA | individual reviewer |
| $rank$ | 46 | CHAT | rank of a house in a given applicant's list |
| $rank$ | 39 | HA | rank of a house in a given applicant's list |
| $rank$ | 19 | HR | rank of an agent in a given agent's list |
| $rank$ | 27 | HRT | rank of an agent in a given agent's list |
| $rank$ | 33 | SRI | rank of an agent in a given agent's list |
| $rank$ | 209 | SRTI-GRP | rank of an edge given by $rank : E \longrightarrow \mathbb{R}$ |

| Symbol | Page | Context | Meaning |
|---|---|---|---|
| $\blacktriangleright$ | 41 | HA | "more popular than" relation |
| $\blacktriangleright$ | 368 | SRI | "more popular than" relation |
| $\blacktriangleright$ | 366 | WHA | "more popular than" relation |
| $\succ_L$ | 383 | CHAT | left-domination relation on profiles |
| $\mathcal{S}$ | 21 | HR | set of stable matchings |
| $\mathcal{S}$ | 152 | HRT | set of strongly stable matchings |
| $\mathcal{S}$ | 152 | HRT | set of super-stable matchings |
| $\mathcal{S}$ | 34 | SRI | set of stable matchings |
| $\mathcal{S}(a_k)$ | 258 | WF-2 | acceptable sets of partners of $a_k \in W \cup F$ |
| $S$ | 261 | SPA | set of students |
| $s(I)$ | 131 | SMI | size of a stable matching in $I$ |
| $s^+(I)$ | 137 | SMTI | max. size of a weakly stable matching |
| $s^+(I)$ | 200 | SRTI | max. size of a weakly stable matching |
| $s^-(I)$ | 144 | SMTI | min. size of a weakly stable matching |
| $s^-(I)$ | 200 | SRTI | min. size of a weakly stable matching |
| $s_i$ | 253 | HRS | size of a resident |
| $s_i$ | 261 | SPA | individual student |
| $s_i(S)$ | 283 | SR-TR | $a_i$'s least-preferred member of $S$ |
| $s(a_i)$ | 360 | CHA | $a_i$'s most-preferred $s$-house not equal to $f(a_i)$ |
| $s(a_i)$ | 361 | CHAT | $a_i$'s most-preferred houses that are even in $G_1$ |
| $s(a_i)$ | 335 | HA | first non $f$-house on $a_i$'s list |
| $s(a_i)$ | 345 | HAT | $a_i$'s most-preferred houses that are even in $G_1$ |
| $t(I)$ | 137 | SMTI | number of preference lists with ties |
| $t(a)$ | 279 | 3DSM-CYC | next agent in $t$ where $a \in t$ |
| $t(a)$ | 274 | 3GSM | pair formed by removing $a$ from triple $t$ |
| $t(a_i)$ | 281 | 3PSA | $t \backslash \{a_i\}$ where $a_i \in A$ |
| $t(a_i)$ | 284 | 3WKT | next element of triple $t$ containing $a_i$ |
| $\mathcal{U}$ | 15 | $G$ | set of unreachable vertices |
| $U$ | 274 | 3GSM | set of men |
| $U$ | 22 | SMI | set of men |
| $U_k$ | 234 | HR-CQ | common quota of bounded set $H_k \in \mathcal{H}_k$ |
| $U_M$ | 23 | SMI | men who are assigned in $M$ |
| $u(M)$ | 349 | HAT | unpopularity factor of $M$ |
| $u_j$ | 230 | HR-LQ | upper quota of $h_j \in H$ |
| $u_j^k$ | 239 | HR-CR | upper quota of $R_j^k \in \mathcal{R}_j$ |

| Symbol | Page | Context | Meaning |
|--------|------|---------|---------|
| $ut$ | 40 | HA | utility of applicant–house pair |
| $ut(M)$ | 40 | HA | utility of matching $M$ |
| | | | |
| $\vee$ | 21 | HR | join of two stable matchings |
| $V$ | 14 | $G$ | vertices of $G$ |
| $V$ | 19 | HR | vertices in underlying graph |
| $v$ | 410 | RA | valuation function |
| | | | |
| $\wedge$ | 21 | HR | meet of two stable matchings |
| $W$ | 274 | 3GSM | set of women |
| $W$ | 40 | HA | largest weight of an applicant–house pair |
| $W$ | 22 | SMI | set of women |
| $W$ | 255 | WF | set of workers |
| $w_i$ | 255 | WF | individual worker |
| $w_j$ | 274 | 3GSM | individual woman |
| $w_j$ | 22 | SMI | individual woman |
| $wt$ | 40 | HA | weight assigned by an applicant to a house |
| $wt$ | 23 | SMI | weight assigned by one agent to another |
| $wt(M)$ | 40 | HA | weight of $M$ |
| $wt(M)$ | 23 | SMI | weight of $M$ |

# Index